Handbook of
Drug Screening

DRUGS AND THE PHARMACEUTICAL SCIENCES

Executive Editor
James Swarbrick
AAI, Inc.
Wilmington, North Carolina

Advisory Board

DRUGS AND THE PHARMACEUTICAL SCIENCES

A Series of Textbooks and Monographs

1. Pharmacokinetics, *Milo Gibaldi and Donald Perrier*
2. Good Manufacturing Practices for Pharmaceuticals: A Plan for Total Quality Control, *Sidney H. Willig, Murray M. Tuckerman, and William S. Hitchings IV*
3. Microencapsulation, *edited by J. R. Nixon*
4. Drug Metabolism: Chemical and Biochemical Aspects, *Bernard Testa and Peter Jenner*
5. New Drugs: Discovery and Development, *edited by Alan A. Rubin*
6. Sustained and Controlled Release Drug Delivery Systems, *edited by Joseph R. Robinson*
7. Modern Pharmaceutics, *edited by Gilbert S. Banker and Christopher T. Rhodes*
8. Prescription Drugs in Short Supply: Case Histories, *Michael A. Schwartz*
9. Activated Charcoal: Antidotal and Other Medical Uses, *David O. Cooney*
10. Concepts in Drug Metabolism (in two parts), *edited by Peter Jenner and Bernard Testa*
11. Pharmaceutical Analysis: Modern Methods (in two parts), *edited by James W. Munson*
12. Techniques of Solubilization of Drugs, *edited by Samuel H. Yalkowsky*
13. Orphan Drugs, *edited by Fred E. Karch*
14. Novel Drug Delivery Systems: Fundamentals, Developmental Concepts, Biomedical Assessments, *Yie W. Chien*
15. Pharmacokinetics: Second Edition, Revised and Expanded, *Milo Gibaldi and Donald Perrier*
16. Good Manufacturing Practices for Pharmaceuticals: A Plan for Total Quality Control, Second Edition, Revised and Expanded, *Sidney H. Willig, Murray M. Tuckerman, and William S. Hitchings IV*
17. Formulation of Veterinary Dosage Forms, *edited by Jack Blodinger*
18. Dermatological Formulations: Percutaneous Absorption, *Brian W. Barry*
19. The Clinical Research Process in the Pharmaceutical Industry, *edited by Gary M. Matoren*
20. Microencapsulation and Related Drug Processes, *Patrick B. Deasy*
21. Drugs and Nutrients: The Interactive Effects, *edited by Daphne A. Roe and T. Colin Campbell*
22. Biotechnology of Industrial Antibiotics, *Erick J. Vandamme*

23. Pharmaceutical Process Validation, *edited by Bernard T. Loftus and Robert A. Nash*
24. Anticancer and Interferon Agents: Synthesis and Properties, *edited by Raphael M. Ottenbrite and George B. Butler*
25. Pharmaceutical Statistics: Practical and Clinical Applications, *Sanford Bolton*
26. Drug Dynamics for Analytical, Clinical, and Biological Chemists, *Benjamin J. Gudzinowicz, Burrows T. Younkin, Jr., and Michael J. Gudzinowicz*
27. Modern Analysis of Antibiotics, *edited by Adjoran Aszalos*
28. Solubility and Related Properties, *Kenneth C. James*
29. Controlled Drug Delivery: Fundamentals and Applications, Second Edition, Revised and Expanded, *edited by Joseph R. Robinson and Vincent H. Lee*
30. New Drug Approval Process: Clinical and Regulatory Management, *edited by Richard A. Guarino*
31. Transdermal Controlled Systemic Medications, *edited by Yie W. Chien*
32. Drug Delivery Devices: Fundamentals and Applications, *edited by Praveen Tyle*
33. Pharmacokinetics: Regulatory • Industrial • Academic Perspectives, *edited by Peter G. Welling and Francis L. S. Tse*
34. Clinical Drug Trials and Tribulations, *edited by Allen E. Cato*
35. Transdermal Drug Delivery: Developmental Issues and Research Initiatives, *edited by Jonathan Hadgraft and Richard H. Guy*
36. Aqueous Polymeric Coatings for Pharmaceutical Dosage Forms, *edited by James W. McGinity*
37. Pharmaceutical Pelletization Technology, *edited by Isaac Ghebre-Sellassie*
38. Good Laboratory Practice Regulations, *edited by Allen F. Hirsch*
39. Nasal Systemic Drug Delivery, *Yie W. Chien, Kenneth S. E. Su, and Shyi-Feu Chang*
40. Modern Pharmaceutics: Second Edition, Revised and Expanded, *edited by Gilbert S. Banker and Christopher T. Rhodes*
41. Specialized Drug Delivery Systems: Manufacturing and Production Technology, *edited by Praveen Tyle*
42. Topical Drug Delivery Formulations, *edited by David W. Osborne and Anton H. Amann*
43. Drug Stability: Principles and Practices, *Jens T. Carstensen*
44. Pharmaceutical Statistics: Practical and Clinical Applications, Second Edition, Revised and Expanded, *Sanford Bolton*
45. Biodegradable Polymers as Drug Delivery Systems, *edited by Mark Chasin and Robert Langer*
46. Preclinical Drug Disposition: A Laboratory Handbook, *Francis L. S. Tse and James J. Jaffe*
47. HPLC in the Pharmaceutical Industry, *edited by Godwin W. Fong and Stanley K. Lam*
48. Pharmaceutical Bioequivalence, *edited by Peter G. Welling, Francis L. S. Tse, and Shrikant V. Dinghe*

49. Pharmaceutical Dissolution Testing, *Umesh V. Banakar*
50. Novel Drug Delivery Systems: Second Edition, Revised and Expanded, *Yie W. Chien*
51. Managing the Clinical Drug Development Process, *David M. Cocchetto and Ronald V. Nardi*
52. Good Manufacturing Practices for Pharmaceuticals: A Plan for Total Quality Control, Third Edition, *edited by Sidney H. Willig and James R. Stoker*
53. Prodrugs: Topical and Ocular Drug Delivery, *edited by Kenneth B. Sloan*
54. Pharmaceutical Inhalation Aerosol Technology, *edited by Anthony J. Hickey*
55. Radiopharmaceuticals: Chemistry and Pharmacology, *edited by Adrian D. Nunn*
56. New Drug Approval Process: Second Edition, Revised and Expanded, *edited by Richard A. Guarino*
57. Pharmaceutical Process Validation: Second Edition, Revised and Expanded, *edited by Ira R. Berry and Robert A. Nash*
58. Ophthalmic Drug Delivery Systems, *edited by Ashim K. Mitra*
59. Pharmaceutical Skin Penetration Enhancement, *edited by Kenneth A. Walters and Jonathan Hadgraft*
60. Colonic Drug Absorption and Metabolism, *edited by Peter R. Bieck*
61. Pharmaceutical Particulate Carriers: Therapeutic Applications, *edited by Alain Rolland*
62. Drug Permeation Enhancement: Theory and Applications, *edited by Dean S. Hsieh*
63. Glycopeptide Antibiotics, *edited by Ramakrishnan Nagarajan*
64. Achieving Sterility in Medical and Pharmaceutical Products, *Nigel A. Halls*
65. Multiparticulate Oral Drug Delivery, *edited by Isaac Ghebre-Sellassie*
66. Colloidal Drug Delivery Systems, *edited by Jörg Kreuter*
67. Pharmacokinetics: Regulatory • Industrial • Academic Perspectives, Second Edition, *edited by Peter G. Welling and Francis L. S. Tse*
68. Drug Stability: Principles and Practices, Second Edition, Revised and Expanded, *Jens T. Carstensen*
69. Good Laboratory Practice Regulations: Second Edition, Revised and Expanded, *edited by Sandy Weinberg*
70. Physical Characterization of Pharmaceutical Solids, *edited by Harry G. Brittain*
71. Pharmaceutical Powder Compaction Technology, *edited by Göran Alderborn and Christer Nyström*
72. Modern Pharmaceutics: Third Edition, Revised and Expanded, *edited by Gilbert S. Banker and Christopher T. Rhodes*
73. Microencapsulation: Methods and Industrial Applications, *edited by Simon Benita*
74. Oral Mucosal Drug Delivery, *edited by Michael J. Rathbone*
75. Clinical Research in Pharmaceutical Development, *edited by Barry Bleidt and Michael Montagne*

76. The Drug Development Process: Increasing Efficiency and Cost Effectiveness, *edited by Peter G. Welling, Louis Lasagna, and Umesh V. Banakar*

77. Microparticulate Systems for the Delivery of Proteins and Vaccines, *edited by Smadar Cohen and Howard Bernstein*

78. Good Manufacturing Practices for Pharmaceuticals: A Plan for Total Quality Control, Fourth Edition, Revised and Expanded, *Sidney H. Willig and James R. Stoker*

79. Aqueous Polymeric Coatings for Pharmaceutical Dosage Forms: Second Edition, Revised and Expanded, *edited by James W. McGinity*

80. Pharmaceutical Statistics: Practical and Clinical Applications, Third Edition, *Sanford Bolton*

81. Handbook of Pharmaceutical Granulation Technology, *edited by Dilip M. Parikh*

82. Biotechnology of Antibiotics: Second Edition, Revised and Expanded, *edited by William R. Strohl*

83. Mechanisms of Transdermal Drug Delivery, *edited by Russell O. Potts and Richard H. Guy*

84. Pharmaceutical Enzymes, *edited by Albert Lauwers and Simon Scharpé*

85. Development of Biopharmaceutical Parenteral Dosage Forms, *edited by John A. Bontempo*

86. Pharmaceutical Project Management, *edited by Tony Kennedy*

87. Drug Products for Clinical Trials: An International Guide to Formulation • Production • Quality Control, *edited by Donald C. Monkhouse and Christopher T. Rhodes*

88. Development and Formulation of Veterinary Dosage Forms: Second Edition, Revised and Expanded, *edited by Gregory E. Hardee and J. Desmond Baggot*

89. Receptor-Based Drug Design, *edited by Paul Leff*

90. Automation and Validation of Information in Pharmaceutical Processing, *edited by Joseph F. deSpautz*

91. Dermal Absorption and Toxicity Assessment, *edited by Michael S. Roberts and Kenneth A. Walters*

92. Pharmaceutical Experimental Design, *Gareth A. Lewis, Didier Mathieu, and Roger Phan-Tan-Luu*

93. Preparing for FDA Pre-Approval Inspections, *edited by Martin D. Hynes III*

94. Pharmaceutical Excipients: Characterization by IR, Raman, and NMR Spectroscopy, *David E. Bugay and W. Paul Findlay*

95. Polymorphism in Pharmaceutical Solids, *edited by Harry G. Brittain*

96. Freeze-Drying/Lyophilization of Pharmaceutical and Biological Products, *edited by Louis Rey and Joan C. May*

97. Percutaneous Absorption: Drugs–Cosmetics–Mechanisms–Methodology, Third Edition, Revised and Expanded, *edited by Robert L. Bronaugh and Howard I. Maibach*

98. Bioadhesive Drug Delivery Systems: Fundamentals, Novel Approaches, and Development, *edited by Edith Mathiowitz, Donald E. Chickering III, and Claus-Michael Lehr*

99. Protein Formulation and Delivery, *edited by Eugene J. McNally*

100. New Drug Approval Process: Third Edition, The Global Challenge, *edited by Richard A. Guarino*

101. Peptide and Protein Drug Analysis, *edited by Ronald E. Reid*

102. Transport Processes in Pharmaceutical Systems, *edited by Gordon L. Amidon, Ping I. Lee, and Elizabeth M. Topp*

103. Excipient Toxicity and Safety, *edited by Myra L. Weiner and Lois A. Kotkoskie*

104. The Clinical Audit in Pharmaceutical Development, *edited by Michael R. Hamrell*

105. Pharmaceutical Emulsions and Suspensions, *edited by Francoise Nielloud and Gilberte Marti-Mestres*

106. Oral Drug Absorption: Prediction and Assessment, *edited by Jennifer B. Dressman and Hans Lennernäs*

107. Drug Stability: Principles and Practices, Third Edition, Revised and Expanded, *edited by Jens T. Carstensen and C. T. Rhodes*

108. Containment in the Pharmaceutical Industry, *edited by James P. Wood*

109. Good Manufacturing Practices for Pharmaceuticals: A Plan for Total Quality Control from Manufacturer to Consumer, Fifth Edition, Revised and Expanded, *Sidney H. Willig*

110. Advanced Pharmaceutical Solids, *Jens T. Carstensen*

111. Endotoxins: Pyrogens, LAL Testing, and Depyrogenation, Second Edition, Revised and Expanded, *Kevin L. Williams*

112. Pharmaceutical Process Engineering, *Anthony J. Hickey and David Ganderton*

113. Pharmacogenomics, *edited by Werner Kalow, Urs A. Meyer, and Rachel F. Tyndale*

114. Handbook of Drug Screening, *edited by Ramakrishna Seethala and Prabhavathi B. Fernandes*

Handbook of
Drug Screening

edited by

Ramakrishna Seethala

Bristol-Myers Squibb Company
Princeton, New Jersey

Prabhavathi B. Fernandes

Ricerca, LLC
Concord, Ohio

MARCEL DEKKER, INC.   NEW YORK · BASEL

ISBN: 0-8247-0562-9

This book is printed on acid-free paper.

Headquarters
Marcel Dekker, Inc.
270 Madison Avenue, New York, NY 10016
tel: 212-696-9000; fax: 212-685-4540

Eastern Hemisphere Distribution
Marcel Dekker AG
Hutgasse 4, Postfach 812, CH-4001 Basel, Switzerland
tel: 41-61-261-8482; fax: 41-61-261-8896

World Wide Web
http://www.dekker.com

The publisher offers discounts on this book when ordered in bulk quantities. For more information, write to Special Sales/Professional Marketing at the headquarters address above.

Current printing (last digit):
10 9 8 7 6 5 4 3

PRINTED IN THE UNITED STATES OF AMERICA

Preface

Screening is an essential early step in drug discovery during which lead molecules are identified for chemical synthetic programs. Because of a low rate of success in obtaining approval of new drugs from regulatory agencies, there has been a heightened demand for increasing the number of drug candidates that enter clinical trials. Combinatorial chemistry was developed to meet the need for increasing the number and diversity of samples available for screening. The completion of the human genome sequencing program has promised to deliver many thousands of new targets for screening. To meet the challenge of the increased number of targets, chemical diversity, and demand for new lead compounds for drug discovery, screening technology has taken great strides in recent years, evolving from low-throughput to high-throughput and ultrahigh-throughput methods. During the 1990s, partnership with engineers, spectroscopists, biologists, biochemists, and others led to the development of new techniques, high-capacity instrumentation and automation for high-throughput screening. *Handbook of Drug Screening* addresses these advances that have been made to enhance screening and drug discovery.

Chapter 1 presents an overview of screening, discussing the importance of screening, the history of the development of screens to present techniques, and the future technologies that will be expected to drive screening. In Chapter 2, the reader is drawn into a general description of targets, the art and science of developing screens, the requirement for automation of repetitious steps, and maintenance of chemical sample records, as well as construction of a coordinated chemistry–biology database. These considerations affect the rate at which effective compounds are found and help in making a successful screen.

Every pharmaceutical and agricultural company has a chemical compound collection gathered from many decades of synthetic programs. In addition, compounds that are available from academic sources have been systematically collected and distributed to screening laboratories by chemical vendors. Combinatorial chemistry has added thousands to millions of compounds to these collections as well. All of these compounds are a source of chemical diversity for new screening programs. Various assay platforms are available to screen such a large library of compounds against several classes of targets in a cost-effective fashion. Various screening strategies that have general applications for many types of screens are discussed in Chapter 3.

Screening has been revolutionized and the aim is to screen tens of thousands of compounds each day. Each step in a screen adds expense and time as well as increases the errors. Homogeneous technologies that use mix-and-read capabilities without additional sample handling steps are favored; these are described in Chapter 4. Homogeneous screens are reliable and automatable for increased throughput. Fluorescence, chemiluminescence, and radioactive-based technologies including fluorescence polarization, homogeneous time-resolved fluorescence, and scintillation proximity, which are amenable to high-throughput and ultrahigh-throughput screening, are discussed with several examples.

The various types of screens—cell-based (microbial and yeast) and cell-free (enzyme, receptor, and ion channel)—are discussed in Chapters 5–10, including the pros and cons of each method, which will enable the reader to select the method best suited to the screen being developed. Microbes have been used as surrogate cellular systems for heterologously expressed targets because they provide an inexpensive means of running cell-based screens. In addition, microbes are genetically well characterized and are relatively easily manipulated using current molecular biology tools. Isolated enzymes and cloned receptors allow the development of targeted screens. However, a ligand or inhibitor that is identified in such screens must alter the function of the target in living cells. Many compounds identified in isolated systems are unable to cross the cell membrane barrier or somehow act differently against the target in its native configuration and cellular location. This lack of conversion of screening of hits identified in cell-free systems to leads that display whole-cell activity has led to the development of functional screens using cellular systems. In addition functional assays are used to evaluate and confirm the positive samples, called ''hits,'' from the primary screening.

In addition to functional screens using cell-based systems, some functional assays can be developed by coupling the activity to transcriptional readouts. The study of gene expression has led to the identification of many transcriptional factors that are good drug targets. Chapter 11 discusses functional screening methods for inhibitors of transcription factors.

Historically, a large source of chemical diversity for screening has been derived from natural products from microbes and plants. Combinatorial chemistry has decreased the reliance on these natural products for sample diversity. However, compounds made by combinatorial synthesis are relatively simple. Large, complex compounds, such as those originating from natural products, have not been amenable to combinatorial chemistry. Chapter 12 discusses how engineering genes involved in complex biosynthetic pathways have allowed the synthesis of new complex products, the so-called unnatural natural products, by combinatorial biology. These newly derived natural products are adding to the chemical diversity available to high-throughput screening laboratories.

Until recently, the identification of a lead drug candidate was followed by many years of toxicology, metabolism, and pharmacology to predict future safety, absorption and drug pharmacokinetics. These costly and time-consuming steps in drug development have now been supplanted to some extent by screening methods for in vitro toxicology and metabolism. High-throughput absorption, metabolism and toxicity screening that is integrated into drug design is described in chapters 13 and 14.

Mass-production methodologies facilitated the sequencing of the human and microbial genomes. The next challenge is the identification of the function of these genes. And, the race to identify genes that will be useful as drug targets has started. The conversion of the new genomic information to screens and then drugs is discussed in chapters 15–17. Chapters 15 and 16 describe identification of new drug targets from the genomic database using proteomics and other innovative strategies to identify function and validate new genes as drug targets. Chapter 17 explores how the large databases of information in the fields of genomics, combinatorial chemistry, high-throughput screening, and clinical trials can be used in drug discovery and development.

The availability of several liquid-dispensing systems, detection systems in microtiter plate-reader format, and the advances in robotics and chip based technology have made it possible to integrate and automate several types of screens. The cost issues, robust nature, and sensitivity of the automated platforms are discussed in Chapters 18 and 19. Until recently, screening was mostly at the macro level and at low-throughput using milliliters of reagents. Currently, it is at the micro-level in 96- and 384-well microtiter plates using 1 to 2 microliters of reagents and enabling high-throughput screening. Technology for nano-level screening is in the immediate future. The miniaturization will enable ultrahigh-throughput screening of more than 100,000 compounds per day. Issues surrounding screen miniaturization, such as sensitivity, and reliability are detailed in Chapters 20 and 21.

Handbook of Drug Screening aims to be a reference text for either new scientists who want to make screening a career and for research scientists who want to understand the process of drug discovery or who want to identify ligands for their target of interest. The book addresses state-of-the-art screening techniques and discusses the advances to be made in the future. The need for such a reference book of screening is felt by both academic and industrial researchers. This book attempts to cover the major aspects of screening, sharing the many years of experience of the authors to benefit beginners and professionals in screening and to increase the efficiency and productivity of drug discovery.

Ramakrishna Seethala
Prabhavathi B. Fernandes

Contents

Preface   *iii*

Contributors   *xi*

1. Moving Into the Third Millennium After a Century of
    Screening   1
    Prabhavathi B. Fernandes

2. Basic Considerations in Designing High-Throughput Screening
    Assays   5
    Duane Bronson, Nathaniel Hentz, William P. Janzen, Mark D.
    Lister, Karl Menke, Jeffrey Wegrzyn, and G. Sitta Sittampalam

3. Screening Platforms   31
    Ramakrishna Seethala

4. Homogeneous Assays for High-Throughput and Ultrahigh-
    Throughput Screening   69
    Ramakrishna Seethala

5. Microbe-Based Screening Systems   129
    Prabhavathi B. Fernandes

6. Molecular Genetic Screen Design for Agricultural and
   Pharmaceutical Product Discovery   153
   Donald R. Kirsch, William Baumbach, Julia N. Heinrich,
   Margaret Lai, Mark H. Pausch, Laura Sarokin, Sanford
   Silverman, and James C. Walsh

7. Receptor Screens for Small Molecule Agonist and Antagonist
   Discovery   189
   Ramakrishna Seethala

8. Functional Assay Screens   265
   Maria L. Webb, Robert A. Horlick, Bassam Damaj, and Kirk
   McMillan

9. Enzyme Screens   283
   Thomas D. Y. Chung and Dennis J. Murphy

10. Screening Strategies for Ion Channel Targets   313
    Thomas H. Large and Martin W. Smith

11. High-Throughput Screening Assays for Detection of
    Transcription   335
    Mohanram Sivaraja

12. Screening of Combinatorial Biology Libraries for Natural
    Products Discovery   357
    Christopher J. Silva, Paul Brian, and Todd Peterson

13. Higher-Throughput Screening Assays With Human Hepatocytes
    for Hepatotoxicity, Metabolic Stability, and Drug–Drug
    Interaction Potential   383
    Albert P. Li

14. High-Throughput Screening for Metabolism-Based Drug–Drug
    Interactions   403
    Vaughn P. Miller and Charles L. Crespi

15. The ATCG of Drug Discovery   415
    Prabhavathi B. Fernandes

Contents

16. Genomics/Functional Proteomics for Identification of New
    Targets   433
    A. Donny Strosberg

17. Bioinformatics: Identification of Novel Targets and Their
    Characterization   447
    Chandra S. Ramanathan

18. The Evolution of Laboratory Automation   477
    Jack Elands

19. Robotics and Automation   493
    Robert F. Trinka and Franz E. Leichtfried

20. Assay Miniaturization: Developing Technologies and Assay
    Formats   525
    *Kevin R. Oldenburg, Ilona Kariv, Ji-hu Zhang, Thomas D. Y.
    Chung, and Siqi Lin*

21. Screening in the NanoWorld: Single-Molecule Spectroscopy and
    Miniaturized High-Throughput Screening   563
    Rodney Turner, Dirk Ullmann, and Sylvia Sterrer

Index   *583*

Contributors

William Baumbach   Morphochem, Inc., Monmouth Junction, New Jersey

Paul Brian   Department of Combinatorial Biosynthesis, Cubist Pharmaceuticals, Vancouver, British Columbia, Canada

Duane Bronson   Sphinx Pharmaceuticals, Eli Lilly and Company, Research Triangle Park, North Carolina

Thomas D. Y. Chung   DuPont Pharmaceuticals Company, Wilmington, Delaware

Charles L. Crespi   GENTEST Corporation, Woburn, Massachusetts

Bassam Damaj   Pharmacopeia, Inc., Princeton, New Jersey

Jack Elands   IDBS Ltd., Guildford, United Kingdom

Prabhavathi B. Fernandes   Ricerca, LLC, Concord, Ohio

Julia N. Heinrich   Wyeth-Ayerst Research, Princeton, New Jersey

Nathaniel Hentz   Sphinx Pharmaceuticals, Eli Lilly and Company, Research Triangle Park, North Carolina

Robert A. Horlick   Pharmacopeia, Inc., Princeton, New Jersey

William P. Janzen   Sphinx Pharmaceuticals, Eli Lilly and Company, Research Triangle Park, North Carolina

Ilona Kariv   DuPont Pharmaceuticals Company, Wilmington, Delaware

Donald R. Kirsch   Wyeth-Ayerst Research, Princeton, New Jersey

Margaret Lai   BASF Corporation, Princeton, New Jersey

Thomas H. Large   Neuroscience Discovery Research, Eli Lilly and Company, Indianapolis, Indiana

Franz E. Leichtfried   Sales and Marketing, Robocon GmbH, Vienna, Austria

Albert P. Li   In Vitro Technologies, Inc., Baltimore, Maryland

Siqi Lin   DuPont Pharmaceuticals Company, Wilmington, Delaware

Mark D. Lister   Sphinx Pharmaceuticals, Eli Lilly and Company, Research Triangle Park, North Carolina

Kirk McMillan   Exelixis, South San Francisco, California

Karl Menke   Sphinx Pharmaceuticals, Eli Lilly and Company, Research Triangle Park, North Carolina

Vaughn P. Miller   GENTEST Corporation, Woburn, Massachusetts

Dennis J. Murphy   Corporate Research, Hercules Incorporated, Wilmington, Delaware

Kevin R. Oldenburg   DuPont Pharmaceuticals Company, Wilmington, Delaware

Mark H. Pausch   Wyeth-Ayerst Research, Princeton, New Jersey

Todd Peterson   Department of Technology and Development, Genicon Sciences Corporation, San Diego, California

Chandra S. Ramanathan Bioinformatics, Bristol-Myers Squibb Company, Wallingford, Connecticut

Laura Sarokin BASF Corporation, Princeton, New Jersey

Ramakrishna Seethala Drug Discovery, Pharmaceutical Research Institute, Bristol-Myers Squibb Company, Princeton, New Jersey

Christopher J. Silva Department of Combinatorial Biosynthesis, Cubist Pharmaceuticals, Vancouver, British Columbia, Canada

Sanford Silverman Wyeth-Ayerst Research, Princeton, New Jersey

G. Sitta Sittampalam BioResearch Technology and Product Development, Lilly Research Laboratories, Eli Lilly and Company, Indianapolis, Indiana

Mohanram Sivaraja CURIS, Cambridge, Massachusetts

Martin W. Smith Sphinx Pharmaceuticals, Eli Lilly and Company, Research Triangle Park, North Carolina

Sylvia Sterrer EVOTEC BioSystems AG, Hamburg, Germany

A. Donny Strosberg Hybrigenics, Paris, France

Robert F. Trinka Robocon Incorporated, Plymouth Meeting, Pennsylvania

Rodney Turner EVOTEC BioSystems AG, Hamburg, Germany

Dirk Ullmann EVOTEC BioSystems AG, Hamburg, Germany

James C. Walsh BASF Corporation, Princeton, New Jersey

Maria L. Webb Pharmacopeia, Inc., Princeton, New Jersey

Jeffrey Wegrzyn Sphinx Pharmaceuticals, Eli Lilly and Company, Research Triangle Park, North Carolina

Ji-hu Zhang DuPont Pharmaceuticals Company, Wilmington, Delaware

1

Moving Into the Third Millennium After a Century of Screening

Prabhavathi B. Fernandes
Ricerca, LLC, Concord, Ohio

For many decades, the discovery of drugs has depended on screening. From the previous century until the 1960s, chemical compounds and natural products were tested in whole animal assays, and in the case of antibacterial and antitumor testing, the samples were tested in whole cell assays. Although screening was labor intensive and expensive, it was fruitful. A large number of successful drugs such as the cephalosporins, aminoglycosides, tetracyclines, macrolide antibiotics, doxyrubicin, etoposide, and other anticancer agents, steroids and drugs for use in the central nervous system and cardiovascular areas were discovered by screening [1,2]. In many cases, identification of the lead molecule, such as penicillin, cephalosporin, macrolides, captopril, and mevacor, spawned a large number of analogs through chemistry programs [3–7]. A single lead molecule identified in one company has, in many cases, kept several hundred industrial chemists and biologists busy for several years.

Screening natural products and a limited number of compound libraries fell out of favor in the early 1980s because very few useful new chemical entities were being identified. Advances in molecular biology in the 1980s contributed to dependence on structural biology, and it was thought that new drugs could be designed [9–11]. At this time, analytical methods were being refined for structural biology. The concept of screening chemicals made in academic and industrial laboratories began to complement the screening of natural products. A large pool of compounds became available through the opening of Russia's borders. Academic chemists, who worked to establish new synthetic methods, became sources of chemical diversity for screening, as the compounds on their shelves, in many cases, had never been tested for biological activity. This chemical diversity was

1

increased tremendously by chemical stores of large numbers of compounds that were made for past programs at pharmaceutical and chemical companies. At the same time, progress in biotechnology allowed isolation of large amounts of pure proteins that were drug targets. Thus testing chemical and natural product libraries against isolated enzymes and proteins became the preferred method for screening [12]. Combinatorial synthesis of thousands of molecules became possible in the 1990s, and many companies tied combinatorial chemistry with high throughput screening [13–20]. The use of microtiter plates made it possible to screen large numbers of samples while keeping reagent costs down. Those companies that were the first to use these methods had products in their portfolios in the 1990s.

Did the use of isolated enzymes and cells speed drug discovery, or did it simply delay the testing of drug leads in whole cell assays or animals? Many drug leads did not show activity in whole cells, which gave employment to many chemists to improve the cell permeability of leads for soluble protein screens. Time gained in identifying a lead molecule could often be lost trying to get whole cell activity [21]. Of course, if the target was at the cell surface, like receptors, screening with isolated membrane systems led to a fair degree of success.

Structural chemistry in the mid-1990s progressed to identify chemical types that could recognize motifs within certain protein classes, such as G-protein coupled receptors and tyrosine kinases [22,23]. Sensitive reporter systems, such as luciferase and fluorescence, which could be used to detect intracellular changes, made whole cell based screening feasible again [24]. Whole cell based screening provided the flexibility of screening against proteins in interacting pathways as well as identifying leads that already acted in cellular systems. Today, screening with isolated proteins as well as whole cell systems are both used in high throughput screening.

Biotechnology has made it possible to identify new drug targets and develop many assays in a short time frame. The identification of many new potential drug targets derived from genomic sequences has increased the drive to win the discovery race by testing hundreds of thousands of samples per day in a large number of screens [24]. Miniaturization of screens has allowed the throughput to increase, and microplates ranging from 384 wells to 9600 wells have been developed [24]. Databases have been built to accommodate millions of data points. During the late 1990s, high throughput screening groups have become ultra high throughput screening groups.

The engineering and automation industries have become major players in the high throughput screening arena. Instrumentation for sample delivery, plate handling, and millions of repetitive motions required for running assays have made robotics and automation an essential component for high throughput screening. Small sample sizes have necessitated the development of very sensitive readers [24–26].

With the need of careful coordination of activities between biologists, chemists, and computer and database specialists and automation for high throughput screening, high throughput screening departments currently consist of a mixture of biologists, chemists, engineers, and computer and networking specialists. High throughput screening departments have become high technology departments. The promise of high throughput screening is that more leads will be identified and that these screening leads will be developed to make new chemical entities and eventually new drugs. During the last few years, the number of new chemical entities entering the marketplace has not increased by much in spite of the increased throughput. The art of screening needs to be coupled to the science of screening again, in order to make high throughput and ultra high throughput screening become *effective* throughput screening.

The 1980s had the promise of structural drug design; the 1990s was the promise of combinatorial chemistry. Future successes may be derived from genomics and new gene sequences that can be identified as drug targets. Identifying the biological role of proteins coded by these new genes as well as their interacting proteins has opened the field of functional proteomics [27]. As analytical and computational tools meet higher resolution needs, structural biology will increase its importance in designing effective blockers of activators of functional domains [28]. Functional proteomics is expected to give hundreds of new targets to drug discovery in the next decade [29]. As the next millennium unfolds, these new protein targets will be tested in high throughput screening. The challenge is to keep high throughput screening an effective means of drug discovery and deliver on the promise of finding new drugs from screening.

REFERENCES

1. S Omura. Philosophy of new drug discovery. Microbiol Rev 50:259–279, 1986.
2. LH Corporale. Chemical ecology: a view from the pharmaceutical industry. Proc Natl Acad Sci USA 92:75–82, 1995.
3. R Wiley, DH Rich. Peptidomimetics derived from natural products. Med Res Rev 13:327–384, 1993.
4. GR Donowitz, GL Mandell. Beta-lactam antibiotics. N Engl J Med 318:419–426, 490–500, 1988.
5. GH Albers-Schonberg, MB Joshua, OD Lopez, JP Hensens, JP Springer, J Chen, S Ostrove, CH Hoffman, AW Alberts, AA Pachett. Dihydromevinolin, a potent hypocholesteremic metabolite produced by *Aspergillus terreus*. J Antibiot 34:507–512, 1981.
6. MD Greenspan, HG Bull, JB Yudkovitz, DP Hanf, AW Alberts. Inhibition of 3-hydroxy-3-methylglutaryl-CoA synthase and cholesterol biosynthesis by beta-lactone inhibitors and binding of these inhibitors to the enzyme. Biochem J 289: 889–895, 1993.

7. CW Cushman, HS Cheung, EF Sabo, MA Ondetti. Development and design of specific inhibitors of angiotensin-converting enzyme. Am J Cardiol 49:1390–1394, 1982.
8. TL Blundell. Structure-based drug design. Nature 384, suppl 23–26, 1996.
9. LM Amzel. Structure-based drug design. Curr Opin Biotech 9:366–369, 1998.
10. GW Bemis, MA Murcko. The properties of known drugs. 1. Molecular frameworks. J Med Chem 39:2887–2893, 1996.
11. DJ Cummins, CW Andrews, JA Bentley, M Cory. Molecular diversity in chemical databases: comparison of medicinal chemistry knowledge bases and databases of commercially available compounds. J Chem Inf Comput Sci 36:750–763, 1996.
12. SP Manly. In vitro biochemical screening. J Biomol Screening 2:197–199, 1997.
13. T Carell, EA Wintner, AJ Sutherland, J Rebek Jr, YM Dunayevskiy, P Vouros. New promise in combinatorial chemistry: synthesis, characterization, and screening of small-molecule libraries in solution. Curr Opin Chem Biol 2:171–183, 1995.
14. MG Bures, YC Martin. Computational methods in molecular diversity and combinatorial chemistry. Curr Opin Chem Biol 2:376–380, 1998.
15. JJ Burbaum, MH Ohlmeyer, JC Reader, I Henderson, LW Dillard, G Li, TL Randle, NH Sigal, D Chelsky, JJ Baldwin. A paradigm for drug discovery employing encoded combinatorial libraries. Proc Natl Acad Sci USA 92:6027–6031, 1995.
16. A Persidis. Combinatorial chemistry. Nature Biotech 16:691–693, 1998.
17. JA Ellman, MA Gallop. Combinatorial chemistry. Curr Opin Chem Biol 2:317–319, 1998.
18. JC Hogan. Directed combinatorial chemistry. Nature 384, suppl 17–19, 1996.
19. KD Janda. Tagged versus untagged libraries: methods for the generation and screening of combinatorial chemical libraries. Proc Natl Acad Sci USA 91:10779–10785, 1994.
20. S Borman. Combinatorial chemistry. Chem Eng News April 6, 47–67, 1998.
21. JR Broach, J Thorner. High throughput screening for drug discovery. Nature 384, suppl 14–16, 1996.
22. GW Milne, MC Nicklaus, S Wang. Pharmacophores in drug design and discovery. SAR QSAR Environ Res 9:2–28, 1998.
23. FA Al-Obeidi, JJ Wu, KS Lam. Protein tyrosine kinases: structure, substrate specificity, and drug discovery. Biopolymers 47:197–223, 1998.
24. PB Fernandes. Technological functions in high-throughput screening. Curr Opin Chem Biol 2:597–603, 1998.
25. A Persidis. High throughput screening. Nature Biotech 16:489, 1998.
26. A Persidis. Biochips. Nature Biotech 16:981–983, 1998.
27. S Fields. The future is function. Nature Genetics 15:325–324, 1997.
28. T Gaasterland. Structural genomics: bioinformatics in the driver's seat. Nature Biotech 16:625–627, 1998.
29. P Roepstorff. Mass spectrometry in protein studies from genomic to function. Curr Opin Biotech 8:6–13, 1997.

2

Basic Considerations in Designing High-Throughput Screening Assays

Duane Bronson, Nathaniel Hentz, William P. Janzen, Mark D. Lister, Karl Menke, and Jeffrey Wegrzyn
Eli Lilly and Company, Research Triangle Park, North Carolina

G. Sitta Sittampalam
Eli Lilly and Company, Indianapolis, Indiana

I. INTRODUCTION

The design of assays for high throughput screening (HTS) is anything but basic. To be effective in HTS, an assay must be robust, reproducible, and automatable. To make matters worse, these techniques must also support complex biological systems often involving engineered cell lines and always requiring comparison of results from hundreds of thousands of assay points. It is not rocket science— if only it were that simple!

Fortunately, the development of these screens is becoming widely recognized as critical expertise and appropriate organizational structures are being implemented in lead discovery groups. The conversion of a biological disease target into a screen requires not only an understanding of the biology and biochemistry underlying both the disease and the readout of the screen but also an understanding of the more factorylike processes employed in the HTS lab. This unique function involves automation, engineering, and high volume data acquisition and analysis and may reside in specialized groups or within a broader scientific function. Careful management of these two divergent viewpoints is critical to the success of any HTS operation. Finally, the selection of appropriate compound diversity for a particular target and the capabilities to store, distribute, and resynthesize should be planned in advance as part of the screen design process.

5

II. TARGET CLASSES

Biological scientists will usually argue that the selection of the target for HTS is the most important decision. "Screeners," on the other hand, may maintain that any target is a viable candidate for lead discovery if a proper screen is developed and it is run correctly. Medicinal chemists believe that appropriate diversity is the most critical and that leads can be found for any target with appropriate compound libraries. The truth lies somewhere among these opinions. The question whether one is testing appropriate chemical diversity to find a lead can only be answered as a binary output: you either found something or you did not; this will be discussed later. The selection of a therapeutically valid target and the assay used to elucidate it is not as clear-cut.

Targets for HTS have historically come from basic scientific research in which years were spent discovering the details surrounding the manifestation of a disease or symptom. From these results one could select key biological responses as steps for intervention. Most successful drugs on the market today are from validated and "druggable" targets of this type [1]. Examples of these targets include enzymes (proteases, polymerases), interacting proteins, and receptors involved in signal transduction and transcriptional processes in the cell. Today more and more targets arrive at screen development with little or no characterization. A target discovered via genomic techniques might have a strong correlation with disease biology but is little more than a DNA sequence when the process of lead discovery is initiated. In such cases, the screen may be creating a tool to elucidate the function of the target and even its structure. Hence HTS is becoming an important tool in finding lead molecules for established targets and new chemical entities that serve to elucidate the pharmacology of novel disease targets.

III. TYPES OF HTS ASSAYS

Two basic assay formats are employed in HTS: in vitro biochemical assays and cell based assays. Targets that are routinely screened using in vitro formats include various enzymes, receptors, and protein–protein interaction assays. Cell based assays are utilized for targets that involve signal transduction pathways (including receptors) and in screens designed to identify antimicrobial agents. In both cases, the screens are generally carried out in 96- or 384-well microtiter plates using colorimetric, fluorescence, luminescence, and radiometric detection techniques to measure assay end points. Commercial plate readers available for HTS applications are reviewed elsewhere [2,3]. Each of these major target types can be subdivided by the detection technique employed. Major types and examples of each are given in Table 1. It should be noted that in the author's laboratories, leads have been discovered in all target types and with all detection tech-

Table 1 Detection Platforms

Radiochemical	Fluorescence detection	Colorimetric or spectrophotometric readouts
Scintillation proximity assay (SPA)	Time resolved fluorescence (TRF)	ELISA
FlashPlate™ radioimmunoassay (RIA)	Fluorescence polarization	Luminescence
	Fluorescence resonance energy transfer (FRET)	Turbidity

nologies (unpublished results), thereby making the selection more difficult but no less critical.

Reporter assays (cell based) are widely employed to detect G-protein coupled receptor (GPCR) signaling and can be employed to detect interruptions in signaling as well as gene transcription and regulation. Functional cellular readouts, as the name implies, are more direct measures of cellular activity. The classic example is the direct measure of ion flux into cells using fluorescent dyes [4]. The simplest assays are crude measures of toxicity in which the ability of a substance to kill a cell type (e.g., a tumor cell) selectively indicates a positive result.

One issue that scientists would like to ignore, but rarely can, is cost. In industrial HTS and screen development operations, the ability and need to track costs becomes critical. Larger operations have the advantage of averaging costs over many screens, but the job of tracking becomes commensurately more complex. Smaller operations have the advantage in tracking but may be more limited in their ability to absorb high cost screens. In either case, cost cannot be used as the only decision point for selecting technologies. Decisions must be made early in the process of developing a screen that will drive the detection technologies that can be employed. Driving screen types based on cost may lead to compromising the biological relevance of the screen. The decision to employ or evaluate a new technology may be driven by the desire to lower costs for supplies and/or manpower. To aid in this type of decision, we have been tracking performance data for several years. The most complete set of data encompasses 26 targets and includes data on cost, time, and throughput. These are roughly divided into cellular and biochemical assay methods. A number of analyses were applied. Among the data collected were cycle time, complexity, cost, and throughput. For this exercise, they were subdivided into the following platforms as shown in Table 2. The results of this analysis are shown in Figure 1.

A. In Vitro Biochemical Assays

Targets frequently screened using in vitro biochemical methods are enzymes, receptors, and protein–protein interactions of significance in disease pathology,

Table 2 Types of HTS Assays

Cellular assays	Biochemical assays
Mammalian cellular	Receptor–ligand binding
Bacterial/fungal	Enzymatic assay (proteases, other)
Calcium channel assay	Kinases and phosphatases
Cellular assay with SPA detection	Protein–protein interactions

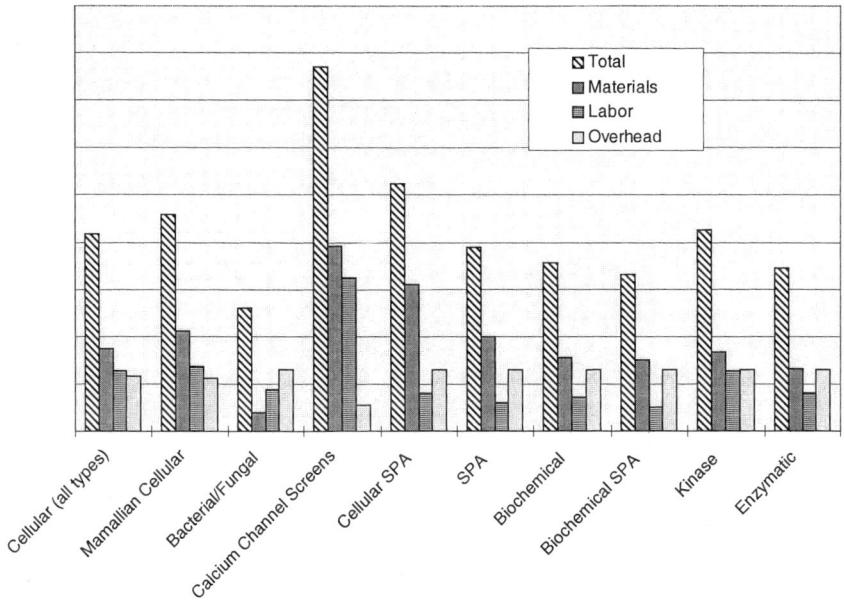

Figure 1 Relative cost per compound in various assay platforms for HTS. This data includes material, labor, and overhead costs.

many examples of which have been described in the literature [1,5,6]. In a typical screen designed to identify inhibitors for enzyme targets, test compounds are mixed with the enzyme followed by the addition of the specific substrate. Product formation is then detected by monitoring a label on the product or a unique spectral property of the product itself (absorbance, fluorescence, chemiluminescence, etc.). In the presence of an inhibitor, the observed signal is significantly attenuated (or enhanced), thus identifying an active compound. Optimization and validation of enzymatic assays for HTS involves several steps that address reaction conditions (buffers, pH, temperature), substrates (natural or synthetic), enzyme kinetics, estimation of signal windows, and the response of known inhibitors (if available). Examples of assay formats applicable to various enzyme targets are shown in Figures 2 and 3. As noted in Table 1, these formats employ colorimetric, fluorescence, and radiochemical methods to follow the conversion of suitable substrates to products. Although the targets shown are proteases, kinases, and phosphatases, the assay formats are applicable to other enzyme types such as DNA ligases and helicases. Clearly, the design of appropriate substrates and the physicochemical properties of the products will determine the assay and detection platforms selected for HTS applications.

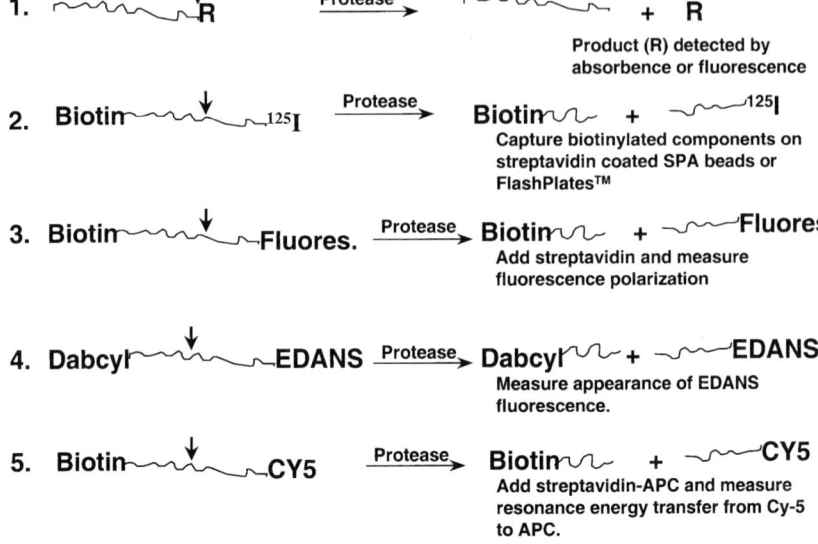

Figure 2 Common assay schemes for protease targets. These schemes are amenable to "mix and measure" formats that are preferable in HTS.

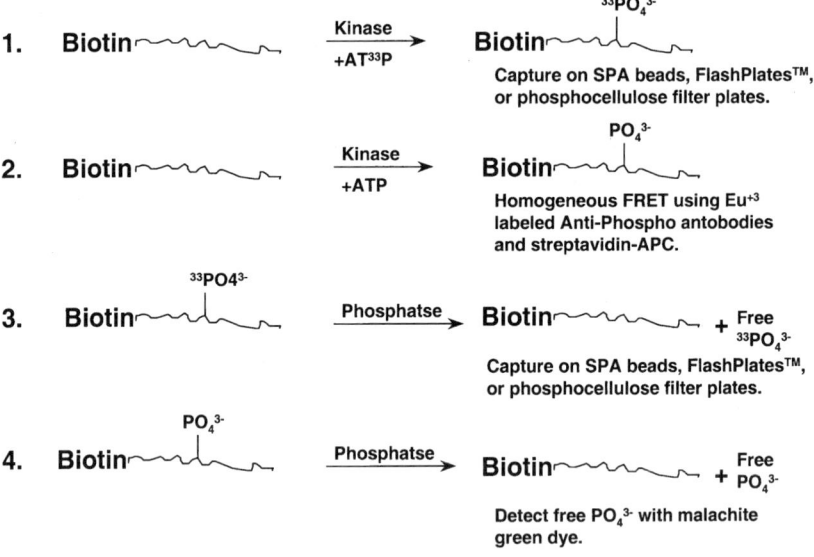

Figure 3 Common assay schemes for kinase and phosphatase targets compatible "mix and measure" formats.

Receptor binding assays are traditionally carried out using natural ligands labeled with radiotracers such as ^3H and ^{125}I [7]. Membrane receptor preparations are made from cells or tissues naturally expressing the receptors, or from engineered cell lines that have been manipulated to overexpress the receptors. In either case, it is critical to establish suitable conditions to generate soluble receptors or membrane suspensions that retain maximum binding affinity for the native ligands [8,9]. It is not unusual to encounter situations in which procedures employed for receptor preparations seriously affect binding affinity under identical conditions. For example, factors such as growth conditions of engineered cell lines, origin and collection of the native tissues, buffers, pH, ionic strength, and buffer additives have to be carefully optimized for each receptor type. In many cases, scaling up of an optimized procedure for HTS is not straightforward and may require reoptimization.

Typical binding experiments involve competitive binding of a labeled ligand and an antagonist or agonist to the receptor. The bound and free ligand is separated and the extent of binding is estimated to determine agonist or antagonist activity. Current assay formats in HTS of receptors employ membrane filtration [7,8] or scintillation proximity assay (SPA) principles [9,10] to measure the ratio of the bound and free labeled ligand. In both cases, assays are performed in 96- or 384-well plates, although the use of a 1536-well assay using an imaging detector has also been reported [11].

Protein–protein interactions play a significant role in protein function and signal transduction and the control of cell cycle [12]. Pathophysiology of many diseases involves abnormal signal transduction processes requiring intervention via inhibitors of specific protein–protein interactions. Screens for such targets can be designed by labeling one or both proteins and measuring the extent of binding using suitable detection techniques. SPA [13], homogeneous time-resolved fluorescence (HTRF) [14], and enzyme-linked immunosorbent assay (ELISA) [15] methods have been used in these screens. Since the apparent binding affinities of protein interactions are considerably lower than those of typical receptor–ligand pairs, homogeneous nonseparation methods (SPA, HTRF) are preferable to ELISA and membrane filtration methods that may alter the protein binding equilibrium [16].

B. Cell-Based Assays

Engineered cell lines overexpressing selected targets have been used extensively in primary HTS applications [17]. GPCRs and proteins involved in signal transduction and transcriptional regulation are common targets screened in this mode, in addition to screens designed to identify antimicrobial agents [18]. Even when the targets are screened using in vitro biochemical methods (receptors and enzymes), more than 95% of the secondary and tertiary assays are in cell-based

Table 3 Signal Transduction by Different Reporter Assays

Reporter gene	Signal	Expression	Cost	HTS compatibility
β-Lactamase	Fluorescence	High, stable. No endogenous enzyme	Medium–high	High
CAT	Radioisotope, ELISA readout	Stable. No endogenous enzyme	Expensive	Poor
Luciferase *FF*	Chemiluminescence. Cell lysis required	High, stable in most mammalian cells	Medium–high	High
Luciferase *Ren*	Chemiluminescence. Cell permeable substrate	High, stable in most mammalian cells	Medium–high	High
SEAP (SPAP)	Colorimetry, chemiluminescence. Fluorescence. Secreted into the medium	High, stable. Endogenous enzyme not secreted	Inexpensive	Good–high
hGH	RIA, ELISA	Secreted	Expensive	Poor
β-Galactosidase	Colorimetry, chemilumnescence	Stable. Endogenous enzyme in some cell lines	Inexpensive	Medium
β-Glucuronidase	Colorimetry, chemiluminescence	Stable. Endogenous enzyme in some cell lines	Inexpensive	Medium
GFP	Fluorescence at multiwavelengths. Nonenzymatic	Stable. Requires post-translational modifications	Inexpensive	Poor

assay formats. These assays are run in medium- or low-throughput formats to elucidate cell permeability, cytotoxicity, and mechanistic linkage of the target to the proposed hypothesis. Both primary and engineered cell lines are employed in secondary and tertiary assays.

Measurement of cellular responses frequently involves the use of reporter genes coupled to the target of interest through specific signal transduction mechanisms or the viability of the cells themselves. Listed in Table 3 are commonly used strategies that utilize colorimetry, chemiluminescence, and fluorescence signal readouts in commercial plate readers.

IV. SCREEN DESIGN ISSUES

Design of all screens will critically depend on (1) the type of target, (2) the available reagents, and (3) the assay readout mode that can be rapidly automated using existing equipment. Equally important is to determine the suitability of in vitro biochemical or cell based assay formats for the target under consideration and the cost of implementing the screen on the chosen platform. In all cases, the simplest assays for HTS are those that are amenable to "mix and measure" formats avoiding complex operations such as washing and plate-to-plate reagent transfers. Although frequently ignored, the assay validation and automation issues should be addressed at this early stage to avoid delays in the implementation phase. These issues are discussed below.

A. Reagents and Assay Validation

In order to ensure that a high-throughput screen is robust and consistent from plate to plate and day to day it is important that the reagents and assay conditions be characterized and optimized. A number of targets have been delayed and even stopped because a reagent was replaced with a different product lot. This includes everything from plates, commercial reagents to in-house generated biological reagents. This is quite possibly the most overlooked step in developing and validating a screen and one of the top causes for delays in screen completion. Prior to running a screen, more of one lot of a given reagent should be obtained than is needed to complete the screen. Obviously, there will be times when this is not possible. Under these circumstances all lots that are needed should be obtained and compared side by side to ensure similar or equivalent results. If reagents cannot be obtained at one time due to availability or stability issues, sufficient amounts of the old reagent should be retained to do direct comparison tests with the new lot to ensure that the different lots are consistent.

Another area that is often overlooked is reagent stability. During development of a high-throughput assay, quite often only a few plates (or even a partial plate) are needed. In these cases reagents are removed from freezers or refrigerators, used, and then replaced in a relatively short span of time. During the actual operation of a screen, however, the amount of time that reagents are out at ambient temperature could be extensive. For example, a receptor-binding assay that uses crude membrane preparations could have substantial consequences if material begins to fall out of suspension under certain conditions or time [19]. Therefore all reagents should be tested for the length of time that they are stable at room temperature. The amount of signal to background will determine how much loss of activity is acceptable (see window determination below). Any acceptable time-dependent loss in activity also assumes that controls for calculating percent inhibition or percent agonist are on each plate. Also critical is the stability of reagents to freeze–thaw. It is best to aliquot into vials and freeze the amount needed for each day's screens. However, this is not necessary if reagents have been tested for repeated freeze–thaws. Again, it is best to keep the number of freezer removals to a number where the loss of activity still gives you the minimum required signal.

Since DMSO is the preferred vehicle for compound/sample delivery, the assay should be tested in a dose–response manner for its tolerance of this reagent. Depending on the final test compound concentration and the assay's tolerance to DMSO, a compromise might have to be reached on running the screen at less than optimal conditions. For this reason the DMSO tolerance should be determined early in the development process so that further optimization can be done under screening conditions.

When running cell based screens one should address specific issues that may cause problems in the HTS mode. One is the scale-up of cells. Perhaps contrary to popular belief, scaling up to the amount of cells needed for a day's run in HTS is not always just a matter of numbers. The increased time of exposure to treatment due to the required handling of many more cell flasks and the larger number of cells washed are a few of the conditions that can alter the cell response in a given assay.

Cells should also be grown under the different conditions that may be required during the screen. That is, if the cells are to be grown over the weekend, then during development this should be tested. In some cases cells grown over the weekend have given different responses from cells grown during the week. The reason may be subtle or it may be as obvious as growing too confluent or seeding flasks with cells too dilute.

If cell washings are required during the screen assay, then automated plate washers should be used some time during the development phase. In some cases cell lines or cell lines that have been transfected may attach loosely to the bottom of the plate well. These cells may be easily disrupted or damaged. Adjusting the wash flow and its direction (pointed toward the plate wall) can reduce cell disturbance. Also, adjusting aspiration depth and length of aspiration time can minimize cell loss as well.

Once an assay has been optimized, it should be validated to ensure its robustness for consistent quantitative determination of active compounds. One method for accomplishing this is to run whole plates at three different concentrations, maximum signal, mid-signal, and background [20–22]. If one is looking for an inhibitor (or signal decrease) there should be a window between the 3 standard deviation (S.D.) line from the maximum and the 3 S.D. line from the mid-signal. The window (or space in between) should be equivalent to or greater than 3 S.D. (e.g., the maximum signal S.D.) (Fig. 4). The same type of determination could be done when looking for increased signal, only using the 3 S.D. minimum and mid-signals. These types of experiments should be done at least three independent times to ensure consistency from screen run to screen run.

Lastly, prior to starting the HTS run, all automated equipment should be used with a small number of plates (~5) to ensure that the assay is comparable to the hand run plate (which should be run in parallel). Data from at least three independent experiments should be compared to establish reproducibility. The next step is to ensure that the assay can be scaled up by running a large number of plates (~50 × 96 well plates or ~20 × 384 well plates) using the appropriate concentration of DMSO as a dispensing solvent (without compounds). Again, three independent experiments should be carried out to ensure robustness. At this point the high-throughput screen should be ready to run.

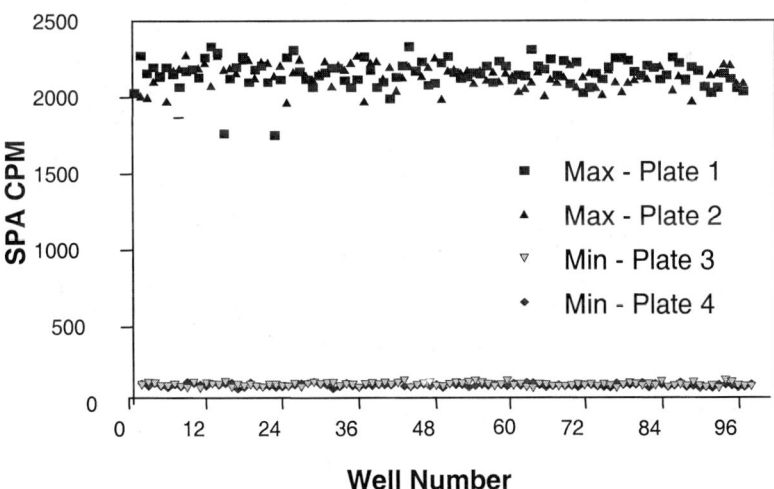

Figure 4   Signal window in a 96-well plate assay showing a coefficient variation < 4% after careful optimization.

B. Assay Readout and Detection Technologies

Recent advances in technology over the past couple of years have led to explosive growth in techniques for use in HTS [2,23–32,49]. Although radioactive assay formats such as scintillation proximity assay (SPA) and filter binding assays are very useful and are still widely used, the current trend is to use nonradioactive labels for HTS. A reason for this trend is that radioactive assays for use in HTS generate enormous quantities of waste and pose serious health and safety risks, not to mention cost.

1. Isotopic Detection Methods

SPA utilizing microbeads has become the standard assay format in many laboratories. SPA provides a "mix and read" approach to measure ligand binding interactions with picomolar to nanomolar K_d values [10,19]. The SPA method employs a scintillant embedded within the microbead matrix. The bead surface is also coated with a capture molecule (e.g., streptavidin, wheat-germ agglutinin, glutathione, protein A). The latter feature is critical, since the bead can capture receptors and biotinylated targets such as other proteins, enzymes, etc. Thus when a radiolabeled molecule (ligand) interacts with the target on the surface beads, it comes within close proximity of the scintillant within the bead. Radioactive decay particles from specific labels (such as ^3H and ^{125}I) interact with the scintillant and generate light signals that can be counted in a manner similar to liquid scintillation counting. A significant advantage is that this feature allows binding measurements to be made without physical separation of the bound and free label. Such physical separations are difficult to automate. SPA can also be carried out by incorporating the scintillant directly onto the surface of the microtiter plate wells (Flashplate®, NEN Life Science Products) [33] or by embedding it directly into the plastic of the microtiter plate (Scintistrip®, Wallac Oy) [34]. Another advance in radioactive HTS is the production of specially designed microtiter plates for cell based assays (Cytostar-T™, Amersham Life Science Inc.) [35]. In this case the scintillant is incorporated into the base of the microtiter plates.

In addition to SPA bead and plate formats, streptavidin-coated membranes that capture the signal of interest have also been developed for high-density (up to 1536) plate footprints (SAM™ Biotin Capture Membrane, Promega Corp.) [36]. In this membrane based technique a filtration step is required.

2. Nonisotopic Detection Methods

Nonradioactive methods are being favored more in HTS laboratories because they tend to offer similar sensitivities and less cost and are safer to use and dispose. In addition, these methods (namely fluorescence) are more amenable to miniaturization than methods such as SPA. For example, a standard 100 μL volume in an

SPA assay generating 4000 CPM would have 40 CPM in a 1 µL volume, requiring a much longer counting time to obtain the same level of precision.

Fluorescence intensity is a very sensitive technique when applied to HTS, but it is not without limitations. Specifically, background fluorescence of biological matrices can dramatically affect the signal. However, several cell based methods employ fluorescence intensity as the detection method, including assays that measure Ca^{2+} flux and membrane potential [4,24].

Another technology that has allowed nonisotopic homogeneous measurements for HTS is fluorescence resonance energy transfer (FRET) [23,24,27,29–31]. The requirements for FRET are that two interacting species of interest are labeled with fluorophores, one with a donor and the other with an acceptor. When the two labeled species are brought to within an appropriate distance (40–50 Å), the donor fluorophore nonradiatively transfers its energy to the acceptor fluorophore. The signal can be measured by monitoring the decrease in the donor fluorescence emission, the increase in acceptor emission, or a ratio of the two. A relatively new reporter system, β-lactamase, measures the disruption of coumarin/fluo-3 acetoxymethyl FRET interaction. It is important to note that any FRET based system can also be measured directly by fluorescence intensity if only one of the fluorophores is monitored.

HTRF [23,24,27,29–31] is another important tool in HTS where fluorescence energy transfer between rare-earth lanthanides such as europium and allophycocyanin [37] or Cy5 dye is used. When the lanthanide is excited (i.e., Eu^{3+} excited at 340 nm) and it is in the vicinity of the acceptor fluorophore, the long fluorescence lifetime of the lanthanide allows time-resolved measurements. Advantages of HTRF are that it has a large Stokes shift, and the background signal due to biological interference is essentially resolved from the signal of interest.

In addition to HTRF, heterogeneous assays can be set up, where the bound and unbound species are separated. The bound europium species is then placed in the presence of dissociation enhanced lanthanide fluorescence immunoassay (DELFIA) enhancement solution [38].

Fluorescence polarization (FP) measures the rotation of a fluorescing species in solution, where the larger the molecule the more polarized the fluorescence emission. FP allows true homogeneous determinations by measuring the ratio of bound to free species [23,27,29–31]. Further, FP allows the "mix and read" format to be utilized for HTS because measurements can be taken without the separation of bound and free species. Another advantage of FP over other techniques, such as FRET, is that only one species needs to be labeled: the smaller entity of the binding pair of interest is usually labeled.

Fluorescence correlation spectroscopy (FCS) has also been recently described for HTS applications [23,27,29,30]. FCS measures binding interactions by observing time-dependent and spontaneous fluctuations in fluorescence intensities in very small volumes (nL). The fluctuations result from the Brownian motion of molecules, which can be directly affected by molecular interactions. Sensitive FCS measurements can be made both in solution and in cellular compartments.

Luminescent assays utilizing reporter genes such as luciferase and β-galactosidase are used to a large extent in HTS laboratories [24,30,31]. Highly sensitive enzyme-linked immunosorbent assays using alkaline phosphatase or horseradish peroxidase are used in HTS as well. Recently, electrochemiluminescence has been reported in which redox-active labels (i.e., ruthenium complexes) have been employed for very sensitive detection [39]. Another very common approach to HTS is by using colorimetric readouts [30,40]. This can be accomplished by either monitoring the production of a colored compound or by measuring the change of one color to another.

New approaches to high and medium throughput screening are being introduced on a regular basis. Very recently, matrix-assisted laser desorption ionization time-of-flight mass spectrometry has been described for multiple-target HTS applications [41]. Using this process, the interactions between the targets and the library of compounds can be monitored. This technique also allows the identification of the individual compounds of interest. Even more radical departure from the plate based methodology is the chip based technology using continuous flow channels. Electrophoretic separation is used to separate colored and fluorescent reaction substrates and products of interest in miniaturized formats [52].

C. Instrumentation: Automation and Detection

The unit operations performed in the overwhelming majority of assays can be divided into four categories: (1) preparing the samples, (2) assembling the assay, (3) detection of the signal, and (4) timing of various operations in the screen.

1. Sample Preparation

Preparation of samples includes the following three steps: (a) transferring an aliquot of sample in DMSO from a storage plate to a sample plate that will be used for a particular assay, (b) diluting the compound in the aqueous assay buffer to a concentration of both compound and DMSO that is compatible with the assay, and (c) transferring an aliquot of diluted sample to the assay plate.

The first step can be performed in advance and stored by a centralized group responsible for compound library management and distribution. The compound dilution and transfer operations are done as part of the assay. These steps are typically the initial operations of most assays and provide a great opportunity for automation and standardization that is reusable across a large number of assays. This is where some of the greatest increases in productivity have occurred in recent years. Only a few years ago, 4- or 8-tip liquid handling robots were being used to transfer samples to the assay plate and throughputs of 15–20 plates/ hr were typical. Because this is the first step of the assay, this operation would commonly determine the daily throughput of the assay. Today, 96 tip pipettors with automated plate handling can achieve throughputs of 100 plates/hr for the same operations.

2. Assembling the Assay

This includes all the remaining steps in setting up the screen. The simplest assays have reagent additions (enzyme, substrates, stop buffers, radiolabel, etc.) uniformly across the entire plate and incubation steps. Incubations may be at room temperature or for cell-based assays may require controlled humidity and elevated temperature and CO_2. More complicated assays may require plate washing, media exchange, filtration, plate-to-plate transfers, and agitation.

3. Detection

Measurement of the signal in the screens is generally carried out in 96- or 384- microtiter plates using colorimetric, fluorescence, luminescence, and radiometric detection techniques to measure assay end points. Plate readers for 96- and 384- well plates are widely available [42,43]. However, new detection devices that permit imaging of plates with high density plates (864, 1536, 3456, and 9600 wells) are becoming available [25,44].

D. Timing

Critical issues from an automation perspective for all of these steps include timing, liquid handling requirements, and plate handling. There are a number of issues related to timing that affect automating an assay for HTS operations:

1. Cycle Time

A multichannel peristaltic reagent dispenser can add a reagent to a stack of twenty-five 96-well plates in less than 6 minutes. However, it can take 6 minutes to read a single plate in a liquid scintillation counter. This large range in throughput rates for the instruments commonly used in HTS causes scheduling difficulties for integrated screening systems. The instruments with short cycle times are underutilized, while the instrument with the longest cycle time limits the throughput of the system. This has been a major factor in the movement from integrated screening systems using robots to move plates from one instrument to another instrument, to stand-alone workstations with stacker based plate handling. By decoupling unit operations, slower process steps can be supported by multiple instruments operating in parallel.

2. Incubation Times

Incubation affects automation in several ways. Short ($<$ 30 minutes), tightly constrained incubation times are difficult to manage unless an integrated system with a scheduler is used. Preferable are incubation periods with sufficient leeway to allow plates to be processed in batches of 25 to 50 without the plate-to-plate variability in timing causing unacceptable variability in the results. Long incubation times (e.g., 48 hr) can increase risk in that a very large quantity of plates will be in process before the first results are read. If there is a quality problem (with a reagent for instance), the total number of plates in the process is in jeopardy.

E. Liquid Handling

There are several types of liquid handling processes. Additions may be as simple as a single reagent added to every well on the plate. More complicated are schemes that require different reagents added to certain wells. These require pipeting instruments that can be single tip, 8 tip, 96 tip, or 384 tip. Some of these instruments aspirate from reagent reservoirs, while syringe pumps feed others. The most complicated liquid handling step would be performing serial dilution of a compound across a row of a plate. Liquid handling instruments can readily perform accurate additions down to 10 μL. 5 μL quantities are achievable with care but require extra attention to instrument setup and maintenance, liquid properties, and fine-tuning of the liquid handling method. Assay miniaturization beyond 384 well plates will require instruments that can accurately dispense in the single- to submicroliter range.

F. Plate Handling

Automated plate handling can be accomplished in several ways. More and more instruments are building in plate handling capability in the form of a "stacker" that moves plates from an input magazine, through the instrument, to an output magazine. In this case, the user only has to load and unload the magazines. A 25-plate magazine of 384 well plates contains 8000 samples that can be manipulated at once. Other instruments without stackers can be integrated to a simple plate handler or to a rotary or rail based robotic system with plate storage carousels.

G. Compound Diversity

The issue of what compounds to test in a screen can be critical to the success or failure of a lead discovery effort. If one had infinite resources in terms of both capacities to screen and compounds, then this would not be an issue. However, in the real world, decisions must be made. One obvious choice is whether to pool compounds and test mixtures of chemical entities [45,46]. The type of target selected may drive this choice. For example, certain cellular assays may show interference from orthogonal pooling techniques. If the libraries are in a pooled format, they may be precluded from considering complex functional cellular assays. Likewise, if the screening technologies are focused on these types of cellular assays, orthogonal pooling may not be an option.

Another method of reducing the number of compounds needed to interrogate a screen is the application of diversity analysis. In most simple terms, this means testing your compound library to see how different they are from one another. To those of us not trained as medicinal chemists, it is a matter of deciding if the compounds look more like clothespins than bunny rabbits. However, the analysis of the diversity of a set of chemicals has become quite complex and controversial [47,48].

H. Data Handling and Analysis

HTS generates large amounts of assay data. Handling this data using unsuitable software systems is frequently a bottleneck in the screening process. The scale of a typical HTS operation (hundreds to thousands of assay plates/day) quickly outstrips the capabilities of the familiar and easy-to-use spreadsheet-like packages. Since regular and timely review of assay data is essential to successful high throughput operations, screening laboratories will need to devote considerable resources to developing data handling systems.

There are several commercially available software packages for HTS (Activity Base, MDL Information Systems, Oxford Molecular, Tripos). These packages vary in their capabilities and should be examined with the point of view that pieces of the process may be best served by a given package, but that supplying a complete solution is unlikely. The various elements in this case will require front- and back-end application support via their programming interfaces or APIs (Application Programmer Interfaces).

The purpose of this section is to examine some essential data-related activities in the screening process, point out properties that the various subsystems should have, and offer some general advice for scientist/developers interested in some or all of their own HTS software. Developing a complete custom package for HTS is a formidable task requiring high levels of cooperation between the various teams that produce and consume information and those that develop the data management tools. Data storage models or objects need to be developed for each of the subsystems, and APIs will have to be developed to permit communication between them. (In this discussion, the term ''object'' refers to the representation of a set of data as a series of text and numerical variables in computer memory.)

There are many factors to be weighed in choosing the best tools and approaches for developing HTS systems. In our own efforts, we have benefited greatly from the use of object-oriented programming methods. These approaches emphasize a modular approach that produces systems that are extensible, easy to maintain, and code-efficient. Every consideration should be given to ease of use and flexibility, but in the end the flexibility desired by the users must be carefully balanced against both the flexibility allowed by the automated assay systems and the IT cost of maintaining overly complex software. A typical model for the operational flow of HTS is shown:

REGISTRATION (Master Library) → Test Sample Inventory (Distribution) → Registered Sample Data + Raw Instrument Data → Calculators → Database.

1. Chemical Registration Inventory and Sample Tracking

All HTS operations require a test sample inventory system for the storage and retrieval of large numbers of natural and synthetic substances. The issues surrounding these activities are beyond the scope of this discussion. However, all HTS operations depend critically on the ability to access sample compounds on a regular basis.

2. Test Sample Registration

The substances to be assayed should have bar codes or other UPC labels on their containers (e.g., a 96-well microtiter plate) and should be positively tracked throughout the screening process. The assay ''run'' is a useful concept for registering the transactions as a unit of work. A run is defined as a group of sample containers that is transferred to a set of assay containers that are then processed, calculated, and loaded to the database as a single transaction. Runs should be identified with a name or index so that the transaction can be queried effectively. A software application, which associates sample containers to assay containers, should be considered the starting point for any screening run. Storage records for the assay data may be created in the database at this point (update model) or in dedicated sample preparation tables for later loading (insert model). The relative merits of the two approaches are discussed below.

3. Data Collection

Assay data for HTS is generated on a wide variety of instruments or ''readers.'' Virtually all these readers are controlled via software provided by their manufacturer. Instrument control software varies widely in complexity but usually has menu options allowing some customization of its exported (e.g., ASCII text) data files. At present, there are no standard data formats adopted by the instrument makers, so developers of an HTS calculating system need to write a number of routines or "data filters" which parse the text file to a format or formats compatible with their system. User intervention in the form of operations such as "cut and paste" should be avoided in these routines. To limit the complexity and variety of the data filters, the automation team can implement process controls to standardize the formats used by the various readers.

4. Calculation and Data Review

Some critically important properties that HTS calculating and review software should have are (1) flexibility—end users should be able to place various control types and choose assay mappings for the calculators from a simple interface. Other specifications needed for proper calculations should be supplied to the calculator from predefined prototypes as well. Caution should be exercised in permitting too much flexibility in the calculating software, as the automated processes themselves are usually the primary reason for limiting flexibility. (2) Graphical interface—numerical trends are best perceived using graphs of the data. If possible, incorporate interactive graphical tools for labeling points, navigating through the plates and subsetting the data. (3) Speed and ease of use— the HTS laboratory is a hectic place. Any tool used by screening personnel should be fast, intuitive to use, and robust. Wherever possible, precompute variables and store them in RAM to allow efficient random navigation through the data. Choices from the interface should be from list controls rather than edit fields. Storing the configuration of the calculator settings with the data object generated by the calculator allows users to correct for minor changes in the protocol (e.g., the location of the controls was accidentally reversed) at run time.

The principal calculations done on the raw data in HTS are of two types. The first of these is the anecdotal or "rapid" type, in which diverse samples are tested for their activity at a single, fixed concentration in one or a few assay wells. To facilitate cross-screening target queries in the results database tables, these results are generally expressed as a percent activity of an assay "window" [20] established by the control samples. The other common modality is the potency assay, where several wells are used to generate a "calculated result" such as an IC_{50} (for inhibitors). The accuracy of this latter value varies widely depending on the methodologies used (i.e., replication, curve fitting method). While all scientists desire high accuracy, it should be remembered that accuracy costs money and consumes precious resources. For most purposes, the confirmatory nature of a 5–10 point dilution of a test sample is sufficient for screening purposes in its early phases. The more advanced assay methodologies should be reserved for promising lead molecules. Most forms of secondary or other "value-added" assays are potency assays that should be run in a high-accuracy mode.

The review process should provide its users the ability to pass or fail sets of data and add comments at appropriate levels. Detailed understanding of the results in terms of the chemical diversity analyzed is not needed at this stage, but process errors within acceptable ranges need to be noted, and false positives are often spotted and removed from the follow-up queue at this stage. Generally, the entire process of calculating, reviewing, and uploading to the results database should take approximately 5–10 minutes per hundred assay plates. Advances in automation and miniaturization [51,52] indicate that these rates and higher will be needed to keep pace effectively in the modern screening laboratory.

5. Long-Term Storage and Data Mining

The results and/or data are the products from HTS. Careful organization and indexing of these data in a relational database is of the utmost importance for maximizing its value. This database is the screening operations "memory" and should be able to provide information rapidly on a wide variety of operational activities, in addition to performing its primary role as a repository of linked chemo- and bioassay information. Developers should be wary of storing too much extraneous information at too high a level. Any memory can be clouded by too much information, and while the batch number of a lead compound needs high-level exposure, the one from the NaCl in the assay does not.

The performance difference in query times between a data model that has been optimally designed and enhanced for performance and one that has not can be many orders of magnitude in size. Databases that take excessively long periods to return an answer, no matter how right that answer may be, defeat the very purpose for which they were created because users tend to quit using them. Instead, they will pull out subsets of the data and maintain them locally and separate from the rest of the data. Decisions will then be made on data that can become out of date or out of context and vary from one copy to the next. The number of users logged onto a database application server and the types of queries they run can have a dramatic impact on the performance of even a well-tuned database, and efforts should be made to shift processing tasks to maximize database availability. Finally, users must be educated on at least a basic level about the types of data stored by their organization and the relative cost of retrieving it. Users of a system who are unaware of these issues often make inefficient use of the database, to the detriment of all its customers.

Fortunately, the sound application of existing technologies in the area of relational database design can and should be adequate to handle the data storage needs of any screening laboratory for the next several years. These same technologies currently allow for the extraction of that data using SQL at acceptable rates. However, enough tools to adequately place this data in context and make it comprehensible to biologists and chemists are not readily available. Too many database access tools consider the records to be the desired output from a database query. In reality, the user often spends a considerable amount of time computing summary statistics and preparing charts of the data. The tools developed for HTS should take the user straight to these summaries and charts to save time. Additionally, the full power of the chemical database needs to be exploited to build structural models to guide follow-up synthetic work.

Tremendous gains in the ability to generate HTS data have created new challenges for the methods used to store and retrieve data from bioassays. To provide the best tools for their operations, many HTS groups develop software for at least some stages of the operation. By applying sound principles of system design, fast and efficient systems can be developed in-house. Although the initial investment in creating these systems is probably higher than that for a packaged system, the returns over the long run are substantial.

V. FUTURE TRENDS

Presently, HTS operations are based on performing the assays within a microtiter plate footprint. The density of the wells within the footprint may vary but the basic mode of operation remains consistent. Simply put, compounds and reagents are mixed within the wells of a plate. Once the reaction is complete some form of detection (e.g., fluorescence, absorption, photon counting) is used to determine the compound's interaction with the reagents. This mode of operation has evolved because of two important factors. The first is a need to have an ordered array so that identification of compounds can be accurately maintained throughout the assay process. Microtiter plates provide a convenient platform for isolating an array of compounds while providing a container to conduct the assay. The second factor is that a large number of commercially available instruments, from liquid handling equipment to detectors, have been designed and standardized to work with the microtiter footprint. Therefore many screens can be conducted with off-the-shelf instrumentation. Although the instruments continue to improve in speed and accuracy, manipulation of a large number of plates can create some bottlenecks within the process.

Emphasis is now being placed on gaining efficiency by screening in plates of higher well density (e.g., 384, 1536, or 9600 wells) [42–44]. Due to the smaller well volumes, conducting screens in higher density plates have the real potential to decrease screening costs by dramatic reduction of reagents consumed. In some cases, miniaturization may enable screening targets, since an enzyme or substrate may be difficult to obtain in quantities required for microliter scale screening. The speed of the process can increase, since time spent dispensing liquids is reduced, and the number of plate manipulations for a fixed number of compounds is significantly smaller. Even some compound dilution steps may be eliminated if the nanoliter dispensing technology proves to be accurate enough in dispensing stock solutions of compounds. The substantial benefits that can be obtained with nanoscale screening do come at a price. Most instrumentation currently used in HTS laboratories cannot work with plates at densities higher than 384. Therefore significant new capital investment in liquid handling and detector instrumentation is required. Liquid dispensing will move from positive displacement syringe driven systems to piezo or solenoid valve dispensing systems. These devices require a much higher level of particle control to avoid obstruction of the flow paths, but the technology should prove reliable enough for most screening labs. Experimental conditions such as liquid volume variations due to evaporation, that were not a concern in 96-well plates, are critical in 1536-well plates and those of higher densities [44,49]. There is also concern with cell based screens in this format. Primarily, will enough viable cells make it through the dispensing heads into each well and produce an adequate detection window? However, even with the considerable challenges associated with miniaturization screening systems several factors will continue to push the industry to smaller assay volumes. The productivity of many combinatorial chemistry efforts continues to expand the diversity and size of the compound library [47,48]. In addition, large programs are under way to increase the number of targets discovered through genomics. These two factors are creating pressure to increase the number of assays per screen, while increasing the number of targets. Therefore, in an effort to control costs associated with screening, the industry will continue to move towards higher density plates.

A term often linked to the development of high-density plates is UHTS. The number of assays commonly affiliated with UHTS is 100,000 per day. Therefore it would take processing of a relatively few high-density plates to obtain UHTS throughput levels. However, introduction of nanoliter dispensing technology is not necessarily a requirement for UHTS. The introduction of 96- or 384-channel liquid handling devices and rapid scanning fluorescence detection have enabled many laboratories to obtain the 100,000 compounds/day level of throughput for a select group of screens without the introduction of microfluidics dispensing techniques and the associated problems.

There are additional new technologies being developed that may evolve into the next generation of UHTS platform. One of the technologies on the horizon is a system that is capable of performing biochemical analysis or even chemical synthesis within flowing microchannels. These devices are often referred to as biochips or genechips [50,51]. The heart of the system begins with a small (about 2 square inches) flat substrate, such as glass, quartz, or possibly some polymers that have small channels etched within the block [52]. Typical channel size is 50 microns wide by 10 microns deep, creating an internal volume of 0.5 nL per every 1 mm of channel length. Channels are etched into the substrate by means of a mask produced by photolithography, and etching occurs with exposure to an acid. The photolithography technology is a proven production technique for integrated circuits within the electronics industry. Fluid movement through the microchannels occurs by electrokinetic forces. Microscopic versions of pumps, valves, reactors, extractors, and advanced separations systems can be created within the mirochip. For a screen application, channels would be etched into a pattern to simulate the process flow of reagent additions. For example, entrance ports and microchannels can be created for compound stock solutions and diluent. These channels would merge to form a single channel. From this point, channels supplying substrate, enzymes, or stop agents are added downstream. Areas for incubation and detection can also be added. After the reaction occurs an area of a microchannel will be used as a flow cell for spectroscopic analysis. Alternately, the chip effluent could be directed into a mass spectrometer for analysis. These systems seem particularly well suited for biochemical assays, especially if the reaction kinetics are fast.

VI. CONCLUSION

Drug screening for identifying novel chemical structures for different targets is moving to HTS and UHTS. The assays have to be simple, homogeneous, robust, reproducible, and fully automatable to adapt to HTS. The design of a cell based or an in vitro biochemical assay that is radioisotope based or nonradioisotope based takes into consideration the type of target, available reagents, assay readout, and availability of automation equipment. With the advent of homogeneous fluorescence based assays and other novel signal detection technologies and miniaturization strategies (384-, 1536-, 3456-, and 9600-well plates) it is possible to meet the 100,000 assays per day capacity. The microchip technology coupled with microfluidic technology or capillary electrophoresis is being developed to meet the needs of UHTS.

ACKNOWLEDGMENTS

The authors thank the Assay Technology Group for useful discussions and Steven Kahl for Figure 4.

REFERENCES

1. Nature supplement, Intelligent drug design. Nature 384, suppl 1–26, 1996.
2. AJ Kolb, K Neuman. Beyond the 96-well microplate: instruments and assay methods for the 384-well format. J Biomol Screen 2:103–109, 1997.

3. E Zubritsky. Microplate fluorometers reach critical mass. Anal Chem News Views 39A–49A, 1999.
4. RY Tsien. Probes of dynamic biochemical signals inside living cells. In: AW Czarnick, ed. Fluorescent Chemosensors for Ion and Molecule Recognition. American Chemical Society, Washington DC, 1992, pp 130–146.
5. RM Evans. The steroid and thyroid hormone superfamily. Science 240:889–895, 1988.
6. SL Schreiber. The chemistry and biology of immunophilins and their immunosuppressive ligands. Science 251:283–287, 1991.
7. FP Bymaster, DO Calligaro, JF Falcone, RD Marsh, NA Moore, NA Tye, P Seeman, DT Wong. Radioreceptor binding profiles for the atypical antipsychotic olanzapine. Neuropsychopharmacology 14:87–96, 1996.
8. GL Stiles. A1 adenosine receptor G-protein coupling in bovine brain membranes: effects of guanine nucleotides, salt and solubilization. J Neurochem 51:1592–1598, 1988.
9. R Bosse, R Garlick, B Brown, L Menard. Development of non-separation binding and functional assays for G-protein coupled receptors for high throughput screening: pharmacological characterization of the immobilized CCR5 receptor on FlashPlate. J Biomol Screen 3:285–292, 1998.
10. S Udenfriend, I Gerber, N Nelson. Scintillation Proximity Assay: a sensitive and continuous isotopic method for monitoring ligand/receptor and antibody/antigen interactions. Anal Biochem 161:494–500, 1987.
11. Z Li, S Mehdi, I Patel, J Kawooya, M Judkins, W Zhang, K Diener, A Lozada, D Dunnington. An ultra-high throughput screening approach for an adenine transferase using fluorescence polarization. J Biomol Screen 5:31–37, 2000.
12. AW Murray, MW Kirschner. What controls cell cycle. Sci Amer 264:56–63, 1991.
13. LM Sonatore, D Wisniewski, LJ Frank, PM Cameron, JD Hermes, AI Marcy, SP Salowe. The utility of FK506-binding protein as a fusion partner in scintillation proximity assays: application to SH2 domains. Anal Biochem 240:289–297, 1996.
14. G Mathis. Probing molecular interactions with homogeneous techniques based on rare earth cryptates and fluorescence energy transfer. Clin Chem 41:1391–1397, 1995.
15. VA Ellsmore, AP Teoh, A Ganesan. A high throughput enzyme-linked immunosorbent assay for inhibitors of the interaction between Retinoblastoma protein and the Leu-X-Cys-X-Glu motif. J Biomol Screen 2:207–211, 1997.
16. AF Braunwalder, L Wennogle, B Gay, KE Lipson, MA Sills. Application of scintillation microtiter plates to measure phosphopeptide interactions with the GRB2-SH2 binding domain. J Biomol Screen 1:23–26, 1996.
17. TL Messier, CM Dorman, H Brauner-Osborne, MR Brann. High throughput assays of cloned adrenergic, muscarinic, neurokinin, and neurotrophin receptors in living mammalian cells. Pharmacol Toxicol 76:308–311, 1995.
18. KR Oldenburg, KT Vo, B Ruhland, PJ Schatz, Z Yuan. A dual culture assay for detection of antimicrobial activity. J Biomol Screen 1:123–130, 1996.
19. SD Kahl, FR Hubbard, GS Sittampalam, JM Zock. Validation of a high throughput scintillation proximity assay for 5-hydroxytryptamine-1E receptor binding activity. J Biomol Screen 2:33–40, 1997.

20. GS Sittampalam, PW Iversen, JA Boadt, SD Kahl, S Bright, JM Zock, WP Janzen, MD Lister. Design of signal windows for high throughput screening assays for drug discovery. J Biomol Screen 2:159–169, 1997.
21. J-H Zhang, TDY Chung, KR Oldenburg. A simple statistical parameter for use in evaluation and validation of high throughput screening assays. J Biomol Screen 4: 67–73, 1999.
22. M Lister, WP Janzen. ISLAR '96, Proceedings 43.
23. PB Fernandes. Technological advances in high throughput screening. Curr Opin Chem Biol 2:597–603, 1998.
24. JE Gonzalez, PA Negelescu. Intracellular assays for high throughput screening. Curr Opin Biotech 9:624–631, 1998.
25. KR Oldenburg, J-H Zhang, T Chen, A Maffia, F Blom, AP Combs, TDY Chung. Assay miniaturization for ultra high throughput screening of combinatorial and discrete compound libraries: A 9600-well (0.2 microliter) assay system. J Biomol Screen 3:55–62, 1998.
26. J Driscoll, R Delmondo, R Papen, D Sawutz. MultiPROBE nL components for drug discovery assay miniaturization. J Biomol Screen 3:237–239, 1998.
27. L Silverman, R Campbell, JR Broach. New assay technologies for high throughput screening. Curr Opin Chem Biol 2:397–403, 1998.
28. KA Giuliano, RL DeBiasio, RT Dunlay, A Gough, JM Volosky, JM Zock, GN Pavlakis, DL Taylor. High content screening: a new approach to easing bottlenecks in the drug discovery process. J Biomol Screen 2:249–259, 1997.
29. MV Rogers. Light on high throughput screening: fluorescence-based assay technologies. Drug Disc Today 2:156–160, 1997.
30. GS Sittampalam, SD Kahl, WP Janzen. High throughput screening: advances in assay technologies. Curr Opin Chem Biol 1:384–391, 1997.
31. JJ Burnbaum, NH Sigal. New technologies for high throughput screening. Curr Opin Chem Biol 1:72–78, 1997.
32. E Litborn, M Stjerstrom, J Roeraade. Nanoliter titration based on piezoelectric drop-on demand technology and laser induced fluorescence detection. Anal Chem 70: 4847–4852, 1998.
33. BA Brown, M Cain, J Broadbent, S Tompkins, G Henrich, R Joseph, S Casto, H Harney, R Greene, R Delmondo, S Ng. FlashPlate™ Technology. In: JP Delvin, ed. High Throughput Screening. New York: Marcel Dekker, 1997, pp 317–328.
34. GR Nakayama, MP Nova, Z Parandoosh. A scintillating microplate assay for the assessment of protein kinase activity. J Biomol Screen 3:43–48, 1998.
35. L Smith, MJ Price-Jones, KT Hughs, NRA Jones. Counting CytoStar-T™ scintillating microplates on the Microbeta® JET. J Biomol Screen 3:227–230, 1998.
36. A Tereba. High density protein kinase assay with sub-attomole sensitivity. J Biomol Screen 3:29–35, 1998.
37. GW Meller, MN Burden, M Preaudat, Y Joseph, SB Cooksley, JH Ellis, MN Banks. Development of a CD28/CD86 (B7-2) binding assay for high throughput screening by homogeneous time-resolved fluorescence. J Biomol Screen 3:91–99, 1998.
38. J Liu, M Gallagher, RA Horlick, AK Robbins, ML Webb. A time-resolved fluorometric assay for galanin receptors. J Biomol Screen 3:199–206, 1998.

39. RO Williams. Electrochemiluminescence: a new assay technology. IVD Technology, November 1995.
40. JCW Comley, T Reeves, P Robinson. A 1536 colorimetric SPAP reporter assay: comparison with 96- and 384-well formats. J Biomol Screen 3:217–225, 1998.
41. F Hsieh, H Keshishian, C Muir. Automated high throughput multiple target screening of molecular libraries by microfluidic MALDI-TOF MS. J Biomol Screen 3: 189–198, 1998.
42. DLT Rose. Challenges in implementing high-density formats for high throughput screening. Laboratory Automation News 2:12–19, 1997.
43. WP Janzen, P Domanico. The 384-well plate: pros and cons. J Biomol Screen 1: 63–64, 1996.
44. JCW Comley, A Binnie, C Bonk, JC Houston. A 384-HTS for Human Factor VIIa: comparison with 96- and 864-well formats. J Biomol Screen 2:171–178, 1997.
45. M Snider. Screening compound libraries consommé or gumbo? J Biomol Screen 3: 169–170, 1998.
46. TDY Chung. Screen compounds singly: why muck it up. J Biomol Screen 3:171–173, 1998.
47. JH Wikel, RE Higgs. Application of molecular diversity analysis in high throughput screening. J Biomol Screen 2:65–67, 1997.
48. RW Spencer. Diversity analysis in high throughput screening. J Biomol Screen 2: 69–70, 1997.
49. J Major. Challenges and opportunities in high throughput screening. J Biomol Screen 3:13–17, 1998.
50. J Moukheiber. A hail of silver bullets. Forbes Magazine, January 26, 1998.
51. BJ Sedlak. Gene chip technology ready to impact diagnostic markets. Gen Eng News, 17 December 1997.
52. JJ Burbaum, M Knapp, S Sundberg, JW Parce. Point-counterpoint. J Biomol Screen 5:5–12, 2000.

3
Screening Platforms

Ramakrishna Seethala
Bristol-Myers Squibb Company, Princeton, New Jersey

I. INTRODUCTION

Drug discovery in the pharmaceutical and biotechnology companies has seen spectacular changes in the last decade mainly due to technological advances in biology, biochemical assays, genomics, proteomics, combinatorial chemistry, miniaturization, automation, computerization, and information technologies. With this technological revolution, drug screening capacities have increased immensely, allowing the pharmaceutical companies to process an increased number of drug targets rapidly by miniaturization and automation using robots for screening the large compound decks and combinatorial compound libraries. To bring a drug to market quickly by reducing the time taken from the target identification to the drug development, pharmaceutical companies have been implementing an assembly-line approach by automation of drug screening.

The screen paradigm is given in Figure 1. From the time a target is selected to screening the compound deck involves the development of assays, optimization of assays, screen development, optimization, and validation before screening the compound deck. The goal of screening groups is to generate a large number of hit compounds that can be advanced to lead compounds that will be further advanced to pre-clinical and clinical development [1]. The aim of the pharmaceutical companies is to achieve first-in-class or best-in-class drugs so that they may become blockbuster drugs. This mandates quality screens and compound decks that enable the generation of quality hits. The screening groups implementing innovative screening technologies, by the use of robots, novel assays, and new miniaturized screening formats, may create a technology gap between the screening groups and the therapeutic area [2]. If the assays developed in therapeutic

31

Screen Paradigm

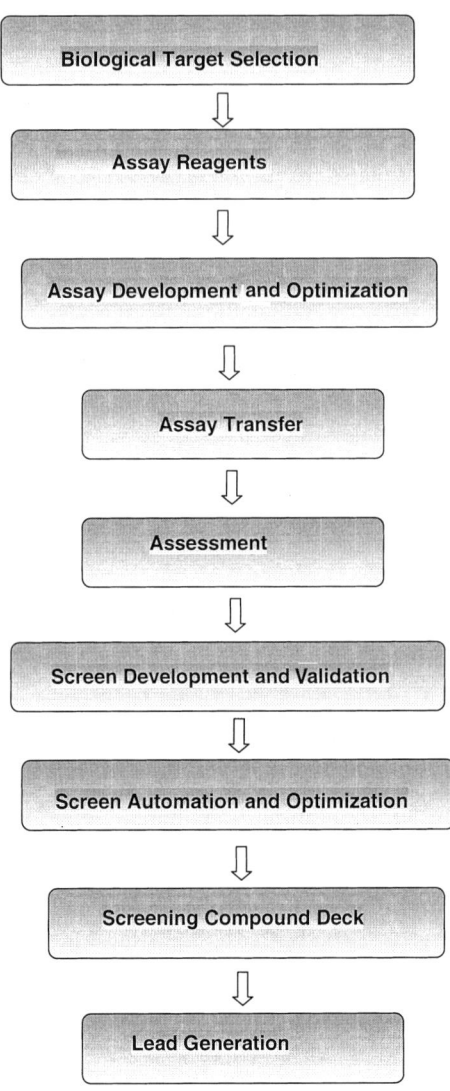

Figure 1 Screen paradigm. General lead generation organizational structure in big Pharma is given here. The target selection, developing assay reagents, and assay development and optimization are usually done in the therapeutic areas in consultation with an HTS group. The therapeutic area transfers the assay to a centralized HTS group where the assay and target priority are reviewed. From the assay transferred a screen is developed and validated. The screen is automated and optimized and then run against the large compound deck (compound collection), which will result in lead generation. Leads will be confirmed, and the dose–response for those will be determined. Lead information is sent to the therapeutic area for followup on the leads.

areas are not compatible with the high-throughput screening (HTS) and ultrahigh-throughput screening (uHTS), these assays have to be modified or restarted to develop into good screens, thus delaying screening and impeding on screening resources. Hence understanding of various screen platforms and thinking very early in a program to develop assays compatible with HTS will enable screening groups to process the most efficient way. The therapeutic area biologist who is familiar with various screen platforms will be in an advantageous position to evaluate the best assay for the target screening with the identification of a target. In consultation with screening groups, the best course of the screen can be determined very early in the conception of the project, which will enable developing appropriate reagents in sufficient quantities required for the screening of enormous compound decks (compound collections) of the big pharmaceutical companies.

The HTS hits are taken to lead generation by testing for proof of concept, and, depending on the pharmacophore, a few of them are taken to lead optimization by synthesis of compounds by medicinal chemists to meet the goals set for a lead compound that can go into preclinical and clinical studies. Lead optimization many times needs synthesis of several hundred compounds that have to be checked for specificity in primary, secondary, and tertiary screens, which necessitates a throughput of medium to high. It is essential to use screens that give robust signals and can be carried with ease as primary screens.

Sometimes, if developing a very good screen takes several weeks due to problems in making the reagents and developing a robust assay, but if it is feasible to assemble a medium-throughput screen with the available reagents, then one can go ahead with such a screen if it saves time and resources. Some of the secondary or tertiary assays may not be easy to develop into good screens and may have to live with the analytical (test tube) assays that are available. Understanding various screen platforms will help in arriving at the best screens with the available equipment and reagents. The quality of leads that will be generated will be only as good as the quality of the screen and the quality of the compound library.

The design of the assay depends on the target, the availability of reagents and plate readers, and the adaptability to miniaturization and automation for screening. It is always advantageous to develop an assay in the format preferred by your HTS group so that the screen is validated and run immediately. Because the deck compounds have to be run in many targets without depleting the compound stocks, it is preferable to develop a screen that can be easily adapted to higher density plates. 96-well plate assays may be the assays of the past; the current screening in HTS groups use at least 384-well plates. Although higher density plates like 1536-, 3456-, and 9600-well plates are in the experimental stage, in a couple of years, when liquid delivery, mixing, and evaporation are better understood and regulated, HTS may be switched to these higher density plate formats.

II. ASSAY FORMATS

Drug discovery until the 1970s depended on low-throughput and single-test-tube assays. With the advent of multiwell plate assays, the throughput has increased, and 96-well plates gained popularity in the last two decades. With multiwell plate availability, spectrophotometers that read 96-well plates have been available for more than two decades, allowing us to develop HTS assays. Slowly, other 96-well plate readers, fluorometers, luminometers, scintillation counters, and multimode plate readers became available. Most of the plate readers now available have stackers that hotel several plates (10–40 plates). The plate readers with stackers are very useful for reading end point reactions in several plates, and these plate readers are compatible with robotic systems to automate fully the screening process.

A. Plate Sizes

A standard 96-well microplate with 8×12 array will have round wells with flat, conical, or round bottoms. The typical assay volume is 100–200 μL, and volume has to be > 40 μL for a good mix of reagents. The outside dimensions of the base, the footprint, recommended by the Society of Biomolecular Screening (SBS) are a length of $5.030 \times 0.01''$ and a width of $3.365 \pm 0.01''$ with a $\pm 0.004''$ tolerance [3,4]. Three heights of the plate recommended for 96-well plates are a single height ($0.565 \pm 0.01''$), a double height ($1.130 \pm 0.01''$), and a deep well ($1.695 \pm 0.01''$) height. The deep well plates can have square wells to increase the capacity of each well. The lid footprint is length 5.030 ± 0.01 and width $3.365 \pm 0.01''$. Since plate manufacturers followed the standard dimensions of 96-well plates, there has been tremendous development in liquid dispensing, automation, and plate readers.

To increase the throughput (screening capacity) and save on the reagents used for screening, higher density plates (384-well plates and the like) are being tested for HTS in the last 5 years. The conversion from 96-well plates to higher density plates has been very slow initially. The 384-well plates with a 16×24 array have outside dimensions (recommended by the SBS) of $5.030''$ and $3.365''$ with a $\pm 0.01''$ tolerance. The typical assay volume is 25–50 μL. Most of the liquid dispensing systems and plate readers used for 384-well plates are the same as those that are used for 96-well plate assays. There are not that many dedicated 384-well plate dispensing and signal reading systems. Hence there is not much time to be saved, as it takes $4\times$ more time in liquid dispensing and signal reading compared to 96-well plates, which led to a lack of enthusiasm for 384-well plates initially. The driving force to go to the 384-well format is miniaturization of the assay, thus saving on reagents and precious deck library compounds, increasing the capacity of assays for continuous automated screens, and reducing waste by

75% (such as used plates), thus saving about 33% of costs (see Chap. 20). Now most of the drug discovery screening groups have transitioned to screening a plate format of 384-well plates thanks to the development of excellent instrumentation for accurate, low volume handling and sensitive detection.

Further miniaturization of assays in much higher density plates such as 1536-well plates is now in the testing phase. The liquid dispensing systems and signal readers for the 1536-well plate are being developed. 1536-well plates with 32 × 48 arrays have outside dimensions (recommended by the SBS) of 5.030″ and 3.365″ with a ±0.01″ tolerance. The typical assay volume in the 1536-plate is 2–10 µL. The inherent problems of assay volumes at the single-digit microliter and submicroliter level include evaporation, mixing of components, and signal readouts. There is substantial progress in instrumentation of low-volume liquid dispensing and sensitive detection. With the development of assay technologies such as imaging technologies, in a few years screening will probably progress to the 1536-well plate format and to increased throughput to uHTS. There will be substantial savings in costs with the adaptation to 1536-well plate screening, savings in the invaluable library of compounds with several compounds coming from combinatorial chemical synthesis in small quantities, and it will be possible to screen several targets that are going to be available from genomics.

A few labs and biotech companies have been using, sporadically in test mode if not in HTS, 864-well, 3456-well, and 9600-well plates. For these plate formats to gain popularity, the liquid dispensing systems and plate readers have to be developed.

There is a new trend to go from plate format to chip-based technologies where in liquid handling is controlled electronically. Chip technology is now currently used in genomics and gene expression screening. Although it may take a few years before this technology becomes a routine screening reality, then the screening will be very inexpensive, reach uHTS, and deliver hits for lead optimization, helping in go-no-go decision of a target for further development in a very short time.

B. Plate Types

Microtiter plates are made of either polystyrene or polypropylene. Polylysine-coated multiwell polypropylene plates sterilized by UV irradiation are used for tissue culture. 96- and 384-well microtiter plates are available with conical, flat, and round bottoms. Depending on the assay, the plate format is selected. For cell-based assays in adherent cell format, flat bottomed tissue culture plates are used. For cell-free assays, either round bottomed or conical bottomed plates are used, which allows a good mix of compounds in smaller volumes. Plates can be clear, white opaque, black opaque, black with clear bottom, or white with clear bottom, depending on the application. It is a common practice to use clear plates for plating the cells in cell-based assays, and treating the cells with compounds, and aliquots are transferred to read plates, which are black or white plates. In cell-based assays, with the availability of tissue culture treated white and black plates with clear bottom, these plates are utilized for growing cells, treating cells with compounds, and developing the signal and reading the signal directly; or often the signal is read after covering the clear bottom with an opaque stick-on paper. Now tissue culture treated white plates with opaque bottom (Packard) are also available, and cell-based assays can be readily done in these plates. In radioactive and luminescence- and fluorescence-based assays, opaque plates are advantageous, because the crosstalk (contribution from the surrounding wells to the signal in the well being read) is minimized in these plates.

C. Plate Layout

The layout depends on the type of assay, the screening or dose response, the number of replications, the test compound distribution plate layout, and the type of readout. If the assay is robust and there are no edge effects, screening is run in single point with 80 compounds per 96-well plate. The typical plate layout is identical to the plate layout of the compound plate with the empty wells being used for standards, controls, and blanks. The first assay plate layout consists of test compounds in all wells from A1 to H10, whereas A11–H11 and A12–H12 are used for positive and negative controls (Fig. 2A). In another layout, test compounds are in A1–H2, A4–H5, and A7–H12, whereas A3–H3 and A6–H6 are used for controls and blanks (Fig. 2B). In another layout, test compounds are in A1–G12 and standards in H1–H12 (Fig. 2C). This layout is mostly used for dose response, with 6 concentrations in duplicate or 12 concentrations in a row and duplicates in the next row to determine IC_{50} or EC_{50} of the hits. When 8- or 10-point dose response is run, the Figure 2D plate layout, either columnwise or rowwise, respectively, can be used. Duplicates can be run, either in adjacent wells in the same or the next row or in two separate plates in the identical position. In a 384- or 1536-well plate, suitable layouts can be used depending on the compound plate layout; and if dose response is run, the layout depends on the number of points and replications. Depending on the plate layout, programs for liquid dispensing protocols and data processing programs are selected.

Figure 2   Different plate layouts for screening in 96-well plates. (A) In each 96-well plate, 80 test samples (T) are in A1 to H10, and the standards (St) and blanks (Bl) are in A11–H11 and A12–H12, respectively. (B) To avoid edge effect the standards and blanks are placed in A3–H3 and A6–H6, respectively. (C) Dose–response layout with duplicates in adjacent wells in the same row with 6 concentrations of the compound. (D) Dose–response at eight concentrations column-wise with duplicates in adjacent wells in the same row.

A

	1	2	3	4	5	6	7	8	9	10	11	12
A	T	T	T	T	T	T	T	T	T	T	St	Bl
B	T	T	T	T	T	T	T	T	T	T	St	Bl
C	T	T	T	T	T	T	T	T	T	T	St	Bl
D	T	T	T	T	T	T	T	T	T	T	St	Bl
E	T	T	T	T	T	T	T	T	T	T	St	Bl
F	T	T	T	T	T	T	T	T	T	T	St	Bl
G	T	T	T	T	T	T	T	T	T	T	St	Bl
H	T	T	T	T	T	T	T	T	T	T	St	Bl

B

	1	2	3	4	5	6	7	8	9	10	11	12
A	T	T	St	T	T	Bl	T	T	T	T	T	T
B	T	T	St	T	T	Bl	T	T	T	T	T	T
C	T	T	St	T	T	Bl	T	T	T	T	T	T
D	T	T	St	T	T	Bl	T	T	T	T	T	T
E	T	T	St	T	T	Bl	T	T	T	T	T	T
F	T	T	St	T	T	Bl	T	T	T	T	T	T
G	T	T	St	T	T	Bl	T	T	T	T	T	T
H	T	T	St	T	T	Bl	T	T	T	T	T	T

C

	1	2	3	4	5	6	7	8	9	10	11	12
A	1 10	1 10	1 3	1 3	1 1	1 1	1 0.3	1 0.3	1 0.1	1 0.1	1 .03	1 .03
B	2 10	2 10	2 3	2 3	2 1	2 1	2 0.3	2 0.3	2 0.1	2 0.1	2 .03	2 .03
C	3 10	3 10	3 3	3 3	3 1	3 1	3 0.3	3 0.3	3 0.1	3 0.1	3 .03	3 .03
D	4 10	4 10	4 3	4 3	4 1	4 1	4 0.3	4 0.3	4 0.1	4 0.1	4 .03	4 .03
E	5 10	5 10	5 3	5 3	5 1	5 1	5 0.3	5 0.3	5 0.1	5 0.1	5 .03	5 .03
F	6 10	6 10	6 3	6 3	6 1	6 1	6 0.3	6 0.3	6 0.1	6 0.1	6 .03	6 .03
G	7 10	7 10	7 3	7 3	7 1	7 1	7 0.3	7 0.3	7 0.1	7 0.1	7 .03	7 .03
H	St	St	St	St	St	St	St	St	St	St	St	St

D

	1	2	3	4	5	6	7	8	9	10	11	12
A	1 10	1 10	2 10	2 10	3 10	3 10	4 10	4 10	5 10	5 10	St	Bl
B	1 3	1 3	2 3	2 3	3 3	3 3	4 3	4 3	5 3	5 3	St	Bl
C	1 1	1 1	2 1	2 1	3 1	3 1	4 1	4 1	5 1	5 1	St	Bl
D	1 0.3	1 0.3	2 0.3	2 0.3	3 0.3	3 0.3	4 0.3	4 0.3	5 0.3	5 0.3	St	Bl
E	1 0.1	1 0.1	2 0.1	2 0.1	3 0.1	3 0.1	4 0.1	4 0.1	5 0.1	5 0.1	St	Bl
F	1 .03	1 .03	2 .03	2 .03	3 .03	3 .03	4 .03	4 .03	5 .03	5 .03	St	Bl
G	1 .01	1 .01	2 .01	2 .01	3 .01	3 .01	4 .01	4 .01	5 .01	5 .01	St	Bl
H	1 .001	1 .001	2 .001	2 .001	3 .001	3 .001	4 .001	4 .001	5 .001	5 .001	St	Bl

III. ASSAY TECHNIQUES

Assays measure either activation or inhibition. In activation assays, the basal (control) activity is minimal; some compounds do not increase the activity (inactive compounds), and compounds increase to an expected level (usually greater than 3σ with defined kinetic values) (active compounds). A compound achieves complete activation when it reaches the maximal activity similar to the activation seen with a natural activator such as hormone or ligand (used as gold standard), and those compounds that do not activate fully are partial activators. In inhibition assays, the activity of an enzyme, receptor–ligand binding, or receptor functional activity in the presence of a ligand or an activator is not effected by inactive compounds, and compounds that show expected inhibition may be taken as potential leads; with some compounds, complete inhibition (the activity reaches background) is achieved. Depending on the target, and the biological approach to the disease, an assay may be developed to identify activators or inhibitors.

The quality of an assay is loosely expressed as a signal-to-noise ratio (S/N) or signal-to-background ratio (S/B).

$$S/N = \frac{\text{Mean signal } - \text{ mean background}}{\text{Standard deviation of background}} \tag{1}$$

$$S/B = \frac{\text{Mean signal}}{\text{Mean background}} \tag{2}$$

However, the S/N ratio does not contain all the information needed to evaluate the quality of assay. On the other hand, the S/B ratio does not contain any information regarding data variation. Thus neither ratio is appropriate in the evaluation of an assay. Recently, Zhang et al. [5] reported a new parameter, the z' factor, for the evaluation of the quality of a biological assay for HTS (for details see Chap. 20).

$$z' = 1 - \frac{(3\sigma_3 + 3\sigma_c)}{(\mu_s - \mu_c)} \tag{3}$$

where σ_s is the standard deviation of the sample and σ_c is the standard deviation of the control, μ_s is the mean of the samples, μ_c is the mean of the controls, and $(\mu_s \times \mu_c)$ defines the usable dynamic range of the screen. The z' factor is a useful tool for the evaluation, comparison, and validation of any bioassays in general and HTS assays in particular. The z' factor around 1 is an ideal assay for HTS, $< 1–0.5$ is a good assay, $< 0.05–0.1$ is a moderate assay, and 0 or < 0 is a very poor assay that cannot be used for HTS.

The assay methods can be broadly divided into two groups: (1) in vitro (cell-free) biochemical assays that are used for enzyme assays and receptor binding assays, and (2) cell-based assays. The assays in these two groups may be either heterogeneous or homogeneous types. The traditional assays about two decades before were single tube analytical assays with low throughput. Although the compound deck size was relatively small, the screening of each target took several months, and screening groups were able to handle only a handful of screens. The gradual transformation of screening to the plate format and improved engineering, spectroscopic, and biochemical techniques enabled workers to develop very sensitive homogeneous and heterogeneous assays with increased throughput and automation of the screening, allowing them to screen increased sizes of compound libraries, with reductions in the costs of materials and screening time. With the development of highly sensitive assays and the availability of higher density plates and sensitive plate readers that can read high-density plates, screening is progressing to uHTS.

IV. IN VITRO (CELL-FREE) ASSAYS

Cell-free assays include simple to very complex systems (Fig. 3). These biochemical assays include enzyme assays, protein–protein interactions, and membrane receptor–ligand and soluble receptor–ligand binding assays. The advantages of in vitro biochemical screening include more ready accessibility of the compounds

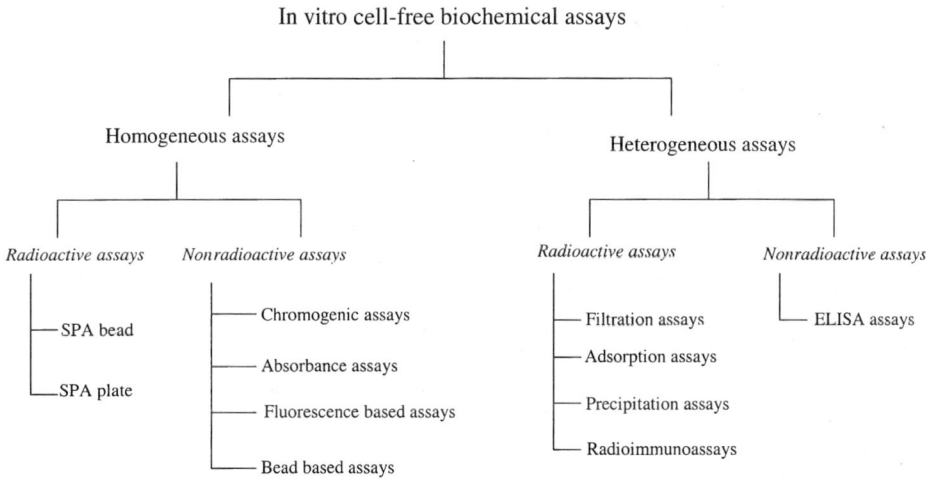

Figure 3 Cell-free biochemical assays. The in vitro assays are mainly classified into heterogeneous and homogeneous assays and subdivided into radioactive and nonradioactive assays.

to the target, easy identification of the target of the compound without any ambiguity, a well-defined mechanism of action, the possibility of developing inexpensive screens, the easy adaptability to newer technologies, amenability to miniaturization, and more ready automation [6].

A. Heterogeneous Assays

Heterogeneous assays are multistep assays that involve multiple additions, incubations, washings, transfers, filtrations, and readings of the signal. These assays are labor intensive, with complicated steps, and generally are difficult to automate and run as HTs. Radioactive and nonradioactive heterogeneous methods have been in use.

1. Nonradioactive Heterogeneous Assays

Enzyme immunoassays are very widely used *in vitro* assays.

Enzyme Linked Immunosorbent Assay (ELISA). ELISAs are the most commonly used *in vitro* assays in multiwell format. ELISAs are heterogeneous assays that typically use a detection system composed of a primary antibody for antigen recognition coupled to a secondary antibody conjugated with an enzyme (horseradish peroxidase or alkaline phosphatase) for signal amplification with an appropriate substrate. In the ELISA the substrate is coated on the surface of the plate, and an enzyme reaction is performed that converts the substrate to the product; a product-specific antibody couples to the product on the surface and then there is interaction with the secondary antibody–enzyme conjugate. After each step, the plates are washed to remove excess reagents. The bound secondary antibody–enzyme conjugate is measured after incubation with substrate that yields a color, fluorescence, or chemiluminescence signal, the last being most sensitive. For example, in the protein tyrosine kinase assay, a poly(Glu-Tyr 4:1) peptide substrate is coated to the microplate wells, and after a protein kinase reaction, tyrosine phosphorylated (PY) peptide (the product) is coupled to PY monoclonal antibody, which in turn is coupled to peroxidase-labeled goat antimouse IgG secondary antibody; color is developed by incubation with *o*-phenylenediamine, and this color is read at 492 nM in a plate reader [7].

A simple enzyme immunoassay consists of an enzyme substrate attached to a microplate well surface; the product formed is coupled to a product-specific antibody labeled with an enzyme detection system or a fluorescent tag. For example, in a protein tyrosine kinase assay a microplate is coated with poly(Glu-Tyr 4:1) peptide substrate; after enzyme reaction, the phosphorylated peptide is coupled to the Eu-labeled PY antibody, washed, treated with enhancement reagent, and read in a fluorometer (excitation 320 nM and emission 615 nM) [8].

2. Radioactive Heterogeneous Assays

Assays utilizing radiolabeled compounds are very commonly used because of their high sensitivity and robustness, despite handling hazards and radioactive waste generation. Receptor–ligand binding assays, until recently, have utilized radioactive ligand binding to receptor membranes and cells. Protein phosphorylation assays mainly used $^{33}PO_4$ or $^{32}PO_4$ transfer from [^{33}P or ^{32}P]-ATP. The classical methods of separation of radioactive products from radioactive substrates consist of filtration, adsorption, and precipitation, and the throughput by these methods is considerably low.

Filtration Assays. In the conventional method, the radioactive product is separated from the radioactive substrate by filtration through glass-fiber filters followed by washing several times. After drying the filter at room temperature, the filter is transferred into a vial, scintillant is added and counted in a scintillation counter. About 15 years back, filter manifolds (Millipore, Bedford, MA) that can hold multiple filters were in use for filtration. Later, with the availability of 8×12 array filtration systems (Tomtec, CT), the reaction contents from a 96-well microtiter plate are filtered through a large filter, sandwiched between an 8×12 block, and washed. After adding scintillant, the filter is countered in a Beta plate counter (Wallac, Turku, Finland). Now 96-well filter plates (Packard, Meriden, CT; Polyfiltronics and Millipore) and filtration units (Packard, Millipore) have facilitated quick separation and washing. After washing, the plate is air-dried, the bottom of the plate is sealed, scintillant is added, the top of plate is sealed, and the plate is counted in a MicroBeta (Wallac) or a Topcount (Packard) scintillation plate counter. The throughput of filtration assays increased considerably with 96-well filter plates, filtration units, and plate counters.

Adsorption Assays. In protein kinase reactions the phosphorylated product (acidic) by ionic interaction is captured on phosphocellulose filters; filter washed, air-dried and transferred into a vial; scintillant is added, and the vial is counted in a scintillation counter. Recently, biotinylated peptide has been used as a substrate, and the phosphorylated product is captured on a streptavidin filter or a streptavidin-coated filter plate; the free [^{32}P or ^{33}P]-ATP is removed by washing, the filter is transferred into a vial; scintillant is added, and the vial or plate is counted in a scintillation counter [9].

Precipitation Assays. In the traditional enzyme assays, the radiolabel from the substrate is transferred to a protein acceptor, and the radiolabeled product is isolated by precipitation with trichloroacetic acid (TCA); the precipitate is collected by filtration and washing, the filter is transferred into a vial, and the vial is counted after the addition of scintillant. In the farnesyl-transferase assay, the transfer of ^3H-farnesylpyrophosphate to farnesyl-transferase is assayed by precipitation with TCA; then the TCA extract is filtered onto a filter, washed, and counted after the addition of scintillant [10].

Radioimmunoassays. The radioimmunoassay is a classical method for measuring hormones, ligands, and other biomolecules. In the insulin radioimmunoassay, insulin in samples is measured by incubating sample with ^{125}I-insulin tracer and insulin antibody and protein-A coated beads into test tubes or wells of microtiter plates; the beads are settled by centrifugation, the supernate is removed, and the radioactivity associated with the beads is assayed by direct counting of the tubes in a gamma counter or after adding scintillant microtiter plates are counted in a scintillation plate counter [11].

Variations of the radioimmunoassay protocol have been used with biotin-streptavidin systems, wherein biotin-labeled substrates are used in enzyme reactions and the radioactive-labeled products are captured either on streptavidin-coated plates or on streptavidin-coated membranes [9].

B. Homogeneous Assays

Homogeneous assays are one-pot assays with no transfer or wash steps. In the classical assay, all the reagents are added in one step or in multisteps, and the signal is read in a plate reader. The assays are radioactive, chromogenic, absorbance, fluorescence, or luminescence assays. These assays are either radioactive or nonradioactive assays. The homogeneous assays are liquid-phase, solid-phase or bead-based assays (some of them are discussed in detail in Chap. 4).

1. Radioactive Homogeneous Assays

The radioactive homogeneous assays are based on scintillation proximity assays (SPA) with either SPA beads or scintillant-coated plates. The principle of SPA is that when a radioactive molecule binds to a scintillant-coated solid phase (bead or microplate), the radio isotope is brought in close proximity to the scintillant, giving a signal that can be measured in a scintillation counter. Radioactive molecules that are not bound will not be in close proximity and will fail to give signal and need not be separated from bound radioactive molecules.

SPA Bead Technology. When a tritium atom decays, it releases a β-particle with an energy of 6 KeV and a mean pathway of 1.5 μm. If this β-particle meets with a suitable scintillant molecule within 1.5 μm of being released, the particle will emit light that can be measured in a scintillation counter [12]. But if the particle travels a greater distance than 1.5 μm, it will not have enough energy to cause scintillation. The path lengths for other isotopes, ^{35}S, ^{33}P, and ^{125}I, are 66, 126, and 17.5 μm, respectively, and these isotopes will produce light if the radioactive particles come in contact with scintillant molecules within their respective distances. Two types of beads, polyvinyl toluene and yittrium silicate, area used in SPA. In SPA bead technology, scintillant is incorporated into small fluoromicrospheres (beads) that are derivatized to bind specific molecules. The radioactive molecule bound to the SPA bead is in close proximity to the scintillant on the bead and stimulates the scintillant to emit light, whereas the unbound radioactive molecule will be too distant, and the particle energy will not reach the scintillant on the bead nor emit light. SPA bead technology can be applied to enzyme assays, receptor binding assays, radioimmunoassays, protein–protein interactions, and PCR quantitation assays [13].

SPA bead technology is the proprietary technology of Amersham (Arlington Heights, IL). In a protein tyrosine kinase assay, biotin-labeled peptide substrate is incubated with ^{33}P-ATP, protein tyrosine kinase, and streptavidin-coupled SPA bead. The phosphorylated peptide binds to the streptavidin SPA bead, and the radioactive label is in close proximity to the scintillant on the bead and emits light that can be measured in a scintillation plate counter [14]. Receptor–ligand binding assays with membrane receptors have used wheat germ agglutinin coated SPA beads for the detection of receptor-bound ligand [15]. Nuclear receptors (soluble receptors) engineered as fusion proteins with the His_6 tag or derivatized with biotin have been used to develop homogeneous receptor–ligand binding assays with SPA copper beads or SPA streptavidin beads, respectively.

SPA Plate Assays. The SPA plate assays consist of 96-well microplates containing scintillants either incorporated in the plastic (ScintiStrps, Wallac) or layered on the bottoms of the wells (FlashPlate, NEN, Boston, MA). The scintillant plate surface is coated with a protein (streptavidin, antibody, or secondary antibody) that binds the radiolabeled product, bringing the radiolabel in close proximity to the scintillant and giving a signal (detailed in Chap. 4). A nonradioactive substrate can be coated onto the plate, and reaction produces radiolabeled product that will be in close proximity to the scintillant and produce a signal. The signal is proportional to the radioactive molecules bound to the SPA medium. Some biomolecules like streptavidin-precoated generic SPA plates are commercially available. The SPA plates have been used for enzyme assays, receptor binding assays, and radioimmunoassays [16].

2. Nonradioactive Homogeneous Assays

The majority of the nonradioactive homogeneous assays are fluorescence-based assays. Bead-based methods are becoming increasingly popular for screening assays.

Chromogenic Assays. Screens based on chromogenic assays have been used for enzyme assays. The substrate containing chromophore is colorless, but when the chromophore breaks away from the substrate it yields color, and the absorbance at λ_{max} is read in a plate reader. The assay can be easily automated and is inexpensive. The limitation of this type of assay is the lack of sensitivity compared to fluorescence and radioactive assays, e.g., in the β-glucuronidase assay, the substrate p-nitrophenyl β-glucuronide is colorless, but the p-nitrophenol formed in the reaction under alkaline conditions is colored, and absorbance at 415 nM is measured [17]. For enzymes that generate phosphate or pyrophosphate such as phosphatases, malachite green dye method (absorbance at 660 nm) can be used.

Absorbance Assays. In enzyme assays, where the substrate absorbs light in the UV region and the reaction product has no absorbance, or vice versa, absorbance assays could be used for quantitation of the reaction by reading in a plate reader. In some reactions, though neither the reaction substrate nor the product has UV absorbance, the substrate or product could be coupled to another enzyme assay that can be monitored by absorbance, e.g., in the ATP–citrate lyase enzyme reaction, oxaloacetate, ADP, and acetyl CoA are formed from the reactants ATP, citrate, and coenzyme A. The oxaloacetate formed is utilized as a substrate for malic dehydrogenase that uses NADH, which has absorbance at 340 nm, giving rise to NAD, which has no absorbance [18]. Thus with the coupled enzyme assay, the ATP–citrate lyase reaction is followed by the disappearance of absorbance of NADH at 340 nm.

Fluorescence Assays. Fluorescence assays are 100 to 1000 times more sensitive than colorimetric or spectrophotometric assays. Fluorescence methods are becoming the popular nonradioactive methods for HTS with the availability of 96-, 384-, and 1536-well plate readers. Several fluorescence detection methods are being used for HTS that include fluorescence intensity, fluorescence resonance energy transfer, homogeneous time-resolved fluorescence resonance energy transfer, fluorescence correlation spectroscopy, fluorescence polarization, and fluorescence life-times.

FLUORESCENCE INTENSITY ASSAYS. These assays include fluorogenic assays and fluorescence quench assays. In fluorogenic assays, the reactants are nonfluorescent, and the reaction produces a fluorescent product; the reaction is monitored as an increase in fluorescence signal, e.g., in the β-glucuronidase assay, umbelliferyl β-glucouronide, which is nonfluorescent, serves as a substrate, and umbelliferone produced in the reaction is fluorescent in alkaline conditions (see Chap. 4) [17]. An improved fluorogenic substrate, 6,8-difluro-4-methylumbelliferylphosphate, was used as a substrate for acid phosphatase with a 10-fold higher fluorescence signal [19]. Nucleic-acid-specific dyes, when bound to nucleic acid, produce fluorescence signals. PicoGreen dye binds specifically to double-strand DNA, OliGreen to single-strand DNA, RiboGreen to RNA [19]. These dyes have been used for quantitative estimation of the respective nucleic acids and in assays monitoring the synthesis or hydrolysis of nucleic acids.

The fluorescence of a fluorescent group covalently linked to substrate is quenched and the reaction cleaves the substrate, freeing the fluorescent group and giving rise to a fluorescence signal in fluorescence quench assays, e.g., the peptidase assay with bisamide of R110 as substrate and enzyme action produces R110 monoamide and R110, which are fluorescent [19].

FLUORESCENCE POLARIZATION. Fluorescence polarization (FP) and anisotropy are interchangeable terms and measure the same process. When a fluorophore is excited with polarized light, the emitted light is also polarized. The polarization value of a molecule is proportional to the rotational relaxation time and depends on the molecular volume under defined conditions (see Chap. 4 for more details). In FP assays, the molecular size is proportional to the polarization of the fluorescence emission. The FP assays can be classified into three different modes (Fig. 4).

1. Size increase: When a small fluorescent molecule binds or transfers fluorescent group to a large molecule results in an increase in the size of the fluorescent molecule adduct, resulting in increase of FP signal.

A variation of this will be a competition assay in which a drug compound competes with the tracer fluorescent small molecule for binding or transferring to a large molecule, thus reducing the FP signal.

2. Size reduction: A bigger fluorescent molecule is degraded to a smaller fluorescent molecule as in the case of proteases or nucleases, resulting in loss of polarization signal.

3. Indirect assays: (a) The direct immunoassay, in which the fluorescent substrate is converted to a fluorescent product and immunocomplexed with product-specific antibody, thus forming a bigger adduct, resulting in gain of FP signal [21].

(b) A competitive immunoassay, wherein the nonfluorescent product competes with the fluorescent product (used as a tracer) in coupling to the product-specific antibody, reducing the FP signal [22, 23].

(c) Similarly, a peptide or nucleotide substrate with biotin and fluorescent tags at the opposite ends of the molecule; when cleaved by protease or nuclease activity, the fluorescent portion is separated from the biotin-containing region, resulting in a reduction of the fluorescent molecule binding to avidin; consequently the FP signal is reduced.

FP is a well-known technique in diagnostics. FP is a simple homogeneous assay format adaptable to a variety of assays: enzyme assays including protein tyrosine kinases [21–23], protein serine and threonine kinases [24], adenosine transferase [25], immunoassays for cAMP [26], and receptor–ligand assays [27].

FLUORESCENCE RESONANCE ENERGY TRANSFER (FRET) ASSAYS. A donor and an acceptor fluorophore are requiredto be in close proximity for FRET to occur. The emission of the donor has to overlap with the excitation of the acceptor, and in many cases the donor and acceptor fluorophores are different (see Chap. 4 for details). FRET assays include quench or quench relaxation assays. When the donor and the acceptor fluorophores are in close proximity to each

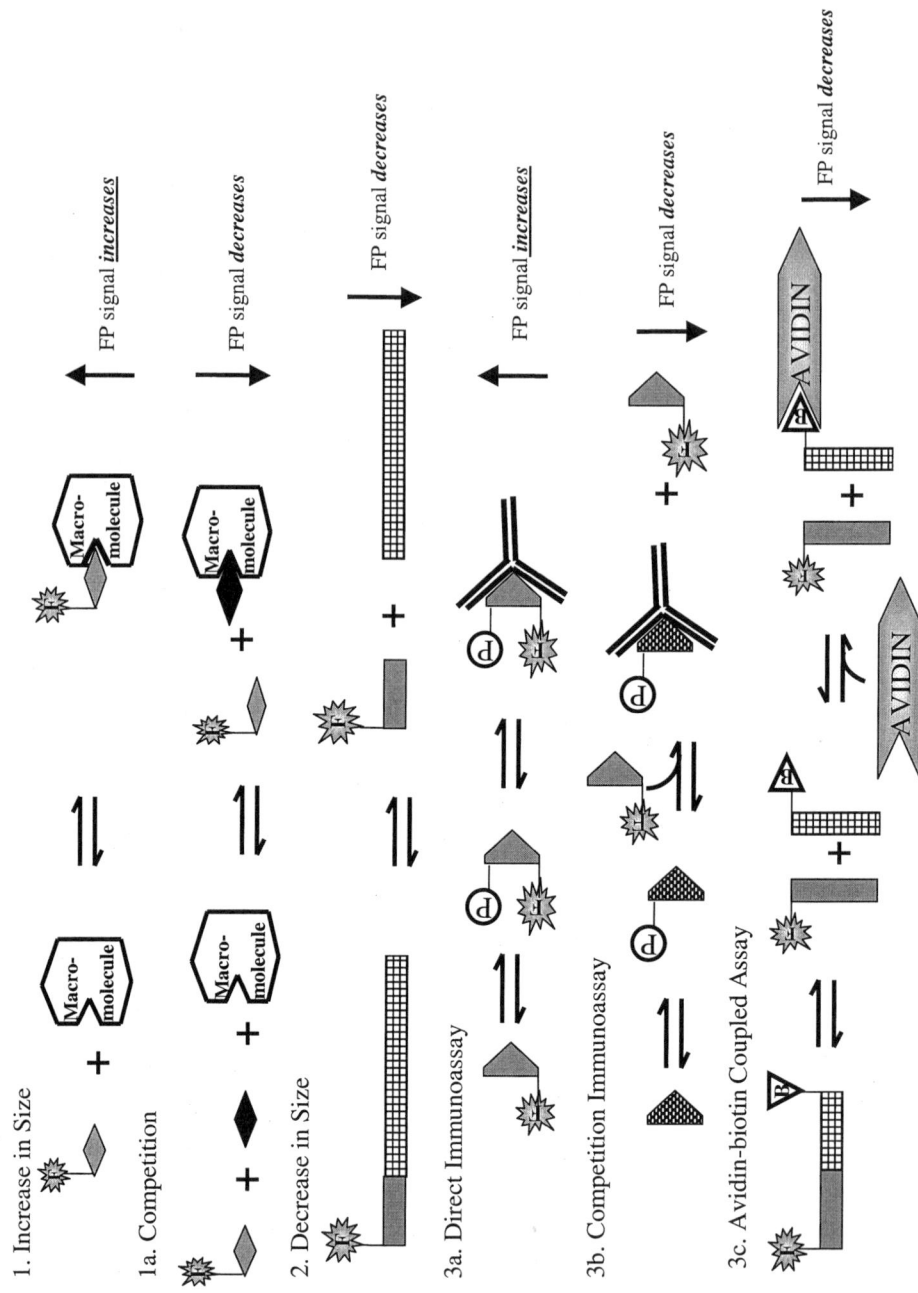

other, the fluorescence is quenched through intermolecular or intramolecular resonance energy transfer. If this intermolecular resonance energy transfer is prevented between the fluorophores, or if intramolecular resonance energy transfer is disrupted due to cleavage of the substrate, the result is an increase of the fluorescence intensity of the donor fluorophore. The FRET assay is widely used for protease assay with peptide substrates containing FRET pair of fluorophores such as 5-(2-aminoethylamino) naphthalene-1-sulfonic acid (EDANS) as donor at the C-terminus, which is quenched by the acceptor 4-(4-dimethylaminophenylazo) benzoic acid (DABCYL) at the N-terminus. Protease action cleaves the peptide, separating the donor and acceptor fluorophores, disrupting the intramolecular quenching, and increasing the fluorescence intensity [28].

HOMOGENEOUS TIME-RESOLVED FLUORESCENCE (HTRF)/TIME-RESOLVED FRET (TRFRET). Time resolution in fluorescence signal detection is gaining importance in HTS. In time-resolved fluorescence (TRF), lanthanide chelates with long lifetimes of 100 to 1000 μs are used, and the fluorescence emission is measured after a time delay. The advantages of TRF in HTS include sensitivity, robustness, and minimization of interference from background prompt fluorescence. FRET is the most common format of TRF used in HTS, wherein lanthanide serves as a donor and allophycocyanine (APC) (a phycobiliprotein from red al-

---

Figure 4 Different modes of fluorescence polarization assays. 1. Increase in size. When a small fluorescently labeled substrate or ligand molecule binds to a macromolecule, due to rotational constraints it orients in the plane of polarization; the emitted light is polarized, resulting in increased FP value. 1a. A variation of this: a test compound competes with the fluorescent molecule, thus reducing the association of the small fluorescent molecule with the macromolecule. 2. Decrease in size. When a large fluorescent molecule (high FP value) is hydrolyzed, releasing a small fluorescent molecule, which rotates freely in solution and orients randomly in the plane of polarization, it results in a decrease of FP value. 3a. Direct immunoassay. A fluorescent small molecule substrate is converted to product, and the fluorescent product conjugates with product-specific antibody, resulting in a constrained large molecule, and the FP value increases. The amount of product formed is proportional to the increase in FP signal. 3b. Competition immunoassay. A small molecule substrate is converted to product, which competes with fluorescent product (as tracer) for product-specific antibody, thus reducing the amount of fluorescent product conjugated to the antibody and hence reducing the FP value. The amount of product formed is proportional to the decrease in FP signal. 3c. When a molecule with a fluorescent group at one end and a biotin group at the other end is used as a substrate for a hydrolyzing enzyme, the fluorescent group is separated from the biotin-containing product by enzyme reaction. The biotinyl product without a fluorescent group will bind to avidin but not the fluorescent product. The substrate containing both biotin and the fluorescent group binds to avidin with high FP value. As the substrate is converted to product there will be a loss of FP signal because the fluorescent group binding to avidin is decreased.

gae) as an acceptor, resulting in long-lived fluorescence of the acceptor. HTRF® is a proprietary technology of Cis-Bio International (France) and Packard Instruments Company (Meriden, CT), wherein Europium cryptate is used as a donor and XL665, a modified APC, as an acceptor [29]. Lance™ is another proprietary technology of Wallac (Turku, Finland), wherein lanthanide chelates are used as donors [30]. With Eu^{3+} chelate as a donor, Cy5 (an APC) is used as an acceptor, and with Tb^{3+} chelate as a donor, rhodamine serves as an acceptor. The applications of HTRF/Lance assays include protein–protein, DNA–DNA, DNA–RNA, and DNA–protein interactions, enzyme assays, immunoassays to quantitate biomolecules, and receptor–ligand assays.

FLUORESCENCE CORRELATION SPECTROSCOPY. In fluorescence correlation spectroscopy (FCS) measurements, temporal fluctuations in the fluorescence signal detected from the diffusion of individual fluorescent molecules into and out of a small tightly focused confocal element are analyzed by autocorrelation techniques [31]. The average number of fluorophore molecules in the detection volume, i.e., concentration, the average brightness per molecule, and the diffusion time of the components can be obtained from the FCS autocorrelation data. FCS assays can be miniaturized to uHTS without losing sensitivity. Mass-dependent FCS screens are applied for ligand–receptor assays, enzyme assays like protease assays, and protein kinase assays. Mass-independent FCS assays based on fluorescence intensity changes include protease assays in which a quenched substrate is cleaved with increased fluorescence.

FLUORESCENCE LIFETIME ASSAYS. The fluorescence lifetime of a fluorophore is a characteristic value and is not influenced by spurious background signals. Fluorophore lifetimes are measured using pulsed or phase-modulation techniques. In the pulsed technique, the sample is illuminated with a pulsed laser, and the fluorescence intensity decay is measured as a function of time. In the phase-modulation technique, the sample is illuminated with a modulated continuous light, and the fluorescence lifetime is measured from the shift of the modulation of the excitation light and the fluorescence emission. A microplate-compatible reader for the measurement of lifetimes is available at present, and an HTS-compatible plate reader is being developed by LJL Biosystems.

Bead-Based Technologies. Beads have been used as solid supports for a long time. Considerable improvements in bead technology permit a variety of beads varying from 20 nm to several millimeters in diameter. Beads are used extensively in HTS assays, since they are a convenient and cost-effective means of performing separations, localizing interactions, and labeling binding events [32]. The SPA bead produced by Amersham is the first bead-based assay used in HTS. SPA bead technology uses radioisotopes and is described in Sec. IV.B.1.a. Here other bead technologies using nonradioactive methods will be discussed.

AMPLIFIED LUMINESCENCE PROXIMITY HOMOGENEOUS ASSAY (ALPHA). Alpha™ screen (Biosignal, a Packard company, Meriden, CT) uses a donor bead containing a photosensitizer that absorbs light at 680 nm and converts ambient molecular oxygen to generate chemiluminescence at 370 nm if the acceptor bead is in close proximity, within 200 nm [33]. The chemiluminescence energy immediately excites a flurophore in the acceptor bead that emits long-lived fluorescence at 520–620 nm. The long lifetime (0.3 s) of this fluorescence signal allows measurements in time-resolved mode reducing the background. Alpha screen applications include enzyme assays such as protein kinases and proteases, immunoassays such as cAMP, and protein–protein and protein–DNA interactions.

FLUORESCENCE MICROVOLUME ASSAY TECHNOLOGY (FMAT). FMAT™ (PE Biosystems, Foster City, CA) is a homogeneous fluorescence imaging assay format used for cell-based and bead-based assays; it measures the fluorescence of Cy5-based flurophores associated with cells or beads [34]. A 633 nm He/Ne laser scans a 1×1 mm area of the cells or beads settled at the bottom of the well. The effective concentration of the fluorophore on a bead or cell is greater than the background fluorescence and is discriminated from unbound background fluorescence. FMAT employs a unique macro confocal imaging system with a He/Ne laser that automatically focuses on and scans fluorescently labeled cells or beads resting on the bottom of the inner surface of the multiwell plate. FMAT has been used for protein kinase, protein–protein interaction, and receptor binding assays [34]. This technology has also been applied to ELISA type assays called fluorescence linked immunosorbent assays (FLISA). Assays can be multiplexed using different size beads and different fluorophores, as the reader with two PMTs is capable of reading two fluorophores simultaneously. IL-6 and IL-8 peptides in the growth media of cytokine stimulated HUVEC cells in the same wells of a 384-well plate have been detected successfully [35].

MICROVOLUME FLUOROMETRY. Affymax (Palo Alto, CA) has developed microvolume fluometry (MVF), which is similar to FMAT. In the MVF, receptor-coated particles (whole cells expressing the receptor can also be used as the receptor-coated particles) are stained when a Cy5- or Cy5.5-labeled ligand binds; the localized signal is measured in a modified capillary fluorometer that is designed to detect fluorescence from Cy5 and Cy5.5 [36]. The MVF scans a 1 mm^2 area, and the emitted light is collected by two PMTs, one for measuring Cy5 and the other for measuring Cy5.5. In multiplex receptor binding assays for IL-1R and IL-5R, each receptor was immobilized on a different size bead, and the binding affinities obtained for different ligands in the 864-well plate were similar to those of the radioligand binding assay [36].

LASER-SCANNING IMAGING. This is similar to FMAT in principle and has been used to assay cell or bead fluorescence; it has been used for receptor assays and protein–protein interactions [37]. It also can be used for multiplexing with different size beads and fluorophores.

MICROVOLUME TWO-PHOTON EXCITATION. Two-photon systems can suppress background effectively by optical filtering. In two-photon excitation, the excitation of fluorescent molecules is in three-dimensional focal volume [38]. Amino-modified polystyrene microspheres (3.1 μm) covalently coated with antibody and the antigen bound can be quantified from individual microparticles by the use of two-photon excitation of fluorescence. The infrared 1.064 μm two-photon laser beam is not absorbed by biological materials, making it possible to assay blood samples. Whole cells can also be used in place of microparticles. The sensitivity and dynamic range obtained by this method suggests that this method will be an inexpensive method for measuring biomolecules in solution.

ELECTROCHEMILUMINESCENCE (ECL). ECL is based on an amplified signal using ruthenium chelate; it is called the Origen system and is from IGEN (Gaithersburg, MD). In ECL, low voltage applied to an electrode triggers an oxidation–reduction reaction of ruthenium. Tripropylamine is consumed in the oxidation process, and the ruthenium chelate is recycled, enabling the label to go through several redox cycles in the read-time. ECL technology has been applied to immunoassays, enzyme assays, protein–protein, and nucleic acid quantitation [39].

MAGNETIC BEAD-BASED SCREENING. Receptor (binding region of the receptor protein) IgG fusion protein is bound to anti-IgG-coated magnetic beads (Chugai Pharmaceuticals, San Diego, CA) and screened with a phase-peptide library, and the nonbinding phase is washed. This method identifies compounds that bind to specific receptors, unlike in the traditional cell-based assays, where the interaction could be to a nonspecific receptor. The bead-based assays are more sensitive (by at least an order of magnitude) than the cell-based assays.

FLOW CYTOMETRY. This is a microparticle-based flow cytometric method described by Becton Dickinson (San Jose, CA). It can be used for simultaneous measurement of multiple cytokines in the test samples. Polystyrene beads are dyed with fluorescence dyes to different intensities, and each intensity bead is coupled with a different antibody and has been used for multiplexing in a FAC-Scan flow cytometer. In the sandwich assay (bead-Ab-cytokine-Ab), a second detector antibody coupled to fluorescent dye phycoerythrin that emits at 585 nm has been used to quantitate the cytokines [40]. The bead population can be separated by flow cytometry, and multiple cytokines can be simultaneously detected.

V. CELL-BASED ASSAYS

Cell-based assays closely mimic the environment of a living cell and have been used for confirmation of leads coming from primary in vitro biochemical screens. Cell-based assays are used for targets where biochemical assays are not available.

Cell-based assays also give information about cellular interactions with the target and also shed light on the stability of compounds [41]. Traditionally, the cell-based assays have been low throughput or medium throughput due to the cumbersome steps involved. With the advances in molecular, assay, and instrumentation technologies, now homogeneous high-throughput cell-based assays are available for primary screening. Like biochemical assays, cell-based assays can be divided into heterogeneous and homogeneous assays (Fig. 5).

A. Heterogeneous Assays

The heterogeneous assays consist of radioactive and nonradioactive assays. Nonradioactive assays are mainly ELISA assays.

1. ELISA Assays

Cells are treated with compounds, and the cellular changes are assayed by ELISA assays. Cell-based assays for screening receptor-mediated changes in signal transduction like cAMP can be measured in the cell extracts by ELISA assays.

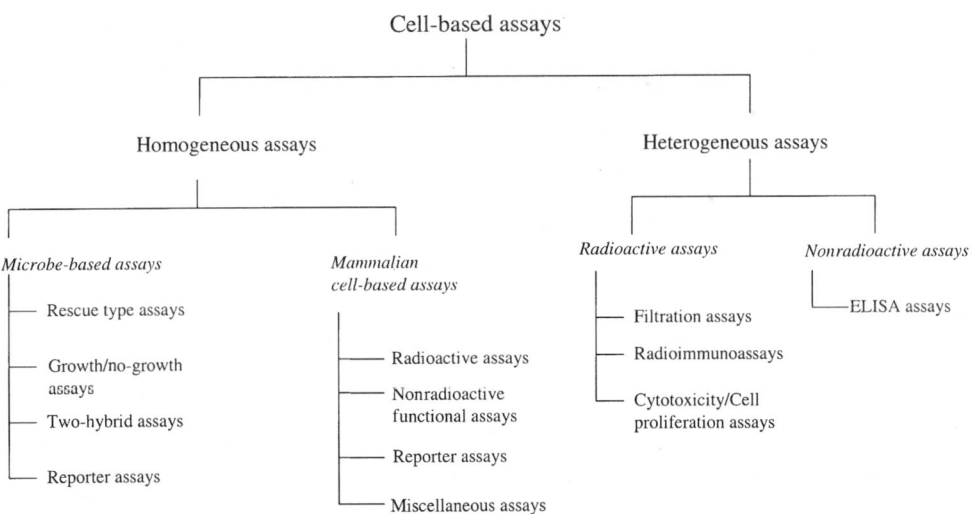

Figure 5 Cell-based assays. Cell-based assays are mainly dealt with as heterogeneous and homogeneous assays. Heterogeneous assays are divided into nonradioactive and radioactive assays. Homogeneous assays are further divided into microbe-based and mammalian cell-based assays.

2. Radioactive Assays

The assays involving radioisotopes are generally limited to receptor-binding assays and quantitation of biomolecules like hormones in cell extracts by radioimmunoassays.

Filtration Assays. Cell-based receptor-binding assays can be screened by filtration assays as described in Sec. IV.A.2., *Filtration Assays.*

Radioimmunoassays. Whole-cell assays in which changes in cell contents due to the treatment with compounds are assayed by radioimmunoassays as described in Sec. IV.A.2., *Radioimmunoassays.*

Cell Proliferation Assays. The classical cell proliferation assays consist of ^{3}H-thymidine incorporation into DNA or ^{51}Cr release. The assay consists of incubating the cells with ^{3}H-thymidine, removing the extracellular radiolabel by washing, and counting the cell-associated ^{3}H-thymidine. This assay at best is medium throughput, and alternate dye-binding HT methods have replaced this assay.

B. Homogeneous Assays

The homogenous cell-based assays can be done in microbes, yeast, or mammalian cells. These assays consist of growing the cells, treatment of the cells with compound, and developing and reading the signal. The homogeneous cell-based assay refers to the assay in a single step or multiple step additions in the same well of a microtiter plate.

1. Microbe-Based Assays

Microbe-based screening has been used to find antibacterial agents and cytotoxic anticancer agents (see Chap. 5). Mammalian proteins can be expressed in microbial cells at high levels often as in insoluble aggregates as inclusion bodies. Mammalian proteins undergo posttranslational modifications in the cells that are essential for biological activity of some proteins, and these modifications are absent in microbial systems limiting the use of expression to those proteins that are not affected by posttranslation modifications. Microbial-based screens are less expensive than the mammalian cell-based assays, as the reagents used for cell growth medium are inexpensive compared to tissue culture reagents, and microbe growth conditions are simpler than for the mammalian cells.

Antibacterial Activity. In HTS for finding new antibiotics, the antibacterial activities of compounds are tested by diffusion experiments in agar medium. Two gram-positive and two gram-negative strains are seeded in the medium, and compounds are spotted on the surface medium. Following overnight incubation, the size of the inhibition zone is determined by measuring the zone diameter. These assays have been automated with robots under sterile environment in major Pharma [42].

Growth/No Growth Assays. With functional expression of homologous or heterologous targets in microbial systems like *E. coli or Saccharomyces*, rendering the cell dependent on the target expressed, a growth or no growth (of the microbe) type of screen can be developed (see Chap. 5).

Rescue Type Assays. High-level expression of a protein that induces cell lysis in a microbe can be rescued by compounds that inhibit the function of that protein. Thus a functional protease screen can be designed by expression of the protease and by insertion of a peptide sequence of a protease substrate within an important protein. When this protein is cleaved by the protease loses its functional activity, and growth is inhibited. Compounds that inhibit protease can rescue the growth [43].

Reporter-Based Assays. SOS response involves the activation of several genes, and when coupled to a reporter the SOS response can be monitored by the reporter activity. For inhibitors of topoisomerase and gyrase, a screen with β-galactosidase receptor was used in which SOS response leads to the overexpression of β-galactosidase [42]. In the reporter assays, a target protein is coupled to a promoter (transcriptional factor) that in turn is coupled to a reporter protein like β-galactosidase, luciferase, or chloramphenicol acetyl transferase. Thus a target protein is engineered in the extracellular domain of the ToxR protein in *E. coli.* When compounds bind to the target protein, promote dimerization of the extracellular domain of hybrid ToxR protein, which activates the *toxR* promoter and consequently activates the expression of reporter and can be easily read in a plate reader [44].

Yeast Expression. As yeast offers null background for human receptors, human GPCRs along with appropriate mammalian G-proteins can be expressed in yeast coupled to the pheromone signaling pathway to screen for agonists and antagonists [45,46]. A yeast-based functional transcription assay for human progesterone receptor in 384- and 1536-plate format has been reported [47].

Two-Hybrid Systems. The classical yeast two-hybrid system in *Saccharomyces cerevisiae* is frequently used to identify protein–protein interactions, to screen libraries for isolating and identifying genes that encode interacting partners for a protein of interest [48]. Yeast two-hybrid systems have been used for detection and analysis of protein–protein interactions to identify interacting proteins. The two-hybrid system has been used for identification of interaction partners for the protein sequences generated from human genome sequencing, thus finding the functional activity (see Chap. 5). In the yeast two-hybrid screening method a particular protein of interest is used as a "bait" to discover the interacting proteins [49]. Yeast two-hybrid screens have been used for functional receptor–ligand interactions [50]. Yeast two-hybrid screens can be used for identifying drugs that disrupt two-hybrid interactions. In a variation of the yeast two-hybrid method, a third protein is expressed along with the usual activating domain (AD) and binding domain (BD). This third protein may promote an interaction by interacting with both hybrid proteins, or the third protein may prevent the formation of a two-hybrid complex [48].

2. Mammalian Cell-Based Assays

Functional cell-based assays and receptor binding assays can be performed with intact cells. Advances in assay technologies have made it possible to miniaturize the assays and use these assays as primary assays. Unlike biochemical assays, cell-based assays take several days before initiation of the assays and from initiation to completion may take several hours. The readouts of homogeneous format are radioactive, luminescence, or fluorescence. In designing and evaluating cell-based assays for screening it is important to have a good cell line and to develop robust and reliable assays.

Cells recombinantly engineered with target protein gene, stable expression of the target protein is the preferred way to develop cell-based assays. Sometimes transiently expressed target genes also are used in secondary functional assays, but not for screening large libraries. Transient expression can vary batch to batch and needs a large amount of DNA for transfection of a large number of cells; it is not practical for HTS because it is laborious, requires large amounts of reagents, and cannot maintain quality control day to day. When recombinant cells cannot be used because of intellectual property issues, a natural cell line with higher expression of the target protein can be used. Some of the requirements for mammalian cells used in HTS are that the cell line should only require common tissue culture reagents; it should only require standard incubation conditions such as 5% CO_2 and 37°C; the doubling time of the cells is preferably less than 48 h; the signal should be stable for at least 20 passages; the assay should not require too many cells ($< 100,000$/well); it should be free of mycoplasma; cells from frozen stock should be viable without alteration in the growth curve; and the target protein should be expressed at a high enough level in the cells. The assay should be robust, with a signal-to-noise ratio of at least 4:1 and little fluctuation in replicates; the assay performance should be consistent from day to day; ideally the assay signal should not vary with passage and cell density; and the assay should be amenable for miniaturization.

Radioactive Assays. Cell-based radioactive homogeneous assays have been used for functional assays and receptor–ligand assays.

RECEPTOR BINDING ASSAYS. Receptor binding assays with membrane receptors can also be assayed with whole cells, either adherent or suspension cells, with a radioligand. A homogeneous SPA assay can be developed using WGA-SPA beads and cells in suspension. SPA assays also can be performed in Cytostar plates (Amersham) with adherent cells, or suspension cells in FlashPlates using WGA-coated FlashPlates (NEN Life Sciences).

GTP-γ-S BINDING ASSAYS. Agonist binding to a GPCR results in conformational change in the receptor activating G-protein. The oligomeric G-protein with bound GDP is inactive, and activation involves exchange of GTP for GDP followed by dissociation of subunits to the α subunit (which binds GTP) and the βγ complex, which activate or inactivate downstream effector molecules like adenyl cyclase, protein kinases, and phospholipase C. G-protein activation can be assessed by measuring the agonist-induced binding of a nonhydrolyzable GTP analog [^{35}S]GTPγS to cells. The earlier [^{35}S]GTPγS binding assay consisted of filtration and washing followed by counting after addition of scintillant. Recently, homogeneous assays using SPA FlashPlates [51,52] have shown that these assays are comparable to the conventional filtration assay. The SPA [^{35}S]GTPγS binding assay can be used for characterization of ligands for orphan receptors whose ligands and signal transduction pathways are not known.

SIGNAL TRANSDUCTION ASSAYS. Some signal transduction pathways of receptors can be monitored using radioisotopes. Adenyl cyclase is regulated by GPCRs that couple to G-proteins $G_{\alpha S}$, $G_{\alpha i}$, and $G_{\alpha q}$, and Ca^{2+}-calmodulin signals as a result the intracellular cAMP concentration changes. The old methods involved laborious, inconvenient, and time-consuming extraction procedures. Several improved assays have been in use to determine the cAMP levels (described in Chap. 7). Homogeneous cAMP assay kits by different vendors for HTS are available.

In the SPA-bead-based one-step Biotrak cAMP assay (Amersham), the extraction and measurement are done in the microtiter plate used for culturing cells. The SPA assay is based on competition between the cellular cAMP and exogenously added tracer of [^{125}I]-cAMP. In this method, the radiolabeled cAMP (tracer) binds to cAMP-specific antibody, which binds to the SPA bead coated with secondary antibody or Protein A. The signal is detected due to the close proximity of the radiolabel to the scintillant on the bead. The unbound radioligand is not detected, as it is not in close proximity to the scintillant on the bead. The cells cultured in the microtiter plate or cells in suspension in the well are lysed after treatment with drug, and the cellular cAMP is measured by competition radioimmunoassay after addition of the SPA beads and other reagents and incubation overnight. This procedure involves few reagent addition steps and no transfer steps. This is a sensitive, automatable, robust, and highly reproducible assay.

Intracellular cAMP also can be measured by scintillation proximity assay using FlashPlate (NEN Life Sciences) in which cells in suspension are incubated with the drug in a cAMP FlashPlate coated with anti-cAMP antibody. Following incubation, cells are lysed and [^{125}I]-cAMP is added; they are incubated overnight, and the FlashPlate is counted [15]. The endogenous cAMP competes with [^{125}I]-

cAMP and is quantitated from a standard curve. These assays are very robust and reproducible and can be miniaturized to 384-well plates.

Nonradioactive Functional Assays. In response to activation of many membrane receptors by ligand binding, the signal transduction mechanisms are activated. Activation of GPCRs that couple to $G_{\alpha q}$ stimulates phosphoinositol hydrolysis, increasing the accumulation of inositol phosphate and cytosolic Ca^{2+} transients. Similarly, other GPCRs that couple to G-proteins $G_{\alpha s}$, $G_{\alpha i}$, $G_{\alpha q}$, and Ca^{2+}-calmodulin signal by regulating adenyl cyclase activity. Different classes of receptors signal through different signal transduction pathways, and here only the very well characterized functional assays will be discussed.

CA^{2+} ASSAYS. Ca^{2+} concentrations can be measured in a fluorescence imaging plate reader (FLIPR) using the fluorescent dye Fluo-3 with adherent or suspension cells [53]. In a FLIPR assay, the Ca^{2+} signal is measured in all the wells of 96- or 384-well plates simultaneously; reading is done every second, so kinetic measurements of calcium oscillations (described in detail in Chap. 7) can be made.

Aequorin has been used as a Ca^{2+} indicator and validated for many GPCRs. The apoaequorin gene has been expressed in a variety of cell types. A typical aequorin signal in mammalian cells occurs within 30 s as a flash. The receptor of interest is expressed stably in a cell line constitutively expressing apoaequorin. In this assay system, cells expressing apoaequorin are charged with the prosthetic group coelenterazine forming holoprotein. Activation of the receptor with ligand binding increases $[Ca^{2+}]_i$, which binds to coelenterazine, oxidizing to coelenteramide, generating photons, which can be quantitated in a luminometer [54,55]. The signal occurs within 30 s and is measured in a luminometer equipped with an injector. A luminometer with six injectors and six PMTs is used in HTS. However, the signal can be more efficiently measured in a CCD-based luminescence imaging plate reader. There are several commercial luminescent imaging systems: the Chemiluminescence Imaging Plate reader (CLIPR) (Molecular Devices), the LeadSeeker (Amersham Pharmacia Biotech), the Wallac ViewLux (Wallac, Turku, Finland), and the NorthStar (PE Biosystems, Foster City, CA); these have to be adapted to the aequorin bioluminescent assay.

Ca^{2+} can also be measured as a reporter gene assay by coupling a reporter like luciferase to transcription factors, the nuclear factor of activated T cells (NFAT), or the cAMP response element binding protein (CREB) [54].

CYCLIC AMP ASSAYS. A high efficiency fluorescence polarization (HEFP) cAMP assay is a homogeneous cAMP assay that can measure cAMP levels in whole cells and is based on competition between cAMP produced in the cell and exogeneously added fluorescent cAMP as tracer (LJL BioSystems). In this assay, cells are incubated with the drug, cells are lysed, fluorescent cAMP tracer and cAMP-specific antibody are added to the lysate, and the FP signal is measured [26,56,57].

A chemiluminescent cAMP immunoassay developed by PE Biosystems is a competitive immunoassay wherein the endogenously produced cAMP competes with cAMP-AP conjugate [58]. Cells treated with drugs are lysed, and cAMP-AP conjugate and an anti-cAMP antibody are added. The cellular cAMP reduces the cAMP-AP conjugate in the immunocomplex, reducing the signal. After washing, the chemiluminescent substrate for AP is added, and signal is measured in a Luminometer, TopCount, or MicroBeta plate counter. Signal reduction is proportional to the amount of cAMP present. In ECL cAMP assay (IGEN), the endogeneous cellular cAMP competes with ruthenylated cAMP for immunocomplex formation with anti-cAMP antibody that complexes with antirabbit antibody coated beads [59]. Cyclic AMP measurements can also be done with AlphaScreen technologies [60]. A luciferase-reporter-based cAMP assay involves expression of reporter plasmids containing luciferase gene under transcriptional control of 6 or 12 cAMP response elements (CREs) in CHO cells stably expressing human β2-AR [61]. The CRE-directed luciferase reporter gene assay has been used for the functional assay of GPCRs that involve the activation of AC and for GPCR agonists that inhibit AC activity.

A melanophore-based rapid functional assay for GPCRs and tyrosine kinase receptors (TKRs) has been developed and has been extended to receptors that dimerize upon ligand binding. DNA encoding GPCRs and TKRs expressed in melanophores and respond to the endogeneous signaling system within the melanophore to mediate cell darkening and lightening. Ligand binding to G_S-linked receptors in melanophores activates AC and pigment dispersion and cell darkening, which can be read as an increase in absorbance [62,63]. Stimulation of melanophores expressing G_i-linked receptors with an agonist results in the inhibition of AC, which causes pigment aggregation and cell lightening and is measured as a decrease in absorbance.

To monitor fluctuations in cAMP in living cells, protein kinase A (PKA) subunits are fused with two appropriate green fluorescent proteins (GFPs) that can be used as donor and acceptor for FRET [64]. Dissociation of PKA subunits by cAMP disrupted FRET between the GFPs, and the fluorescence signal is dependent on the levels of cAMP.

Reporter-Based Assays.   Most of the transcription factors are modular, consisting of a DNA-binding domain and activation domain. These domains can be interchanged between different factors and still retain their functional properties. A reporter gene construct consists of an inducible transcriptional control element driving the expression of a reporter gene. Reporter genes code for proteins that possess unique enzyme activities, and the activity assays are adaptable to HTS. The test DNA is ligated upstream of the coding region of the reporter gene to generate a chimeric gene in which the regulatory element controls the expression of a reporter gene [65]. Functional enhancer elements that can augment transcription are placed upstream of the promoter. The reporter gene construct with an inducible transcriptional element driving the expression of the reporter gene will regulate the reporter protein synthesis in response to receptor activation. Several reporter gene vectors are commercially available with a promoter or enhancer, proper cloning sites, reporter gene, intron, poly A site, antibiotic resistance, and prokaryotic origin of replication (Fig. 6). The fusion reporter gene constructs are expressed into mammalian cells by a variety of transfection methods, the most popular being calcium phosphate, DEAE-dextran, lipofectamine, and electroporation methods [65].

Several reporter genes are available with a choice of many different vectors for appropriate cell line. The popular reporter systems are firefly luciferase, secreted alkaline phosphatase (SEAP), chloramiphenicol acetyltransferase (CAT), β-galactosidase, β-lactamase, and green fluorescent protein (GFP).

The firefly luciferase is a 62-kDa protein that is active as a monomer. The overall reaction catalyzed by luciferase is

$$\text{D-luciferin} + \text{ATP} + \text{O}_2 \rightarrow \text{oxyluciferin}$$
$$+ \text{AMP} + \text{CO}_2 + \text{PPi} + \text{light (562 nm)} \tag{4}$$

Luciferin is oxidized producing a flash of light ($t_{1/2} = 0.3$ s) with highest bioluminescence quantum yield of 0.82. The signal is read in a luminometer wherein the reagent is injected into each well before reading the signal. Luminometer with single injector and single well reader takes a long time to read a 96-well

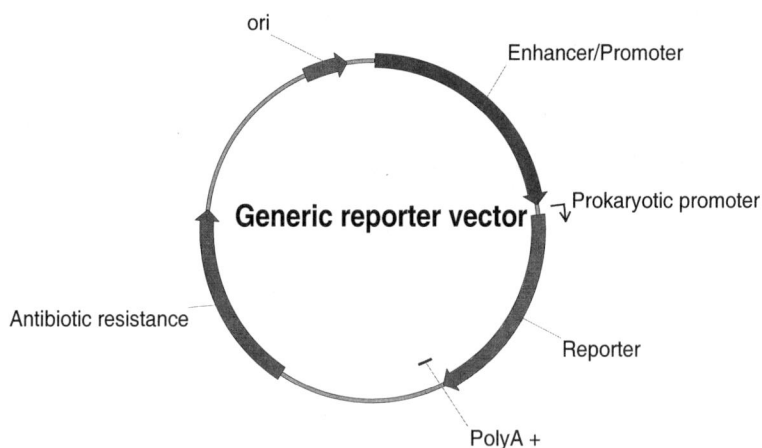

Figure 6 Map of a generic reporter vector. Ori represents the bacterial origin of replication. Promoter and enhancer attached to the reporter gene are cloned into the cloning site. A poly (A) sequence is positioned downstream of the reporter sequence to allow in vitro synthesis.

plate and hence is not fit for HTS. Now luminometers with 6-injectors and 6-PMTs are available that shorten the read time dramatically. The luminescence signal if converted from flash ($t_{1/2}$ = s) to glow ($t_{1/2}$ = h) can be read in a scintillation counter though with less sensitivity than in a luminometer [66]. Luciferase assay kits are available from several vendors (Luc-Lite from Packard; Steady-Glo from Promega; Luc-Screen from Tropix) that are amenable for homogeneous assays. The cells are incubated with the drugs in tissue culture treated Optiplates (Packard); a cell lysis reagent and luciferase assay reagents are added, and after 10 to 15 min incubation plates are read in a scintillation plate reader. This assay can be easily miniaturized into a 384-well plate and can be completely automated and further miniaturized into a 1536-well plate [67].

Secreted alkaline phosphatase (SEAP) is a unique human placental alkaline phosphatase that is secreted from the cell into the medium and is present in very few cell types. SEAP is heat resistant and homoarginine resistant, and endogenous alkaline phosphatase activity is suppressed with either treatment. Though colorimetric assays have been used in the past, the assays are not sensitive enough for use in screening. Now extremely sensitive chemiluminescent and fluorescent assays have been developed that are as sensitive as luciferase. Because SEAP is secreted into the medium, the assay can be done by adding directly the required reagents and reading the signal in a plate reader.

CAT was a widely used reporter because it is a stable enzyme from *E. coli* and is not present in mammalian cells. CAT-catalyzed acetylation occurs at the 3-hydroxyl position [Eq. (5)] which by nonenzymatic rearrangement is converted to 1-acetylchloramphenicol [Eq. (6)] and by further acetylation at the 3-position produces diacetylchloramphenicol [Eq. (7)].

$$\text{Chloramphenicol} + \text{acetyl CoA} \rightarrow \text{3-acetylchloramphenicol} + \text{CoA} \qquad (5)$$

$$\text{3-acetylchloramphenicol} \rightarrow \text{1-acetylchloramphenicol} \qquad (6)$$

$$\begin{aligned}\text{3-acetylchloramphenicol} &\rightarrow \text{acetyl CoA} \\ &\rightarrow \text{1,3-diacetylchloramphenicol} + \text{CoA}\end{aligned} \qquad (7)$$

Earlier CAT detection methods used radioactive- or fluorescent-labeled acetyl CoA and required either thin layer chromatography separation of products or differential solvent extraction of diacetyl product from acetyl CoA substrate. These methods are complex, labor-intensive, and time-consuming and hence are not suitable for screening. Recently a homogeneous assay was reported using FlashPlate. Biotinyl chloramphenicol is coated on a streptavidin FlashPlate, and the cell extract containing the enzyme is incubated with ^{3}H or ^{14}C-coenzyme A. The plate is counted in a scintillation counter or, to increase the sensitivity after aspiration of the contents of the well, the plate is counted. It is possible to use this FlashPlate CAT assay for screening [68].

β-Galactosidase catalyzing the hydrolysis of β-galactosides is a widely used reporter gene. Endogeneous β-galactosidase (optimal pH 3.5) in mammalian cells is either very low or absent, hence *E. coli* β-galactosidase (optimal pH 7.3) is often used as an internal standard to monitor the transfection efficiency in transient transfections. The common β-galactosidase assay is a colorimetric assay using chromogenic substrate *o*-nitrophenyl β-galactopyranoside. More sensitive chemiluminescent assays using Galacto-Star or Galacto-Light-Plus™ (Tropix) and fluorescent assays have been developed that are useful in HTS. The *E. coli* β-glucouronidase reporter gene is mainly used in plant cells and also in mammalian cells to a limited extent. The optimal enzyme activity for *E. coli* β-glucouronidase is obtained at pH 7 compared to pH 4.5 for the mammalian enzyme. β-Glucouronidase can be assayed with a chromogenic substrate *p*-nitro-β-glucuronide. The fluorescent assay with 4-methylumbelliferyl *D*-glucuronide [17] and the chemiluminescent assay using GUS-Light™ (Tropix) are more sensitive assays.

β-Lactamase from *E. coli* is gaining popularity as a reporter system [69]. Recently, the soluble membrane-permeable substrates 6-chloro 7-hydroxy coumarin and fluorescein-coupled cephalosporin have been used for developing a very sensitive homogenous FRET assay (see Chap. 7). Though this system is sensitive, robust, and simple, due to the proprietary nature of this assay a license has to be obtained to use this reporter system.

GFP is a protein from *Aequorea victoria*. The GFP system is becoming a very popular reporter for HTS in drug discovery. The wild type protein emits green light (λ_{max} = 509 nm) when excited at 395 nm, and the signal is very weak in mammalian cells. Several GFP variants with different spectral properties and improved fluorescence properties are available [70]. Vectors for the GFP reporters are available from several vendors. The GFP reporter system can be used for monitoring the functional activity of cytokine and nuclear hormone receptors that rapidly regulate transcription. Robust, sensitive, functional GFP-reporter-based assays have been developed with GPCRs that stimulate adenyl cyclase upon agonist binding using a GFP reporter construct with 5XCRE in the promoter.

3. Miscellaneous Assays

Lead compounds and compounds of interest for lead optimization are routinely tested for cytotoxicity, inhibition of cytochrome P450 isoenzymes (CYPs), compound permeability in Caco-2 cells, and specificity in other related assays. Thus compounds from an estrogen receptor screen have to be tested in other nuclear hormone receptor assays like those for androgen, progesterone, glucocorticoid, and TSH. Profiling of the compounds at the time of the lead optimization is very helpful for selecting compounds to in vivo studies.

Cell Proliferation and Cytotoxicity Assays.   The effect of compounds on the cell is generally assessed with nonspecific cytotoxicity assays. In a cell viability test using tetrazolium compound, MTT is added to cells, and living cells reduce the tetrazolium to a highly colored formazen salt and can be read in a plate reader as an end point reading. However, MTT is relatively insoluble, and other tetrazolium compounds such as MTS and XTT with improved solubility properties have been developed to assess the viability of living cells [71]. The Alamar Blue assay is a widely used homogeneous cell proliferation/cytotoxic assay that can be performed in 96- or 384-well plates in HTS mode [72]. When Alamar Blue is added to either adherent or suspension cell culture, the dye becomes reduced to a highly fluorescent red form that can be read in a spectrophotometric or fluorescence reader. The intensity of color or fluorescence is proportional to the cell viability. Alamar Blue is nontoxic to cells and allows continuous monitoring of cells at various periods of time.

A new cytotoxicity and proliferation assay based on fluorescent Oxygen BioSensor technology has been developed (BD PharMingen, San Diego, CA) [73]. This is a rapid homogeneous assay for monitoring cell viability, proliferation, and death. The fluorescent compound, tris 4,7-diphenyl-1, 10 phenanthroline ruthenium (II) chloride, is used as oxygen biosensor in a 96-well plate. Test compounds and cells in growth medium are added to the biosensor coated plate and incubated in the tissue culture incubator, and fluorescence was read in a fluorimeter using the bottom plate reading configuration at different times for kinetic measurement. This assay was successfully used with a variety of eukaryotic and prokaryotic cells. There is a good correlation between cell number and signal (the intensity) in the absence of toxic compounds.

CYPs. During the lead optimization studies, several lead compounds will have to be tested in absorption, distribution, metabolism, and excretion, and pharmacokinetics (ADME/PK) has to be studied to promote the early leads to compounds. Liver metabolism is one of the major determinants in the half-life of a drug in the bloodstream. CYPs are important in drug clearance. Xenobiotic metabolism is divided into phase I and phase II. Compounds are generally oxidized by CYPs in phase I and are conjugated with glucuronic acid or sulfuric acid in phase II. CYPs are a superfamily of several related proteins. At least six major forms of CYPs are involved in the metabolism of pharmaceuticals in man: CYP1A2, CYP2D6, CYP2C9, CYP2C19, CYP3A4, and CYP 2E1 [74,75]. Each isozyme has a specific preferred substrate. Routinely the compounds are tested for CYPs with human hepatocytes, isolated microsomes, human hepatoblastoma cell lines, or expressed human CYPs using a specific substrate for the subtype of CYP. A fluorescent homogeneous cell-based assay for CYP1A2 isozyme in the HepG2/C3A cell line using ethoxyresorufin as substrate in a 96-well plate has been reported (Amphioxus Cell Technologies, Inc., Houston, TX) [76]. Also, an assay for CYP3A4 has been developed [76].

Chip Technologies. Microfabrication and microfluidics-based chip technology is emerging and may replace HTS with further miniaturization. Microchip technology has become a powerful technology and is widely used in DNA analysis. DNA chips are glass surfaces bearing arrays of DNA fragments at discrete addresses, at which the fragments are available for hybridization [77]. Two DNA chip formats currently in wide use are the cDNA format and the in situ synthesized oligonucleotide array format. DNA chips have about 10,000 genes on an area of 3.6 cm². The Incyte cDNA chips have capillary arraying of cDNAs and two-color hybridizations scanned separately for each color. The dual-color oligonucleotide chips contain 65,000–400,000 oligonucleotides on a 1.6 cm² glass surface (Affymetrix, Santa Clara, CA). The drug effects and drug metabolism can be evaluated by monitoring the expression of drug-metabolizing and toxicology-marker genes in liver and in cultured hepatocytes. Other applications of DNA chips are the determination of the expression of each gene in human tissues and cell lines and gene expression in diseased tissues. Microchip technology also has been used for enzyme assays, receptor binding assays, and binding antigen to monoclonal antibodies. With advances in piezoelectric pulse ink jet delivery and microfluidic capillary electrophoresis and electro-osmosis, microchip technology will be a promising platform for HTS in few years.

VI. CONCLUSIONS

The human genome consisting of 3 billion base pairs and about 30,000 genes, whose sequence is being completed, will generate an increased number of molecular targets. This coupled with increasing compound collections necessitates the adaptation of HTS assays that can be run rapidly with minimal volumes to preserve the compounds and rapidly identify lead compounds. In the early 1990s, screening capacity was a few thousand samples per day with few targets (< 20) per year. This improved to several thousand samples per day with about 30 targets per year in the late 1990s in big Pharma. With technological advances in assay methods and miniaturization capabilities it is possible to achieve in the 2000s the capacity of 100,000 assays per day. The number of targets that are being screened at present range between 50 and 60 targets per year, and each target has to be screened with a million compounds in big pharmaceutical industries. Both the number of targets and the compound deck size will be increasing due to advances in genomics and combinatorial chemical synthesis, respectively, thus requiring the development of robust, homogeneous HTS assays. To decide on an appropriate screening assay, it is important for the target team to understand the screening platforms available and the screens HTS groups prefer for automation of the screen. Although several assay methods are available for a particular target, for example, tyrosine kinases and cAMP assays (see Chap. 7), to pick a robust assay for development one needs to know the liquid handling systems, detection systems, and automation equipment available. Homogeneous assays are preferred over heterogeneous assays for HTS, but one cannot compromise if the assay is not robust or the signal-to-noise ratio and CV are not acceptable. Sometimes functional cell-based assays are preferred as primary assays compared to in vitro biochemical assays. The availability of several fluorescence technologies and rapid advances in imaging technologies make it possible to assay at a single cell or a single molecule in the near future.

REFERENCES

1. PB Fernandes. Technological advances in high throughput screening. Current Opinion Chem Biol 2:597–603, 1998.
2. S Fox, S Farr-Jones, MA Yund. High throughput screening for drug discovery: continually transitioning into new technology. J Biomol Screen 4:183–192, 1999.
3. Working group updates. Standards in robotics and instrumentation. Microplate standardization, report 3, August 16, 1996. J Biomol Screen 1:163–168, 1996.
4. SBS proposed microplate specification. Revised July 1999. J Biomol Screen 4:169–174, 1999.
5. J-H Zhang, TDY Chung, KR Oldenburg. A simple statistical parameter for use in evaluation and validation of high throughput screening assays. J Biomol Screen 4:67–73, 1999.
6. SP Manly. In vitro biochemical screening. J Biomol Screen 2:197–199, 1997.
7. I Lazaro, M Gonzalez, G Roy, LM Villar, P Gonzalez-Porque. Description of an enzyme linked immunosorbent assay for the detection of protein tyrosine kinase. Anal Biochem 192:257–261, 1991.
8. AF Braunwalder, DR Yarwood, T Hall, M Missbach, KE Lipson, MA Sills. A solid-phase assay for the determination of protein tyrosine kinas activity of c-src using scintillating microtitration plates. Anal Biochem 234:23–26, 1996.
9. Y Reiss, MC Seabra, JL Goldstein, MS Brown. Purification of ras farnesyl:protein transferase. Methods 1:241–245, 1990.
10. E Schaefer, K Hsaio, S Guimond. Cell signaling. Detection and quantitation of protein tyrosine kinases. Promega Notes 59:2–9, 1999.
11. Biotrak cellular communication assays. Rat insulin [125I] assay system with magnetic separation. Amersham Code RPA 547.
12. S Udenfriend, L Gerber, N Nelson. Scintillation proximity assay, a sensitive and continuous isotopic method for monitoring ligand–receptor and antigen–antibody interactions. Anal Biochcm 161:494–500, 1987.
13. ND Cook. Scintillation proximity assay—a versatile high throughput screening technology. Drug Discov Today 1:287–294, 1996.
14. D Pocius, K Amrein. Detection of the activities of tyrosine(Y) kinase and tyrosine phosphatase (PTPs) that regulate the phosphorylation status of SRC family of tyrosine kinases utilizing a novel scintillation proximity assay (SPA) system. FASEB J 10:1458, 1996.
15. SD Kahl, FR Hubbard, GS Sittampalam, JM Zock. Validation of a high throughput scintillation proximity assay for 5-hydroxy tryptamine1E receptor binding activity. J Biolmol Screen 2:33–39, 1997.

16. I Kariv, ME Stevens, DL Behrens, KR Oldenburg. High throughput quantitation of cAMP production mediated by activation of seven transmembrane domain receptors. J Biomol Screen 4:27–32, 1999.

17. RA Jefferson. Assaying chimeric genes in plants: the GUS gene fusion system. Plant Mol Biol Reporter 5:387–405, 1987.

18. R Seethala, WB Benjamin. An essential arginyl residue at the active site of ATP– citrate lyase. Biochemical J 195:735–743, 1981.

19. KR Gee, W-C Sun, MK Bhalgat, RH Upson, DH Klaubert, KA Latham, RP Haugland. Fluorogenic substrates based on fluorinated umbelliferones for continuous assays of phosphatases and β-galactosidases. Anal Biochem 273:41–48, 1999.

20. RP Haugland. Handbook of Fluorescent Probes and Chemical Research. 6th ed. Molecular Probes, 1996.

21. R Seethala, R Menzel. A homogeneous, fluorescence polarization assay for src-family tyrosine kinases. Anal Biochem 253:210–218, 1997.

22. R Seethala. Fluorescence polarization competition immunoassay for tyrosine kinases. Methods 22:61–70, 2000.

23. R Seethala, R Menzel. A fluorescence polarization competition immunoassay for tyrosine kinases. Anal Biochem 255:257–262, 1998.

24. JJ Wu, DR Yarwood, Q Pham, MA Sills. Identification of a high-affinity antiphosphoserine antibody for the development of a homogeneous fluorescence polarization assay for protein kinase C. J Biomol Screen 5:23–30, 2000.

25. Z Li, MI Patel, J Kawooya, M Judkins, W Zhang, K Dieener, A Lozada, D Dunnington. An ultra-high throughput screening approach for an adenine transferase using fluorescence polarization. J Biomol Screen 5:31–38, 2000.

26. JR Sportsman, LJ Leytes. Miniaturization of homogeneous assays using fluorescence polarization. Drug Disc Today: HTS suppl 1:27–32, 2000.

27. M Allen, J Reeves, G Mellor. High throughput fluorescence polarization, a homogeneous alternative to radioligand binding for cell surface receptors. J Biomol Screen 5:63–69, 2000.

28. S Grahn, D Ullmann, HD Jakubke. Design and synthesis of fluorogenic trypsin peptide substrates based on resonance energy transfer. Anal Biochem 265:225–231, 1998.

29. G Mathis. HTRF® technology. J Biomol Screen 4:308–310, 1999.

30. I Hemmila. Lance™: Homogeneous assay platform for HTS. J Biomol Screen 4:303–307, 1999.

31. M Auer, KJ Moore, FJ Meyer-Almes, R Guenther, AJ Pope, KA Stoeckli. Fluorescence correlation spectroscopy: lead discovery by miniaturized HTS. Drug Disc Today 3:457–465, 1998.

32. MB Meza. Bead-based HTS applications in drug discovery. Drug Discovery Today: HTS suppl 1:38–41, 2000.

33. Application Notes. AlphaScreen. Bio-signal, a Packard BioScience Company.

34. S Miraglia, EE Swartzman, J Mellentin-Michelotti, L Evangelista, C Smith, I Gunawan, K Lohman, EM Goldberg, B Manian, P-M Yuan. Homogeneous cell- and bead assays for high throughput screening using fluorometric microvolume assay technology. J Biomol Screen 4:193–204, 1999.

35. EE Swartzman, SJ Miraglia, J Mellentin-Michelotti, L Evangelista, P-M Yuan. A

homogeneous and multiplexed immunoassay for high-throughput screening using fluorometric microvolume assay technology. Anal Biochem 271:143–151, 1999.

36. C Martens, A Bakker, A Rodriguez, RB Mortensen, RW Barrett. A generic particle-based nonradioactive homogeneous multiplex method for high-throughput screening using microvolume fluorimetry. Anal Biochem 273:20–31, 1999.

37. P Zuck. Ligand–receptor binding measured by laser-scanning imaging. Proc Natl Acad Sci USA 96:11122–11127, 1999.

38. P Hanninen, A Soini, N Meltola, J Soini, J Soukka, E Soini. A new microvolume technique for bioaffinity assays using two-photon excitation. Nat Biotech 18:548–560, 2000.

39. R Williams. Electrochemoluminescence: A new assay technology. IVD Technology 28–31, 1995.

40. R Chen, L Lowe, JD Wilson, E Crowther, K Tzeggai, JE Bishop, R Varro. Simultaneous quantification of six human cytokines in a single sample using microparticle-based flow cytometric technology. Clin Chem 45:1693–1694, 1999.

41. Z Parandoosh. Cell-based assays. J Biomol Screen 2:201–202, 1997.

42. M-H Beydon, A Fournier, L Drugeault, J Becquart. Microbiological high throughput screening: an opportunity for the lead discovery process. J Biomol Screen 5:13–21, 2000.

43. JO McCall, S Kadam, L Katz. A high capacity microbial screen for inhibitors of human rhinovirus protease Sc. Biotechnology 12:1012–1016, 1994.

44. R Menzel, ST Taylor. Periplasmic membrane-bound system for detecting protein–protein interactions. US patent 5,521066, 1966.

45. C Klein, JI Paul, K Sauve, MM Schmidt, L Arcangeli, J Ransom, J Trueheart, JP Manfredi, JP Broach, AJ Murphy. Identification of surrogate agonists for the human FPRL-1 receptor by autocrine selection in yeast. Nat Biotechnol 16:1334–1337, 1998.

46. J Zhu, CR Kahn. Analysis of a peptide hormone–receptor interaction in the yeast two-hybrid system. Proc Natl Acad Sci USA 94:13063–13068, 1997.

47. M Berg, K Undisz, R Thiericke, T Moore, C Posten. Miniaturization of a functional transcription assay in yeast (human progesterone receptor) in the 384- and 536-well plate format. J Biomol Screen 4:9–14, 1999.

48. BL Drees. Progress and variations in two-hybrid and three-hybrid technologies. Curr Opi Chem Biol 3:64–70, 1999.

49. L McAlister-Henn, N Gibson, E Panisko. Applications of the yeast two-hybrid system. Methods 19:330–337, 1999.

50. BA Ozenberger, KH Young. Functional interaction of ligands and receptors of the hematopoietic superfamily in yeast. Mol Endocrinol 9:1321–1329, 1995.

51. R Bosse, R Garlick, B Brown, L Menard. Development of nonseparation binding and functional assays of G-protein coupled receptors for high throughput screening. Pharmacological characterization of the immobilized CCR5 receptor on FlashPlate. J Biomol Screen 3:285–292, 1998.

52. J Watson, JV Selkirk, AM Brown. Development of FlashPlate technology to measure [^{35}S]GTPγS binding to Chinese hamster ovary cell membranes expressing the cloned human 5-HT1B receptor. J Biomol Screen 3:101–105, 1998.

53. KS Schroeder, BD Neagle. FLIPR: a new instrument for accurate high throughput optical screening. J Biomol Screen 1:75–80, 1996.

54. LC Mattheakis, LD Ohler. Seeing the light: calcium imaging in cells for drug discovery. Drug Disc Today: HTS suppl 1:15–19.
55. MD Ungrin, LMR Singh, R Stocco, DE Sas, M Abramovitz. An automated aequorin luminescence-based functional calcium assay for G-protein-coupled receptors. Anal Biochem 272:34–42, 1999.
56. YZ Klompus, J Daijo, M Zhang, P Panfili, R Sportsman. Cellular cAMP detection with high efficiently fluorescence polarization (HEFP). 6th Annual Conference of the Soc Biomol Screen, Vancouver, Canada, Sept 6–9, 2000.
57. P Kasila, D Duyka, P Banks. Florescence polarization cAMP assays for high throughput screening in 96- and 384-well formats. Product Bulletin, NEN Life Science Products.
58. AC Chiulli, K Trompeter, M Palmer. A novel high throughput chemiluminescent assay for the measurement of cellular cyclic adenosine monophosphate levels. J Biomol Screen 5:239–247, 2000.
59. IGEN Homepage on World Wide Web. URL: *http://www.igen.com/*
60. ALPHAScreen™ Technical Note ATN-002. BioSignal, a Packard Bioscience Company.
61. SS Vansal, DR Feller. An efficient cyclic AMP assay for the functional evaluation of β-adrenergic receptor ligands. J Receptor Signal Transduction Res. 19:853–863, 1999.
62. ME Nuttall, JC Lee, PR Murdock, AM Badger, F Wang, JT Laydon, GA Hofman, GR Pettman, JA Lee, A Parihar, BC Van Wagenen, J Fox, M Gowen, RK Johnson, MR Mattern. Amphibian melanophore technology as a functional screen for antagonists of G-protein coupled 7-transmembrane receptors. J Biomol Screen 4:269–277, 1999.
63. MN Potenza, GF Graminski, MR Lerner. A method of evaluating the effects of ligands upon Gs protein coupled receptors using a recombinant melanophore-based bioassay. Anal Biochem 206:315–322, 1992.
64. M Zaccolo, F DeGiorgi, CY Cho, L Feng, T Knapp, PA Negulescu, SS Taylor, RY Tsien, T Pozzan. A genetically encoded, fluorescent indicator for cyclic AMP in living cells. Nat Cell Biol 2:25–29, 2000.
65. J Alam, JL Cook. Reporter genes: application to the study of mammalian gene transcription. Anal Biochem 188:245–254, 1990.
66. AJ Kolb, K Neumann. Luciferase measurements in high throughput screening. J Biomol Screen 1:85–88, 1996.
67. AM Maffia III, I Kariv, KR Oldenburg. Miniaturization of a mammalian cell-based assay: luciferase reporter gene readout in a 3 microliter 1536-well plate. J Biomol Screen 4:137–142, 1999.
68. BA Brown, M Cain, J Broadbeat, S Tompkins, G Henrich, R Joseph, S Casto, H Harney, R Greene, R Delmondo, S Ng. FlashPlate™ Technology. In: JP Devlin, ed. High Throughput Screening. New York: Marcel Dekker, NY 1997, pp 317–328.
69. G Zlokarnik, PA Negulescu, TE Knapp, L Mere, N Burres, L Feng, M Whitney, K Roemer, RY Tsien. Quantitation of transcription and clonal selection of single living cells with β-lactamase as reporter. Science 279:84–88, 1998.
70. RY Tsien. The green fluorescent protein. Ann Rev Biochem 67:509–544, 1998.

71. GR Nakayama. Microplate assays for high-throughput screening. Cur Opin Drug Disc Develop 1:85–91, 1998.

72. GR Nakayama, MC Caton, Z Parandoosh. Assessment of the Alamar Blue assay for cellular growth and viability in vitro. J Immunol Methods 204:205–208, 1997.

73. M Wodnicka, RD Guarino, JJ Hemperly, MR Timmins, D Stitt, JB Pitner. Novel fluorescent technology platform for high-throughput cytotoxicity and proliferation assay. J Biomol Screen 5:141–152, 2000.

74. FP Guengerich. Cytochrome P-450 3A4: regulation and role in drug metabolism. Ann Rev Pharmacol Toxicol 39:1–17, 1999.

75. DA Smith, MJ Ackland, BC Jones. Properties of cytochrome P450 isoenzymes and their substrates. Part 2: properties of cytochrome P450 substrates. Drug Disc Today 2:479–486, 1997.

76. JH Kelly, NI Sussman. A fluorescent cell-based assay for cytochrome P-450 isozyme 1A2 induction and inhibition. J Biomol Screen. 5:249–253, 2000.

77. D Gerhold, T Rushmore, CT Caskey. DNA chips: promising toys have become powerful tools. Trends Biochem Sci 24:168–173, 1999.

4

Homogeneous Assays for High-Throughput and Ultrahigh-Throughput Screening

Ramakrishna Seethala
Bristol-Myers Squibb Company, Princeton, New Jersey

I. INTRODUCTION

Though drug therapy in the 1990s is based on less than 500 molecular targets, the number of potential therapeutic targets has been increasing exponentially with the advances in genomics, combinatorial chemistry, and competition to increase drug discovery processes in pharmaceutical industries [1]. With the recent completion of human genome sequencing (3 billion nucleotides) and understanding disease processes at the genetic level it is estimated that the potential molecular targets for drug discovery will be between 5,000 and 10,000. Consequently, the number of screens and compounds being screened will be increasing. This rapid growth in targets necessitates an increase in the pace and productivity of high-throughput screening (HTS). HTS is becoming an important strategy in finding large numbers of new leads that will eventually be developed into new drugs. All the assays must be screened with large compound libraries (> million compounds) to obtain novel lead compounds. More than 90% of HTS screens were run in 96-well plates by the end of 1998, and it is predicted that about 50% of HTS will be in 384-well plates or higher density plates in the year 2000. HTS has been growing into ultrahigh-throughput screen (uHTS) with the development of high density plates and nanotechnology. To increase the throughput of screening, the assays must be very simple and robust and preferably homogeneous (add reagents and read), and amenable to miniaturization, liquid dispensers, and robots (automation), thus decreasing the time and manpower required for completion of screening the compound deck. There have been constant improvements in liquid handling systems, detection instruments, data analysis, high density plates, and assay technologies to be able to automate drug screening completely.

Homogeneous assays mostly involve in vitro biochemical assays. A few homogeneous cell based assays have also been developed recently. In vitro assays are advantageous because the assay can be targeted for a specific enzyme, receptor, or protein. The other advantages of in vitro assays include speed, less expense, and more reproducibility of the assays as large amounts of the reactants and protein from the same lot can be used for the screen. In cell based assays the concentration of test compounds in the cell depend on the permeability and efflux of the compound. The less permeable compounds in mammalian cells may not be detected in a whole cell screen but may be an excellent chemotype affecting the target in in vitro biochemical assays. Some chemotypes may be degraded in the cell to nonactive compounds and may not be detected. These chemotypes can be detected in the in vitro biochemical assays and have the potential of being amenable to chemical modifications to increase permeability or stability in the cell to obtain more potent compounds. Also, with the recombinant cells, the signal and response may change with the number of passages of the cell line and cell biology reagents.

Conventional assays such as receptor–ligand binding, enzyme, and cell based assays are multistep assays which in the past involved filtration, precipitation, or adsorption and several wash steps to separate bound ligand from free ligand or product from substrate. ELISA assays, which gained popularity in the last two decades, involve several incubation and wash steps that are labor intensive and can achieve at best medium throughput ($< 5K$ samples per day). Assays involving radioactive substrates and ligands generate large volumes of radioactive waste that create handling problems. In addition, these assays are not adaptable for complete automation. Each additional processing step in an assay will increase the time for completion of the assay, reduce the signal, and thus reduce the throughput of a screen. In HTS mode, at least 10K samples need to be processed per day. There is a dire need for HTS to develop homogeneous assays that are robust and consist of reactions that are very rapid, taking minutes to reach equilibrium. They should also have reduced processing steps, faster in terms of sample readout, adaptable to high density plates, and easy to automate.

In homogeneous assay technology, all the reaction components either in solution or with some reactants attached to solid phase are incubated and read in a plate reader with no further processing steps. Most of the homogeneous assays are fluorescence based assays because fluorescence is very sensitive and can easily be quantitated with fluorescence detectors and miniaturized. Environmentally safe fluorescence assays can be developed because these do not generate hazardous waste. Major research efforts in the last few years produced several new homogeneous assays that were based on fluorescence, chemiluminescence, or radioactive assays. New nanotechnologies including DNA array chips, quantum dots, and nanoparticle technologies are evolving that will be suitable for uHTS in a very high density format. The fluorescence based assays include (a) fluorescence intensity assays that include fluorogenic, fluorescence quench, and quench relaxation assays; (b) fluorescence polarization (FP) assays in which the signal is proportional to molecular volume (size); (c) fluorescence resonance energy transfer (FRET); (d) homogeneous time-resolved fluorescence (HTRF), homogeneous time-resolved fluorescence resonance energy transfer (TR-FRET) or Lance assay in which fluorescence resonance energy from a donor molecule is transferred to an acceptor molecule; (e) confocal fluorescence microscopy; (f) fluorescence imaging analysis; and (g) fluorescent reporter systems such as green fluorescent protein (GFP) and β-lactamase reporter systems. Homogeneous assays have also been developed with electrochemiluminescence and chemiluminescence. The homogeneous radioisotope assays include (a) scintillation proximity assay (SPA), now a widely used technology, in which the isotope labeled biomolecule bound to appropriate protein or antibody coated on SPA bead (scintillant coated bead) comes in close proximity of the scintillant on the bead and produces a scintillation signal; (b) FlashPlate; (c) Cytostar assay technologies, in which the radioisotope bound to the antibody or receptor coated on to the scintillant coated microtiter plate wells produces a scintillation signal; the unbound radioisotope need not be separated from the bound isotope; and (d) Leadseeker technology, which combines SPA and imaging technologies. Other technologies, such as surface plasma resonance, are being developed for higher throughput assays. These homogeneous assays are discussed in greater detail in this chapter.

II. FLUORESCENCE SCREENS

The fluorophores are either intrinsic or extrinsic fluorophores. Several biological molecules contain naturally occurring fluorophores (intrinsic fluorophores). Tryptophan is the most highly fluorescent amino acid in proteins and contributes more than 90% of the fluorescence of a protein. Proteins absorb at 280 nm, and the emission maximum ranges between 320 and 350 nm. Though tyrosine is fluorescent in solution, its emission in a protein is weaker. Nucleotides and nucleic acids are not fluorescent except yeast t-RNA[phe], which contains a highly fluorescent Y base. NADH is highly fluorescent, with absorption and emission maxima at 340 nm and 450 nm, respectively, but NAD$^+$ is nonfluorescent. The quantum yield of NADH increases 4-fold when bound to a protein. FMN and FAD are fluorescent with absorption and emission maxima at 450 nm and riboflavin at 515 nm [2].

When the intrinsic fluorescence of a macromolecule is not adequate, external fluorophores are conjugated to them to improve spectral properties. In most of the biological assays extrinsic fluorophores conjugated with biomolecules have been used. Fluorescein, rhodamine, Texas Red, and 4,4-difluro-4-bora-3a,4a-diazo-s-indacene (BODIPY) dyes have been widely used for labeling proteins and nucleic acids. These fluorescent dyes have longer wavelengths of excitation and emission that minimizes the background fluorescence of biological samples. Some fluorophores such as coumarin derivatives, and dansyl chloride are excited with shorter wavelengths. Dansyl chloride is a very widely used fluorophore to label proteins and amines. Various derivatives of these fluorescence dyes are available that react with a primary amino group, the ϵ-amino group of lysine, —SH group of cysteine, —OH, or —COOH groups of amino acids, and —OH or phosphate group of sugar moiety in nucleotides. ATP derivatives with etheno bridge (ϵ-ATP derivatives) or lin-benzo AMP are highly fluorescent and retain hydrogen bonding properties and can be used for labeling nucleotides. 1-anilino-8-naphthalenesulfonic acid (1,8-ANS), *bis*-anilino-8-naphthalenesulfonic acid (bis-ANS), and 2-*p*-toluidinyl-naphthalene-6-sulfonic acid (2,6-TNS) are non-fluorescent in water but highly fluorescent in nonpolar solvents and when bound to proteins at hydrophobic pockets. Lipids in the membranes can be labeled with 9-vinyl anthracene, 1,6-diphenylhexatriene, or perylene. The probes are insoluble in water and partition into the lipid layer of membranes [2].

A. Fluorescence Intensity Screens

The readout of these assays is either an increase or a decrease in the fluorescence intensity. Fluorogenic assays and fluorescence quench relaxation assays are manifested in the increase of fluorescence intensity, whereas fluorescence quench assays show a decrease [3].

1. Fluorogenic Assays

The reactants (such as methylumbelliferyl derivatives, ANS, bisANS, nucleic acid specific dyes) of the assay are not fluorescent, but the products generated are fluorescent and measured as an increase in fluorescence intensity. β-Glucuronidase, nuclease, and polymerase assays are discussed here.

Instrumentation. Several fluorescence plate readers from different manufacturers that can read fluorescence of 96- and 384-well microtiter plates with stackers are available. A few being used in the author's lab are: Victor 1420 (Wallac, Gaithersburg, MD), Analyst and Acqueyst (LJL BioSystems, Sunnyvale, CA), CytoFluor 4000 (PerSeptive Biosystems, Framingham, MA), Titertek (Flow Laboratories, AL), Spectramax Gemini (Molecular Devices), and HTS 7000 (Perkin-Elmer Corp., Norwalk, CT). The fluorescence assay can be fully automated using a robot, a liquid handler system, and a plate reader, and the throughput can be increased to 50,000 to 100,000 compounds per day.

β-Glucuronidase Assay.   In the homogeneous fluorogenic assay for human β-glucuronidase the fluorogenic substrate, 4-methylumbelliferyl-D-glucuronide (MUG) is hydrolyzed to fluorescent 4-methyl umbelliferone (4-MeU) (Fig. 1) [4,5]. The product 4-MeU is fluorescent only when the hydroxyl group is ionized (the pKa of this hydroxyl group is 8–9), and maximal fluorescence is obtained at pH > 10. The enzyme activity is proportional to the fluorescence signal and is measured in a fluorescence plate reader with excitation at 355 nm and emission at 465 nm. This fluorogenic assay was found to be 100-fold more sensitive than the colorimetric methods. The β-glucuronidase assay can be minia-

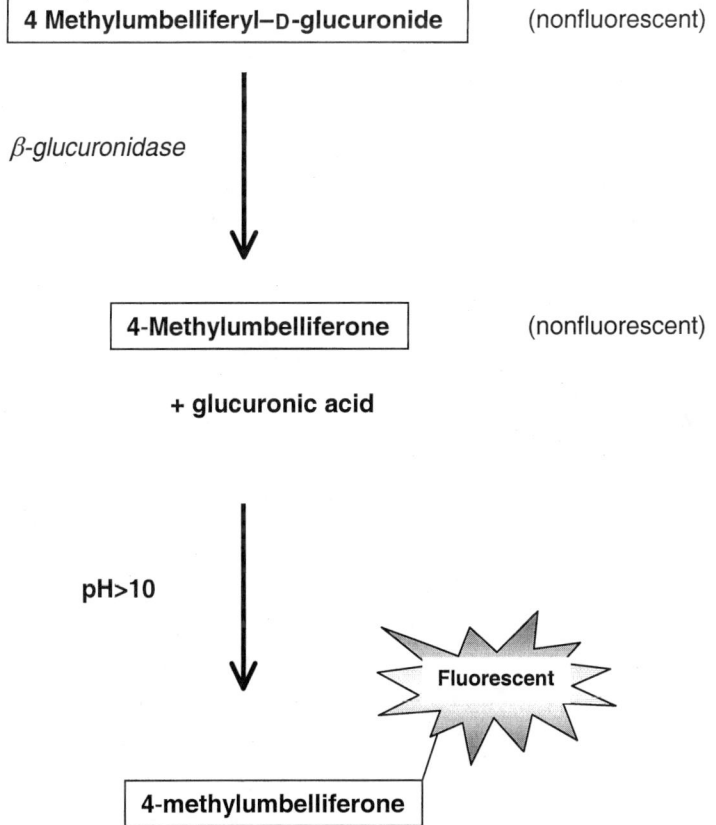

Figure 1   Fluorogenic β-glucuronidase assay. The substrate 4-methylumbelliferyl-D-glucuronide is nonfluorescent. The product 4-methylumbelliferone formed is also non-fluorescent but becomes fluorescent in alkaline solution. Fluorescence can be read in a fluorescence plate reader with excitation at 355 nm and emission at 465 nm.

turized into 384-well plate, and the signal is comparable to that of 96-well microplate (Fig. 2). The fluorimetric β-glucuronidase assay is a simple, homogeneous, robust assay that does not involve any separations and washes. The assay is amenable for miniaturization in high density plates (384- and 1536-well plates).

Nuclease and Polymerase Assays. Molecular Probes developed several dyes (nonfluorescent or low intrinsic fluorescence) that specifically interact with double strand DNA (dsDNA), single strand DNA (ssDNA) or RNA and, upon binding to the nucleic acids, exhibit several-fold fluorescence enhancement and quantum yield increases. PicoGreen interacts with dsDNA and produces a fluorescence signal while ssDNA and RNA in the sample do not contribute to this fluorescence signal. PicoGreen is used for detection of picogram level of dsDNA in solution [6]. OliGreen specifically binds to ssDNA and oligonucleotides and produces a fluorescence signal. RiboGreen binds specifically to RNA and has

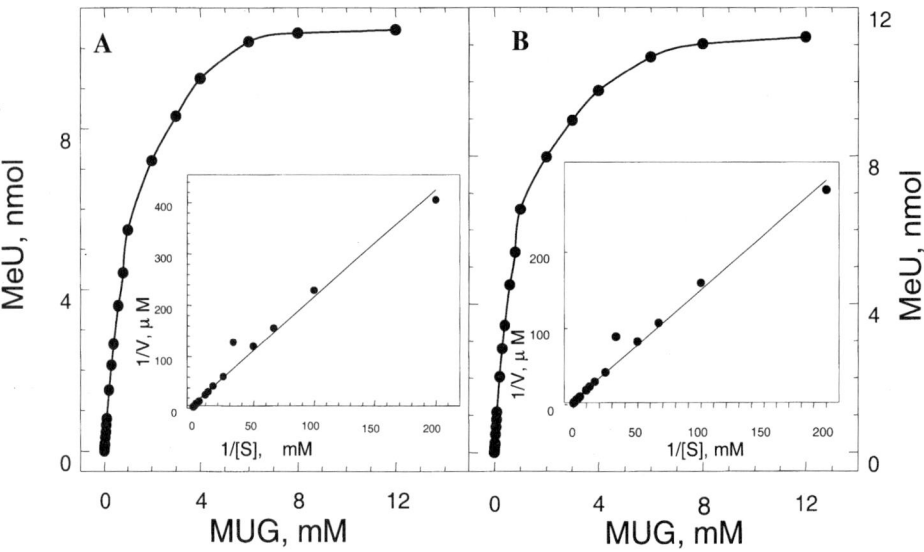

Figure 2 Comparison of β-glucuronidase screen in (A) 384-well and (B) 96-well format. Increasing concentrations of MUG were incubated with 3 or 10 ng β-glucuronidase in acetate buffer, pH 4.8, at room temperature in a total volume of 15 or 50 μL in a 384- or 96-well micro plate, respectively. The reaction was terminated with the addition of 15 or 50 μL stop buffer (0.4 M Na_2CO_3, pH 10) to each well in a 384- or 96-well micro plate respectively. The 96-well plate was read in Titertek fluorescence reader and the 384-well plate was read in Victor 1420 (Wallac). The saturation binding curves were similar and the K_ms obtained were also similar.

been used for quantitation of RNA in solution [7]. Nuclease or polymerase assays using the specific nucleic acid binding dyes can be developed.

RNase H hydrolyzes the RNA strand from the RNA-DNA hybrid substrate. A homogeneous fluorescence RNase H assay can be developed with PicoGreen dye. In this assay, the substrate poly r(A)-d(T)$_{12-18}$ is incubated with the RNase H enzyme at 37°C for 60 min. The reaction is terminated with the addition of PicoGreen dye solution and read in a fluorescence plate reader. The dye binds to DNA in the intact DNA-RNA duplex, resulting in a fluorescence signal. As the RNA strand is hydrolyzed with the enzyme action, the double strand hybrid is depleted, resulting in a decrease in the dye bound to the substrate. Thus with the RNase H activity the fluorescence intensity is decreased.

A DNA polymerase assay can be developed wherein the fluorescence signal increases with enzyme activity. In this assay, the dsDNA synthesized in the enzyme reaction binds the dye and increases the fluorescence signal, whereas the ssDNA substrate does not bind. These assays can be performed in 96- or 384-well plate format or even in higher density plates. A limitation of these assays is the identification of false positives with compounds that interact with the DNA-RNA by intercalation and interference from colored compounds. Nevertheless, this nonradioactive, simple, homogeneous assay can be used for rapid primary screening, and the false negatives can be eliminated in the secondary screening.

2. Fluorescence Quench Assays

When a fluorescent molecule is constrained by covalent modification of the reactive groups (substrate), the fluorescence is quenched. During the assay reaction, if the covalent bond of the substrate is cleaved, free fluorescent dye is released, and the fluorescence intensity increases due to relaxation of quenching.

Peptidase Assay. Several amine containing dyes, 7-amino-4-methylcoumarin, 7-amino-4-chloromethylcoumarin, 6-aminoquinoline, rhodamine 110 (R-110), N-(4-chloromethyl)benzoyl rhodamine 110, 5-(and 6-)chloromethylrhodamine 110, and 6-amino-6-deoxyluciferin (Molecular Probes), when covalently linked to amino acids or peptides, change the spectral properties and cause the fluorescence to be quenched [6].

Rhodamine 110 exhibits spectral properties similar to fluorescein with excitation and emission λ of 496 and 520 nm, respectively. Several bisamide derivatives of rhodamine 110 have been used as specific substrates for protease activity in solution and living cells [6]. These substrates contain peptides covalently linked to the two amino groups of rhodamine, thereby suppressing its absorption and fluorescence; thus the fluorescence of R-110 is quenched in these substrates (Fig. 3). The nonfluorescent bisamide R-110 substrate is cleaved to give fluorescent monoamide, and further cleavage yields more fluorescent R-110 [6]. The

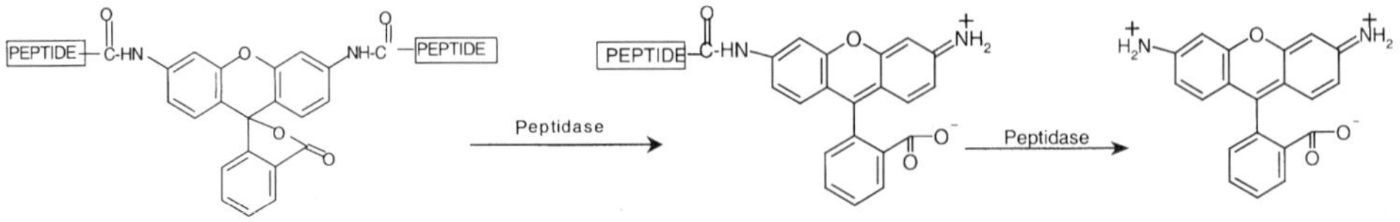

Bisamide   Monoamide   Rhodamine 110

(Nonfluorescent)   (Fluorescent)   (Fluorescent)

Figure 3 Fluorogenic peptidase assay. In this peptidase assay the substrate, a nonfluorescent bisamide of rhodamine 110, was cleaved to the fluorescent monoamide and then
to the highly fluorescent free rhodamine 110.

fluorescence intensity of the monoamide derivative and R-110 is constant between pH 3 and pH 9.

3. Comments

Fluorescence intensity assays are simple assays that can be readily miniaturized.
Total fluorescence is measured, and these assays give little information for design
of assay and quality control. These assays have strong interference from test
compound inner-filter and autofluorescence effects, which are difficult to detect
and to correct for.

B. Fluorescence Polarization Screens

Fluorescence polarization (FP) is a homogeneous technique in which FP signal
is proportional to the molecular size of the fluorescent molecule in defined conditions.

1. Theory of FP

When a fluorescent sample is excited with a polarized light, the emission from
the sample is polarized. Polarization of a fluorophore is the result of its orientation
relative to the direction of the polarized excitation. When fluorescent molecules
in solution are excited with polarized light, the degree to which the emitted light
retains polarization reflects the rotation that the fluorophore underwent during
the interval between absorbance and subsequent emission. In 1926 Perin first
described the utility of FP to study the molecular interactions in solution [8]. FP
instrumentation was developed by Weber [9] and adapted to homogeneous assays
by Dandliker [10]. FP is a powerful technology for the determination of molecular
interactions in solution [9–15]. The polarization value of a molecule is proportional to the molecule's rotational relaxation (correlation) time and is described
by the Stokes equation,

$$\rho = \frac{3\eta V}{RT} \qquad (1)$$

where ρ is rotational relaxation time (the time required to rotate through 68.5°), η is the viscosity of the medium, V is the molecular volume of the molecule, R is the gas constant, and T is the temperature. Therefore if viscosity and temperature are held constant, polarization is directly proportional to molecular volume. Polarization and anisotropy are commonly used to describe the molecular interactions in solution.

The fluorescence polarization (P) is defined as

$$P = \frac{I_\parallel - I_\perp}{I_\parallel + I_\perp} \qquad (2)$$

Anisotropy (r) is defined as

$$r = \frac{I_\parallel - I_\perp}{I_\parallel + 2I_\perp} \qquad (3)$$

where $I_\parallel$ is the intensity of emission parallel to the plane of excitation light and $I_\perp$ is the intensity of emission perpendicular to the plane of excitation. Polarization and anisotropy can be interconverted using the following equations [2]:

$$P = \frac{3r}{2 + r} \qquad (4)$$

$$r = \frac{2P}{3 - P} \qquad (5)$$

Polarization and anisotropy use the same measurements [Eqs. (2) and (3)]. Although anisotropy values are preferred, FP is the most common technology used in HTS, hence only FP will be discussed here. All the FP readers measure FP in milli P (mP) (1 polarization unit = 1000 mP). The total fluorescence intensity (I) can be determined from the same data from the equation $I = I_\parallel + 2I_\perp$. Polarization ($P$), being a ratio of emission light intensities, is a dimensionless entity and does not depend on the intensity of the emitted light or on the concentration of the fluorophore. If the fluorophore is small, it rotates or tumbles faster, and the resulting emitted light is random with respect to the plane of polarization (depolarized) and will have lower FP value (Fig. 4). If the fluorophore is large, it remains relatively stationary, and the emitted light will remain polarized and have higher FP signal [13,14]. Theoretically, the minimum FP possible is 0 and the maximum is 500 mP for fluorescein. However, the real experimental minimum observed is 40–80 mP for small molecules and the experimental maximum for large molecules is 100–300 mP [14,15]. This window of FP signal between the minimum and the maximum is sufficient because the ratiometric FP signal

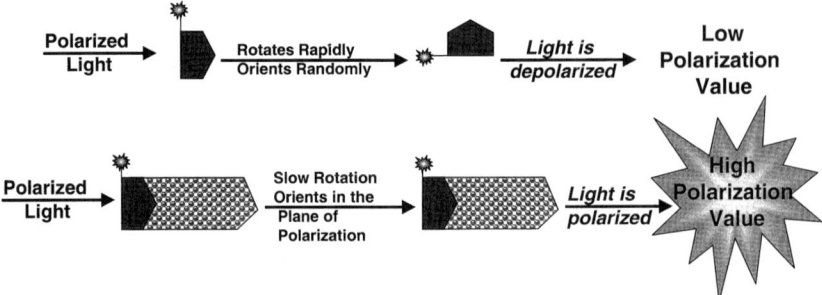

Figure 4 Principle of fluorescence polarization. Top panel: A small fluorescent molecule in solution rotates rapidly and orients randomly, the emitted light is depolarized, and the polarization value is low. Bottom panel: A small fluorescent molecule binds to a macromolecule, the resulting fluorescent complex rotates slowly in solution, it orients in the plane of polarization, and the emitted light is polarized giving a high polarization value.

is highly reproducible. Unlike other assays where the robustness of the assay depends on the magnitude of the signal-to-noise ratio, in FP assay ΔP (the highest signal–lowest signal) is important. A ΔP of 100 mP in a FP assay is considered a good assay and $\geq$ 150 mP is considered a robust assay.

FP is a homogeneous technology consisting of simple mix reagents and read format. FP can be easily automated. The reactions are very rapid, reaching equilibrium very quickly. Fluorescein derivatives are the most common fluorescent derivatization reagents for covalent labeling. Fluorescein is a well-studied molecule that has high absorptivity, excellent fluorescence quantum yield, and good water solubility, and its excitation maximum closely matches the 488 nm spectral line of the argon-ion laser. A number of other fluorophore (BODIPY®, Texas Red™, Oregon Green®, Rhodamine Red™, Rhodamine Green™, etc.) derivatives are commercially available with various chemistries for making fluorescent bioconjugates [6]. Many biotechnology labs have been synthesizing fluorescent derivatives of custom peptides and nucleic acids, and a few ready made fluorescent compounds are available with some vendors. The reagents are stable, and large batches required for a screen can be prepared at one time. In FP assays only one tracer is needed. FP is free from interferences, independent of intensity, and insensitive to colored compounds because it is a ratiometric measurement.

2. Instrumentation

The first single-tube bench-top instrument, the FPM-1, was developed by Jolley Consulting and Research Inc. (Grayslake, IN) for measuring FP signal for homogeneous assays for drug screening. Beacan 2000 (PanVera Corp., Madison, WI) is another single-tube bench-top instrument. Now FP plate readers from several manufacturers are available. FPM-2 from Jolley Consulting and Research Inc. reads 96-well plates; Polarion from Tecan (Research Triangle Park, NC), Polarstar from BMG LabTechnologies (Durham, NC), and Analyst from LJL Biosystems (Sunnyvale, CA) read both 96- and 384-well plates, and Acqueyst from LJL BioSystems reads 1536-well plates. These FP instruments are very sensitive in measuring fluorescence intensity and FP signals. Polarstar and Analyst are multimode signal detection systems that can read fluorescence intensity, fluorescence polarization, time-resolved fluorescence, and luminescence. All the FP plate readers are robot-friendly for transport and communication.

A simplified schematic diagram for Analyst is given in Fig. 5. In microplate readers, the excitation and emission energy can be focused through the meniscus of the samples in the microplate wells. Analyst achieves almost identical performance in both 96- and 384-well plates by using SmartOptics, which selects and configures light sources, filters, detectors, and optical paths and analyzes light from the same sensed volumes. SmartOptics can place the sensed volume in the middle of the well or at the bottom of the well for cell based assays.

3. Types of FP Assays

FP assays can be classified into three groups. The first group represents assays that acquire FP signal by binding a small molecule containing a fluorophore to a macromolecule and forming a larger fluorophore adduct, e.g., protein–DNA, antigen–antibody, DNA–DNA, DNA–RNA, and protein–protein interactions [12,14–17] and receptor–ligand binding [18–20]. The second group represents

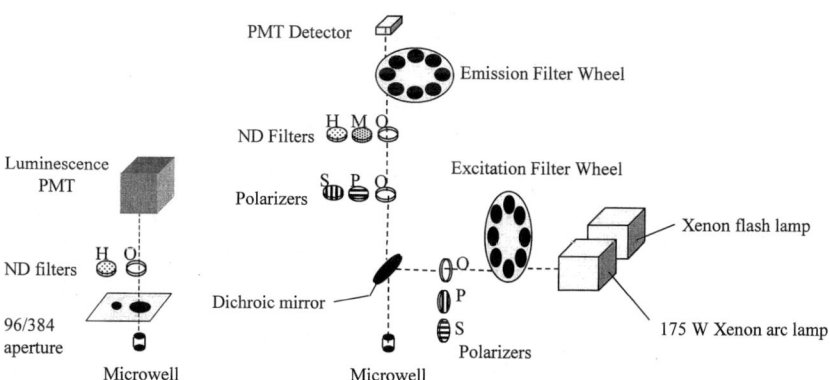

Figure 5 Schematic of optical pathway of Analyst™. The optical path for Analyst™ in fluorescence polarization mode is given. Monochromatic light passes through filter and polarizer and the polarized light excites the sample. The emitted light is passed through horizontal and vertical polarizers and an emission filter and measured in a PMT detector. (Courtesy of LJL Biosystems.)

the assays that incur loss of polarization signal due to cleavage of relatively larger fluorescent molecules (fluorophore containing macromolecules) to smaller fluorescent molecules, e.g., nuclease, helicase, and protease assays [13,21–23]. The third group represents facilitated (indirect) assays in which the fluorescent reactant molecule is coupled to antibody with acquisition or loss of FP signal, e.g., protein tyrosine kinases (PTKs) and protein tyrosine phosphatases (PTPs). FP has been extensively used in the last decade in clinical laboratories in competitive immunoassays for the detection of drugs and hormones [13–15]. FP assays representative of each of these groups are discussed below in detail.

FP Protein Tyrosine Kinase Assay.   Seethala and Menzel have developed the first robust FP-PTK assay [14–16]. In the direct FP-PTK assay, the phosphorylated fluorescein–peptide (fl-phos-peptide) product formed in the reaction is measured by immunocomplexing with the antiphosphotyrosine antibody (PY antibody) which results in an *increase in the FP signal* (Fig. 6). The direct FP-PTK assay can only be used with a peptide substrate and requires large amounts of antiphosphotyrosine antibody. To overcome these problems, the FP-PTK competition immunoassay was developed [15,16]. In this assay, phosphorylated peptide or protein produced by kinase reaction competes with the fluorescent phos-

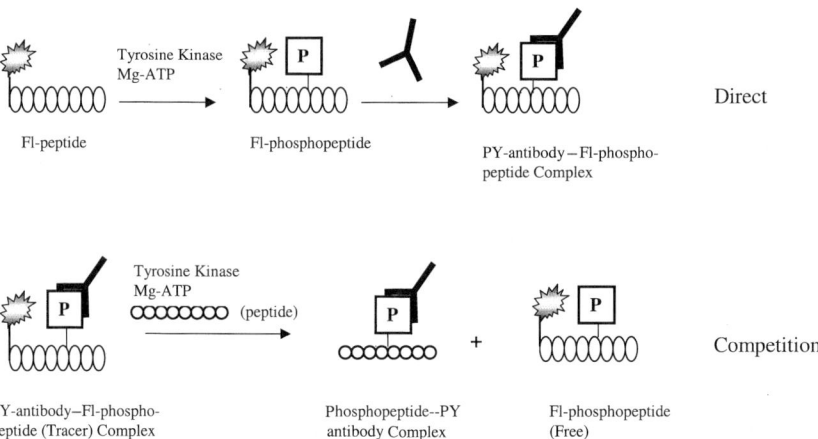

Figure 6   FP protein tyrosine kinase assay methods. In the direct FP-PTK assay, a fluorescein-peptide substrate is incubated with PTK, Mg^{2+}-ATP, and antiphosphotyrosine (PY) antibody [14]. The phosphorylated fluorescein-peptide (fl-phos-peptide) product is immunocomplexed with the PY antibody, resulting in an *increase in the FP signal*. The signal is proportional to the phosphorylated product formed. In the FP-PTK competition immunoassay [15,16], phosphorylated peptide or protein produced by a kinase reaction will compete with a fluorescent phosphopeptide used as a tracer for immunocomplex formation with PY antibody. In this format kinase activity results in a *loss of the FP signal*, and the FP signal is inversely proportional to the phosphorylated product formed in the reaction.

phopeptide tracer for PY antibody. In this format, kinase activity results in a *loss of the FP signal.*

In the FP-PTK competition assays, a peptide substrate (at about K_m concentration) is incubated with PTK (Lck), 0.1 mM ATP, 5 nM fl-phosphopeptide as tracer, and PY-54 phosphotyrosine antibody (0.25–1.00 µg) in the assay buffer in a final volume of 25 or 100 µl in a 384- or 96-well plate, respectively. After incubation at room temperature for 30 min, the plate is read in a FP reader. The affinity of fl-phosphopeptide was highest with PY antibody ascites fluid and monoclonal antibody PY54 [14–16] among different monoclonal and polyclonal anti-PY antibodies tested (Fig. 7A). The fl-phos-peptide binding to PY 54 is rapid and reached equilibrium in less than 5 min, and it remained unchanged for at least 60 min (Fig. 7B). The dissociation of PY antibody-fl-phosphopeptide com-

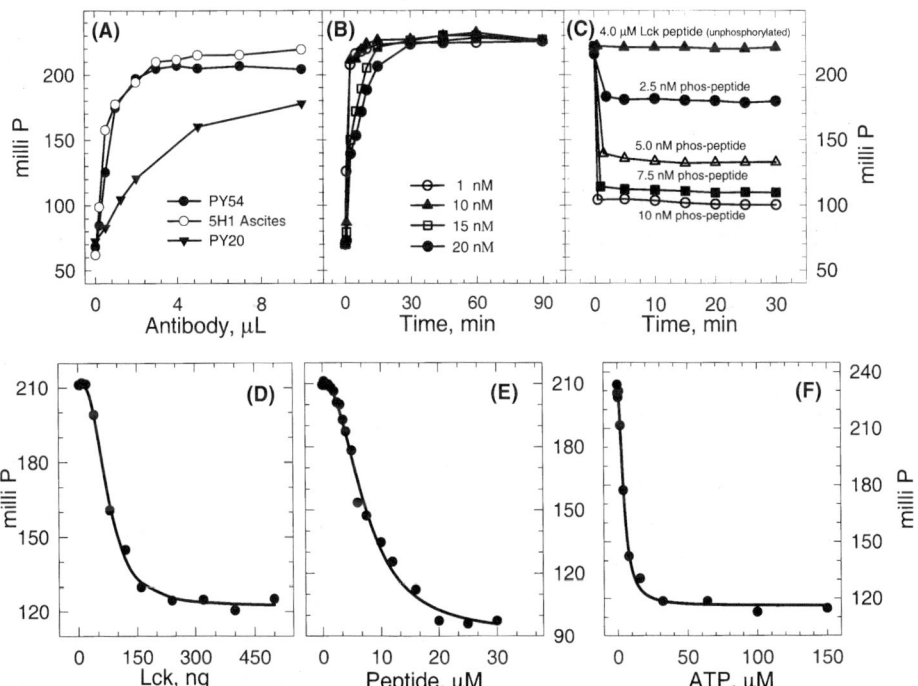

Figure 7 Optimization of FP protein tyrosine kinase assay. (A) FP signal with different PY antibodies. (B) Association of fluorescein labeled tyrosine–phosphorylated Lck-peptide (fl-phos-peptide) with 10 µg of PY antibody at different concentrations of peptide substrate. (C) Dissociation of PY antibody-fl-phos-peptide complex by phos-peptide as a function of time. Dependence of Lck activity on (D) Lck, (E) Lck peptide, and (F) ATP concentrations. (From Ref. 16.)

plex was rapid, and complete dissociation was achieved with phospho-Lck peptide within 5 min (Fig. 7C). PTK activity showed a good concentration dependence on Lck, ATP, Lck-peptide, and enolase (Figs. 7D–F).

Inhibition by staurosporine, a potent nonspecific PTK inhibitor, and by PP, a specific Lck/Fyn inhibitor competitive with ATP [24], were evaluated both at 5 and 20 μM ATP (Fig. 8). The IC_{50} for staurosporine was 30 and 110 nM and for PP was 70 and 300 nM at 5 and 20 μM ATP, respectively, and these values are comparable to the values obtained by parallel $^{32}PO_4$ transfer assay and FP direct assay. In addition, results with a panel of proprietary PTK inhibitors suggested that the FP competition immunoassay can successfully detect inhibitors of Lck with the same rank order of potency.

This FP-PTK assay is very simple, nonradioactive, and highly sensitive and does not involve separation of substrate and product. A variation of this method would also be suitable for the assay of phosphatases and has been successfully so used. The simplicity and speed of this method makes FP-PTK and FP-phosphatase assay ideal for HTS. The advantages of this assay over other more commonly used kinase assays such as $^{32}PO_4$ transfer assay, ELISA, or DELFIA

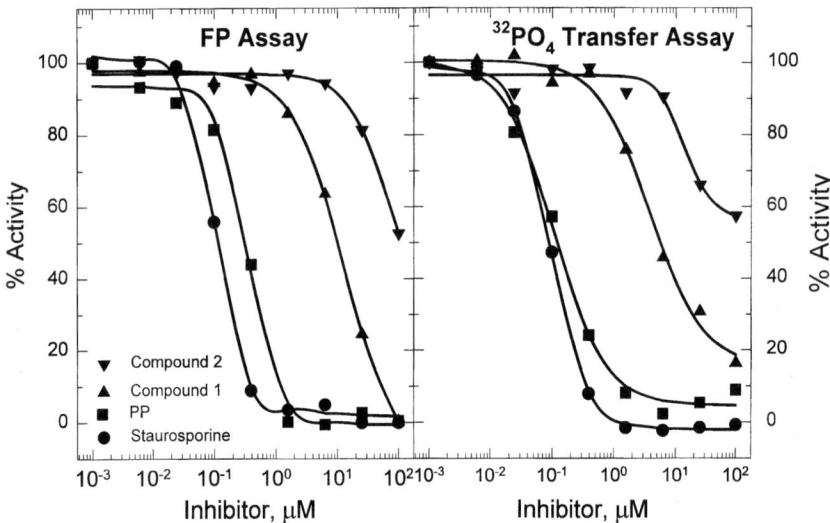

Figure 8 FP-PTK assay validation. To validate the FP-PTK competition immunoassay inhibition by staurosporine, a potent nonspecific protein kinase inhibitor, and PP, a specific Lck/Fyn competitive inhibitor, inhibition was evaluated at 5 and 20 μM ATP and compared with a concurrently run $^{32}PO_4$ transfer assay. The IC_{50}s obtained by both the FP competition assay and the $^{32}PO_4$ transfer assay were similar. (From Ref. 16.)

include the use of nonisotopic substrates and its being a simple one-step assay, without separation, precipitation, washing, or processing steps after incubation. This method can easily be automated for high throughput drug discovery screening. PTK assay kits based on this FP assay described here are now commercially developed by LJL BioSystems and PanVera Corp.

FP Serine/Threonine Kinase Assay. Several nonradioactive PTK assays based on FP, HTRF, DELFIA, ELISA, and ECL formats have been described that utilize the high affinity anti-PY specific antibody [14–16]. These approaches cannot be applied for serine/threonine kinase assays due to the unavailability of a specific, high affinity antibody for phosphoserine and phosphothreonine. Recently, a FP assay for general kinase assay has been described [17] in which the peptide substrate is thiophosphorylated with ATPγS by protein kinase, followed by biotinylation of the thiophosphate group with iodoacetyl LC-biotin, and finally incubation with streptavidin and determination of the FP signal (Fig. 9). The FP signal increased and was directly proportional to the product formed. The Ki obtained for H-89, an ATP-competitive inhibitor of protein kinase A, was calculated to be about 60 nM, which is similar to that reported. The drawback of this FP method is that biotinylation of the thiophosphate group of the peptide takes a long time ($\sim$ 8 hr). Nevertheless, the assay can be adapted to the microtiter plate and can be used as a homogeneous assay after some optimization. It is a better alternative to the $^{32}PO_4$ transfer assay.

Recently, a FP assay for protein kinase C using a specific monoclonal antiphosphoserine antibody has been described [18]. A selective substrate peptide containing ser or thr is phosphorylated with ATP using PKC along with diacylglycerol and phosphatidylserine. After incubation, the reaction is stopped with the addition of EDTA, fluoresceinylated phosphopeptide tracer and antipeptide antibody and read in a plate reader. Phosphorylated product competes with the tracer, and the enzyme activity is inversely proportional to the FP signal. FP-PKC kits are available from LJL BioSystems and PanVera Corp.

FP-RNase H Screen. RNase hydrolizes RNA strands of RNA-DNA hybrid. Fl-RNA-DNA-biotin hybrid substrate is prepared by first synthesizing a 52 mer fluorescein labeled RNA transcript from Hind III linearized pSP65 cDNA using fluorescein 12-UTP. A complimentary 5′-biotin-52-mer oligo DNA is synthesized (bio-DNA) and annealed to the prepared pSP65 fl-RNA. The FP signal of fl-RNA-bio-DNA hybrid, upon binding to avidin, increased from 105 to 350 mP. The ΔP of 245 mP is a robust signal for a FP assay. RNase H reaction was carried out by incubation of fl-RNA-bio-DNA hybrid with HIV RNase H (HIV-reverse transcriptase/RNase H, Worthington) at pH 8.0 and 37°C in a total volume of 50 μL. The reaction was terminated with the addition of stop buffer (50 μL) containing 20 mM MOPS pH 7, 20 mM EDTA and 10 μg of avidin, incubated for a further 15 min, and read in a FP reader. With the enzyme action, the

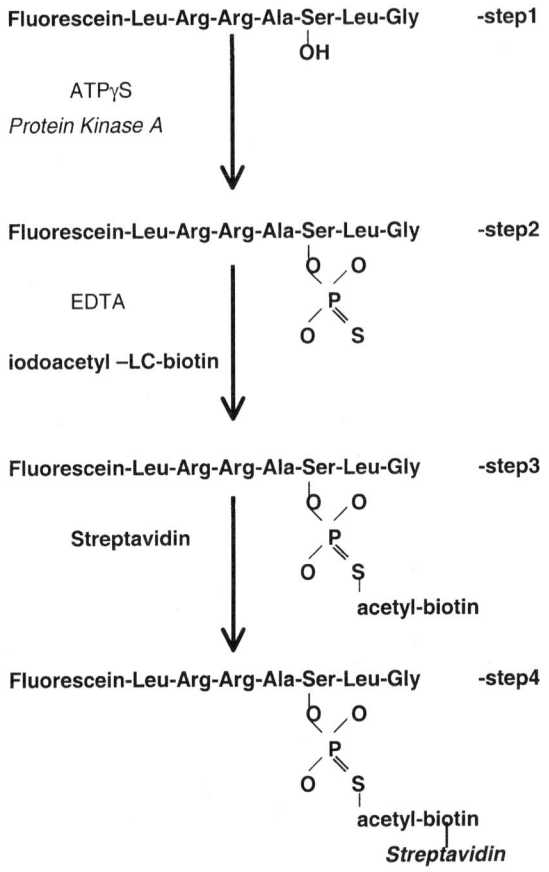

Figure 9   Schematic of a FP-serine-threonine kinase assay. A fluorescein labeled peptide is thiophosphorylated (step 2) by incubation with ATPγS and kinase. Biotin is attached to the sulfur by incubation with iodoacetyl-LC-biotin (step 3), and the binding of streptavidin to thio-biotinyl peptide (step 4) produces a FP signal.

RNA of the RNA-DNA hybrid is hydrolyzed. The fl-RNA will no longer be attached to the avidin bound biotin-DNA strand and hence *the FP signal will decrease* with increasing enzyme activity in this assay. The FP signal decreased to 110 mP from 306 mP. There was no enzyme activity in the absence of reaction mixture (Mg^{2+} and Mn^{2+}).

The time dependence of the FP signal with HIV-RNase H showed that 60 min incubation is optimal (Fig. 10A). RNA hydrolysis was measured as a function of RNase H concentration; the FP signal showed that 2 units of RNase H was

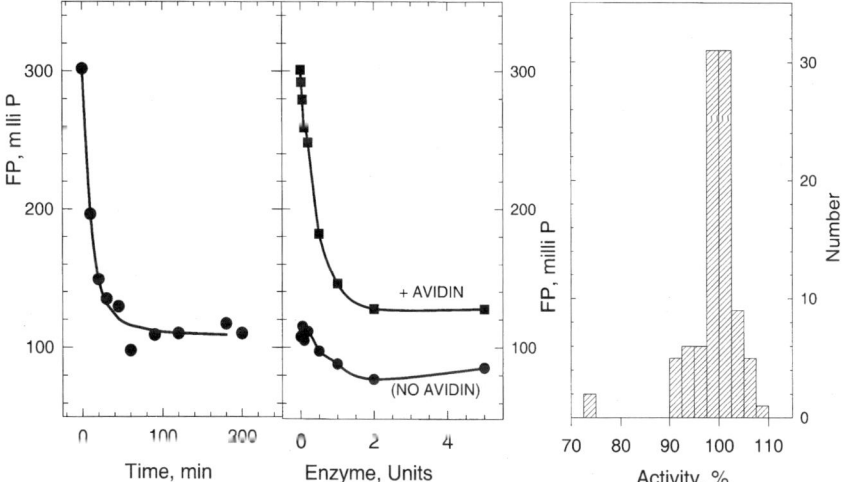

Figure 10 FP RNase H assay. (A) Time-course of RNase H reaction. (B) RNase H activity as a function of RNase H enzyme concentration. (C) Validation with random compounds (10 μM). The distribution of the activity was normal with > 95% of compounds falling between 90 and 110%.

optimal (Fig. 10B). Evaluation of FP RNase H assay with a plate of random compounds at 10 μM (Fig. 10C) showed that the distribution of activity was very tight, with most of the activity between 90 and 110%. This suggests that the assay does not give spurious results and that synthetics can be tested at 10 μM or higher concentrations. The results with FP RNase H assay suggest that the signal is robust and reproducible and can be used for HTS.

Protease Assay. Standard FP protease assay kits using fluorescein labeled α-casein as substrate are available (Panvera and Molecular Probes). The FP signal of the relatively large fluorescein substrate is high and when cleaved by protease action produces smaller labeled fragments that have low FP signal. Fl-casein substrate, however, is not stable at lower pHs below 7. To assay various proteolytic enzymes by FP, a new pH-independent substrate BODIPY-α-casein was used between pH 2 to 11 [21]. A peptide substrate derivatized by biotinylation of a γ-aminobutyric acid modified amino terminus and labeled with 5-(4,6-dichlorotriazinyl)aminofluorescein at the carboxy terminus was used for cytomegalovirus protease [22]. The substrate binds to avidin and increases FP signal. Enzyme action cleaves the substrate generating a fluorescein containing peptide separated from biotin that will not be able to interact with avidin thereby resulting in the loss of FP signal.

FP Receptor Binding Assay. The bulk of HTS targets are receptors such as G-protein coupled receptors (GPCRs) and other membrane receptors, and nuclear receptors. For successful application of the FP method, a membrane receptor has to be expressed in a high copy number ($\sim$ 100,000) in each cell. The ligand has to have very high affinity for the receptor, and a substantial amount of fluorescent ligand has to bind to the receptor for FP assay. Previously, FP membrane receptor binding assays were limited to analytical methods using large amounts of membranes [19]. Recently, FP assay has been used for both peptide (vasopressin V_{1a} and δ-opioid) and nonpeptide (β$_1$-receptor and 5-HT3) receptors in 96- and 384-well formats [20].

The FP method has been successfully used for nuclear receptors [18]. FP assays amenable for HTS using purified nuclear receptor protein and a fluorescent ligand have been developed for estrogen receptors ERα and ERβ [17] and other nuclear receptors. ERα and ERβ FP assays have been developed using full length receptors and Fluormone™ ES1 (a natural fluorescent ligand) and ES2 (a fluorescein labeled estradiol). The ligand is incubated with the receptor protein for 1 hr, and the FP signal is measured in a plate reader. The assay produced a robust signal of 200 mP. The FP ligand binding assay can be extended to all other nuclear receptors. The FP nuclear receptor and membrane receptor assays are discussed in greater detail elsewhere in this volume (Chap. 7).

4. Comments

FP is a simple and reasonably predictive technology and can be used for a variety of different assays. FP is a robust homogeneous assay that can be applied to higher density 384- and 1536-plates. Only one component of the assay is required to be labeled with a fluorophore and is suitable for small ligands ($<$ 15 kDa). FP is insensitive to inner-filter effects. Since the FP signal is a ratiometric measurement it is less susceptible to quenching from colored compounds. However, autofluorescence compounds can interfere with a FP assay. Sometimes, propeller effects on the fluorescent tag may restrict the use of the FP assay. FP plate readers from several manufacturers are available in different price ranges.

C. FRET Assays

Fluorescence resonance energy transfer (FRET) is a phenomenon wherein excitation energy is transferred from a donor molecule to an acceptor molecule without emission of a photon. FRET has been used for measuring the distances between interacting molecules under physiological conditions with near angstrom resolution [25]. For FRET to occur, donor and acceptor molecules have to be in close proximity (10–100 Å), excitation of the acceptor must overlap with the emission of donor, and donor and acceptor transition dipole orientations must be parallel.

The rate at which the energy is transferred from donor to acceptor is governed by the Förster equation [26]

$$k_T = \frac{1}{\tau_D} \times \left(\frac{R_0}{R}\right)^6 \tag{6}$$

$$R_0^6 = 8.79 \times 10^{-5} \left(\kappa^2 \phi_D \frac{J}{n^4}\right) \tag{7}$$

The rate of Förster energy transfer k_T is dependent on τ_D, the fluorescence lifetime of donor (D) in the absence of acceptor molecules. R is the distance between donor and acceptor, and R_0 is the Förster radius, the distance at which energy transfer is 50% efficient (typically between 10 and 50 Å). Thus $k_T = 1/\tau_D$ when $R_0 = R$. This equation applies to a single configuration of donor and acceptor, and distributions of donor and acceptor have to be averaged appropriately over distance and orientations. The Förster distance is dependent on κ^2, the dipole orientation factor between donor and acceptor; ϕ_D is the quantum yield of the donor in the absence of acceptor; n, the refractive index of the medium (which is 1.4 in aqueous solution), and J, the overlap integral between the donor and acceptor. $J = \int F_D(\lambda) \cdot \varepsilon_A(\lambda) \cdot \lambda^4 \cdot d\lambda$ where $F_D(\lambda)$ is the peak-normalized fluorescence intensity of donor and ε_A is the molar absorption coefficient of the acceptor at wavelength λ [3,25–27].

In most cases the donor and acceptor dyes are different, and FRET can be detected as the quenching of donor fluorescence by the acceptor or appearance of the sensitized fluorescence of the acceptor. Typical R_0 for the donor–acceptor pair of fluorescein and tetramethylrhodamine is 55 Å, 5-(2-aminoethylamino) naphthalene-1-sulfonic acid (EDANS) and 4-(4-dimethylaminophenylazo)benzoic acid (DABCYL) is about 33 Å, and 5-(2-iodo-acetyl aminoethylamino) naphthalene-1-sulfonic acid (IAEDANS) and fluorescein is 46 Å [2,5,25,26].

1. FRET Applications

FRET applications include quench and quench relaxation assays.

Protease Assay. In a peptide containing C-terminus EDANS and N-terminus DABCYL, the fluorescence of EDNAS is quenched by the acceptor DABCYL. Cleavage of this internally quenched fluorogenic substrate leads to an enormous increase in fluorescence intensity as the donor is separated from the acceptor. The fluorescence intensity is proportional to the hydrolysis of the substrate. Thus the protease activity can be monitored by the fluorescence intensity. Several protease assays (e.g., trypsin, HIV-1 protease, renin, hepatitis C virus protease, human cytomegalovirus protease) using FRET have been described [27].

D. HTRF™/Lance™ (TR-FRET) Assays

Time-resolved fluorescence (TRF) is based on the long lifetime properties of lanthanides, europium (Eu), samarium (Sm), terbium (Tb), and dysprosium (Dy). Homogeneous time-resolved fluorescence (HTRF), as the name implies, is a homogeneous assay method that uses fluorescence resonance energy transfer from europium cryptate (EuK) donor to an acceptor fluorophore provided they are at a distance less than 10 nm from one another and that the emission energy of donor overlaps with the excitation of the acceptor. HTRF™ technology is a trademark of Packard and CIS/Bio, and this phenomenon is variously called time-resolved fluorescence resonance energy transfer (TR-FRET) or Lance™ (Lanthanide Chelate Excitation) technology, a trademark of Wallac. For the energy transfer, the donor and acceptor have to be brought into close proximity. The fluorescence lifetime of most of the fluorescent compounds is very short (typically a few nanoseconds). The interference from assay components, microtiter plates, light scatter, and biological samples and compounds is short-lived fluorescence (prompt fluorescence). To avoid this interference, lanthanide chelates with long lifetimes, 100–1000 µs (no polarization), are used as fluorescence energy donors. The common acceptor fluorophores used are modified allophycocyanine, a phycobiliprotein from red algae (XL665, Cy 5), fluorescein, or tetramethylrhodamine. The long-lived fluorescence of lanthanide due to Förster dipole–dipole energy transfer to the acceptor results in a long-lived fluorescence of the acceptor. The energy transfer emission has a decay time directly proportional to donor decay time and inversely to the distance between acceptor and donor. Time resolution in HTRF/Lance reduces scattering and prompt decay background interference from short-lived fluorescence due to delay in the time of measurements of fluorescence.

The rare-earth elements, Eu^{+3}, Su^{3+}, Tb^{3+}, and Dy^{3+}, are poor fluorophores by themselves, and to measure the lanthanide fluorescence, the lanthanides have to be complexed. Europium fluorescence is protected from decay by conversion to the macropolycyclic compound europium cryptate [Eu]K to enhance their fluorescence, and cryptate protects from fluorescence quenching (Fig. 11) [26,28–30]. [Eu]K has convenient linker arms to complex covalently with peptides, proteins, and nucleic acids. Also, selected generic reagents, streptavidin, biotin, WGA, ConA, protein A, anti-DNP antibody labeled with XL665 and biotin, streptavidin, and antiphosphotyrosine antibody labeled with [Eu]K are available from Packard. New chelates of Eu^{3+} and Tb^{3+} are available from Advant (a joint venture between Xenova and Wallac) that are used in Lance technology (Fig. 11) [30]. The fluorescence resonance energy is transferred from Eu chelates emitting

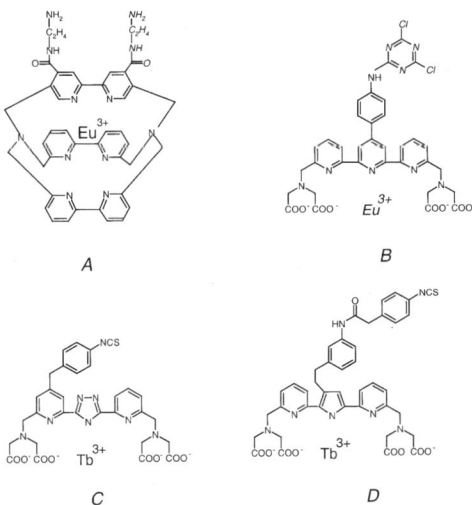

A

B

C

D

Figure 11 (A) Structure of Eu^{3+} cryptate. In Eu^{3+} cryptate (EuK) the Eu^{3+} is tightly bound within the trisbipyridine cryptate cage that protects Eu^{3+} from quenching, enhances Eu^{3+} fluorescence, and provides convenient linkers to attach biomolecules. (B) Structure of Eu^{3+} chelate. (C) and (D) Structure of Tb^{3+} chelates.

maximally at 613 nm to longer wavelength acceptor molecules such as allophycocyanins XL665 or Cy5 and from Tb chelates emiting at 492 and 545 nm to a variety of acceptors such as tetramethylrhodamine, xanthine dyes, or fluorescein [25,26,28]. Some of the generic reagents available include Eu^{3+} or Tb^{3+} labeled antihuman, rabbit or mouse IgG, anti-GST or anti-PY antibody, protein G and streptavidin, and APC labeled streptavidin (Advant).

1. Instrumentation

The dual-wavelength time-resolved fluorometer that can measure time-resolved fluorescence and the ratio of donor fluorescence and acceptor fluorescence is the Discovery HTRF microplate analyzer (Packard) [28]. The schematic for the Discovery is given in Ref. 28. The Discovery simultaneously measures the emission of (Eu)K at 620 nm and the XL665 at 665 nm and the ratio is calculated. The other fluorometers that can measure TRF at two wavelengths one after the other are the 1420 Victor plate reader (EG&G Wallac, Turku, Finland), the Analyst plate reader (LJL Biosystems), and the PolarStar (BMG Technologies). The Victor 1420 schematic diagram of optics is given in Figure 12.

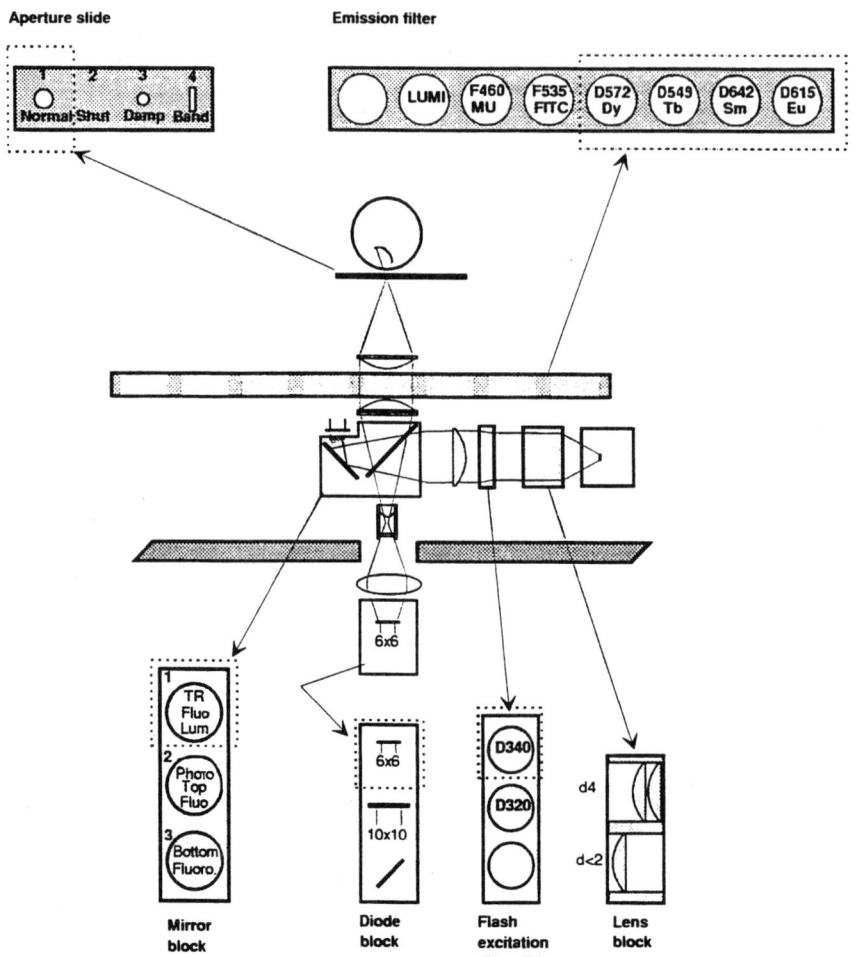

Figure 12 Optical pathway of Victor 1420. Excitation light from UV Xenon flash tube (1500 V) is focused through a filter, conventional lenses, and a dichroic mirror into the sample, and the emitted light passes through a narrow band filter into a R1527 PMT for measurement. The dual window TRF option allows us to record two separate windows in one flash cycle. (Courtesy of Wallac.)

2. Applications of HTRF

Essentially three types of labeling of macromolecules (direct, indirect, and semi-direct) can be done for HTRF/Lance assays similar to labeling for FP [27]. In direct labeling, the donor lanthanide chelate and the acceptor allophycocyanin, rhodamine, or fluorescein are directly labeled on molecule(s) involved in the reaction. In indirect labeling, these donor and acceptor fluorophores are labeled to macromolecules involved in the secondary interactions such as antibody–secondary antibody binding or biotin–streptavidin interaction. Semidirect labeling is a combination of direct and indirect labeling. HTRF/Lance assays can be successively developed for reactions involving macromolecular interactions such as DNA–DNA, RNA–DNA, DNA–protein, protein–protein, and receptor–ligand, for reactions involving hydrolysis of macromolecules such as nucleases or proteases, for reactions involving synthesis of macromolecules such as polymerases, and for other protein modification reactions such as kinases [28–30].

PTK Assay. In the PTK assay, a biotin–peptide substrate is phosphorylated by PTK action. Eu-PY antibody binds to the phosphorylated biotin-peptide [32,33]. The biotin-peptide binds to XL665 or CY5 labeled streptavidin (APC-streptavidin) (Fig. 13). Due to the close proximity of the two fluorophores, the energy from Eu^{3+} is transferred to APC-streptavidin. Typically, with the best PY antibody, a signal-to-noise ratio of 20 or more is obtained [35]. In the indirect method, PY antibody (not labeled with Eu^{3+}) binds to the phosphorylated biotin-peptide and the complex binds to a generic reagent, Eu-protein G or Eu-protein A (Fig. 13). FRET can occur from Eu to APC conjugated to streptavidin, though the signal will be somewhat lower (signal-to-noise ratio $\sim$ 10) than the above direct method [34]. Nevertheless, the use of generic reagents provides an easy assay development with comparable signal response.

The applications for HTRF/TR-FRET technology have been rapidly growing. They include competitive immunoassays to measure hormones like prolactin, βhCG [35], tumor necrosis factor (TNF) α [36], receptor–ligand binding interactions, e.g., TNF receptor 1 (TNFR1)-TNFα binding in which TNFR1 is labeled with lanthanide chelate and the ligand with acceptor. When TNFR1 is labeled with Eu chelate biotin-TNFα is used as ligand and APC-streptavidin as acceptor (Fig. 14), and with TNFR1 labeled with Tb chelate, rhodamine labeled TNFα is used as ligand. The IC_{50} obtained for TNFα is 3.9 and 3.5 nM, respectively. IL-2 receptor α–IL-2 interaction in which IL-2 is labeled with Eu^{3+} and a monoclonal antibody for IL-2 receptor α is labeled with Cy5 [36]; EGF-EGF receptor interaction in which EGF is labeled with Eu^{3+} and a monoclonal antibody against the nonbinding region of EGF receptor is labeled with XL-665 [29]; protein–protein interactions, e.g., *jun-fos* heterodimerization is measured by incubating biotinylated *jun* peptide with XL665-*fos* peptide and (Eu)K-streptavidin [29]; DNA-hybridization assay in which one oligo nucleotide strand was labeled with Eu^{3+}

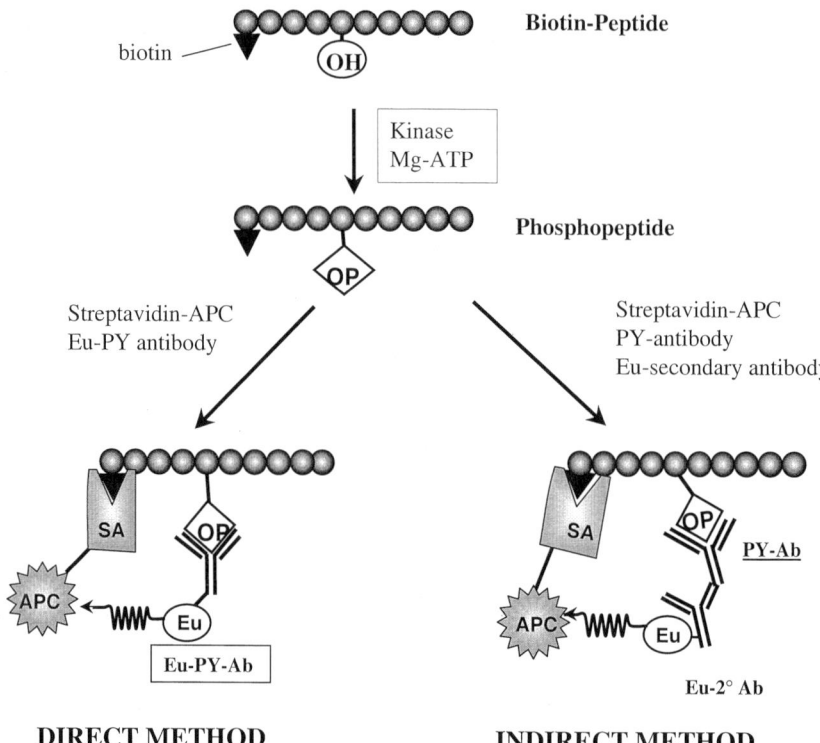

Figure 13 Schematic representation of TR-FRET protein tyrosine kinase assays. In the direct assay, phosphorylated biotinyl peptide is conjugated to Eu^{3+}-PY antibody and energy is transferred to APC-streptavidin bound to the phosphorylated biotinyl peptide. In the indirect assay, Eu^{3+}-protein A or Eu^{3+}-secondary antibody binds to the PY antibody conjugated to the phosphorylated biotinyl peptide and transfers energy to APC-streptavidin bound to the phosphorylated biotinyl peptide. (Courtesy of Wallac.)

at the 5′ end and the complementary strand was labeled with biotin at the 5′ end and coupled to XL-665 labeled streptavidin [38]; helicase assay in which one DNA oligonucleotide strand is labeled at the 3′ end either with CY5 or tetramethylrhodamine and the other strand is labeled at the 5′ end with Eu or Tb chelate, respectively [39].

3. Comments

The HTRF/Lance assay is a nonradioactive, homogeneous, sensitive assay without any separation steps. With appropriate delay time in HTRF/Lance, the back-

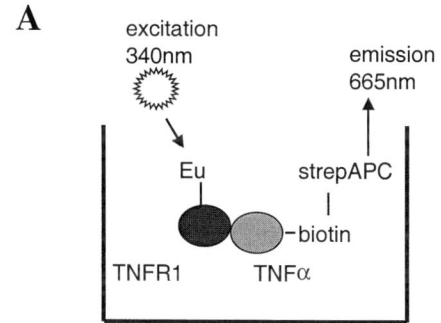

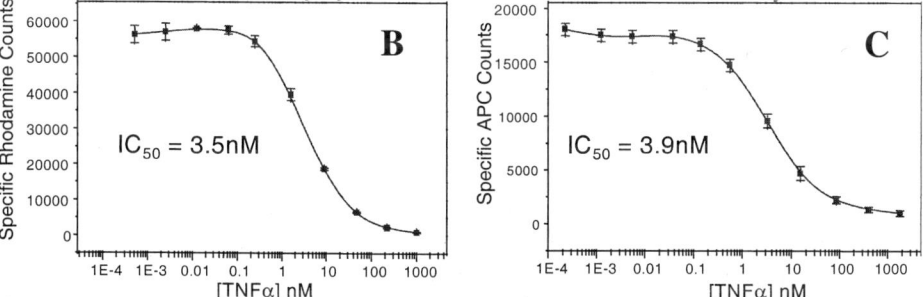

Figure 14 TR-FRET TNFα binding assay. (A) Schematic representation of energy transfer in the TNFα-Eu^{3+} APC Lance assay system. TNFR1 was labeled with Eu, TNFα was biotinylated, and APC bound streptavidin was used in this assay. (B) TNFα competition binding curve using Tb/Rh Lance assay system gave an IC$_{50}$ for a TNFα of 3.5 nM. The TNFα competition binding curve using the Eu/APC Lance assay system gave an IC$_{50}$ for a TNFα of 3.9 nM, which is very close to that obtained with the Tb/Rh system. (Courtesy of Wallac.)

ground interference can be minimized, and the emission of acceptor excited during the excitation of donor can be eliminated, and the pure energy transfer signal from acceptor can be measured. Energy transfer from lanthanide donor to allophycocyanin acceptor occurs when they are in proximity, giving long-lived decay at acceptor emission, the signal being specific for the biomolecular interaction. When acceptor (XL665) is not in proximity to the donor [Eu]K or lanthanide chelates a short-lived signal at 665 nm occurs that is discriminated by a time-delayed measurement. The interference from colored and fluorescent compounds in the assay can be eliminated by measuring the ratio of specific acceptor signal to the donor (Eu)K or lanthanide chelate signal. HTRF/Lance is a homogeneous assay technology that can be used in 96-, 384-, and 1586-well plate format and is amenable to automation, thus suited well for HTS and uHTS. The reagents are environmentally safe and are stable. One of the limitations of this assay is that two interacting biomolecules have to be conjugated to a donor and an acceptor fluorophore if the generic labeled reagents cannot be used, which can be complex and problematic. There is a limited choice of donors and acceptors.

E. Fluorescence Correlation Spectroscopy

Fluorescence correlation spectroscopy (FCS) is a statistical physics based new analytical technology that extracts quantitative information from the spontaneously fluctuating fluorescence molecules of small molecular ensembles [40]. FCS monitors interactions of molecules present at minuscule concentrations in femtoliter volumes and thus offers the highest potential as the detection technique in the nano scale in the determination of molecular interactions in solution, on cell surfaces, or in the cells using homogeneous assays. In FCS, a sharply focused laser beam illuminates a very small volume element (typically femtoliter). Single molecules diffusing through the illuminated confocal volume produce bursts of fluorescent light quanta during the entire course of their journey (Brownian motion), and each individual burst is recorded in a time-resolved manner by a highly sensitive single-photon detector and analyzed using autocorrelation techniques [41,42]. This FCS autocorrelation function gives information on concentration, the diffusion time of all the individual molecules (related to the size and shape of the molecule), and the brightness of each molecule. Thus autocorrelation of the time-dependent fluorescence signal allows differentiation of slow and faster diffusing particles, and binding and catalytic activity can be directly calculated from the diffusion times and the ratio of faster and slower molecules.

1. Instrumentation

The Confocor is a commercial instrument developed by a joint venture between Evotec and Carl Zeiss. EVOscreen™ platform is a modular, miniaturized uHTS based on FCS and Evotec's proprietary FCS-related single molecule detection technology (FCS plus) and reader Confocor™ (discussed in greater detail in this volume, Chapter 22).

2. FCS Applications

Evotec in parternership with SmithKline Beechem and Novartis have developed several FCS based assays representing various classes of target proteins [42]. FCS can be used for mass-dependent and mass-independent fluorescence assays.

Mass-Dependent Assays. Ligand–receptor binding assays can be performed by FCS at the molecular level with membrane bound receptors in live cells on cell surfaces, membrane preparations, or cell derived vesicles and nuclear receptors based either on fluorescence intensity or on FP [41,42]. Screens based on FCS have been used for EGF receptors, acetylcholine receptors, and thyroid receptors [41,42]. FCS protease assays with fluorescent casein substrate based on total intensity and with fluorescent biotinylated peptide substrate based on FP have been reported [41,42]. Enzyme assays such as PTK, PTP [42], protein–protein interactions such as SH2–phosphotyrosine binding [42], DNA/protein interactions such as topoisomerase–DNA binding, thyroid hormone receptor–DNA binding [41,42], and DNA/DNA interactions as in template-primer association have been investigated with FCS [42].

Mass-Independent Assays. Assays based on fluorescence intensity changes have also been developed by FCS. In these assays, the fluorescence of the substrate is quenched, and in the product of the reaction the quenching is relieved, increasing the molecular brightness and the total fluorescence intensity. A quenched fluorescent protease substrate, tetramethylrhodamine (TMR)-quenched fluorescence peptide, or streptavidin-quenched rhodaminegreen (RhGn)-peptide, when cleaved by protease, leads to the relief of quenching, which results in an increase in fluorescence intensity, apparent particle number, and mean confocal intensity [42]. In the RNA–ligand binding assay, association of a RhGn labeled ligand to RNA quenches the RhGn fluorescence of the ligand due to environmental effects on the dye and reduces the confocal fluorescence intensity and apparent particle number [42]. The binding of TMR labeled chemokine to chemokine receptor membrane vesicles increases the cumulative brightness of TMR-chemokine bound vesicles, which can be monitored by FCS [42]. FCS can also be used for Ca^{2+} uptake functional assays for 7-transmembrane receptor [42].

A dual-color fluorescence cross-correlation spectroscopy method suitable for binding and fast catalytic rate study was developed that does not depend on diffusion properties as with conventional FCS [43,44]. Based on dual-color fluorescence cross-correlation spectroscopy, RAPID FCS (rapid assay processing by integration of dual-color FCS) has been developed, which combines short analysis times with the development of fast and flexible assays resulting in sensitive, homogeneous, fluorescence based assays to measure molecular fragmentation and assembly resulting from a reaction/interaction [44]. This further extends the scope of FCS for uHTS.

3. Comments

FCS can be used in a microvolume configuration (1–10 μL assay volumes) and is adaptable for uHTS. The read times for vesicle assays are 5–10 sec and for solution assays 1–2 sec. FCS assays are based on the analysis of the molecular dynamics and the reaction kinetics of fluorescence labeled molecules that undergo temporal changes in their diffusion properties and determine the concentration of interaction parameters. A large number of simultaneously derived molecular parameters are obtained in FCS that allow selection of the most robust signal change for an assay. FCS can be used for several types of assays and for detection techniques monitoring intensity, particle number, polarization, energy transfer, and lifetimes, increasing the scope of this technology. FCS can thus be used for most of the target classes of assays encountered in drug discovery. Though FCS was first described more than 25 years ago, the application of FCS technology to HTS is very recent, and there are no commercial FCS readers available other than that being developed by Evotec (Confocor). Access to the technology is semiexclusive to the consortium partners, limiting the accessibility of this technology. Also, the limited screening data that is in the public domain is from presentations at conferences and a few review articles [41,42]. FCS could prove to be a promising technology for uHTS in the future.

F. Fluorescent Reporter Assays

Cell based assays are either used in conjunction with in vitro assays or in place of in vitro biochemical assays to examine output of specific cellular process. Reporter genes have been used in drug discovery for transcriptional studies as well as for characterization of receptor function and metabolic regulation [45]. Agonist activation of a receptor or a ligand-gated ion channel produces changes in the transcription of a number of genes that can be readily measured by using gene fusions. A reporter gene construct consists of an inducible transcriptional element that controls the expression of a reporter gene. Generally, a strong promoter (constitutively not active) that is controlled by a desired response element that is regulated by receptor activation is fused to the coding region of a reporter protein such as green fluorescent protein, β-lactamase, luciferase, β-galactosidase, chloramphenicol acetyltransferase, or secreted alkaline phosphatase (see Chapter 10).

1. Green Fluorescent Protein

Green fluorescent protein (GFP), a fluorescent protein originally isolated from the jellyfish *Aequoria victoria,* is a 238 amino acid protein that attains the fluorescent state spontaneously; the fluorescence is stable [46]. The wild GFP is relatively low in fluorescence intensity, has multiple absorption and emission maxima, and has about 4 hr lag time between the expression of protein and attaining full fluorescence. These drawbacks of GFP are remedied with the development of new and improved GFP mutants with different spectral properties, increased brightness of fluorescence, and mammalian cell compatible cloning vectors. Because of different spectral properties of mutant GFPs, it is possible to follow two GFPs in the same cell, and also they can be used in FRET assays to study protein–protein interactions and other FRET based assays. A FRET protease assay between two linked variants of the GFP was described [46]. The C-terminus of a red-shifted variant of GFP (RSGFP4) was fused to the N-terminus of a protein linker containing a Factor Xa protease cleavage site, and the N-terminus of a blue-variant of GFP (BFP5) was fused to the C-terminus of the protein linker. In the gene product, energy transfer occurs from BFP5 to RSGFP4. With Factor Xa protease action the protein linker is cleaved and the two GFPs dissociate, resulting in a decrease in energy transfer. The emission ratio of the BFP5 (450 nm) and RSGF4 (505 nm) increases with the protease activity because FRET decreases. A similar chymotrypsin assay using G-protein FRET peptide in 3546-well Nano well assay plates was reported [47].

2. β-Lactamase Reporter Assays

β-lactamase from *E. coli* is a 29 kD product of the ampicillin-resistant gene *Amp* that hydrolyzes pencillins and cephalosporins. The recent report of a membrane-permeant ester derivative CCF2/AM (6-chloro 7-hydroxy coumarin and fluorescein conjugated at 7 and 3′ positions of cephalosporin, respectively), a β-lactamase substrate, provides a quantitative measure of gene transcription and thus allows the development of a homogeneous transcription activation HTS assay with β-lactamase reporter [48]. Excitation of coumarin at 409 nm by FRET results in the emission from fluorescein of a green fluorescence (520 nm). When β-lactamase cleaves fluorescein on the 3′ position, it disrupts FRET and results in the emission of coumarin at 447 nm, i.e., blue fluorescence appears while the green fluorescence of fluorescein is quenched. Thus each molecule of β-lactamase attacks many substrate molecules and changes the fluorescence of substrate molecules from green to blue by disrupting FRET. The response element or promoter with β-lactamase reporter has to be introduced into a clonal cell line with stable expression of the receptor of interest. Jurkat cells stably transfected with M_1 muscarinic receptor and nuclear factors of activated T-cell (NF-AT)-β-lactamase reporter gene with a cytomegalovirus promoter, when treated with carbachol, induced β-lactamase activity as measured by the conversion of green fluorescence to blue fluorescence after loading cell-permeable β-lactamase substrate CCF2/AM. The induction of β-lactamase activity (the appearance of blue cells) depended on carbachol concentration and time of incubation [48]. The measurement of β-lactamase activity as an emission ratio at 450/530 nm improves accuracy. A cell based G-protein coupled receptor assay using the β-lactamase reporter gene system in a 3456-well Nano well assay plate was described [48]. This functional activity assay with β-lactamase reporter holds great promise for screening for receptors and ligand-gated ion channels. The main drawback of reporter gene technology is the limitation on using reporter genes in recombinantly expressed cells (heterologous expression is available to consortium partners of Aurora Bioscience Corp. and other institutions that license the reporter assay technologies).

G. Fluorescence Imaging

In fluorescence imaging technology, unlike in a plate reader wherein the fluorescence signal is read one well at a time, all the wells of a microtiter plate (96-, 384-, 1536-well plates) are read simultaneously by imaging in a CCD camera capable of recording kinetics in the subsecond range.

1. High Content Screening

High content screening (HCS) is analysis of cells using fluorescence based reagents with the ArrayScan system to extract spatial and temporal information of target activities within cells [49]. HCS yields information that will permit more efficient lead optimization before the in vivo testing. There are two types of HCS: (1) using fixed cells with fluorescent antibodies, ligands, and/or nucleic acid probes, and (2) using live cells with multicolor fluorescent indicators and biosensors. Two additional parameters can be measured simultaneously with the availability of two more channels of fluorescence in the ArrayScan system.

Instrumentation. The ArrayScan™ System is being developed by Cellomics along with Carl Zeiss, Jena. It is a tabletop instrument (optics with a spatial resolution of 0.68 µM) with subcellular resolution from many cells in a field within the well of a microtiter plate. The ArrayScan II automatically scans a microtiter plate, acquiring multicolor fluorescence image datasets of fields of cells at a preselected spatial resolution. The CellChip™ System is a miniaturized chip-based screening platform being developed by Cellomics.

Applications. HCS can be very effectively applied to study drug-induced dynamic redistribution of intracellular constituents. In cells transfected with green fluorescent protein coupled to human glucocorticoid receptor (GFR-hGR), drug-induced cytoplasm to nuclear translocation of GFR-hGR was studied by HCS [49]. With HCS, translocation in each cell can be quantified. In addition, the availability of two more fluorescence channels in ArrayScan II allows the determination of two additional parameters in parallel such as other receptors or cellular processes. A high content screen has also been explored for multiparametric measurement of apoptosis, which provides information on parameters such as nuclear size and shape changes, nuclear DNA content, mitochondrial potential, and actin-cytoskeletal rearrangements during drug-induced programmed cell death [49].

Comments. HCS is a promising technology that can be applied for measurement of molecular events such as signal transduction pathways and effects on cell functions. It can be used with fixed cells to measure end points, with live cells continuous monitoring of the activities is possible. Availability of up to two additional fluorescence channels in ArrayScan II allows the measurement of two additional parameters simultaneously. HCS allows subcellular measurements, and data can be obtained for each individual cell in the field. The technology is in the early stages, and there is no published screening data.

2. Fluorescence Confocal Microscope Imaging

With the development of the confocal laser scanning microscope and many technological advancements in laser scanning techniques and digital imaging methods and photostable fluorescent dyes, it is possible to do multiple fluorescence labeling of biological specimens, live cell imaging, and multidimensional microscopy in addition to imaging fixed and fluorescently labeled biological specimens in single and multiple wavelength modes [50]. Confocal laser scanning microscopy, multiple fluorescence labeling, together with immunofluorescence and fluorescence in situ hybridization, have become powerful techniques to map gene expression, for the detection of DNA and RNA and the expression of proteins.

A fast fluorescence confocal microscope optimized for homogeneous cell based assays has been designed by SEQ Ltd. [51]. The instrument is capable of simultaneous two-color laser excitation and detection of three-color imaging. The cells (e.g., expressing the receptor of interest) adhering to the bottom of the plate are probed with a ligand labeled with fluorescent dye such as fluorescein and fluorescent dye LDS 751, a nonspecific nucleic acid stain. A cell by cell analysis of the binding activity can be made using a mask generated by LDS 751 emission. The overlay of this binary mask with a binding activity image yields a pseudocolor map of receptor activity. Real-time data processing for most assays can be obtained by image analysis. Other assays tested include trafficking of a transcription factor and agonist activated transient Ca^{2+} levels. This technology is under development, and so screening data are not available.

3. Fluorescent Imaging Plate Reader

The Fluorescent Imaging Plate Reader (FLIPR) was developed to perform high throughput quantitative optical screening for cell based fluorescent assays [52]. FLIPR measures fluorescence signals in all the wells of a microtiter plate simultaneously, with kinetic updates in the subsecond range. This permits the determination of transient signals such as the release of intracellular calcium using calcium indicators, calcium Green-1 and Fluo-3 [53]. It has also been used for measuring luminescence based luciferase reporter assays [54].

Instrumentation. FLIPR[96] and FLIPR[384] by Molecular Devices will perform optical measurements on all wells of a 96- or 96-and-384-well plate, respectively, at rates of up to once per second. FLIPR uses water-cooled argon-ion laser (5 W) illumination to excite the fluorescent dyes, and the emitted light is detected with a cooled charge coupled device (CCD) camera. The laser provides many discrete spectral lines spaced from 350 to 530 nm. If broader spectral coverage is needed, FLIPR can be fitted with a broad-band xenon arc lamp (300 to 700 nm). FLIPR[384] contains interchangeable pipettor heads for simultaneous dispensing to either 96- or 384-well plates, and FLIPR[96] contains pipettor head for a 96-well plate with an integrated washing station. Typical fluid volumes are 50–100 µL for FLIPR[96] and 2–30 µL for FLIPR[384]. The instrument has precise temperature control. FLIPR has three primary configurations, manual, robot line, and stacker fed.

Applications. FLIPR can be used for measurements of intracellular calcium, intracellular pH, intracellular sodium, and membrane potential. FLIPR assays will be discussed in more detail in Chapter 7.

Comments. FLIPR, with its ability to take readings in all the wells of a plate simultaneously, enables the study of real-time kinetics. Real-time kinetic data gives additional pharmacological information for ranking relative potencies of drugs and gives information on the kinetics of the drug–receptor interaction. Functional response can be measured, thus providing the affinity, efficacy, and function of each drug; it can also distinguish full agonists, partial agonists, and antagonists within a single assay. FLIPR is a complex instrument. The user has to be familiar with all fine tunings needed to get the best results.

4. Fluorometric Microvolume Assay Technology

Fluorometric Microvolume Assay Technology (FMAT™), developed by PE Biosystems, uses Cy5 based fluorophores, and multiwell plates are scanned with a red, 633 helium/neon laser. In FMAT™, the laser is focused on the bottom of the well of a multiwell plate, and the fluorescence associated with each cell or bead is detected over the unbound and background fluorescence. The analysis algorithm ignores background fluorescence. Specific signal is detected as areas of concentrated fluorescence surrounding a cell or a bead, and the remaining background fluorescence is ignored in the final processing of the image data. FMAT is a homogeneous format for intact cell and bead based assays and does not require washing to remove unbound fluorophore. The FMAT™ system can be used with 96-well as well as higher density, 384- and 1536-well plates. The FMAT™ assays can be performed in a two-color format with two PMTs to determine more than one receptor binding assay on a single cell or multiple markers on a cell or a bead.

Instrumentation. The FMAT™ instrument is a fluorescence imager of a single-well plate of multiwell plates developed at Biometric Imaging. The excitation is from a 633 helium/neon laser and focused on a 1×1 mm area at the bottom of the well. The emission from the well is read in two PMTs at different wavelengths to measure the emission of different fluorophores to quantitate two different events on the same cell or bead. The FMAT™ system's optical platform yields population data in image format.

Applications. FMAT™ uses nonradioactive fluorescent tags and has been applied for cytotoxicity assays, functional assays such as ICAM-1 regulation by cytokines, and G–protein coupled receptor binding assays, nuclear hormone receptor assays, tyrosine and serine/threonine kinases, protein–nucleic acid interactions, and protein–protein interactions [55]. ELISA assays that are not HTS compatible, due to the number of wash steps and incubation steps involved, can be reformatted to bead based homogeneous assays with FMAT™. In a typical IL-8 fluorescent-linked immunosorbent assay (FLISA), secondary antibody (goat antimouse IgG) coated beads are complexed with monoclonal anti-IL-8 antibody and incubated with sample, biotinylated polyclonal anti-IL-8 antibody, and streptavidin labeled Cy5 (Fig. 15). IL-8 in the sample forms a sandwich by the matched antibody pair, and Cy5 labeled streptavidin binds to the biotin of the polyclonal antibody. The fluorescence associated with the bead complex is determined in the FMAT™ system over the unbound streptavidin Cy5 and background. FLISA uses 1% of capture antibody that is used in ELISA (coated on plate), thus reducing reagent. FLISA is a one-step incubation assay compared to the multiple wash and incubation steps with ELISA, with equal sensitivity as ELISA. FMAT™ can be used to develop multiplexed assays (simultaneous multiple assays) with different bead sizes or fluorophores within a single well. Though FMAT™ is a homogeneous assay because the imager reads one cell at a time, it will take several minutes reading high density 384- and 1536-plates, restricting throughput.

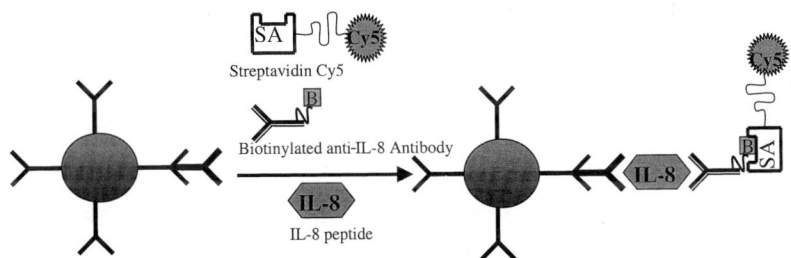

Streptavidin Cy5

Biotinylated anti-IL-8 Antibody

IL-8 peptide

Goat antimouse IgG beads coated with monoclonal anti-IL-8 antibody

Bead bound fluorescence is measured; unbound fluorophores are ignored

Figure 15 Schematic representation of FMAT interleukin-8 immunoassay. Goat antimouse beads coated with IL-8 monoclonal antibody bind IL-8 in the sample; then the second biotinylated IL-8 antibody (to a different epitope) will immunocomplex with IL-8. Streptavidin labeled with Cy5 will bind to biotin residue, and bead bound fluorescence is measured. The unbound fluorophone is ignored and does not give a signal. (Courtesy of PE Biosystems.)

III. CHEMILUMINESCENCE

In electrochemiluminescence, ruthenium chelate attached to a biological substance, when stimulated by applied potential, emits light at a particular wavelength due to a redox reaction to signal the presence of a bioanalyte. This ORIGEN™ technology is a trade mark of IGEN International and has developed a multichannel detection platform. Chemiluminescence in biological systems utilizes the firefly luciferase system. Luciferase catalyzes the reaction of luciferin with ATP, resulting in the emission of photons. The half life of the chemiluminescence has been progressively increased from a flash (seconds) to a glow (hours) luminescent signal, enabling the development of homogeneous HTS assays.

A. Electrochemiluminescence

Electrochemiluminescence (ECL) assay technology developed by IGEN is marketed as ORIGEN technology. ECL utilizes stable ruthenium chelate (TAG) which in the presence of tripropylamine participates in a luminescent reaction that is triggered by the application of low voltage [56]. ECL is capable of quantitating the specific binding of two molecules. In ECL, light is generated when low voltage is applied to an electrode, triggering a cyclical oxidation–reduction reaction of the ruthenium metal ion (Ru^{2+}). The Ru^{2+} is bound in a chelate of tris-(bipyridine). Tripropylamine (TPA) present in molar excess is the second reaction component that is consumed in the oxidation process, recycling the ruthenium chelate (Fig. 16A). The tracer molecule, Ru-chelate, is conjugated to the antibody. The Ru^{2+} labeled component is captured on the surface of polystyrene magnetic beads and brought to the surface of an electrode by applying a magnetic field through a movable magnet. TPA is introduced into the flow cell and a voltage is applied that oxidizes both ruthenium and TPA simultaneously (Fig. 16B). TPA by losing a proton becomes a reducing agent and transfers electrons to the ruthenium ions in an excited state and then decays to the ground state, releasing a photon in the process, which is quantitated at 620 nm [56,57]. The reduced Ru^{2+} is recycled, which along with TPA will intensify and amplify the ECL signal and is read in the ORIGEN® analyzer.

1. Instrumentation

The ORIGEN® 1.5 analyzer by IGEN® International, Inc., for analytical assays and the ORIGEN® ECLM8 electrochemiluminescent reader with 8 ECL modules

---

Figure 16 Electrochemiluminescence assay technology. (A) Structure of ruthenium (II)tris-(bipyridine)NHS ester. (B) Schematic representation of electrochemiluminescence reaction. (C) Schematic representation of ECL process. (Courtesy of IGEN, Inc.)

A

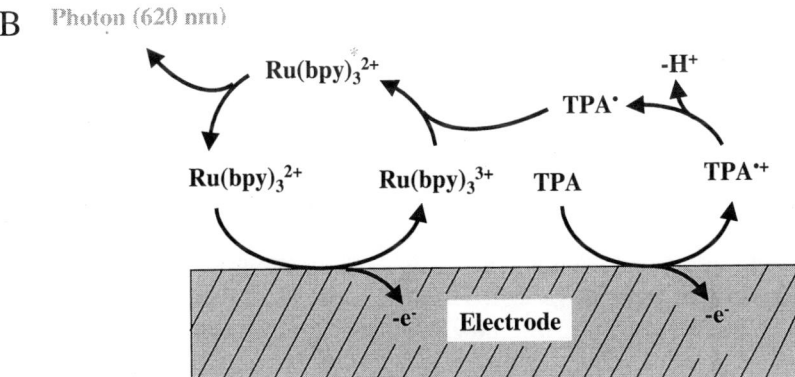

B

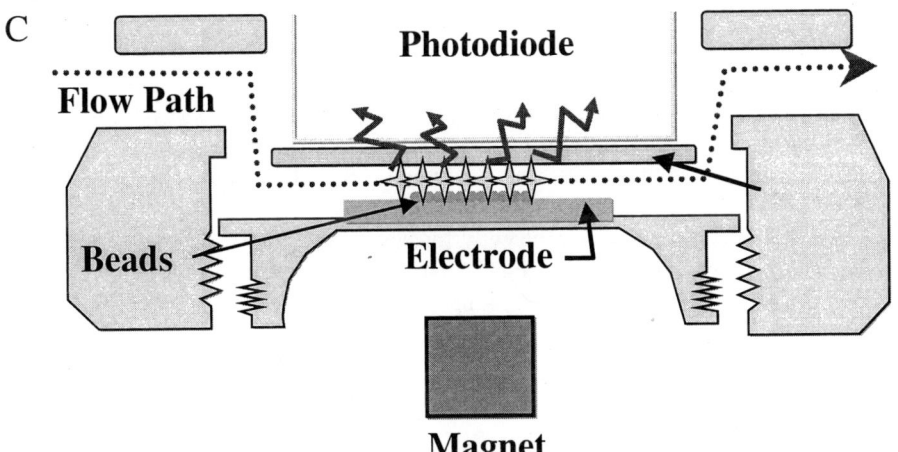

TPA = Tripropylamine

C

for HTS can read a 96-well plate in 10 min. The schematic diagram of Origen Flow system is given in Fig. 16C.

2. Applications

Four derivatives of ruthenium chelate, TAG-amine, TAG-hydrazide, TAG-maleimide, and TAG-phosphoramidite, are available for convenient covalent linking to biomolecules. ECL can be used for large- and small-molecule immunoassays such as for hormones, second messengers, drugs [58]; for enzyme–substrate interaction; for receptor–ligand interactions; and for DNA–DNA, DNA–protein, and protein–protein interactions [59], quantification of nucleic acids [60], and PCR products [61].

PTK Assay. In ECL PTK assay, the peptide substrate is end labeled with biotin, and the phosphotyrosine (PY) antibody is labeled with ruthenium (II) tris-(bipyridine) NHS ester (Fig. 17). The biotinylated peptide is incubated with ATP, Mg^{2+}, PTK, ruthenyl PY antibody, and streptavidin coated magnetic beads (M280 Dynabeads) in 100 μL for 30 min; a 200 μL quench buffer is added and read in the ORIGEN analyzer. An automatic sipper aspirates a specified volume (125–1000 μL) of the reaction mixture and pumps into a flow cell; the magnet moves close to the platinum electrode, which captures the streptavidin magnetic beads (due to the magnetic field) with bound tyrosine phosphorylated biotin-peptide complexed to Ru^{2+}-anti-phosphotyrosine antibody. Voltage (< 2 V) is applied to the electrode to begin the oxidation reaction to produce light and is measured in a PMT in less than 1 sec. The flow cell is then washed with buffer and the next sample is read.

3. Comments

The ECL is a promising technology that can be used in a wide spectrum of assay formats. Screening can be performed using whole blood, culture medium, crude extracts, and membrane preparations. The dynamic range is very broad. The ECL method utilizes nonisotopic small-molecule labels. In this assay the sample has to pass through the flow cell and is washed to remove free Ru^{2+} labeled molecules, resulting in a high signal-to-noise ratio (the background is low because the beads are concentrated on the electrode and washed with TPA). However, ECL cannot be used for measuring real-time kinetics. It is suitable for end point determinations. As the signal produced 30–50 nm above the electrode is measured, ECL can not be used efficiently for whole cell assays. At least four derivatives of Ru^{2+} chelates are available that can be coupled to a variety of biomolecules, and also a custom labeling service is available from IGEN. ECL instruments are one- to eight-module instruments able to read 1–8 samples at a time. The ECL homogeneous assays can be adapted to HTS with the eight-module instrument, which reads each plate in 10 min.

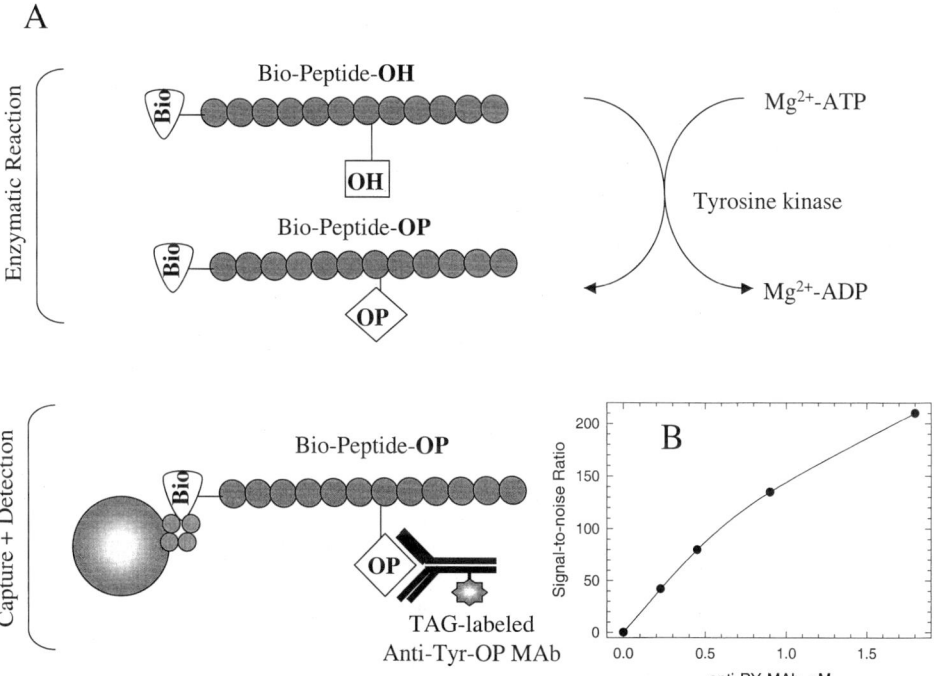

Figure 17 Origen PTK assay. (A) Schematic representation of PTK assay. A biotin-peptide is phosphorylated by PTK. The phosphorylated biotin-peptide is bound to strepta-vidin coated magnetic beads and will give signal by association with TAG labeled PY antibody. (B) Titration with anti-PY antibody in the Origen PTK assay. The S/N of 200 is observed at about 1.8 nM of PY antibody. (Courtesy of IGEN, Inc.)

B. Chemiluminescence

Chemiluminescence is the production of light by a chemical reaction in which a substrate molecule is converted to product with concomitant release of a single photon of light energy. A rapid homogeneous chemiluminescent telomerase hybridization protection assay amenable to HTS was described [62].

1. Luciferase Reporter Assays

Bioluminescence is chemiluminescence that occurs within an organism. Bioluminescent reporters of genetic transcription provide rapid quantitative analysis of many cellular events that are important in drug discovery, such as receptor function, signal transduction, gene expression, and protein–protein interaction. Firefly luciferase, a 62 K protein, is the most widely used luminescent reporter enzyme for studying gene regulation and expression. As no endogenous luciferase activity is found in mammalian cells, the background is very low. Bioluminescence is generated at 560 nm when luciferase converts the substrate luciferin to oxyluciferin in the presence of ATP, and the half-life of the signal is a few seconds. Developments in the luciferase assay increased the half-life from seconds (flash) to minutes (enhanced flash) and finally to hours (glow) [62]. Homogeneous luciferase assays have been developed by adding reagent containing lysis buffer, enzyme stabilizer, and luciferin directly to the cells in culture media and measuring the chemiluminescence signal [63]. Luciferase kits for homogeneous assays are available as Steady-Glo from Promega, LucScreen from Tropix, Luc Lite from Packard and also from other manufacturers. (Luciferase assays will be elaborated in Chapter 7.)

2. Instrumentation

Chemiluminescence can be measured in dedicated microplate luminometers (Wallac, Tropix, and other manufacturers) or the scintillation counters TopCount NXT (Packard) and MicroBeta (Wallac) or CLIPR (Molecular Devices) or the multimode readers Victor 1420, Analyst, and Polarstar.

3. Other Chemiluminescence Reporters

Some other common reporter genes, such as β-galactosidase and secreted alkaline phosphatase, are assayed conventionally with colorimetric reagents, and the sensitivity has been increased with the use of fluorescence substrates. The sensitivity of these enzymes has been remarkably enhanced (by at least 3 orders) with chemiluminescent reporter gene assay reagents, galacto-star and galacto-light/plus for β-galactosidase, and phospha-light for secreted alkaline phosphatase from Tropix.

C. Alpha Screen™

The amplified luminescent proximity homogeneous assay screen (Alpha Screen™) developed by BioSignal, a Packard BioScience Company, is a nonradioactive proximity assay using two proprietary 200 nm diameter latex donor and acceptor beads [64]. The beads are coated with hydrogel to prevent nonspecific interactions and self-aggregation and to provide a functionalized surface for covalent attachment and to retain dyes. The donor beads contain a photosensitizer that absorbs light at 680 nm (laser excitation) and then converts ambient molecular oxygen to the excited singlet state that has a short lifetime of 4 μsec, allowing it to diffuse up to 200 nm in aqueous solution. When the acceptor bead is brought into close proximity (within 200 nm) by a molecular binding event, the singlet oxygen reacts with the thioxene derivative of the acceptor beads, generating chemiluminescence at 370 nm, which is immediately transferred to fluorescent acceptors in the same beads. The fluorophores then emit light at 600 nm with high yield (Fig. 18). The half-life of the decay reaction is 0.3 sec, allowing to operate in time-resolved mode. The singlet oxygen does not react with the unbound acceptor beads; they do not emit light.

1. Instrumentation

The AlphaQuest-AD microplate analyzer from Packard is a single fiber-optic detector that can be used for AlphaScreen assays in 96-well plates as well as 384- and 1536-well high density plates (Fig. 19). The AlphaQuest-HTS is a high speed detection system with four simultaneous detectors. Using efficient fiber-optic bundles, AlphaQuest-HTS provides high speed measurement of 96-, 384-, and 1536-well microplates with a throughput of up to 100,000 samples per day.

2. Applications

AlphaScreen™ is a nonradioactive homogeneous assay technology. AlphaScreen has been applied to several different classes of assays including protein tyrosine kinases, serine-threonine kinases, proteases, helicases, receptor binding assays, protein–protein interactions, protein–DNA interactions, immunoassays for de-

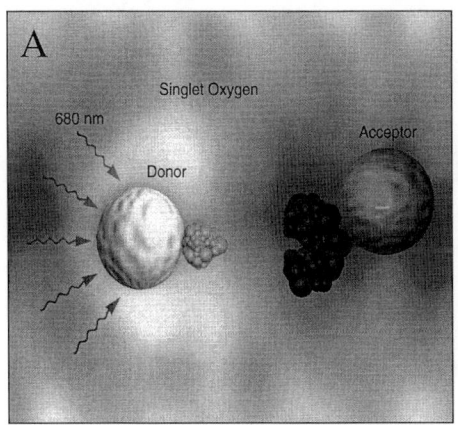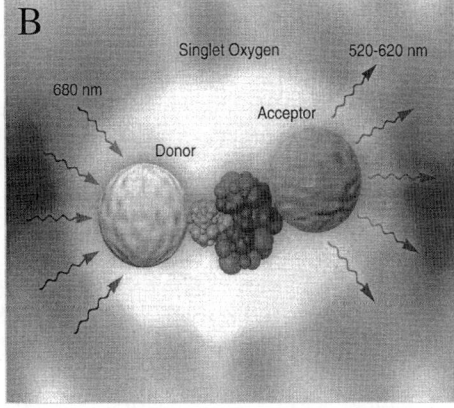

Figure 18 Principle of amplified luminescent proximity homogeneous assay (Alpha) screen. (A) On laser excitation, the chemical signal generated by the donor bead cannot be detected if the acceptor bead is not in close proximity. (B) When acceptor and donor beads are brought into close proximity by specific biological interaction, amplified signal is generated in AlphaScreen. (Courtesy of BioSignal, a Packard BioScience company.)

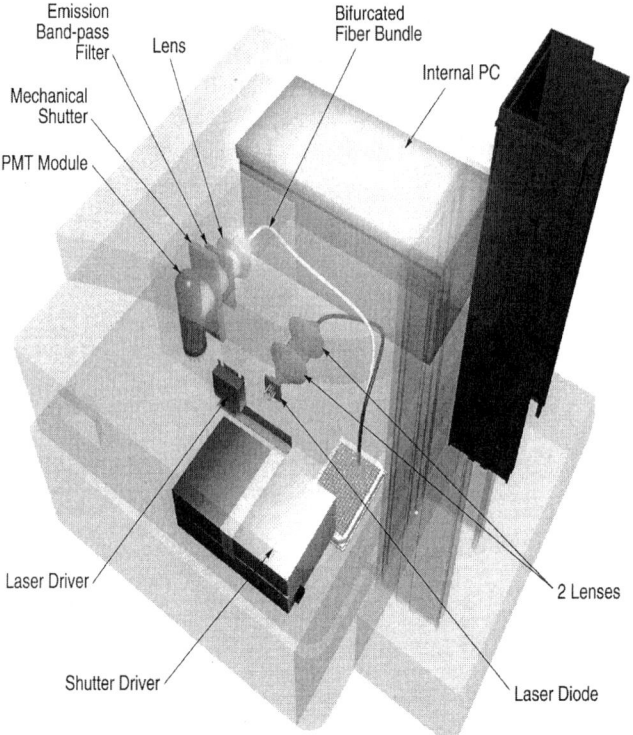

Figure 19   Line drawing of AlphaQuest AD, an AlphaScreen microplate reader. (Courtesy of BioSignal, a Packard BioScience company.)

termining small molecules and hormone levels, and cAMP functional assays for GPCRs (Fig. 20).

3.   Comments

As each donor bead contains a high concentration of photosensitizers, each donor bead emits up to 60,000 singlet oxygen molecules per second, which results in a very high amplification. Excitation at a longer wavelength (680 nm) and emission at a lower wavelength (500–600 nm) minimize nonspecific excitation and reduce background. The beads are the optimal size, too small to settle in aqueous buffers and large enough to be centrifuged. The effective distance between the donor and acceptor (R_0 value) is large, ~ 200 nm, which overcomes the assay limitations of other FRET based systems with weak interactions. Because of the long lifetime (0.3 s) of the AlphaScreen fluorescence signal, measurements are

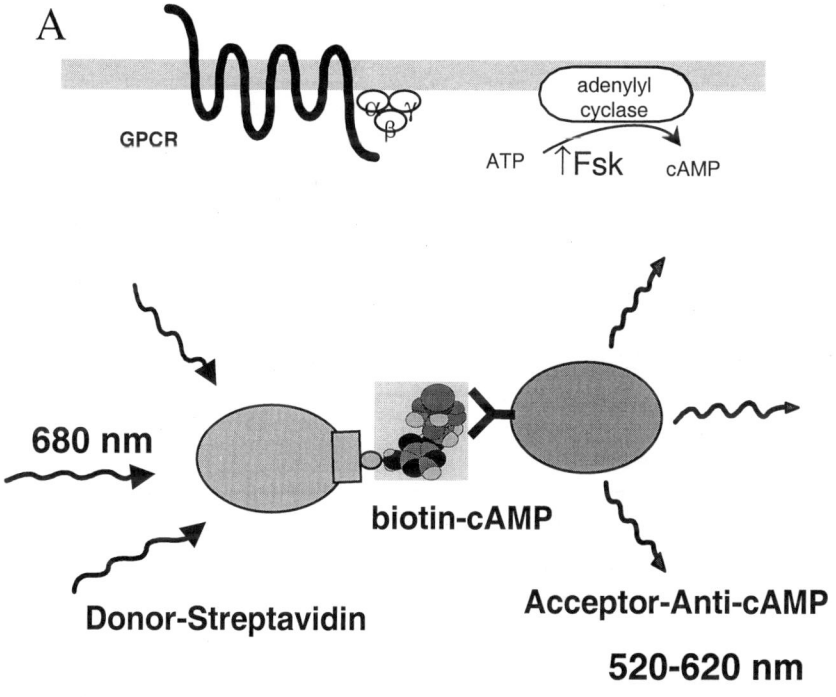

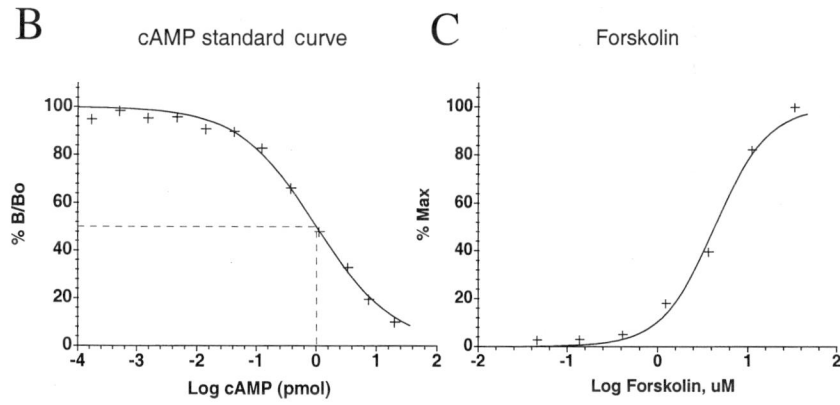

Figure 20 AlphaScreen cyclic AMP assay. (A) Schematic view of AlphaScreen cAMP assay (a whole cell GPCR functional assay) is based on the competition of endogenous cAMP with biotin-cAMP for cAMP antibody coated acceptor beads. (Courtesy of BioSignal A Packard BioScience Company.) (B) Standard curve of cAMP in AlphaScreen assay. (C) Activation of adenyl cyclase by forskolin in CHO-k1 cells expressing a GPCR. EC_{50} obtained for forskolin in this assay was 4.4 µM.

made in time-resolved mode, reducing the background. There is a good potential for this technology, which is still in development; the plate readers are being beta-tested. The unconjugated and labeled beads and the plate readers are only available from Packard. The AlphaQuest instruments are specifically for AlphaScreen technology and limit the use of it.

IV. RADIOACTIVE ASSAYS

In the conventional radioactive assays, the product or bound ligand has to be separated from the radioactive substrate or free ligand by gel filtration, precipitation, adsorption, or filtration and then washing. These procedures are not amenable for HTS and in addition generate large volumes of radioactive waste. Nevertheless, radioactive assays are very sensitive and robust. Reliable new homogeneous radioactive assays have been developed for HTS that do not require the separation of bound radioactivity from free, thus increasing the throughput and greatly reducing radioactive waste.

A. FlashPlate™ Technology

The FlashPlate™ (exclusively licensed to DuPont from Packard Instrument Company) is a white 96-well polystyrene microplate with plastic scintillator coated wells [65]. The FlashPlate like other polystyrene microplates has a hydrophobic surface for the adsorption of protein. Since the scintillant is coated on the microplate, additional scintillant is not required for counting (Fig. 21). The FlashPlate has to be precoated with a substrate, ligand, antibody, or secondary antibody before being used for an assay. The FlashPlate coated with antibodies, proteins, or peptide substrates has been used to develop one-step assays, reducing the amount of radioactive waste and the time for the assay. Generic precoated Flash-Plates (FlashPlate plus) with antibody, protein A, streptavidin, and myelin basic protein are available from DuPont. For further miniaturization of the HTS assays, the 384-well FlashPlate in basic form (without any other precoats) is now available.

1. Instrumentation

Topcount-NXT (Packard) and MicroBeta (Wallac) scintillation plate counters capable of measuring 96- and 384-well microtiter plates are available.

2. Applications

FlashPlate™ technology can be used in many assay formats: (1) enzyme assays such as protein kinase, chloramphenicol acetyl transferase (CAT), helicase, and reverse transcriptase [66]; (2) receptor–ligand binding assays with soluble recep-

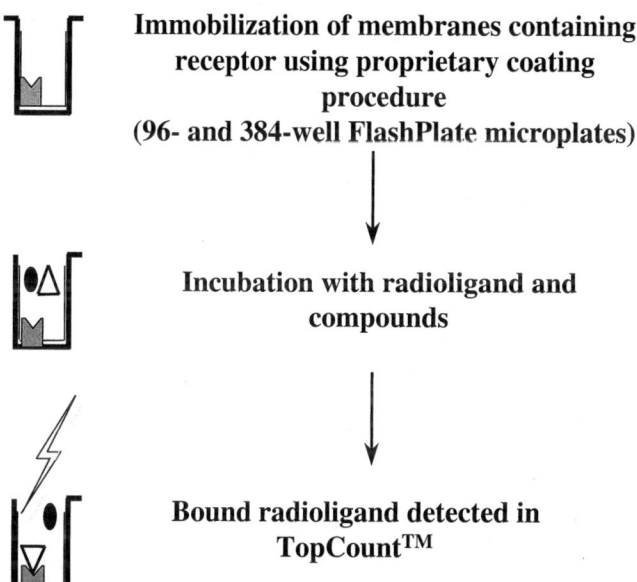

Immobilization of membranes containing receptor using proprietary coating procedure (96- and 384-well FlashPlate microplates)

↓

Incubation with radioligand and compounds

↓

Bound radioligand detected in TopCount™

Figure 21 Schematic representation of FlashPlate technology. Receptor immobilized to a 96- or 384-well FlashPlate is incubated with radioligand and compounds. Receptor bound radioligand is in close proximity with Scintillant and is detected in a TopCount. (Courtesy of Packard, a Packard BioScience company.)

tors (e.g., human estrogen receptor) [67] interleukin-1 receptor, and G-protein coupled receptors (e.g., endothelin receptors) [68]; (3) radioimmunoassays (e.g., cyclic AMP, cyclic GMP, prostaglandin E2) [69]; (4) functional assays with live cells (e.g., adenyl cyclase assay) [69]; and (5) molecular biology techniques including sandwich hybridization assay and translation systems [69]. For CAT assay, biotinylated chloramphenicol is coated on streptavidin coated FlashPlate. The reaction is done in the chloramphenicol coated streptavidin FlashPlate™ by adding ^{14}C- or ^{3}H-acetyl CoA in the reaction buffer with the addition of CAT, incubated at 37°C and counted after incubation. The plate is counted again after aspiration of the liquid and washing with buffer. The counts detected in chloramphenicol are similar either with or without aspiration of the reaction medium, suggesting that the assay can be done without aspiration and washing in a homogeneous mode [66].

3.   Comments

FlashPlate technology can be used for a wide variety of applications such as enzyme, receptor binding, functional and immunoassays and in live cells. Commonly used radioisotopes (i.e., ^{3}H, ^{14}C, ^{35}S, ^{125}I, ^{32}P, ^{33}P, and ^{45}Ca) can be used with FlashPlates. With low energy beta emitters such as ^{3}H, ^{14}C, and ^{35}S isotopes, FlashPlates can be used in homogeneous mode. However, with higher energy beta emitters such as ^{32}P and ^{33}P, the unbound radioactivity should be removed by aspiration and rinsing, because these radioisotopes can be detected due to the long distance traveled by the strong beta particles and may interfere in the assay with high background. Availability of generic FlashPlates precoated with commonly used proteins such as streptavidin, protein A, antibody and MBP and nickel chelate in assay ready format is an added advantage. Preparation of a custom biomolecule bound to FlashPlate involves binding the compound and several washings, which is labor intensive, and batch-to-batch variations may occur.

B. Cytostar-T™ Technology

Cytostar-T™ scintillating microplates from Amersham International plc are standard 96-well format, sterile, tissue culture treated microplates. The Cytostar-T plate is a polystyrene plate with a transparent base coated with scintillant. Upon addition of radioactive tracer to cells grown in the bases of the wells, the scintillant generates light when the tracer is bound to the cell membranes or taken up by the cell due to the close proximity of the radioactive isotope to the scintillant at the base of the plate and can be counted in a standard plate counter [70]. The free radiolabel in the medium is physically too far from the scintillant to trigger a light reaction. Homogeneous cell based assays can be done in the Cytostar-T plate because there is no need for separation of the free radiolabel from the cell bound radiolabel.

1. Instrumentation

The Topcount (Packard) and MicroBeta (Wallac) scintillation plate counters capable of measuring 96- and 384-well microtiter plates are available. The signal is enhanced considerably in these instruments with new counting modes, high efficiency count mode for the Topcount, and paralux counting for the MicroBeta.

2. Applications

Cytostar-T plate cell-based assays can be done in homogeneous mode with radioisotopes (beta-emitters, ^{3}H, ^{14}C, ^{35}S, and ^{45}Ca). Some of the cell-based homogeneous assays that have been done in Cytostar-T include receptor radioligand binding assays with intact cells, amino acid uptake into cells, DNA synthesis monitoring in cells in response to drug treatment, and apoptosis measurements [69].

3. Comments

Cytostar-T can be used for assays with live cells plated on the bottom of the well. As with the FlashPlate assays, the Cytostar-T assays can be used with weak beta emitters in homogeneous assay mode. When strong beta emitters or ^{125}I are used, the plates need to be washed to remove unbound radioactivity.

C. Scintillation Proximity Assay

The Scintillation Proximity Assay (SPA) was first described by Hart and Greenwald in an immunoassay using two polymer beads coated with antigen, one coated with fluorophore and the other with ^{3}H [71]. Antibody agglutination brings many of the ^{3}H beads into close proximity to the fluorophore beads and excites them, and after very long incubations they can be counted in a scintillation counter. Udenfriend et al. have improved on this with microbeads containing a fluorophore and coated with antibody [72]. ^{125}I labeled antigen binds to the antibody on the beads and by its proximity the emitted short-range electrons of the ^{125}I excite the fluorophore in the bead, which can be measured in a scintillation counter without separation of unbound antigen. Amersham International further developed SPA radioisotopic assay technology. SPA is a homogeneous radioisotope assay technology that can be utilized for a variety of biological assays. In SPA the target of interest is immobilized to a small scintillant containing microspheres or fluoromicrospheres (SPA beads) approximately 5 μm in size. The fluoromicrosphere consists of a solid scintillant–polyvinyltoluene core coated with polyhydroxy film, which reduces the hydrophobicity of the particle. Generic SPA beads to which proteins such as antibodies, streptavidin, receptors, and enzymes or small molecules such as glutathione or copper ions are chemically linked to the coating on the bead are available with Amersham.

Assays are done in aqueous buffers with beta emitters such as ^{3}H, ^{14}C, ^{35}S, and ^{33}P isotopes and ^{125}I. When ^{3}H atom decays, it releases a β-particle with an average energy of 6 keV and a mean path length of 1.5 μm in water. The path lengths of ^{14}C, ^{35}S, and ^{33}P isotopes and ^{125}I are 58, 66, 126, and 17.5 μm, respectively. If a ^{3}H β-particle meets a scintillant molecule within 1.5 μm of the particle being released, it will have sufficient energy to excite the scintillant into emitting light. On the other hand, if the β-particle of ^{3}H travels longer distances of more than 1.5 μm, it will not have enough energy to cause scintillation. In SPA, if a radioactive molecule is bound to the SPA bead directly or through a molecule coupled to the bead, it is brought into close proximity for the emitted radiation to stimulate the scintillant to emit light (Fig. 22). Unbound radioactive molecules are too far away from the scintillant of the bead, the energy released is dissipated in the solution before reaching the bead, and no light is produced [73]. The amount of light produced is proportional to the amount of radioactive molecules bound to the SPA bead and is easily measured in a scintillation counter. Thus the bound radioactive molecules only produce the scintillation signal but not the unbound, hence there is no need for separation of free radioisotope in the assay.

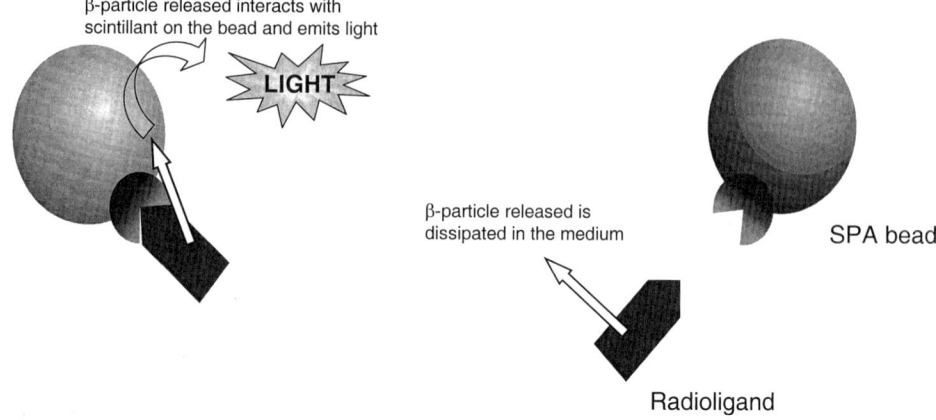

Radioligand bound is in close proximity stimulating the bead to emit light

Unbound radioligand does not stimulate the bead

Figure 22 Principles of SPA. Radiation energy from the radiolabeled ligand bound to the acceptor molecule on the SPA bead is absorbed by bead fluor and generates a signal on the bead, whereas unbound radioligand does not stimulate the bead. (Courtesy of Amersham.)

1. Instrumentation

Topcount NXT (Packard) and MicroBeta (Wallac) scintillation plate counters capable of measuring 96- and 384-well microtiter plates are available.

2. Applications

SPA microsphere beads have been prepared from hydrophobic polymers such as poly(vinyl toluene) (PVT) and inorganic scintillators such as yttrium silicate (YSi) beads [73]. The capacity of the YSi bead is higher than that of the PVT bead. The YSi bead is more dense than the PVT bead. Before developing an assay the compatibility of radioligands with SPA beads has to be tested. Also, the microplates have to be screened for low nonspecific binding of the radioligand. SPA has been applied to a wide variety of different assays including radio-immunoassays (RIAs), receptor binding assays, protein–protein interaction assays, enzyme assays, and DNA–protein and DNA–DNA interaction assays [74]. SPA beads coated with protein A or secondary antibody captures the antibody–antigen complex and is quantitated in the radioimmunoassays (RIAs).

RIAs are used in clinical and pharmacological studies to measure drugs, the second messengers, prostaglandins, steroids, and other serum factors.

SPA beads with wheat germ agglutinin (WGA), polyethylimine WGA-PVT beads, or polylysine coated YSi beads have been used for several membrane receptors including neuropeptide Y, galanin, endothclins, ncrve growth factor, TGFα, TGFβ, Ach, EGF, insulin, angiotensins, β-adrenoceptors, somatostatin, bFGF, dopamine, and interleukin receptors [66,67]. SPA also has been utilized in designing screens for nuclear receptor binding assays with the ligand binding domain of the nuclear receptor expressed as a fusion protein with His-tag (His_6 or His_{10}), radioligand, and nickel-SPA beads. The protein–protein interaction assays that have been developed using SPA consist of SH2 and SH3 binding domains, Fos-Jun, Ras-Raf, selectin, and integrin adhesion assays [65,66]. SPA has been used for protein–DNA binding interaction assays such as binding of transcription factor NF-κB to DNA.

Enzyme assays can be grouped into three main formats (Fig. 23). *(1) Signal removal.* As in the case of hydrolytic enzymes such as proteases, nucleases, phospholipases, and esterases, the radiolabeled substrate is linked to the streptavidin–SPA bead via biotin, and the enzyme action cleaves the radiolabel from the biotinylated portion of the molecule, resulting in a decrease in the signal. *(2) Signal addition.* As in the case of synthetic enzymes such as transferases, kinases, and polymerases, the acceptor substrate is linked to the SPA bead through biotinylation, and the donor substrate is radiolabeled. The action of the enzyme transfers the radiolabel to the acceptor molecule on the bead from the donor, resulting in an increase in the signal. *(3) Product capture.* In this assay format, the radiolabeled product of the reaction is captured by biospecific recognition to antibody as in the case of PTK. In the PTK assay the phosphorylated product but not the substrate is captured specifically by antiphosphotyrosine antibody that binds to protein A or secondary antibody coated onto the bead [73]. SPA has also been used for quantification of PCR using biotinylated PCR primers and [³H] dNTPs. The biotinylated [³H] DNA produced is captured onto streptavidin coated beads [73].

3. Comments

SPA is a very widely used homogeneous assay format for many biological assays. SPA is applicable to HTS and can be adapted for automation. Assays are routinely done in 96-well plates and for increased throughput in 384-well plates. Assays in 384-well plates reduce the radioactive waste generated and the cost of the reagents. The most commonly used radioisotopes in SPA are ³H and ¹²⁵I. Recently, ³³P is also being used in protein kinase assays. Though the path lengths of ³⁵S and ¹⁴C are similar, due to the low specific activity ¹⁴C-labeled compounds (~ 60 mCi/mmol) have not been utilized that much in SPA. When ³⁵S or ³³P is

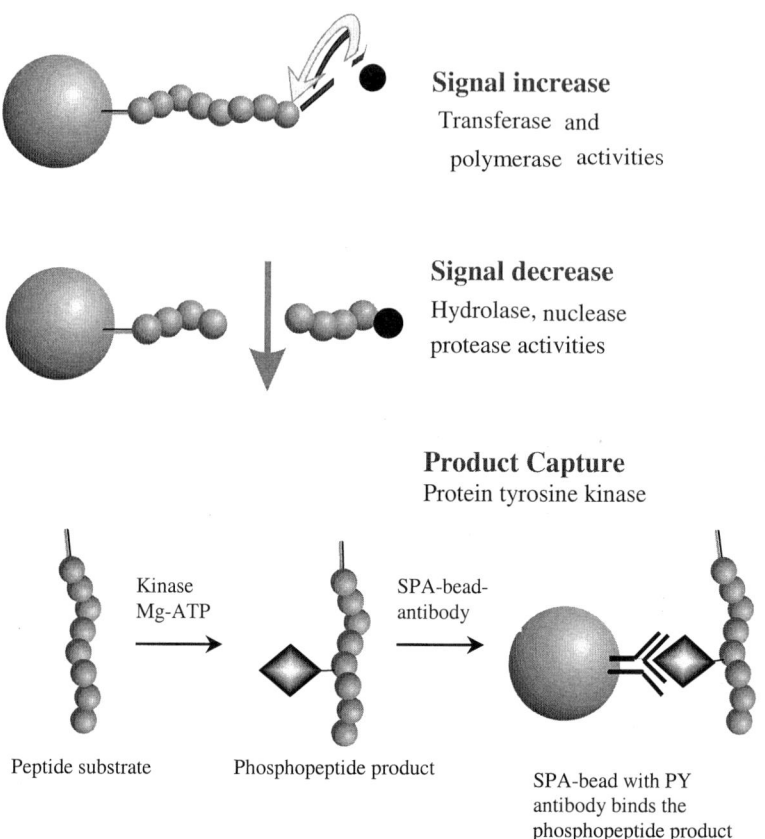

Signal increase
Transferase and
polymerase activities

Signal decrease
Hydrolase, nuclease
protease activities

Product Capture
Protein tyrosine kinase

Kinase
Mg-ATP

SPA-bead-
antibody

Peptide substrate   Phosphopeptide product

SPA-bead with PY
antibody binds the
phosphopeptide product

Figure 23 SPA enzyme assay formats. Schematic representation of the three enzyme assay formats. (1) When a radiolabeled compound is added to a substrate attached to a SPA bead, the result is a signal increase. (2) When a radiolabeled substrate attached to a SPA bead is cleaved, releasing the radiolabel will result in a signal decrease. (3) When a substrate forms a radiolabeled product in the reaction, the product can be captured onto the antibody coated SPA bead, resulting in signal increase. (Courtesy of Amersham.)

used in SPA, for best results the samples are centrifuged to bring the beads to the bottom of the well or CsCl is added to increase the density of the reaction medium, allowing the beads to float to the top for reducing background. The count time in scintillation counters for a 96- or 384-well plate is approximately 10 or 40 min, respectively, and this may restrict the throughput to some extent. The signal-to-noise ratio is generally lower than with the conventional assays but may be adequate for use in HTS. Other critical issues associated with SPA are color quench and detection efficiency of scintillation counting. The availability of many protein and other biomolecule coated generic SPA beads and SPA assay kits from Amersham makes assay development for HTS convenient. The radioactive waste is reduced and does not require any special equipment. However, for SPA use in HTS, a technology transfer agreement has to be obtained from Amersham, which is expensive.

D. LEADseeker™ Homogeneous Imaging System

The LEADseeker homogeneous imaging system is being developed by Amersham in collaboration with Imaging Research Inc. This proprietary system combines imaging instrumentation and specialized software with radioactive proximity reagents that are at least 10 times more sensitive than SPA [75]. The LEADseeker™ Radiometric Imager system consists of a CCD camera and uses europium yttrium oxide (YO:Eu) or europium polystyrene (PST:Eu) particles, which exhibit emission at 615 nm. The LEADseeker™ imaging beads have an emission maximum of 615 nm and show very little quenching with colored (yellow) compounds. These beads are available with streptavidin, WGA, glutathione, protein A, and nickel coatings, which produce higher light output than the SPA bead. The assays have been developed for 96- and 384-well plates. This system is capable of capturing the signal from an entire 384-well plate in a single exposure within 10 min. The camera reads the density of the image in grey scales over a total range of 2^{16} levels. The assays are currently developed for higher density plates such as 1536-well plates.

1. Instrumentation

Amersham is developing LEADseeker instrument in collaboration with Imaging Research Inc. for performance partnership partners. The main features of the LEADseeker consists of a camera with a cooled CCD chip that has an imaging area of 1024×1024 pixels for high-resolution imaging. Shadowing, which is a common problem with standard lenses used in imaging, is overcome in the LEADseeker with the use of the telecentric Borealis lens, which captures light more efficiently from the full area of the plate. The LEADseeker™ will be extended to nonradiometric applications with a multimodality imager that will be capable of reading fluorescence, luminescence, and color. Fluorescent assays will be based on proprietary cyanine fluors (500–800 nM) that will be applicable to including steady-state, FP, FRET, and TRF assays.

2. Applications

The LEADseeker™ can be used for assays that can be performed with the SPA, the difference being the use of more sensitive Eu-complexed polystyrene or yttrium oxide beads in place of SPA beads. Thus the assays can be converted from macro to micro assays using LEADseeker™ technology. Some of the assays that have been tested include reverse transcriptase, EGF binding, GTPγS binding, and extracellular response kinase 1 [75].

3. Comments

The LEADseeker is a homogeneous radioactive/nonradioactive imaging technology suitable for HTS. A CCD camera images the signals from a 384-well plate, and imaged for less than 10 min, and with development of a 1536-plate imager, will allow screening > 100,000 compounds a day. This technology is being developed in partnership with major pharmaceutical companies. Color quench is overcome by the new bead types. When available, this technology will be a rapid homogeneous radiometric and nonradiometric uHTS, which will save reagents and shorten screening times.

V. OTHER METHODS

A. Surface Plasmon Resonance

Surface plasmon resonance (SPR) has become a popular method for looking at biomolecular interactions. SPR occurs when surface plasmon waves are excited at the sensor surface consisting of thin metal such as gold coated onto a glass support [76,77]. SPR is a phenomenon that occurs between incoming photons and electrons in the sensor surface. The light energy at a particular wavelength and angle of incidence is transferred to the electrons in the metal surface, causing alterations in the reflected light. The resonance (nonreflectance) angle is dependent on the refraction index in the vicinity of the metal surface, which in turn is dependent on the mass concentration. Molecules attaching to the sensor surface with gold film cause changes in the refractive index close to the surface, resulting in a change in the SPR signal. Biomolecular binding events cause further changes in the refractive index that is detected as changes in the SPR signal. In the BIAcore (biomolecular interaction analysis), the shift in the resonance angle with time is measured. SPR technique can be used to precisely measure the kinetics of macromolecular interactions.

Sensor surfaces can be functionalized either directly to capture different target molecules or to do affinity capture of the target molecules. SPR of macromolecules uses Au SPR film and a flow cell that houses a chip coated with a thin layer of Au colloidal particles. The sensor chip CM5 has a carboxymethylated dextran matrix surface to which ligand can be immobilized through covalent derivatization through amine, thiol, aldehyde, or carboxyl groups. Different types of sensor surfaces are available that can be used for different assays; the sensor chip CM5 (carboxymethylated dextran matrix surface) with immobilized ligand can be used for studying the interaction with target molecules or affinity capture of an alternative molecule that interacts with the target molecule; the sensor chip SA (streptavidin surface), which captures large biotinylated DNA fragments, is used in nucleic acid interactions; the sensor chip NTA (NTA coated sensor surface) through nickel chelation captures histidine tagged biomolecules that can be used in receptor binding assays; the sensor chip HPA (hydrophobic surface) to which membranes or liposomes containing receptors can be coated are used in receptor binding studies.

When light is reflected off the surface of the Au particle, the angle of reflection gives information about the mass bound to the matrix. A solution containing compounds that interact with the molecule bound to the chip (e.g., ligand to receptor or antigen to antibody) is passed over the surface of the chip. As these molecules interact, there is an effective change in surface roughness of a few nanometers on Au-SPR films that is easily detected. This allows a rapid and direct measurement of the binding kinetics of a broad variety of interactions in real time. Despite its sensitivity, this technique is limited, and it is not applicable to small molecule measurements.

1. Instrumentation

The single channel Biacore probe is used for fast detection and concentration of target biomolecules. In the Biacore-probe, SPR occurs in the gold film at the tip of a sensor probe. The multichannel Biacore X, Biacore 2000, and Biacore 3000 can be used to study biomolecular binding events in real time, allowing direct assessment of kinetic constants. Flow cells use as little as 5 μL sample. The Biacore X is a manual system with one continuous flow pump and two flow cells. The Biacore 2000 and Biacore 3000 are automated systems with two autosamplers and continuous flow pumps and four flow cells on one sensor, which allow immobilization of four different molecules; four different interactions can be monitored simultaneously.

2. Comments

SPR based assays are homogeneous assays though lower throughput without angle scanning. The sensitivity of Biacore technology is sufficient for detection and characterization of binding events involving low-molecular-weight compounds and their immobilized protein targets. Biacore systems measure real-time binding events, with accurate determination of kinetic constants. Automation reduces the analysis times and increases throughput. Multiple interactions can be screened on a large array sensor simultaneously using imaging technology with a CCD camera for detection. As this technology develops it will be a powerful tool for HTS to measure the binding of small molecule compounds to their drug targets directly.

B. CLIPR System

The Chemiluminescence Imaging Plate Reader (CLIPR) a product of Molecular Devices, is an ultra high throughput luminometer system for 96-, 384-, 864-, and 1536-well microplates. The instrument can be used in HTS mode for cell based assays and SPA assays in microplates [78].

1. Instrumentation

CLIPR integrates a high sensitivity CCD camera, a telecentric lens, a high precision positioning mechanism, and a computer system with software for instrument and record data. The CLIPR system can be loaded manually, can have a plate stacker, or can be integrated to a linear robot line. The imaging plate reader system reads plates in under a second and it is possible to do kinetic studies.

C. Infrared Thermography

To measure thermogenesis in a cell culture, infrared imaging system thermogenesis was reported [79]. The infrared imaging system was shown to be a rapid, very sensitive ($0.002°C$), and effective method for measuring thermogenesis in cell culture in vitro. Cells grown in a 96- or 384-well plate are equilibrated in an incubator at 37°C, compounds are added, equilibration at 37°C is done for 10 min, and the heat generation is measured by imaging in the infrared thermography system. Thermogenesis increased in yeast expressing the mitochondrial uncoupling protein-2 after treating with an uncoupler of mitochondrial respiration and in adipocytes treated with rotenone, an inhibitor of mitochondrial respiration or β-adrenergic receptor agonists [79].

1. Instrumentation

Commercial systems are not available in the market. A custom-made infrared thermography system consists of a thermo electrically cooled Agema Thermovision 900 Infrared System AB (at a focal distance of 6 cm), equipped with a SW Scanner and a lens ($40° \times 25°$ lens) that detects a 2–5.4 micron spectral response. The data analyzer consists of OS-9 advanced systems and ERIKA 2.00 software from Agema Infrared Systems. The sensitivity of this infrared thermography system is $0.002°C$, and its robustness (96- as well as 384-well plates) makes this system very useful for HTS assays in detection of altered thermogenic responses in various cell types.

D. Nanoparticle Technologies

Highly luminescent semiconductor quantum dots (small nanoparticles made of zinc sulfide capped cadmium selenide) have been covalently coupled to biomolecules such as various antibodies or DNA probes for use in ultrasensitive biological detection [80,81]. The luminescent labels are $\sim$ 20 times brighter and 100 times more stable against photobleaching and one-third wide in spectral line width compared to organic dyes such as rhodamine. Biomolecules are attached to different color nanoparticles. These biomolecule conjugates are water soluble and biocompatible. When cells are exposed to the different colored nanoparticles containing various antibodies, each antibody binds only to its specific antigen on the cell surface. Depending on the presence of types of antigen on the cell surface, those colored nanoparticles are captured and others are washed away. Spectral readings at different wavelengths give information on the types of antigens present and the amount of each antigen. Similarly, nanoparticles with different DNA probes can be used to identify a large number of gene sequences in blood and other biological samples.

Semiconductor nanocrystals labeled with fluorescent probes have a narrow, tunable, symmetrical emission spectrum and a broad continuous excitation spectrum; they are photochemically stable and may prove to be superior to existing fluorophores and may have many applications in several different assays [82]. These water soluble nanocrystals also have a long fluorescence lifetime (hundreds of nanoseconds), which can allow for time-gated detection of autofluorescence suppression. Several companies developing nanoparticle technologies are Quantum Dot, Auspex, Biocrystals, Nanomat, and Nanosphere.

E. Liquid Crystals

Liquid crystals are used to amplify and transduce receptor-mediated binding of proteins at the surface into optical outputs. Liquid crystal sandwiched between two gold films supporting self-assembled monolayers containing ligands, upon binding of proteins to the specific ligands, will change the surface roughness and trigger rapid changes in the orientations of 1–20 µm thick films of supported liquid crystals and changes the intensity of light transmitted through the liquid crystal, which can be further amplified and transduced into optical signals [83]. The orientations of liquid crystals are sensitive to a wide variety of physicochemical properties of surfaces, which suggests that this approach can be used for the detection of binding of small molecules to proteins and protein aggregates to a surface. This approach does not need electroanalytical apparatus, provides spatial resolution of micrometers, and can be extended to assay the effect of spatially resolved chemical libraries on the ligand–receptor binding.

F. Microchip Technology

The HTS and uHTS assays use volumes of a few microliters (5–10 µL in a 1536-well plate) to several microliters (100 µL in a 96-well plate). Fluid dispensing, mixing, and evaporation are some of the major technical problems in reducing assay volumes to a few microliters or to submicroliter levels. Microchips are designed either for single or multiple use and consist of silicon and glass master chips combined with plastic injection molding or embossing produced by micro-fabrication technologies. Fluids are moved through microscopic channels by either electro-osmosis or electrophoresis (microfluidics). Microfluidic capillary electrophoresis has been successfully used in several different types of HTS enzyme assays in microchips.

Microchip technology has been used for enzyme assays and to determine the binding affinity of monoclonal antibody [84–86]. In a microchip based protein kinase A assay, fluorescein labeled Kemptide was used as substrate. The assay reagents were placed in wells on the microchip, aliquots of the reagents were transported by electro-osmosis into the network of etched channels, and enzyme reaction was performed. The phosphorylated fluorescein labeled Kemptide product was separated from the substrate by on-chip capillary electrophoresis, and kinetic constants for ATP and peptide substrates (K_m) and the inhibition constant (K_i) for inhibitor H-89 were determined. This assay demonstrated the usefulness of microchips for performing enzyme assays. Thus microchip technology has potential for applications to immunoassays, nucleic acid assays, enzyme assays, and receptor-binding assays.

Microchip technology has been widely used in DNA analysis. A DNA chip is a small surface specked with thousands of dots of single stranded DNA of a gene or gene segment. Gene activity in the cells or tissues is measured by collecting mRNA from cells or tissues, converting it to cDNA, labeling it with a dye, and incubating it with a DNA chip. The cDNAs hybridize to complementary sequences on the chip and are identified. DNA array technology is widely used in various diseases including cancer. Microchips are now commercially available from Affymetrix Inc. (Santa Clara, CA), Caliper Technologies (Mountain View, CA), ACLARA BioSciences (Mountain View, CA) and others.

VI. CONCLUSIONS

Several fluorescence, chemiluminescence, luminescence, and radioactive based assays that are one-step assays that do not require wash and multiple incubations or filtrations and washes are detailed in this chapter. There is not a single format that can completely replace all the assays for various targets in drug screening. Radioactive assays are generally very sensitive assays, but handling problems and the generation of radioactive waste are big concerns. In the radioactive assays based on SPA, FLASHplate, Cytostar, and LEADseeker assays, the radioactivity generated is reduced. SPA assays have been gaining in popularity in spite of being radioactive methods. Nonradioactive methods like fluorescence based methods are getting more attention. The FP assay format is comparatively simple as only one labeled molecule is needed. HTRF requires two labeled molecules, and if some generic labeled molecules can be utilized, an assay can be developed faster. Both FP and HTRF formats are very robust and sensitive assays and have the added advantage that these are ratiometric methods, hence there is less interference from colored compounds. When tyrosine kinases and serine kinase were compared in FP, FRET, and SPA formats, all three techniques produced very similar IC_{50} values for some peptide substrates, but the FP assay was found to be faster, cheaper, and more sensitive and robust than the SPA assay [87].

The HTS lab should concentrate on a few technologies that are amenable to a wide variety of assays and also adaptable to high density plates rather than diversifying on too many platforms. It will consume a lot of time and money (instrumentation) to test all the technologies available in optimizing each assay. Some assays can be done with equal efficiency by more than one format, and it will be advantageous to go with an assay format that is already tested in the laboratory. Unless the new technologies offer substantial improvements over the existing platforms, it is not cost-effective to switch. Several new assay technologies based on fluorescence confocal microscopy, fluorescence correlation spectroscopy, fluorescence imaging, electrochemiluminescence, AlphaScreen, FMAT, and SPR methods are being developed to increase the throughput to adopt to HTS and uHTS. Microassays using microchip technologies along with microfluidic array and imaging technologies are gaining importance in drug discovery.

REFERENCES

1. J Drews. Drug discovery: a historical perspective. Science 287:1960–1964, 2000.
2. JR Lakowicz. Principles of Fluorescence Spectroscopy. New York: Plenum Press, 1983.
3. AJ Pope, UM Haupts, KJ Moore. Homogeneous fluorescence readouts for miniaturized high-throughput screening: theory and practice. Drug Disc Today 4:350–352, 1999.
4. RA Jefferson. Assaying chimeric genes in plants. The GUS gene fusion system. Plant Molec Biol Reporter 5:387–405, 1987.
5. AG Rao, P Flynn. A quantitative assay for β-D-glucuronidase (GUS) using microtiter plates. Biotechniques 8:38–40, 1990.
6. RP Haugland. Hand Book of Fluorescent Probes and Chemical Research. 6th ed. Molecular Probes, 1996.

7. LJ Jones, ST Yue, C-Y Cheung, VL Singer. RNA quantitation by fluorescence-based solution assay: RiboGreen reagent characterization. Anal Biochem 265:368–374, 1998.

8. FJ Perrin. Polarisation de la lumière de fluorescence vie moyenne des molécules dans l'état excité. J Phys Radium 7:390–401, 1926.

9. G Weber. Fluorescence and Phosphorescence Analysis. New York: John Wiley, 1966.

10. WB Dandliker, HC Chapiro, JW Meduski, R Alonso, GA Feigen, JR Hamrick, Jr. Applications of fluorescence polarization to the antigen-antibody reaction. Immunochem 1:165, 1964.

11. ME Jolley. Fluorescence polarization immunoassay for the determination of therapeutic drug levels in human plasma. J Anal Toxicol 5:236–240, 1981.

12. DM Jameson, WH Sawyer. Fluorescence anisotropy applied to biomolecular interactions. Meth Enzymol 246:283–300, 1995.

13. ME Jolley. Fluorescence polarization assays for the detection of proteases and their inhibitors. J Biomol Screen 1:33–38, 1996.

14. R Seethala, R Menzel. A homogeneous, fluorescence polarization assay for src-family tyrosine kinases. Anal Biochem 253:210–218, 1997.

15. R Seethala, R Menzel. A fluorescence polarization competition immunoassay for tyrosine kinases. Anal Biochem 255:257–262, 1998.

16. R Seethala. Fluorescence polarization competition immunoassay for tyrosine kinases. Methods 22:61–70, 2000.

17. S Jeong, TT Nikiforov. Kinase assay based on thiophosphorylation and biotinylation. Biotechniques 27:1232–1238, 1999.

18. GJ Parker, TL Law, FJ Lenoch, RE Bolger. Development of high throughput screening assays using fluorescence polarization: nuclear receptor-ligand binding and kinase/phosphatase assays. J Biomol Screen 5:77–88, 2000.

19. AP Tairi, R Hovius, H Pick, H Blasey, A Bernard, A Surprenant, K Lundstrom, H Vogel. Ligand binding to the serotonin-5-HT3 receptor studied with a novel fluorescent ligand. Biochemistry 37:15850–15864, 1998.

20. M Allen, J Reeves, G Mellor. High throughput fluorescence polarization: a homogeneous alternative to radioligand binding for cell-surface receptors. J Biomol Screen 5:63–69, 2000.

21. SA Schade, ME Jolley, BJ Sarauer, LG Simonson. BODIPY-α-casein: a pH-independent protein substrate for protease assays using fluorescence polarization. Anal Biochem 243:1–7, 1996.

22. LM Levine, ML Michner, MV Toth, BC Holwerda. Measurement of specific protease activity utilizing fluorescence polarization. Anal Biochem 247:83–88, 1997.

23. R Bolger, D Thompson. A quantitative RNase assay using fluorescence polarization, Am Biotechnol Lab 12:113–116, 1994.

24. JH Hanke, JP Gardner, RL Dow, PS Changelian, WH Brissette, EJ Weringer, BA Pollok, PA Connelly. Discovery of a novel, potent, and src family-selective tyrosine kinase inhibitor. J Biol Chem 270:695–701, 1996.

25. PR Selvin. Fluorescence resonance energy transfer. Meth Enzymol 246:300–334, 1995.

26. RM Clegg. Fluorescence resonance energy transfer. Curr Opin Biotech 6:103–110, 1995.

27. S Grahn, D Ullmann, HD Jakubke. Design and synthesis of fluorogenic trypsin peptide substrates based on resonance energy transfer. Anal Biochem 265:225–231, 1998.
28. AJ Kolb, JW Burke, G Mathis. Homogeneous, time-resolved fluorescence method for drug discovery. In: JP Devlin, ed. High Throughput Screening. New York: Marcel Dekker, 1997, pp 345–360.
29. B Alpha, J Lehn, G Mathis. Energy transfer luminescence of europium (III) and terbium (III) cryptates of macrocyclic polypyridine ligands. Agnew Chem Int Ed Engl 26:266, 1987.
30. G Mathis. HTRF® technology. J Biomol Screen 4:308–310, 1999.
31. I Hemmilä. Lance™: homogeneous assay platform for HTS. J Biomol Screen 4: 303–307, 1999.
32. T Ahola, J Virtanen, A Toivonen, I Hemmilä, P Hurskainen. Use of generic reagent in Lance™ TR-FRET Assays. Fourth Annual Conference of the Soc Biomol Screen, 21–24 September, Baltimore, MD, 1998, poster SDAT-33.
33. AJ Kolb, PV Kaplita, DJ Hayes, Y-W Park, C Pernell, JS Major, G Mathis. Tyrosine kinase assays adapted to homogeneous time-resolved fluorescence. Drug Disc Today 3:333–342, 1998.
34. I Hemmilä, T Ahola. Homogeneous time-resolved fluorometric energy transfer assay (Lance™) for protein kinase. Fourth Annual Conference of the Soc Biomol Screen, 21–24 September, Baltimore, MD, 1998.
35. K Blomberg, P Hurskainen, I Hemmilä. Development of homogeneous time-resolved fluorescence energy transfer immunoassay (Lance™) for the determination of βhCG. Second Annual Conference of the Soc Biomol Screen, 22–25 September, San Diego, CA, 1997.
36. DC Hill. Trends in development of high-throughput screening technologies for rapid discovery of novel drugs. Curr Opin Drug Disc Develop 1:92–97, 1998.
37. K Stenroos, P Hurskainen, S Eriksson, I Hemmilä, CK Blomberg, C Lindqvist. Homogeneous time-resolved IL-2-IL-2 Rα assay using fluorescence resonance energy transfer. Cytokine 10:495–499, 1998.
38. P Hurskainen, J Virtanen, P Liitti, K Blomberg, I Hemmilä. Suitable substrates for homogeneous enzymatic DNA strand separation assays. Second Annual Conference of the Soc Biomol Screen, 22–25 September, San Diego, CA, 1997.
39. DA Earnsaw, KJ Moore, CJ Greenwood, H Djaballah, AJ Jurewicz, KJ Murray, AJ Pope. Time-resolved fluorescence energy transfer DNA helicase assays for high throughput screening. J Biomol Screen 4:239–248, 1999.
40. S Maiti, U Haupts, WW Webb. Fluorescence correlation spectroscopy: diagnostics for sparse molecules. Proc Natl Acad Sci USA 94:11753–11757, 1997.
41. S Sterer, K Henco. Fluorescence correlation spectroscopy (FCS)—a highly sensitive method to analyze drug/target interactions. J Receptor Signal Transduction Res 17: 511–520, 1997.
42. M Auer, KJ Moore, FJ Meyer-Almes, R Guenther, AJ Pope, KA Stoeckli. Fluorescence correlation spectroscopy: lead discovery by miniaturized HTS. Drug Disc Today 3:457–465, 1998.
43. U Kettling, A Koltermann, P Schwille, M Eigen. Real-time kinetics monitored by dual-color fluorescence cross-correlation spectroscopy. Proc Natl Acad Sci USA 95: 1416–1420, 1998.

44. A Koltermann, U Kettling, J Bieschke, T Winkler, M Eigen. Rapid assay processing by integration of dual-color fluorescence cross-correlation spectroscopy: high throughput screening for enzyme activity. Proc Natl Acad Sci USA 95:1421–1426, 1998.
45. NS Finney. Fluorescence assays for screening combinatorial libraries of drug candidates. Curr Opin Drug Disc Develop 1:98–105, 1998.
46. RD Mitra, CM Silva, DC Youvan. Fluorescence resonance energy transfer between blue-emitting and red-shifted excitation derivatives of the green fluorescent protein. Gene 175:13–17, 1996.
47. L Mere, T Bennet, P Coassin, P England, B Hamman, T Rink, S Zimmerman, P Negulescu. Miniaturized FRET assays and microfluidics: key components for ultra-high-throughput screening. Drug Disc Today 4:363–369, 1999.
48. G Zlokarnik, PA Negulescu, TE Knapp, L Mere, N Burres, L Feng, M Whitney, K Roemer, RY Tsien. Quantitation of transcription and clonal selection of single living cells with β-lactamase as reporter. Science 279:84–88, 1998.
49. KA Giulliano, RL DeBiasio, RT Dunlay, A Gough, JM Volosky, J Zock, GN Pavlakis, DL Taylor. High-content screening: a new approach to easing key bottlenecks in the drug discovery process. J Biomol Screen 2:249–259, 1997.
50. SW Paddock. Confocal laser scanning microscopy. Biotechniques 27:992–1004, 1999.
51. RL Hansen, HA Louis, TD Harris, JK Trautman, N Nicklas. Fast confocal fluorescence microscopy for homogeneous cellular assays in HTS. Fourth Annual Conference of the Soc Biomol Screen, 21–24 September, Baltimore, MD, 1998, poster CBA-23.
52. KS Schroeder, BD Neagle. FLIPR: a new instrument for accurate high throughput optical screening. J Biomol Screen 1:75–80, 1996.
53. Application notes. Measuring intracellular calcium with the FLIPR™ system. Molecular Devices.
54. G Wada, B Neagle, KS Schroeder. Luminescence assays using the FLIPR high throughput system. Third Annual Conference of the Soc Biomol Screen, 22–25 September, San Diego, CA, 1997.
55. S Miraglia, EE Swartzman, J Mellentin-Michelotti, L Evangelista, C Smith, I Gunawan, K Lohman, EM Goldberg, B Manian, P-M Yuan. Homogeneous cell- and bead assays for high throughput screening using fluorometric microvolume assay technology. J Biomol Screen 4:193–204, 1999.
56. R Williams. Electrochemoluminescence: a new assay technology. IVD Technology 28–31, 1995.
57. DR Deaver. A new non-isotopic detection system for immunoassays. Nature 377:758–760, 1995.
58. S Bohlm, S Kadey, K McKeon, S Perkins, R Sugasawara. Use of the ORIGEN electrochemiluminescence detection for measuring tumor necrosis factor-α in tissue culture. Biomed Products 21:60, 1996.
59. SR Hughes, O Khorkova, S Goyal, J Knaeblein, J Heroux, N Riedel, S Sahasrabudhe. α2-macroglobulin associates with β-amyloid peptide and prevents fibril formation. Proc Natl Acad Sci USA 95:3275–3280, 1998.
60. K Motmans, J Raus, C Vandevyer. Quantification of cytokine messenger RNA in

transfected human T cells by RT-PCR and an automated electrochemiluminescence-based post PCR detection system. J Immunol Method 1990:107–116, 1996.

61. JH Kenten, J Casade, J Link, S Lupold, J Willey, M Powell, A Rees, RJ Massey. Rapid electrochemiluminescence assays for polymerase chain reaction products. Clin Chem 37:1626–1632, 1991.

62. DB Lackey. A homogeneous chemiluminescent assay for telomerase. Anal Biochem 263:57–61, 1998.

63. AJ Kolb, K Neumann. Luciferase measurements in high throughput screening. J Biomol Screen 1:85–88, 1996.

64. Application notes. Alpha screen. Bio-signal, a Packard BioScience Company.

65. BA Brown. FlashPlate™ platform for assays and high throughput drug screening. DuPont NEN™ Life Science Products Application Notes.

66. BA Brown, M Cain, J Broadbeat, S Tompkins, G Henrich, R Joseph, S Casto, H Harney, R Greene, R Delmondo, S Ng. FlashPlate™ technology. In: JP Devlin, ed. High Throughput Screening. New York: Marcel Dekker, 1997, pp 317–328.

67. J Haggblad, B Carlson, P Kivela, H Siitari. Scintillating microtitration plates as platform for determination of ^{3}H-estradiol binding constants for hER-HBD. BioTechniques 18:146, 1995.

68. J Watson, JV Selkirk, AM Brown. Development of FlashPlate™ technology to measure [^{35}S]GTPγS binding to Chinese hamster ovary cell membranes expressing the cloned human 5HT$_{1B}$ receptor. J Biomol Screen 3:101–105, 1998.

69. S Watson. In vitro measurement of the second messenger cAMP: RA vs Flash-Plates™. Biotech Update 10:11, 1995.

70. R Graves, R Davies, G Brophy, G O'Beirne, N Cook. Noninvasive real-time method for the examination of thymidine uptake events—application of the method to V-79 cell synchrony studies. Anal Biochem 248:251–257, 1997.

71. HE Hart, EB Greenwald. Mol Immunol 16:265–267, 1979.

72. S Udenfriend, LD Gerber, L Brink, S Spector. Scintillation proximity radioimmunoassay utilizing 125I-labeled ligands. Proc Natl Acad Sci USA 82:8672–8676, 1985.

73. ND Cook. Scintillation proximity assay: a versatile high-throughput screening technology. Drug Discov Today 1:287–294, 1996.

74. N Bosworth, P Towers. Scintillation proximity assay. Nature 341:167–168, 1989.

75. D Powell, B Jessop, M Looker, I Davies, R Waythe, J Turner. The Leadseeker homogeneous imaging system. Fourth Annual Conference of the Soc Biomol Screen, 21–24 September, Baltimore, MD, 1998, poster AM-2.

76. F Legay, P Albientz, R Ridder. Bio-analytical applications of BIAcore, an optical sensor. In: JP Devlin, ed. High Throughput Screening. New York: Marcel Dekker, 1997, pp 443–454.

77. Z Salamon, MF Brown, G Tollin. Plasmon resonance spectroscopy: probing molecular interactions within membranes. Trends Bio Sci 24:213–219, 1999.

78. KL Lohman. A novel fluorescent HTS platform using homogeneous miniaturized cell and bead-based assays for drug discovery and development. Fourth Annual Conference of the Soc Biomol Screen, 21–24 September, Baltimore, MD, 1998, poster CBA-14.

79. MA Paulik, RG Buckholz, ME Lancaster, WS Dallas, EA Hull-Ryde, JE Weiel, JM Lenhard. Development of infrared imaging to measure thermogenesis in cell culture:

thermogenic effects of uncoupling protein-2, troglitazone, and β-adrenoceptor agonists. Pharmaceutical Res 15:944–949, 1998.

80. D Rotman. Quantum dot com. Technology Review, Jan–Feb 2000:51–57, 2000.
81. VK Gupta, JJ Skaife, TB Dubrovsky, NL Abbott. Optical amplification of ligand-receptor binding using liquid crystals. Science 279:2077–2080, 1998.
82. M Bruchez, M Moronne, P Gin, S Weiss, AP Alivisatos. Semiconductor nanocrystals as fluorescent biological labels. Science 281:2013–2016, 1998.
83. WCW Chan, S Nie. Quantum dot bioconjugates for ultrasensitive nonisotopic detection. Science 281:2016–2018, 1998.
84. CB Cohen, E Chin-Dixon, S Jeong, TT Nikiforov. A microchip-based enzyme assay for protein kinase A. Anal Biochem 273:89–97, 1999.
85. NH Chiem, DJ Harrison. Monoclonal antibody binding affinity determined by microchip-based capillary electrophoresis. Electrophoresis 19:3040–3044, 1998.
86. V Dolnik, S Liu, S Jovanovich. Capillary electrophoresis on microchip. Electrophoresis 21:41–54, 2000.
87. R Bolger. High-throughput screening: new frontiers for the 21st century. Drug Disc Today 4:251–253, 1999.

5
Microbe-Based Screening Systems

Prabhavathi B. Fernandes
Ricerca, LLC, Concord, Ohio

I. HISTORICAL INTRODUCTION

Microbe-based screening has been used to identify antibacterial agents as well as cytotoxic anticancer agents. Cell death was a phenotype that could easily be followed. Screening against targets in other areas, such as cardiovascular and neuroscience, was done in the past in isolated tissues, cells, or even in whole animals. With the start of biotechnology it became possible to clone and express various target proteins, and this led to in vitro screening in cell-free systems [1]. These screens are relatively simple to run and generally yield a large number of "hits." However, many of the hits from cell-free screens do not demonstrate whole cell activity, and the hits cannot be used to validate the target or to determine the utility of the lead molecule [2]. Thus lead molecules identified in cell-free screens meet the formidable hurdle of requiring chemical modification to derive molecules that can cross the cell membrane barrier while retaining activity against the original target. It is not a simple task to gain cell permeability while retaining selectivity against the target protein with no untoward activity against other cellular components. Thus, the time gained by rapidly identifying hits from a soluble protein screen is lost while using chemical modifications to gain selective activity against the target in a whole cell. Cellular systems simulate the natural milieu of the cell in which the target resides, and target proteins behave more as they do in their native cell than when they are tested in a cell-free environment. Furthermore, disease targets may involve complex protein interactions, and some of these interactions may not be known. These complex protein interactions are better observed in cellular systems than in cell-free assays.

II. ADVANTAGES OF CELL-BASED SCREENING SYSTEMS

During the last few years, it has become apparent that cell-based screens provide significant advantages in that hits identified in these screens can be rapidly developed into drugs [2,3] (Fig. 1). Moreover, as there are many thousands of targets in a cell, hits in a cell-based screen show some degree of selectivity towards the target or they would not be detected in the screen. Hits from a cell-based screen can be shown to have activity in the cell where the target is naturally expressed. Such hits face little delay in moving into lead optimization for preclinical candidate identification. Once the activity of the hits is confirmed in secondary assays, chemical modifications can be focused on optimization of potency and pharmaceutical properties (Fig. 2).

Although many enzymes can be cloned and expressed to produce large amounts of protein, it is not always feasible to provide sufficient amounts of

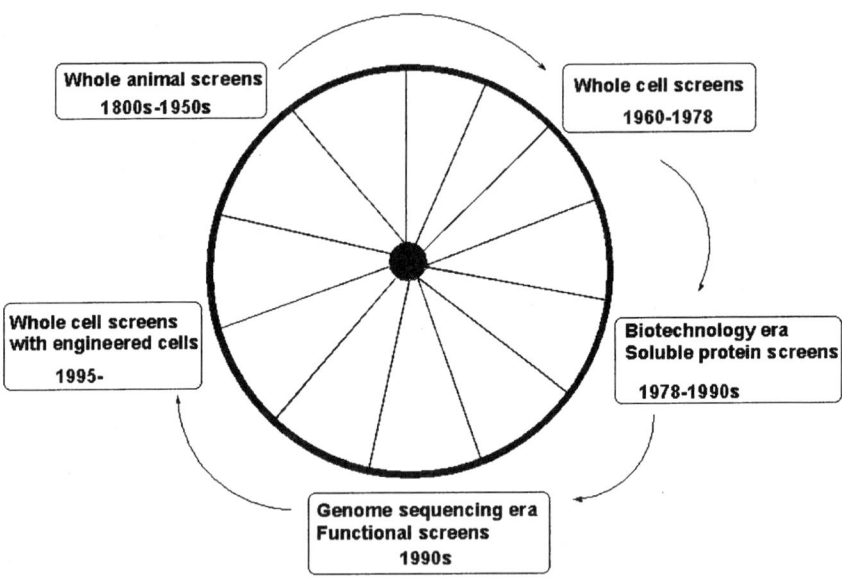

Figure 1   Back to the future: Cell-based screening systems are gaining in popularity because of the quality of the leads generated from these screens and the high information content of cell-based screens.

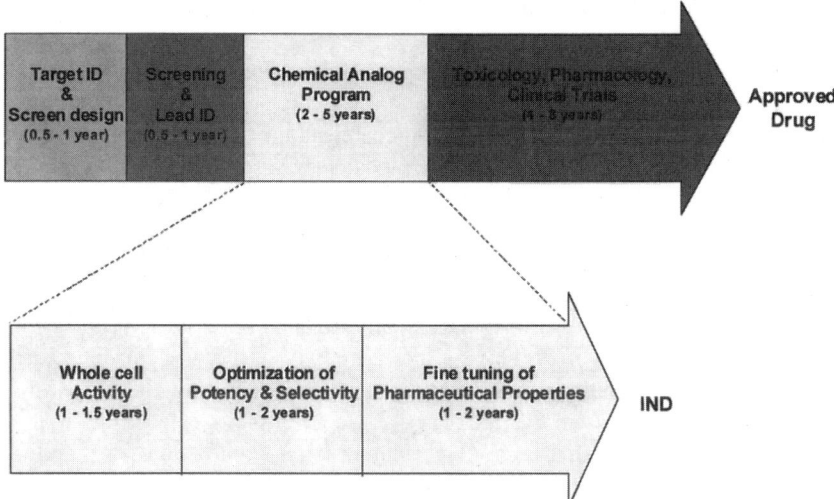

Figure 2 Drug discovery timelines. Chemical analogs from hits in cell-free screens must be modified synthetically to get whole cell activity. This process is time-consuming, and it may not be possible to get whole-cell activity while retaining the enzyme activity. Drug discovery timelines can be shortened by obtaining leads that have activity in whole-cell systems.

enzyme for use in high-throughput screening. Additionally, batch-to-batch reproducibility of activity is important for maintaining the quality of the screen. In cell-based screening systems, once the target protein has been expressed, the reagents are self-generating and the cost of generating reagents is low. A drawback of screening using mammalian cells is the expense of growing them to produce the quantities required for high-throughput screening. The expense and difficulty of producing large numbers of mammalian cells has been decreased to some extent by using miniaturized systems. Miniature systems contain fewer cells and therefore require a robust readout to be detectable. In addition, if there is no interference between the targets, multiple targets can be expressed in the same cell to conserve costs. Such dual-target systems provide a well-controlled screening system in which selectivity can be measured ratiometrically between the targets expressed in the same cell. Drug targets can be used in whole-cell screens without knowing all the interacting proteins necessary for activity such as interacting proteins in signaling pathways, other biochemical pathways, and transcriptional activators and repressors that have not yet been individually characterized [2,4].

III. MICROBIAL-BASED SCREENING

Biotechnology has made it possible to express mammalian proteins in microbial systems. Generally, proteins are heterologously expressed in microbial systems to obtain large quantities of protein for biochemical and structural studies or for producing large amounts of proteins for clinical use. In some cases, foreign proteins expressed at high levels in microbial cells result in inactive protein trapped in inclusion bodies. These insoluble aggregates, in many instances, regain activity after they are isolated, dissolved, and refolded. Also, mammalian proteins that require post-translational modification like glycosylation are not found in microbial systems. These differences between mammalian and bacterial cell expression are not generally limiting to developing screening systems, because, in our experience, some post-translational modifications are not required for the expression of small amounts of protein that are necessary to elicit biological activity in microbial cells. In mammalian cells, glycosylation is required for protein stability and for transport. Microbes offer significant advantages for developing and running screens because of the relative ease with which proteins can be cloned and expressed in microbial cells. In addition, microbial cells are inexpensive to grow and can be manipulated for distribution with ease.

IV. CELL PERMEABILITY

An advantage of cell-free systems is that all active compounds are identified and then secondary cellular systems can be used to differentiate those hits that are cell-permeable and those that are not. Although cloning new targets of interest into microbial systems is easily achieved, microbial cells have evolved an impermeable cell wall and membrane that allow them to survive in the environment. Therefore the target may not be easily accessible to compounds in high-throughput screens that use microbial cells. Mammalian cells, in general, are more permeable than microbial cells, and therefore microbial cell-based screens are likely to miss those compounds that do not penetrate across the cell wall and membrane barrier. In addition, many microbes have very effective efflux systems that pump out compounds. These efflux systems are similar to the multidrug resistance (MDR) transporters found in tumor cells. One of the large classes of efflux systems, or transporters, is called the ATP-binding cassette transporters or ABC transporters. The ABC transporters are conserved from bacteria to man [5].

With such drawbacks, can microbial-based screening be effectively used in screening? Genetic and molecular technology has made it possible to remove some of these barriers and make screen development and screening in microbial systems a viable, inexpensive, and productive alternative to other screening systems. Among microbes, *Saccharomyces cereviseae* or yeast and *Escherichia coli* have been the most popular because of the genetic manipulations possible with these organisms. Since the yeast *S. cerevisiae* is an eukaryote, it is often considered to be a more realistic system for screen development against mammalian targets. However, yeast is slower to grow than *E. coli*, taking 48 hours to grow to measurable densities, while *E. coli* can be used within 6 to 8 hours of growth. Also, genetic manipulations in *E. coli* are considerably less difficult than with yeast, and *E. coli* are more permeable than yeast.

A. Development of Permeable *S. cerevisiae*

Wild-type *S. cerevisiae* are quite impermeable owing to their cell wall and membrane. The cell wall is considered to be latticelike and allows most small molecules to permeate through. *However, the cell membrane is considered to be quite impermeable.* More recently it has been noted that yeast cells are actually permeable, and the lack of drug effect is the result of the activity of multiple efflux systems, belonging to the family of ATP-binding cassette transporters (MDR), called PDR, that rapidly pump out compounds. Transcription factors, pdr1p and pdr3p, down-regulate the expression of hexose transporters, HXT11 and HXT9, which in turn up-regulate the expression of PDR. Thus deleting the hexose transporters, HXT11 or HXT9, confers pleiotropic drug resistance on yeast while over-expression of these transporters results in increased sensitivity to drugs (Fig. 3). Furthermore, deletion of the regulators of the promoter for ATP-binding transporters, PDR1 and PDR3, in HXT11 and HXT9 over-expressing strains, results in supersensitive yeast [6]. These mutant strains are ideal organisms for use as host strains for the development of screens. Improved cell permeability was also

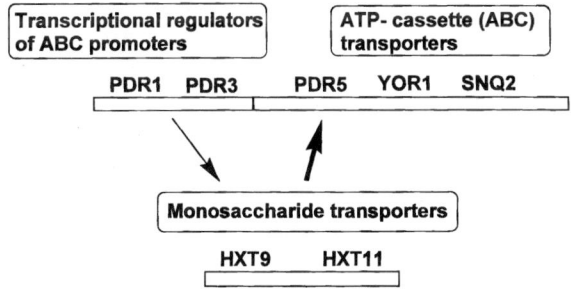

Figure 3 Design of permeable *S. cerevisiae*. The transcription factors pdr1p and pdr3p decrease the expression of the hexose transporters. The hexose transporters increase the expression of the ATP transporters such as PDR5. Deletion of PDR1 and PDR5 increases expression of HXT11, and HXT9 down-regulates the expression of ATP transporters and enhances the susceptibility of *S. cerevisiae* to many compounds.

reported for the yeast strains with deletion of the *YOR1* gene, which encodes another ATP-binding cassette transporter [7]. In addition to transporter mutants, mutants in the ergosterol pathway, such as in *erg6*, are also more permeable to small molecules used in screening than wild-type strains [8]. However, the mutants in the ergosterol pathway have the disadvantage that their growth is affected, and they are difficult to transform.

B. Development of Permeable *E. coli*

Gram-negative bacteria like *E. coli* have an outer membrane, an inner membrane, and a periplasmic space between the membranes. The outer membrane has the lipopolysaccharide (LPS) layer and a number of proteins called porins, which allow molecules that are approximately 600 Kd to enter the periplasmic space [9–11]. The sizing sieve created by the porins is perfect for small molecule discovery programs. *E. coli* permeability can be increased by mutating efflux proteins [12–14] and also by shortening the length of the LPS layer [9]. Mutants in the *rfa2* gene are called rough strains and are more permeable to many drugs [15]. Another gene, called *imp*, has been reported that, when mutated, enhances permeability by altering the surface properties of *E. coli* [16]. As with yeast, mutations in cell wall structures enhance permeability. However, growth and transformation properties could be altered, and the ideal permeable strain must be picked to suit the needs of the particular screen.

V. BIOLOGICAL RELEVANCE OF MICROBIAL SCREENING SYSTEMS

Although microbial systems are less complicated than mammalian cell systems, homologs of many mammalian proteins are found in microbial systems. As of November 1996, 21% (15/70) of the positionally cloned genes that are mutated in human disease have a match in the *Saccharomyces* sequence [17]. In addition, many other mammalian proteins that do not have sequence homology but do have functional homology can be used to complement function in microbial cells. Unlike in higher organisms, the functions of about half of the yeast and *E. coli* genes are known on the basis of amino acid sequence similarity with other proteins of known function [18]. This is an enormous resource that is being used for functional analysis. History has shown us that biological mechanisms revealed from the study of microbial cells will be applicable to higher eukaryotes; in the future, knowledge of the function of microbial proteins will help in elucidating the function of many mammalian proteins. The similarity between living organisms was noted by Jacques Monod when he said "What is true for *Escherichia coli* is true for the elephant, except more so" [19].

VI. USE OF *S. CEREVISIAE* AND *E. COLI* FOR TARGET VALIDATION AND SCREENING

In the simplest approach, functional complementation can be used to derive a screen in which the activity of the heterologous gene is essential for survival. This approach has been successful even in cases where protein homology is limited, as long as the relevant biological activity is complementary. Microbial systems can also be used to define interactions with other proteins. The more difficult approach is to manipulate the heterologously expressed gene to obtain a surrogate phenotype and create "designer microbes." The following are examples of each type of system.

A. Complementation of Homologous Targets

1. *S. cerevisiae*–Based Screens for Immunosuppressants

Immunosuppressants, such as cyclosporin and FK506, inhibit T cell activation and have made tissue and organ transplantation a reality. These drugs were first identified because of their antifungal activity, and their mechanism of immunosuppression long remained a mystery. The mechanism of action of cyclosporin and FK506 was determined using *S. cerevisiae* [20]. Cyclosporin and FK506 bind proteins with peptidyl prolyl isomerase activity (called FK506 binding proteins and cyclosporin binding proteins) and forms a complex that inhibits the calcium-calmodulin phosphatase, calcineurin. The similarity between the activity of the immunosuppressants in T cells and yeast is shown in Fig. 4. In the presence of these immunosuppressants, yeast does not grow and divide after exposure to the yeast pheromone, α-factor. The mechanism of action of Rapamycin, another immunosuppressant, has been elucidated by studies in yeast [21]. The target of rapamycin is known to be TOR, which is involved in phosphorylating protein phosphatase 2A. These phenomena can be used as the basis of screens for new immunosuppressants.

2. Expression of Seven Transmembrane, G-Protein Coupled Receptors in Microbial Systems

Many currently used drug targets are seven transmembrane, G-protein coupled receptors (GPCRs), such as the serotonin receptors, dopamine receptors, and adrenergic receptors. Antagonists have been identified by ligand-displacement assays using mammalian cells or their membrane preparations. Agonists have generally been identified by functional assays. Microbial systems have been adapted to identify agonists and antagonists of GPCRs. The mating factor receptor in *S. cerevisiae* is called Ste2 and is similar in structure to mammalian GPCRs. Mammalian GPCR can be used to replace Ste2, so that the GPCR can signal

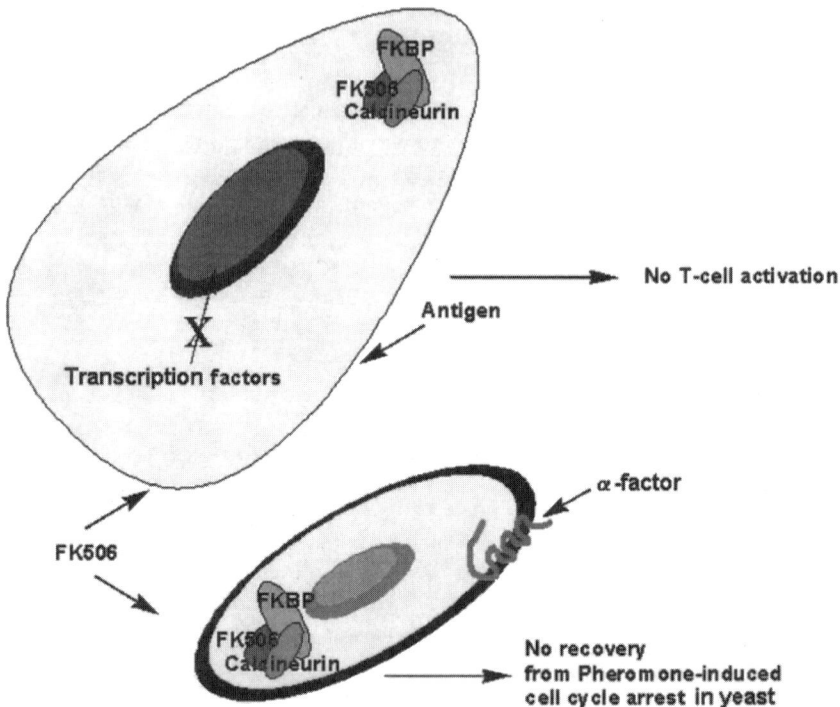

Figure 4   Mechanism of action of FK506 in T lymphocytes and *S. cerevisiae*. The immunosuppressant FK506 binds FK506-binding proteins in T cells and in yeast. Together, these molecules complex with calcineurin and inhibit its activity. In T cells and in yeast, free calcineurin is essential for the activation of transcription factors. The phenotype derived from the activation of transcription factors in T cells is activation and in yeast it is growth after pheromone-induced arrest.

through the mating factor signaling pathway when activated by the GPCR agonist [22]. In order to get efficient downstream coupling with the mating factor pathway kinases, the amino terminal domain of the G_α protein can be replaced by the same region of the human G_α protein [22,23]. The GPCRs expressed in this manner in yeast are useful for screening for agonists. Coupling can also be obtained with the natural yeast G_α protein [24,25].

The natural ligands of GPCRs can be large peptide ligands. In mammalian cells, these GPCRs can be activated by these peptides as well as their derivatives. Antagonists are identified by finding compounds that can displace these peptides. Short peptides can be coexpressed with GPCR in yeast to develop functional antagonist screens as well as for identifying ligands for orphan GPCRs [26].

Orphan GPCRs are those receptors that have been cloned by sequence homology to known GPCRs but whose function and natural ligand are not yet known. Peptide libraries have been expressed with secreted sequence tags that are secreted into the surface of the yeast where they come into contact with the GPCRs. Using these libraries, peptides that specifically interact with and activate the receptor in an autocrine manner can be identified.

GPCRs have been expressed in *E. coli* with limited success. However, *E. coli* has no naturally occurring GPCRs and there is no G-protein signaling pathway. Therefore both the G-protein and the GPCRs have been expressed in *E. coli* to obtain receptors that bind receptor agonists [27,28].

3. Functional Expression of Channels in *S. cerevisiae*

Potassium channels are important targets for cardiovascular, immunological, and neurological diseases. Functional screens for K^+ channel openers and blockers involve expensive equipment and are technically difficult. Therefore simpler assays for developing high-throughput screens have been welcomed. A simple functional screen was developed in *Saccharomyces* using complementation of Trk potassium transporter knockouts [29] (Fig. 5). The inwardly rectifying potassium channel IRK1 has also been expressed in yeast to complement the Trk transporter defect. In the strain expressing the IRK1 channel, the ion channel

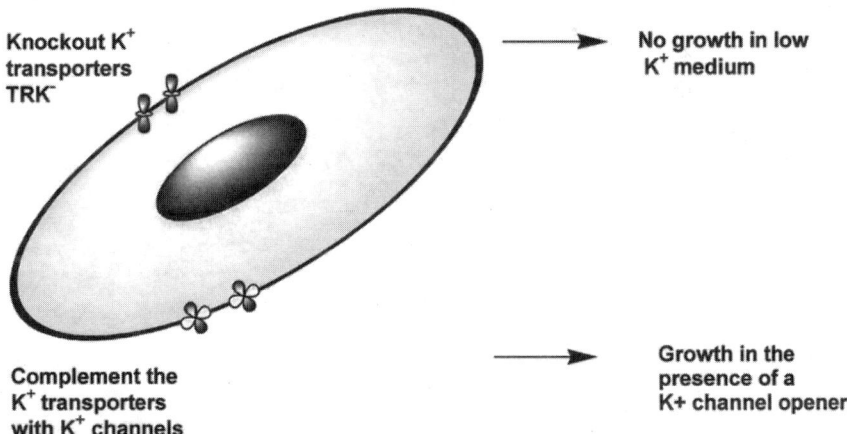

Figure 5 Functional expression of MinK potassium channel in *S. cereviseae*. MinK is a potassium channel found in the heart. It can complement the knockout of the potassium transporters TRK1 and TRK2 in yeast and allows the mutant yeast to grow in a low-potassium medium. Blocking of the MinK channel in yeast will prevent growth of the yeast in a low-potassium medium.

activity correlates well with the growth phenotype and with patch clamp experiments in *Xenopus* oocytes expressing these channels.

The influenza M2 channel has been expressed in *S. cerevisiae* [30]. The influenza M2 channel is a proton channel that is expressed in infected cells: its function is to increase the acidity of the milieu in which the virus sheds its capsid. When expressed in *S. cerevisiae*, the M2 proton channel increases the permeability of yeast membrane to ions resulting in loss of yeast cell viability. In order to develop a screen to find influenza M2 protein inhibitors, it was expressed from a galactose-inducible promoter (Fig. 6). The screen was designed to find compounds that permit growth and rescue the cells from the permeabilizing effects of M2 protein when the growth medium is supplemented with galactose.

Channel screens designed in *S. cerevisiae* have been useful for high-throughput screening. However, yeast is slow growing, and expression of channels in this microbe is difficult and time-consuming. *E. coli* provides an alternative for developing screens to find influenza virus M2 inhibitors. The influenza

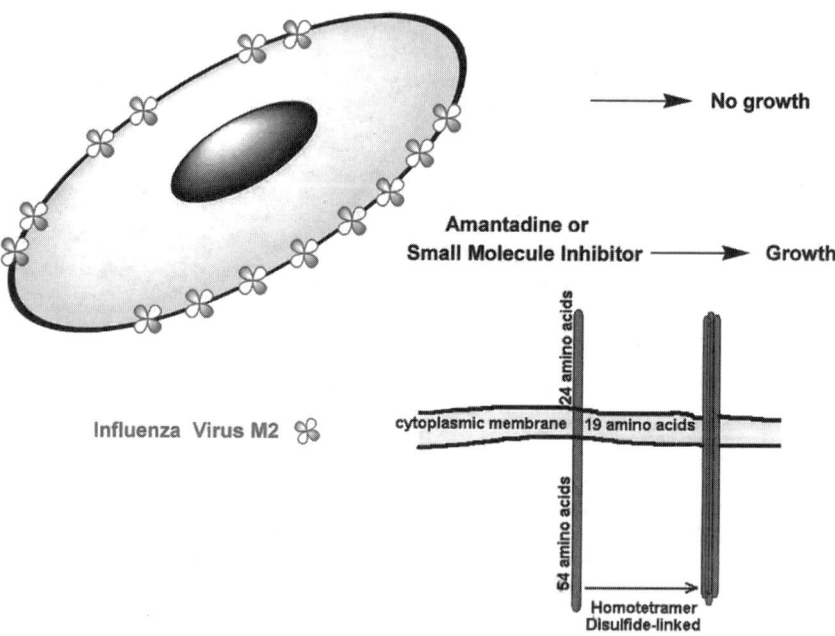

Figure 6 Functional expression of the influenza virus M2 proton pump in *S. cereviseae*. The influenza M2 channel occurs as a tetramer and acts as a proton pump in virus infected cells. Expression of this channel in yeast increases membrane permeability and causes cell death. Blockers of the channel allow growth of yeast cells expressing M2.

virus M2 protein has been expressed under control of the *lac* promoter in *E. coli* [31]. In *E. coli*, M2 protein increases membrane permeability to hydrophilic molecules, such as ONPG, uridine, and hygromycin B, and as in yeast, high level expression of M2 induces rapid lysis in *E. coli*. Thus a rescue screen for compounds that can block influenza virus M2 is easily set up using *E. coli*.

B. Expression of Heterologous Proteins to Get a Phenotype

1. Expression of Steroid Hormone Receptors in *S. cerevisiae*

Screens to find ligands for steroid hormone receptors such as the retinoid receptors have been designed in *S. cerevisiae* [32]. Steroid hormone receptors occur intracellularly. They have a ligand-binding domain, a dimerization domain, and a transactivation domain. When ligands induce these receptors to homo- or heterodimerize, their transactivation domain binds the specific response elements and activates specific promoters. The dimerization of the steroid hormone receptors followed by binding and transactivation of specific promoters can be studied in yeast. Homodimerization has been demonstrated using retinoic acid receptors, thyroid hormone receptors, and estrogen receptors [33,34]. Heterodimerization with the RXR retinoid receptor can also be demonstrated [35]. In this system, the RAR, retinoic acid receptors respond to a number of retinoids, but RXR responds only to the RXR-specific 9-*cis* isomer of retinoic acid [35]. Because all mammalian cells have many representatives of the steroid hormone receptor family expressed naturally, microbial systems offer cells with a "null" background for studying specific interactions.

2. Expression of Human Topoisomerase 1 in *E. coli*

Human topoisomerase 1 binds double-stranded DNA and makes a single covalent phosphotyrosine intermediate, thus relaxing negatively supercoiled DNA. The activity of topoisomerase 1 is increased in dividing cells and is the target for the anticancer agent, camptothecin. Camptothecin traps the covalent phosphotyrosine intermediate, and various cellular activities produce cleavage fragments from this interrupted strand passage reaction. Many companies have sought additional chemotypes that inhibit topoisomerase 1 by using cell-free screening approaches that have been difficult to use in high-throughput screens. In these screens compounds that interact with DNA but do not have enzyme inhibitory activity often appear as false positives. Thus, it is difficult to detect specific inhibitors in these cell-free screens. An *E. coli* based screen has been reported that can easily distinguish between compounds that are only DNA interactive agents and those that produce cleavable complexes made of drug–DNA fragments and topoisomerase 1 [36]. Human topoisomerase 1 and *E. coli* topoisomerase 1 are different in the reaction that they catalyze and have no homology. In the *E. coli*–based screen, human topoisomerase 1 is expressed from the inducible *lac* promoter. If a compound inhibits human topoisomerase 1, it can freeze the DNA cleavage fragments generated during the strand passage reaction. DNA damage is detected in *E. coli* by observing the induction of a *sulA–lac* fusion. General DNA damaging agents, which do not act on topoisomerase 1, are identified by the induction of *sulA–lac* in the absence of topoisomerase 1 induction. Thus compounds that interact with topoisomerase 1 and DNA can be clearly differentiated from those compounds that interact only with DNA in this screen.

3. Development of Screens to Find Protease Inhibitors

Proteases have been used as drug targets for many disease processes ranging from hypertension (angiotensin converting enzyme, ACE) to the human immunodeficiency virus or HIV. Enzyme assays have been useful in screening for protease inhibitors. Microbial systems provide an alternative means for screening against these targets and have the advantage of not requiring the production of large quantities of the enzyme. In addition, for those proteases whose natural environment is the cytoplasm, microbial systems provide a more natural milieu than that of a cell-free screen. It is possible that the difficulty in designing inhibitors against proteases, such as the cytomegalovirus protease and hepatitis C virus proteases, is related to the complex natural substrates that are the targets of these proteases in infected cells. These complex substrates can be engineered into microbial screening systems.

In general, functional screens for proteases are designed by insertion of the protease substrate (peptide sequence) within a protein that, when cleaved by the protease, looses its activity. One screening system that has been described in *S. cereviseae* involves the Gal4 transcriptional activator that induces the expression of several enzymes required to metabolize galactose. The protease substrate can be inserted in the transcriptional activation site of Gal4 [37]. The protease, when expressed from a separate plasmid, cleaves the substrate, destroying the Gal4 transcriptional activator. When the protease is inhibited, the Gal4 transcriptional activator binds the Gal promoter and activates the transcription of galactose metabolizing enzymes. Yeast strains expressing these enzymes do not grow on medium containing 2-deoxygalactose, and this allows the easy detection of protease inhibitors.

A number of protease screening systems have been developed in *E. coli*. The test protease could be made to activate or deactivate a protein that confers a phenotype, for example viability or a scorable phenotype, such as a color reporter or an antibiotic resistance gene. In one system, the transmembrane tetracycline efflux protein, TET, has been used as the indicator and confers tetracycline resistance to the host *E. coli*. TET protein has two multiple transmembrane domains that are connected by a long cytoplasmic loop. The protease substrate can be engineered into the cytoplasmic loop of TET without loosing TET function. Cleavage of TET by the protease that is also heterologously expressed in the same strain makes the *E. coli* sensitive to tetracycline [38,39]. An inhibitor of the protease allows the *E. coli* to grow in the presence of tetracycline as the TET protein remains intact.

Another more intricate system has been developed in *E. coli* for identifying protease inhibitors. This system depends upon the phenomenon that a mutation takes place in the S12 ribosomal protein of the 30S ribosomal subunit that makes *E. coli* resistant to streptomycin [40]. Such a mutation in the S-12 ribosomal protein is simulated by making a S-12 fusion protein with the protease substrate. The *E. coli* expressing the S-12-peptide-substrate chimeric protein is resistant to streptomycin. When the protease is expressed in the same *E. coli* expressing the S12-peptide-substrate chimera, the substrate is cleaved from S-12 and the *E. coli* becomes sensitive to streptomycin. Protease inhibitors preserve the S12-peptide-substrate chimera, and the *E. coli* remain streptomycin-resistant (Fig. 7). This screening system is notable because protease-dependent, dominant phenotypes are more sensitive than recessive phenotypes [40].

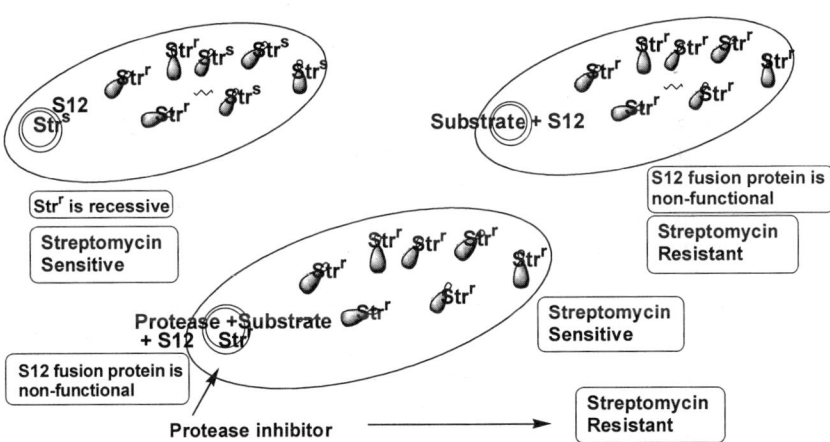

Figure 7 A bacterial protease system. A mutation in the S12 protein in the 30S ribosomal subunit makes *E. coli* streptomycin resistant. A chimeric S12-protease substrate protein mimics a mutation, and *E. coli* expressing S12-protease substrate constructs are streptomycin resistant. When the test protease is coexpressed in the streptomycin resistant *E. coli*, the protease substrate is cleaved from S12 and the *E. coli* reverts to streptomycin sensitivity. Protease inhibitors would prevent the regaining of streptomycin sensitivity.

4. Functional Expression of Tyrosine Kinases and Phosphatases

Tyrosine kinases are important targets for drug discovery for oncology, immunology, and other therapeutic areas. Cell-free assays have been popular for this class of targets since the enzymes are easily produced and phosphorylation of the substrate is simple to detect using labeled ATP. An alternate method using *Schizosaccharomyces pombe* has been published that will find inhibitors that are nontoxic to yeast as well as cell permeable [41]. The prototypic tyrosine kinase, *Src*, was expressed under the control of the inducible promoter. Induction of Src results in cell death, and growth rescue can be used for identifying inhibitors. To adapt the screen for identifying phosphatase inhibitors, this system was modified by co-expressing tyrosine phosphatase on a second plasmid. When the kinase and phosphatase are coexpressed, the cell survives the detrimental effects of kinase expression. Tyrosine phosphatase inhibitors can be identified in this system by looking for agents that selectively kill the strain coexpressing the kinase and phosphatase [41].

5. Methods for Detecting Protein–Protein Interaction

Yeast Two-Hybrid System. Proteins carry out their function in most cases by interacting with other proteins. The yeast two-hybrid system, developed by Fields and Song [42], has revolutionized the study of protein–protein interactions. In this system, the transcription factor, GAL4 from *S. cerevisiae*, is used to set up the assay. GAL4 has two domains, a site-specific DNA binding domain and an acidic region that is required for transcriptional activation. The DNA binding and activation domains can be coded by separate genes as long as they are brought together in a heterodimer to reconstitute a functional transcription factor (Fig. 8). The system is designed so that when GAL4 binds the GAL4 binding domain on the promoter, *LEU2* and/or *HIS3* are expressed. Functionally competent chimeric proteins can be made that consist of the DNA binding domain fused to one protein of an interacting pair and the activation domain fused to the second protein of the interacting pair. Interaction of the proteins that are constructed as chimeras of the activating and DNA-binding domains allow the yeast to grow in the absence of histidine and leucine, thus providing a selective advantage. The yeast two-hybrid system is widely used for identifying homo- and hetero-dimerizing proteins as well as to develop screens to find compounds that can block two proteins from interacting with each other [43].

The yeast two-hybrid system has been modified to measure the dissociation of interacting proteins by using the *URA3* reporter [44]. Yeast cells expressing *URA3* can grow in medium without uracil. When 5-fluoroorotic acid (FOA) is introduced into the medium, *URA3* expressing cells take up FOA and transform it into a toxic compound. Thus the expression of the reporter gene is toxic and

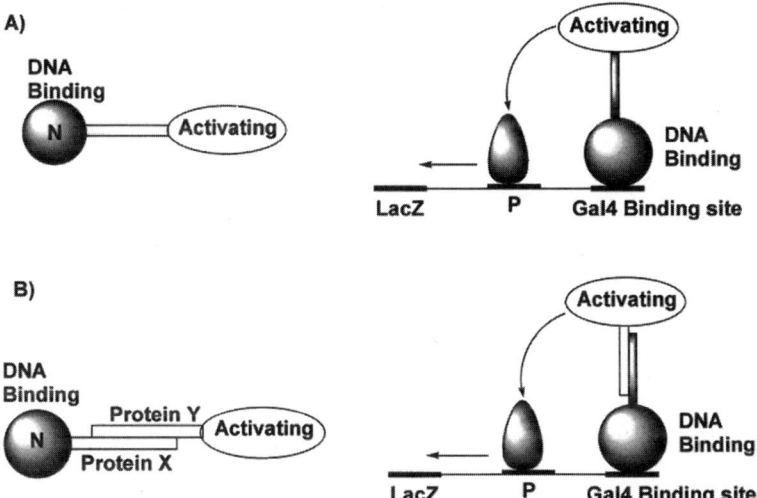

Figure 8 The yeast-two hybrid system. (A) The "wild-type" GAL4 transcription factor has two domains, the activating and the DNA-binding domains. (B) Two interacting proteins X and Y that are chimeras with the DNA-binding domain of GAL4 and the activating domain of GAL4-respectively. When the two proteins interact, the activating domain and the DNA-binding domain are brought together to function as a transcription factor.

provides a powerful selection procedure. This FOA system is used in the "reverse two-hybrid" system, providing a selective growth advantage and a more powerful system for screening. In the reverse two-hybrid system, the interacting protein is expressed inducibly, and only when the interacting proteins are blocked do the cells survive. GAL4 and LexA transcription factors are most often used in the yeast two-hybrid system. In two-hybrid screens, it is useful to have two separate reporter constructs to help in sorting "hits." Reporters such as *LEU2* and *LacZ* can be expressed in the same cell.

The yeast 2-hybrid system has been used to develop screens for ligand–receptor interactions, including peptide hormone receptors and the tyrosine kinase receptors [45–47]. Specific and reversible ligand–receptor interactions between growth hormone and growth hormone receptor, VEGF and KDR, can be studied using the yeast two-hybrid system. Ligand-dependent receptor dimerization can also be studied using three expression plasmids in which the receptor is expressed as a fusion protein with both the DNA binding domain as well as the activation domain. The ligand is expressed from a third plasmid. When the ligand binds the two receptors, the DNA-binding domain and activating domains are pulled together and Gal4 is activated (Fig. 9).

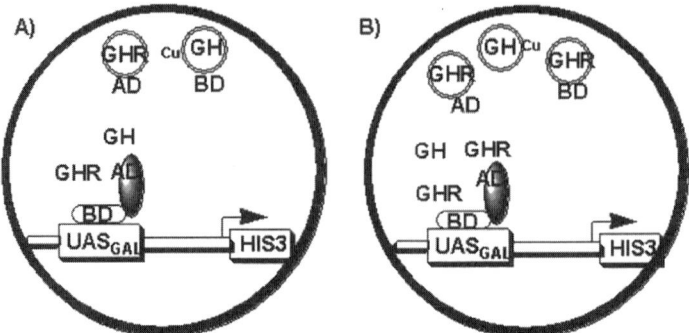

Figure 9 The yeast three-hybrid system. Yeast systems to identify ligand–receptor interactions are shown. In the example shown, (A) growth hormone is expressed as a chimera with the DNA-binding domain of GAL4 under the control of a regulatable promoter (copper). Growth hormone receptor is expressed as a chimera of the activating domain of GAL4. When induced, the activating and binding domains are brought together by the interaction of growth hormone with its receptor. This is a two-hybrid system. (B) In contrast, in the three-hybrid system, both the DNA-binding and activation domains are expressed as chimeras with growth hormone receptor. Growth hormone is expressed from a third plasmid, under regulation of the inducible copper promoter. When uninduced, the activating and binding domains are not brought together. When induced, the excess of growth hormone binds the growth hormone receptor, inducing dimerization and functional activation of GAL4, resulting in the expression of HIS3 reporter.

The yeast two-hybrid system has been adapted to study protein–protein, protein–RNA, protein–DNA, and protein–small molecule interactions [48]. A one-hybrid system has been developed that utilizes *cis*-acting sequences to identify DNA-binding proteins that can initiate transcription [49]. A yeast three-hybrid was developed to study RNA–protein interactions that are especially useful for developing screens against viruses [50]. In this system, the hybrid RNA containing sites recognized by the RNA interacting proteins links the two-hybrid proteins containing the DNA-binding and activation domains, respectively. The yeast two-hybrid system has been recently applied to find inhibitors of the N type calcium channel [51,52]. Alternative screening techniques use mammalian cells to measure calcium channel activity with electrophysiological and spectrophotometric methods to measure calcium influx. These methods are labor intensive, difficult, and not compatible with high-throughput screening. In the yeast two-hybrid system, the interacting, regulatory portion of the $\alpha 1$ subunit of the channel fused to the Gal4 activation domain and the full length $\beta 3$ subunit fused to the yeast Gal4 DNA binding domain were expressed. The system could be adapted to find inhibitors of specific calcium channels by selecting the specific interacting domains.

E. coli *Two-Hybrid System.* The yeast two-hybrid system and its modifications have been widely used for studying protein–protein interactions. Although it is useful for identifying the interacting proteins, the interacting proteins are not easily accessible to inhibitors that must cross the cell wall and membrane and also the nuclear membrane where the two-hybrid constructs are located in the yeast cell. Simpler systems have been developed in *E. coli* that may be more effective for screening. An example of such a system is a protein dimerization system developed in *E. coli.* Dimerization and multimerization is required for the activation of many cell surface proteins, such as single transmembrane receptors and channels. A bacterial system called ToxR has been developed to detect dimerization of cell surface receptors and channels [53]. The *Vibrio cholerae* ToxR gene product is a Type 2 membrane protein that has an extracellular domain, a single transmembrane domain, and a cytoplasmic domain that acts as a transcription factor, binding directly to the Tox promoter to activate toxin secretion. In *V. cholerae*, the extracellular domain of ToxR activates in response to external stimulus and induces the cytoplasmic portion of the receptor to bind the *Ctx* promoter directly to induce transcription of the toxin gene. In the screening system, ToxR has been cloned into *E. coli*, and the extracellular domain of ToxR has been replaced by the extracellular domain of the TrkC, the receptor for the neurotropin, NT3. Dimerization of the extracellular domain of the TrkC receptor activates the ToxR promoter. In the *E. coli* system, a reporter, such as β-galactosidase or the antibiotic resistance gene chloramphenicol acetyl transferase, has been engineered instead of toxin, for obtaining an easily measured readout (Fig. 10).

The ToxR system has been also been used for expressing the influenza virus M2 protein. In contrast to the system described in the previous section, where the M2 protein induces membrane permeability, the M2-ToxR chimera can be used to identify compounds that bind the channel and alter the multimerization state. The anti-influenza drug Amantadine is known to block the M2 channel, and in the M2 ToxR system, it was shown to alter the transcriptional signal produced by the M2-ToxR chimera. Although this screening system does not specifically find blockers of the M2 proton pump, the screen is designed to test a large number of compounds rapidly and identify those compounds that bind and affect aggregation. Some of these compounds could inhibit M2 function. The ToxR system has also been used for studying the formation of homodimers of Immunoglobulin V_L domains [54]. Thus different functional systems can be used to develop screens for the same target.

More recently, a different system using the CadC protein in *E. coli* has been developed for developing protein dimerization screens (unpublished obser-

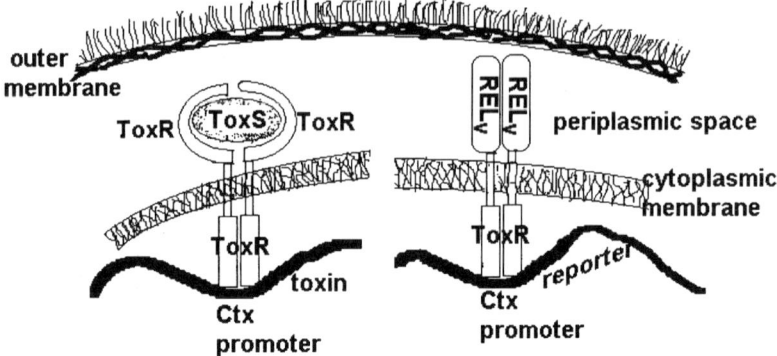

Figure 10 The ToxR, *E. coli* dimerization system. In *Vibrio cholerae*, ToxR is expressed as a cytoplasmic membrane protein that acts as a transcription factor interacting with the Ctx promoter directly with its cytoplasmic domain when it is induced to dimerize by interaction with ToxS. The ToxR protein has been cloned and expressed in *E. coli* to study the dimerization of proteins. The extraceullular dimerization domain of ToxR is replaced with the protein of choice, in this example the immunoglobulin variable domain and expressed in *E. coli*. The Ctx promoter is also cloned into *E. coli* and the toxin gene is replaced by a reporter such as *lacZ*. When the immunoglobulin domains dimerize, the cytoplasmic domain of ToxR is activated and lacZ is expressed.

vations, Facts sheet, Small Molecule Therapeutics, Inc.). In contrast to ToxR, CadC is a naturally occurring single transmembrane protein in *E. coli* that signals in response to pH change. These simple *E. coli* systems are modular in that chimeric proteins can easily be developed with the CadC protein. In addition, these systems can be easily adapted to identify compounds that induce or prevent dimerization.

Proteins carry out functions by interacting with other proteins. The ToxR and CadC protein dimerization systems can be used to measure homodimeric interactions of membrane-associated proteins. There are other *E. coli*–based systems that can be used for measuring heterodimeric interactions. A system developed by Dove et al. [55,56] uses transcriptional activators in prokaryotes that bind near a promoter and contact RNA polymerase. RNA polymerase consists of the β, β^1, σ, and two α subunits. Each α subunit contacts an activator protein, such as λci, two of which are required to occupy the λ-operator sites and activate the promoter. The C-terminal end of the RNA polymerase α subunit is fused to one protein of the interacting pair, and the C-terminal end of the λcl protein is fused to the second protein of the interacting pair. When contact is made between the protein fused to the DNA-bound protein, λcl, and the heterologous protein domain fused to RNA polymerase, transcription is activated. There are many examples of the activation of genes by recruitment of proteins [57], and many of these can be applied to developing protein–protein interaction screens. Another novel *E. coli* two-hybrid system has been developed that takes advantage of the ability of LexA to repress the activator AraC [58]. Chimeras of AraC and LexA were made with the interacting proteins. The LexA-protein 1 chimera interacts with the LexA binding half-site on the DNA, and the AraC–protein 2 chimera binds to the high-affinity AraC binding site. The LexA and AraC operators are separated by an IHF site that is involved in DNA loop formation and possibly helps in repression. Interaction of protein 1 and protein 2 allows LexA to hetero-dimerize with AraC, causing repression of the AraBAD promoter fused to *lacZ*.

Another *E. coli* two-hybrid system that can be used for homo- and hetero-dimeric proteins has been described [59]. This particular system is simple in that reconstitution of enzyme activity is used. The genes of the two interacting pro-teins of interest are fused to the two fragments of the catalytic domain of *Borde-tella pertussis* adenylate cyclase. One fragment with amino acid 1-224 of the cyclase is constructed with a C-terminal fusion and the second with amino acid 225-399 is constructed with an N-terminal protein fusion. Interaction of the pro-teins reconstitutes the adenylate cyclase and results in cAMP synthesis in *E. coli* with a mutation in the *cya* gene. cAMP binds to the catabolite gene activator protein, CAP, and the cAMP/CAP complex can bind specific promoters to turn on certain genes. Reporters of interest, such as lacZ or chloramphenicol resistance genes, can be fused to the cAMP sensitive reporters to obtain an easily measurable readout. Protein–protein interaction can also be measured by fusing the proteins of interest to complementing β-galactosidase deletion mutants [60]. The forced interaction of nonfunctional β-galactosidase units to produce active enzyme is the basis of this technology and provides a second method to measure protein–protein interaction directly without the activation of transcription factors.

Use of Two-Hybrid Systems for Functional Proteomics. Large numbers of potential targets are being identified from human genome sequencing efforts [61]. Many other targets are being identified from sequencing other organisms. The companies that are the first to convert the finite number of targets buried in the sequence information into drug targets and drugs will be successful in the future. Many strategies are being taken to understand the function of new genes as well as their protein products. Homology to known proteins has been useful in designing screens. However, the functions of many other genes will need to be determined. As proteins function by interacting with other proteins, two-hybrid screening methods will be useful for determining interacting proteins. The func-tion of interacting proteins, if known, may give clues to the functions of the new gene. Once the interaction is understood to be important, the two-hybrid pair can be used in a high-throughput screen to find small molecule inhibitors of the interaction by using any of the systems described above. Large numbers of compounds are available for screening from combinational chemistry programs, and cost-effective means of screening these libraries are necessary. Nanodrop technology has been developed to test combinatorial compounds that are available in small amounts [62]. In this method, nanodroplets (100–200 nL droplet) containing the yeast and compound are created. When the compound is released from the bead, it is trapped and cannot diffuse. When the compound is available at high concentrations in the proximity of the yeast with the target interacting proteins expressed in a two-hybrid format. This method combines the power of yeast genetics with combinatorial chemistry for drug discovery. The same droplet technology can be used with any of the microbial screening systems described in this chapter to screen small molecule libraries.

VII. SCREENING IN HIGH-DENSITY FORMATS

As thousands of new potential drug targets from genomic information and protein interaction studies are identified, the future of screening is in using chip technology [63,64]. Microbial systems are especially suited for delivery to microchips. Microbes are robust and are easy to handle and distribute. Thousands of yeast and *E. coli* cells can be deposited on chips in a high-density format. In addition, the surface charge on *E. coli* could be used to array the organisms into the desired format. Currently, the limitation is the sensitivity of reading colorimetric reporters, and alternatives are being investigated. Biosensors and transducers could be used to detect thermal, immunologic, or optical changes [65,66]. Glucose sensing amperometric systems are being used in clinical microbiology and could be developed for HTS [67]. Each of the microbial systems described can be adapted to use the reporter that is suitable for the high-density format on chips for screening.

VIII. CONCLUSIONS

Microbes provide an alternate platform for high-throughput screening. The systems are inexpensive to run, and screens can be developed rapidly. Many modular systems are available that are adaptable to important classes of targets, such as GPCRs, single transmembrane receptors, tyrosine kinase, phosphatases, and ion channels. Functional screens are becoming necessary for developing screens for proteins whose biological function is not yet known as well as on proteins that interact with new proteins. Because of the ease with which new targets can be explored, functional cell-based screens are becoming the preferred method for finding leads for drug discovery. Microbial systems provide simple and cost-effective means of functional screening.

ACKNOWLEDGMENTS

I thank Dr. Rolf Menzel and Dr. Donald Kirsch, who introduced me to many of the concepts described in this chapter.

REFERENCES

1. S Omura. Philosophy of new drug discovery. Microbiol Rev 50:259–279, 1986.
2. Z Parandoosh. Cell-based assays. J Biomol Screening 2:201–202, 1997.
3. JC Walsh. High throughput mechanism-based screening techniques for discovering novel agrochemicals. J Biomol Screening. 3:175–181, 1998.
4. L Guarente. Strategies for the identification of interacting proteins. Proc Natl Acad Sci USA 90:1639–1641, 1993.
5. D Dexter, S Moye-Rowley, S Wu, J Golin. Mutations in yeast *PDR3, PDR4, PDR7* and *PDR9*. Pleiotropic (multiple) drug resistance loci affect the transcript level of an ATP binding cassette transporter encoding gene *PDR5*. Genetics 136:505–515, 1994.
6. A Nourani, M. Wesolowski-Louvel, T Delaveau, C Jacq, A Delahodde. Multiple-drug-resistance phenomenon in the yeast *Saccharomyces cerevisiae*: involvement of two hexose transporters. Mol Cellular Biol 17:5453–5460, 1997.
7. TC Hallstrom, WS Moye-Rowley. Divergent transcriptional control of multidrug resistance genes in *Saccharomyces cerevisiae*. J Biol Chem 273:2098–2104, 1998.
8. JA Prendergast, RA Singer, N Rowley, A Rowley, GC Johnston, M Danos, B Kennedy, RF Gaber. Mutations sensitizing yeast cells to the start inhibitor nalidixic acid. Yeast 11:357–547, 1995.
9. H Nikaido. Structure and functions of the cell envelope of gram-negative bacteria. Rev Infect Dis Jul–Aug 10 Suppl 2:S279–281, 1988.
10. H Nikaido. Porins and specific diffusion channels in bacterial outer membranes. J Biol Chem 269:3905–3908, 1994.
11. H Nikaido, M Vaara. Molecular basis of bacterial outer membrane permeability. Microbiol Rev 49:1–32, 1985.
12. K Lewis. Multidrug resistance pumps in bacteria: variations on a theme. Trends Biochem Sci 19:119–123, 1994.
13. SB Levy. Active efflux mechanisms for antimicrobial resistance. Antimicrob Agents Chemother 36:695–703, 1992.
14. H Nikaido, JA Hall. Overview of bacterial ABC transporters. Methods Enzymol 292:3–20, 1998.
15. L Chen, WG Coleman Jr. Cloning and characterization of the *Escherichia coli* K-12 rfa2 (rfac) gene, a gene required for lipopolysaccharide inner core synthesis. J Bacteriol 175:2534–2540, 1993.
16. PA Sampson, R Misra, SA Benson. Identification and characterization of a new gene of *Escherichia coli* K-12 involved in outer membrane permeability. Genetics 122: 491–501, 1998.
17. A Goffeau. Life with 6000 genes. Science 274:546–567, 1996.

18. FR Blattner, G Plunkett III, CA Bloch, CA Perna, NT Perna, V Burland, M Riley, J Collando-Vides, JD Glasner, CK Rode, GF Mayhew, J Gregor, NW Davis, HA Kirkpatrick, MA Goeden, DJ Rose, B Mau, Y Shao. The complete genome sequence of *Escherichia coli* K-12. Science 277:1453–1462, 1997.

19. SL Wolin. From the elephant to *E. coli*. SRP-dependent protein targeting. Cell 77: 787–790, 1994.

20. ME Cardenas, M Lorenz, C Hemenway, J Heitman. Yeast as model T cells. Perspective Drug Discovery Design 2:103–126, 1994.

21. M Nanhoshi, T Nishiuma, Y Tsujishita, K Hara, S Inui, N Sakaguichi, K Yonezawa. Regulation of protein phosphatase 2A catalytic activity by alpha 4 protein and its yeast homolog Tap42. Biochem Biophys Res Commun 251:520–526, 1998.

22. K King, HG Dohlman, J Thorner, MG Caron, RJ Lefkowitz. Control of yeast mating signal transduction by a mammalian β2-adrenergic receptor and G_s α subunit. Science 250:121–123, 1995.

23. HG Dohlman, J Thorner, MG Caron, RJ Lefkowitz. Model systems for the study of seven transmembrane segment receptors. Annu Rev Biochem 60:653–688, 1995.

24. LA Proce, EM Kajkowski, JR Hadcock, BA Ozenberger, MH Pausch. Functional coupling of a mammalian somatostatin receptor to the yeast pheromone response pathway. Mol Cellular Biology 15:6188–6195, 1995.

25. MH Pausch. G-protein coupled receptors in *Saccharomyces cerevisae*: high throughput screening assays for drug discovery. Trends Biotechnol 15:487–494, 1997.

26. C Klein, JI Paul, K Sauve, MM Scmidt, L Arcangeli, J Ransom, J Trueheart, JP Manfredi. Identification of surrogate agonists for the human FPRL-1 receptor by autocrine selection in yeast. Nature Biotechnol 16:1334–1337, 1998.

27. B Bertin, M Freissmuth, RM Breyer, W Shultz, AD Strosberg, S Marullo. Functional expression of the human serotonin receptor 5HT1A receptor in *Escherichia coli*. J Biol Chem 267:8200–8206, 1992.

28. M Freissmuth, E Selzer, S Marullo, W Shutz, AD Strosberg. Expression of two human β-adrenoceptors in *E. coli*: functional interaction, with two forms of G_s. Proc Natl Acad Sci USA 88:6548–8552, 1991.

29. B Bertin, M Freissmuth, R Jockers, AD Strosberg, S Marullo. Cellular signalling by an agonist-activated receptor/$G_{s\alpha}$ fusion protein. Proc Natl Acad Sci USA 91: 8827–8831, 1994.

30. KM Hahnenberger, M Krystal, K Esposito, W Tang, S Kurtz. Use of microphysiometry for analysis of heterologous ion channels expressed in yeast. Nature Biotechnol 14:880–883, 1996.

31. R Guinea, L Carrasco. Influenza virus M2 protein modifies membrane permeability in *E. coli* cells. FEBS letters 343:242–246, 1994.

32. BL Hall, Z Smit-McBride, ML Provalsky. Reconstitution of retinoid X receptor function and combinatorial regulation of other nuclear hormone receptors in the yeast *Saccharomyces cerevisiae*. Proc Natl Acad Sci USA 90:6929–6933, 1993.

33. DJ Mangelsdorf, C Thummel, M Beato, P Herrlich, G Shultz, K Umesono, B Blumberg, P Kastner, M Mark, P Chambon. The nuclear receptor superfamily: the second decade. Cell 83:835–839, 1995.

34. P Chambon. A decade of molecular biology of retinoic acid receptors. FASEB J 10: 940–954, 1996.

35. BL Hall, Z Smit-McBride, ML Provalsky. Reconstitution of retinoid X receptor function and combinatorial regulation of other nuclear hormone receptors in the yeast *Saccharomyces cerevisiae*. Proc Natl Acad Sci *USA*, 90:6929 6933, 1993.

36. ST Taylor, R Menzel. The creation of a camptothecin-sensitive *Escherichia coli* based on the expression of the human topoisomerase 1. Gene 167:69–74, 1995.

37. BD Kohorn. Isolation of cDNA's encoding proteases of known specificity using a cleavable Gal4 protein. Methods, A Companion to Methods in Enzymol 5:156–160, 1993.

38. TM Block, RH Grafstrom. Novel bacteriological assay for detection of potential antiviral agents. Antimicrob Agents Chemother 34:2337–2341, 1990.

39. JO McCall, S Kadam, L Katz. A high capacity microbial screen for inhibitors of human rhinovirus protease Sc. Biotechnology 12:1012–1016, 1994.

40. RF Balint, I Plooy. Protease-dependent sreptomycin sensitivity in *E. coli*—a system for protease inhibitor selection. Biotechnology 13:507–510, 1995.

41. G Supert-Furga, K Jonasson, SA Courtneidge. A functional screen in yeast for regulators and antagonizers of heterologous protein tyrosine kinases. Nature Biotechnol 14:600–604, 1996.

42. S Fields, OK Song. A novel genetic system to detect protein–protein interactions. Nature 340:245–246, 1989.

43. FJ Germino, ZX Wang, S Weissman. Screening for in vivo protein–protein interactions. Proc Natl Acad Sci USA 90:933–937, 1993.

44. M Vidal, RK Brachmann, A Fattaey, E Harlow, JD Boeke. Reverse two-hybrid and one-hybrid systems to detect dissociation of protein–protein and DNA–protein interactions. Proc Natl Acad Sci USA 93:10315–10320, 1996.

45. BA Ozenberger, KH Young. Functional interaction of ligands and receptors of the hematopoietic superfamily in yeast. Mol Endocrinol 9:1321–1329, 1995.

46. EM Kajkowski, LA Proce, MH Paush, KH Young, BA Ozenberger. Investigation of growth hormone releasing hormone receptor structure an activity using yeast expression technologies. J Recept Signal Transduct Res 17:293–303, 1997.

47. J Zhu, CR Kahn. Analysis of a peptide hormone–receptor interaction in the yeast two-hybrid system. Proc Natl Acad Sc USA 94:13063–13068, 1997.

48. F Tirode, C Malguti, F Romero, R Attar, J Camonis, JM Egly. Conditionally expressed third partner stabilizes or prevents the formation of a transcriptional activator in a three-hybrid system. J Biol Chem 272:22995–22999, 1997.

49. DJ SenGupta, B Zhang, B Kraemer, P Pochart, S Fields, M Wickens. A three-hybrid system to detect RNA–protein interactions in vivo. Proc Natl Acad Sci USA 93: 8496–8501, 1996.

50. S Liberles, ST Diver, DJ Austin, SL Schreiber. Inducible gene expression and protein translocation using nontoxic ligands identified by a mammalian three-hybrid screen. Proc Nat Acad Sci USA 94:7825–7830, 1997.

51. K Young, L Stephen, L Sun, E Lee, M Modi, S Hellings, M Husbands, B Ozenberger, R Franco. Identification of a calcium channel modulator using a high throughput yeast two-hybrid screen. Nature Biotechnol 16:946–950, 1998.

52. WA Catterall. Yeastly brew yields novel calcium channel inhibitor. Nature Biotechnol 16:906, 1998.
53. R Menzel, ST Taylor. Periplasmic membrane-bound system for detecting protein–protein interactions. U.S. patent 5,521,066, May 28, 1996.
54. H Kolmar, C Frisch, G Kleemann, K Gotze, FJ Stevens, HJ Fritz. Dimerization of Bence Jones proteins: linking the rate of transcription from an Escherichia coli promoter to the association constant of REL$_v$. Biol Chem Hoppe-Seyler 375:61–70, 1994.
55. SL Dove, JK Joung, A Hochschild. Activation of prokaryotic transcription through arbitrary protein–protein contacts. Nature 386:627–630. 1997.
56. A Hochschild, SL Dove. Protein–protein contacts that activate and repress prokaryotic transcription. Cell 92:597–600, 1998.
57. M Ptashne, A Gann. Transcriptional activation by recruitment. Nature 386:569–577, 1997.
58. MG Kornacker, B Remsburg, R Menzel. Gene activation by the AraC protein can be inhibited by DNA looping between AraC and LexA repressor that interacts with AraC: possible applications as a two-hybrid system. Mol Microbiol 30:615–624, 1998.
59. G Karimova, J Pidoux, A Ullman, D Ladant. A bacterial two-hybrid system based on a reconstituted signal transduction pathway. Proc Natl Acad Sci USA 95:5752–5756, 1998.
60. F Rossi, CA Charlton, HM Blau. Monitoring protein–protein interactions in intact eukaryotic cells by β-galactosidase complementation. Proc Natl Acad Sci USA 94:8405–8410, 1997.
61. FC Collins, A Patrinos, E Jordon, A Chakravarti, R Gesteland, L Walters, and the members of the DOE and NIH planning groups. New goals for the U.S. human genome project: 1993–2003, 1998.
62. J Huang, SL Schreiber. A yeast genetic system for selecting small molecule inhibitors of protein–protein interactions in nanodroplets. Proc Natl Acad Sci USA 94:13396–13401, 1997.
63. BJ Sedlak. Gene chip technology ready to impact diagnostic markets. Gen Eng News 17:1, 14, 34, 1997.
64. M Leach. Discovery on a credit card? Drug Dis Today 2:263–254, 1994.
65. NH Chiem, DJ Harrison. Microchip systems for immunoassay: an integrated immunoreactor with electrophoretic separation for serum theophylline. Clin Chem 44:591–598, 1998.
66. A Uhlig, E Lindner, C Teutloff, U Schnakenberg, R Hintsche. Miniaturized ion-selective chip electrode for sensor application. Anal Chem 69:4032–4038, 1997.
67. CC Liu, MR Neuman, LT Romankiw, EB Makovos. Microelectronic sensors for simultaneous measurement of PO2 and pH. Adv Exp Med Biol 220:295–298, 1987.

6

Molecular Genetic Screen Design for Agricultural and Pharmaceutical Product Discovery

Donald R. Kirsch, Julia N. Heinrich, Mark H. Pausch, and Sanford Silverman
Wyeth-Ayerst Research, Princeton, New Jersey

William Baumbach
Morphochem, Inc., Monmouth Junction, New Jersey

Margaret Lai, James C. Walsh, and Laura Sarokin
BASF Corporation, Princeton, New Jersey

I. INTRODUCTION

Efforts have continued for many centuries to find substances to treat disease and to combat agricultural pests. The earliest discoveries in both of these fields came about largely by chance, and information regarding these discoveries was then disseminated anecdotally. For example, the medical utility of plant extracts containing cardiac glycosides was initially found by chance observation. Extracts containing these compounds are known to have been used by the ancient Egyptians and the ancient Romans. The first modern description of the therapeutic uses of the most clinically useful cardiac glycoside, digitalis, was written by William Withering in 1785 [1]. Although Withering appreciated the clinical utility of digitalis, the pharmacological basis for the drug's action was not understood at the time, in part because of the serendipitous nature of the compound's discovery.

153

A similar example from the field of agricultural chemistry is the development of Bordeaux mixture as the first effective agricultural fungicide. Although the antifungal properties of copper sulfate had been recognized as early as 1807, copper sulfate mixtures were initially largely used on vines to discourage grape theft [2]. The fungicidal properties of an old copper sulfate preparation, Bordeaux mixture, against downy mildew were first noted in 1882 by Millardet and Gayon. The news of this discovery spread rapidly. Bordeaux mixture was introduced into the United States by 1885.

The discoveries described above were largely accidental. Certainly neither of these inventions nor similar contemporary inventions were based upon a directed research program to find the desired substance. Improvements in the chemical synthesis of organic molecules made it possible by the mid to late 1800s to synthesize molecules that could be exploited as pharmacological agents. Early advances, such as the discovery of the anesthetic properties of ether, were based upon the fortuitous biological activity of relatively simple organic molecules. Other discoveries, such as the development of aspirin, were based on analogue synthesis of plant-derived substances previously known to have therapeutic utility [1]. True large-scale synthesis and screening programs became possible in the twentieth century following major advances in synthetic organic chemistry and the discovery of the therapeutic utility of microbial metabolites.

One classic example of such a chemical screening program is the discovery of DDT. DDT was originally synthesized by Zeidler in 1874 as part of a basic organic chemical synthesis research program. This compound was later rediscovered by Muller and coworkers in 1939 in an early mass chemical synthesis and screening program at Geigy directed toward the identification of improved insecticides [3]. Another example is the mass screening of microbial extracts for antibiotics, which was initiated by several laboratories following the discovery of penicillin by Fleming in 1929 and its later characterization and clinical evaluation by Florey et al. [1]. Since penicillin is derived from a saprophytic fungus, Walksman and coworkers reasoned that degradative soil bacteria might also produce therapeutically useful substances [4]. These workers initiated a program to screen fermentations produced by soil Actinomycetes to identify antibiotics active against tuberculosis which, prior to the discovery of streptomycin by Waksman's group, was a major and largely untreatable infectious disease.

The two preceding examples have much in common. Both discoveries addressed major and highly visible societal needs and were recognized by Nobel Prize citations. In addition, both programs were based upon what are now recognized as traditional screening approaches: the use of whole, laboratory culturable target or surrogate organisms to identify active substances, which were then characterized in a series of increasingly stringent tests (secondary screens) culminating in ''real life'' evaluation (agricultural field or medical clinical trials). While it is currently accepted that most primary screening actives in such programs will not have commercially or clinically useful activity, this screening strategy can be highly successful because of the elimination of undesired molecules via secondary testing. Lastly, novelty was assured in early screening programs because the materials tested had never been previously evaluated for the desired use.

Contemporary screening programs face a number of challenges. It is increasingly difficult to find novel leads from previously tested sources of molecules. New products must show clear superiority to current treatments to be useful and to win commercial acceptance. The development of resistance to established products requires the identification of products acting on novel target sites. Fortunately, new tools are available to meet these challenges. Molecular techniques make it possible to identify biochemical sites of action rapidly and to make early predictions of the usefulness of potential drug targets. Biochemical and molecular biological techniques also make it possible to design screening tests based upon mechanism of action rather than general biological end point targets. Mechanism of action screening increases prospects for novelty and decreases the chance of rediscovering known compounds. Judicious choice of targets increases the prospects for selective compound action and reduces the risk of unintended toxicity. In addition, rapid genome sequencing fully reveals important aspects of biological systems, so that it is no longer necessary to treat a biological system as a "black box."

These tools have been exploited in many ways. This review will focus on one type of screen design approach: the application of molecular genetics to model microbial organisms for screen design. This general strategy has been exploited since the late 1970s and will probably become more valuable with the logarithmically increasing availability of DNA sequence information. The system presented here, the laboratory yeast *Saccharomyces cerevisiae*, is favored for three major reasons: *S. cerevisiae* shows great functional similarity to higher eucaryotes; it has a fully sequenced genome; and it can be manipulated with great efficiency, allowing rapid screen development and analysis of therapeutic targets. Thus molecular-genetic-based screening in yeast provides a direct and important linkage between genes and chemicals to allow rapid exploitation of DNA sequence information for drug discovery.

II. FUNCTIONAL EXPRESSION OF HETEROLOGOUS POTASSIUM CHANNELS IN YEAST

The use of *Saccharomyces cerevisiae* cells as host for heterologous expression of potassium channels was made possible by advances in the understanding of the regulation of potassium ion homeostasis [5,6]. High-affinity potassium uptake in yeast is mediated by a potassium uptake transporter encoded by *TRK1* [7]. *TRK2* encodes a cation transporter with moderate potassium affinity that contributes significantly to potassium uptake [8,9]. Potassium uptake is determined in part by the electrogenic activity of the plasma membrane proton pump Pma1p, which establishes a highly hyperpolarized membrane potential, permitting accumulation of a high intracellular potassium content from growth media containing low potassium ion concentration [6]. Yeast cells lacking high-affinity potassium uptake capacity due to a *trk1Δtrk2Δ* double mutation are incapable of growing on medium lacking sufficient potassium. This conditional phenotype made possible the cloning of potassium channels that supplied potassium uptake capacity.

A variety of potassium channels from heterologous sources have been cloned by complementation of the growth defect of *trk1Δtrk2Δ* double mutant yeast cells. In general the approach to cloning has been the same. Yeast expression vectors were used to construct cDNA libraries that permit production of heterologous proteins. Transformation of *trk1Δtrk2Δ* cells with cDNA expression libraries and plating on low potassium medium permits survival of only those cells containing a cDNA whose product provides high-affinity potassium uptake. The first potassium channels to be cloned by this method were the related plant potassium channels KAT1 and AKT1 [10,11]. These channels, which appear to be composed of six transmembrane domains and a single pore forming domain, were shown by biophysical measurements to confer inward rectifying potassium currents.

More recently, a number of potassium channels from animal species have been functionally expressed in yeast. The *Drosophila melanogaster* ORK1, a unique potassium channel composed of four transmembrane domains and two pore forming domains and the first recognized member of the two pore (2P) potassium channel family [12], was cloned by complementation of the *trk1Δtrk2Δ* defect [13]. Electrophysiological measurements indicate that this channel encodes open rectifier potassium flux capacity. This current-producing property suggests that ORK1 may underlie the long sought-after leak current in myelinated neurons. When measured in yeast cells and *Xenopus* oocytes, the channels retained comparable sensitivity to potassium channel-blocking ions. Due to its localization within the neuromuscular tissues of adult flies, these results are consistent with a role for ORK1 in regulating nerve cell function.

The inwardly rectifying potassium channel IRK1 from guinea pigs is also able to support the growth of potassium uptake-deficient yeast cells [14]. The single transmembrane domain influenza virus M2 nonselective cation channel has also been shown to function in yeast [15]. Expression of the M2 channel in yeast cells interferes with the maintenance of the electrochemical proton gradient, resulting in growth inhibition. This phenotype is reversed by addition of the M2 channel inhibitor amantadine [15]. In addition, yeast cells expressing IRK1 and the M2 channel have been evaluated by microphysiometry [15,16]. The effects of amantadine and BL-1743, another M2 channel blocker, were distinguished using microphysiometer measurements.

In most cases, the heterologous potassium channels expressed in yeast maintained pharmacological responsiveness, suggesting that this configuration may be useful for assembly of high-throughput screening (HTS) assays designed to find small molecule modulators of the potassium channels [17]. Several reports demonstrate the ability of K channel-blocking compounds to inhibit potassium uptake by heterologous channels, thus affecting cell survival [11,13–17]. It is upon this basis that HTS assays designed to identify compounds that modulate the activity of the heterologous potassium channel may be devised. As well, since growth of the potassium-channel-containing yeast cells is dependent on the heterologous protein, genetic approaches to defining potassium channel function may be easily devised. Using such a scheme, the protein structural requirements for KAT1 pore function were assessed [18,19].

Recently, database mining led to the identification of a yeast potassium channel, Tok1p/Ykc1p/Duk1p/York1 [20–23]. *TOK1* appears to encode a protein of eight membrane spanning helices and two pore forming domains. Biophysical measurements indicate that Tok1p represents the previously characterized potassium channels identified by direct biophysical measurement of yeast cells [21,24]. Tok1p mediates outward potassium flux, which is modified by changes in external potassium ion concentration [20]. Deletion of the *TOK1* gene gives rise to no detectable growth phenotype, although in the presence of the yeast plasma membrane H^+-ATPase inhibitor DCCD, a Tok1p-deficient strain exhibits a growth defect [25]. This phenotype might be exploited in the expression of heterologous potassium channels that complement the *tok1* growth defect.

III. CYTOPLASMIC RECEPTORS

Cytoplasmic receptors are ligand binding transcription factors. All except the aryl-hydrocarbon receptor (AhR, a basic helix-loop-helix-PAS [bHLH-PAS] family member) are members of the classical steroid family, which is now part of the nuclear receptor (NR) superfamily (for recent reviews see Refs. 26–29). Because AhR was originally characterized in biochemical studies as a steroid receptor, other putative steroid receptors may actually be bHLH-PAS structural family members. The biochemical features of cytoplasmic receptor ligands are tailor made for drug discovery efforts: they are small lipophilic compounds that average 350 daltons, show exquisite specificity and efficacy, and mediate physiological processes that are vital in arthropods and vertebrates. Steroids and synthetic mimics have medical utility for many conditions including osteoporosis, cancer, and a variety of inflammatory disorders, as well as agricultural utility to enhance production in livestock. The majority of these compounds was discovered by traditional screening approaches. With increasing knowledge of the three-dimensional structure of these receptors and the auxiliary proteins involved in receptor mediated transcription, it is now possible rationally to select drug targets and synthesize ligands (for reviews see Refs. 30–32).

The exploitation of steroid receptor targets in agricultural chemical and veterinary therapeutic discovery has resembled medical discovery strategies, favoring a mechanism-based drug discovery approach. Environmentally favored commercial insecticides, whose mode of action is to interfere with vital biological pathways unique to insects, are referred to as insect growth regulators (IGR). The discoveries of the IGRs RH5849 (subsequently optimized to tebufenozide) and methoprene were serendipitous. Subsequently, it was shown that both are nonsteroid analogues of insect specific steroid hormones: tebufenozide is a bis-acylhydrazine nonsteroid agonist of the ecdysone receptor, and methoprene is a terpenoid, similar in structure to juvenile hormones [33,34]. The ecdysone receptor (EcR) is a member of the steroid receptor family [26]. The *Drosophila melanogaster* methoprene resistance gene (*Met*), which remains after a decade of research as the most likely candidate for the putative nuclear receptor-like component of the juvenile hormone receptor, was recently identified as a bHLH-PAS family member [35,36]. *D. melanogaster* has 18 nuclear receptors [37], six bHLH-PAS family members [36,38–40], and the size of these families may increase several fold with the enlargement of expressed-sequence tag (EST) data bases and complete sequencing of the *Drosophila* genome. In addition, there are a variety of related receptors in other insect species.

The structural features of cytoplasmic receptors (the four classes of steroid/nuclear receptors plus AhR) determine the specific requirements and strategies for screen design. On the primary amino acid level, cytoplasmic receptors have a tripartite structure of independent functional domains. Nuclear receptors (NR) have an N-terminal transactivation domain (TA), a middle DNA binding domain (DBD) and C-terminal ligand binding domain (LBD). AhRs have a similar topology except the TA is moved from the N-to the C-terminus. The DBD of both NR and AhR binds a *cis*-DNA element referred to as a hormone response element (HRE), for which consensus sequences have been identified. The NR superfamily is divided into four groups based on their DNA binding and dimerization properties. Families I and II are ligand binding receptors and families III and IV are orphan receptors. Family I includes the steroid receptors: glucocorticoid receptor (GR), mineralocorticoid receptor (MR), estrogen receptor (ER α and β), progesterone receptor (PR α and β and androgen receptor (AR). In the absence of ligand, these receptors are located in the cytoplasm in association with heat shock proteins (HSPs). The presence of ligand causes the release of the HSP, translocation into the nucleus, binding as homodimers to the HRE, and activation of transcription.

Family II includes the receptors that heterodimerize with the 9-*cis*-retinoid X receptor (RXR) and that bind their cognate HRE in the absence of ligand. Based on their HRE consensus sequence, this family was recently divided into two subfamilies: the original family members, peroxisome proliferator activated receptor (PPAR α, β, and γ), RXR (α, β, and γ), retinoic acid receptor (RAR α, β, and γ), and vitamin D3 receptor, thyroxine receptor (TR α and β), and the newer members whose ligands were recently identified, constitutively active receptor β (CARβ), benzoate X receptor (BXR), pregnane X receptor (PXR), and steroid and xenobiotic receptor (SXR). The classification into families serves only as a guideline, since ER functions as a hybrid of family I and family II receptors.

In family III, the orphan receptors bind the HRE as monomers (e.g., nerve factor induced orphan receptor), and in family IV as dimers (e.g., chicken oval-bumin upstream promoter transcription factor) [26,41,42]. The ability to hetero-dimerize with RXR is a signature feature that an orphan receptor may have a ligand [27]. Novel ligands for orphan receptors are intracrine rather than being endocrine, which may explain why they have been more difficult to identify. The ligand mediated transcriptional activity of some receptors is regulated by a grow-ing list of cofactors that act as corepressors or coactivators. A number of coactiva-tors are histone acetyltransferases (HATs) and members of the bHLH-PAS fam-ily. These cofactors function by recruiting multicomponent complexes that promote chromatin remodeling: corepressors are associated with multiple histone HATS and coactivators with histone deacetylases (HDACs) [31]. Recently, a novel cofactor that is both a corepressor and a coactivator called NR-binding SET-domain containing protein has been identified [43].

AhR shares properties of both families I and II ligand activated NRs. Like family I members, unligated AhR is in the cytoplasm associated with HSP. Upon ligand binding, the HSP dissociate and aryl hydrocarbon receptor nuclear trans-locator (ARNT) associates and transports the complex to the nucleus, where it binds to the xenobiotic response elements (XRE) and activates transcription [29]. ARNT was the first bHLH-PAS member identified and like RXR and the NR family II members, functions as a heterodimeric partner to other bHLH-PAS members. The extent to which NRs and bHLH-PASs interact with each other and share common regulator pathways is being intensely investigated [30,44].

Since yeast has no endogenous cytoplasmic receptors and has a limited metabolic capability, it is a model system for reconstituting ligand mediated cyto-plasmic receptor function. The ligand inducible transactivation activity of all the known mammalian steroid receptors and a growing list of RXR receptors and their heterodimeric partners has been successfully reconstituted in yeast by the classic cis–trans assay (Table 1) [45–53] (for reviews see Refs. 42, 54–58). Yeast is transformed with a receptor expression plasmid(s) and a reporter plasmid driven by hormone-responsive promoter fused to a tractable marker gene. The approaches we routinely use are based on the standard strategies for expression of steroid receptors in yeast [45,54,59] (see Ref. 58 for alternative approaches). The salient feature of the yeast expression vector is that it codes for a ubiquitin-receptor fusion protein, which is cleaved by endogenous yeast protease at the junction to generate an authentic receptor. In addition, either a copper inducible metalothionein promoter is used to regulate the expression level of the receptor or the triosephosphate dehydrogenase promoter is used to obtain constitutive ex-pression. With the AhR system, an expression vector for yeast (pBEVY) that has

Table 1 Cytoplasmic Receptors Expressed in cis/trans Assay in Yeast

Ligand inducible nuclear receptors	Reference
Estrogen receptor alpha	46,47,51–53,57,58
Glucocorticoid receptor	62
Mineralocorticoid receptor	
Androgen receptor	53
Progesterone receptor	45,51,53
Retinoic X receptors (RXR)	164
Vitamin D receptor	
Retinoic acid receptors	164
Oncogenic thyroid hormone receptor (c-erbA)	
Thyroid hormone receptor (c-erbA) and RXR	46
Vitamin D receptor and RXR	
Retinoic acid receptors and RXR	54,164
Peroxisome proliferator-activated receptor and RXR	48,49

Constitutively active nuclear receptors	
Chicken ovalubumin upstream promoter transcription factor (COUP)	46
Hepatic nuclear factor 4	165
Ecdysone receptor (EcR)	59,61
Androgen receptor/EcR chimera and ultraspiracle	61

bHLH-PAS receptors	
Aryl hydrocarbon nuclear receptor	60

a bidirectional galactose inducible promoter was used [60], and the recombinant protein was not a fusion product. Expression vectors with bidirectional promoters are becoming more important because several proteins must be expressed to reconstitute or optimize the activity of some receptors. One note of caution for the use of galactose inducible promoters is that some galactose media are contaminated with bioactive steroid (i.e., estrogen and progesterone [51]).

A frequently used yeast reporter plasmid, YRpC2, has an *XhoI* cloning site in which to insert the HRE upstream of the yeast cytochrome C promoter and *lacZ* reporter gene. The HRE, which typically consists of a duplicate copy of the binding sequence, is critical and may need to be optimized for specific needs. Although *lacZ* is the most frequently used reporter and its expression is quantifiable, it is not a selectable maker. The auxiotrophic markers *URA3* [52] and *HIS3* [46,57] were both used as reporter genes with the ER. The advantage of *URA3* is that its activity is quantifiable, and it can be used as a counter-selectable marker:

the gene product orotidine-5'monophosphate decarboxylase can be measured in a liquid *OMPdecase* assay, and kills yeast by conferring sensitivity to the toxic antimetabolite 5-fluoro-orotic acid (5-FOA). In a Usp antagonist screen, we use the canavanine permease gene (*CAN1*) as a counter-selectable marker [61] (see Table 2). The protease-deficient yeast strain BJ2168 is frequently used for cytoplasmic receptor expression. While in our experience the EcRs and Usp showed comparable activity in BJ2168 and a strain expressing the normal complement of proteases [61], others have observed that the high levels of receptors produced in strain BJ2168 were not always desirable and perhaps were even detrimental for particular assays [54]. Typically, yeast is grown for 4 to 24 h in the presence of ligand to detect reporter gene expression, and β-galactosidase assays are performed in a 96-well format [53,54]. In Usp screen, which employs a *CAN1* reporter, cells are seeded in agar media and test compounds are applied on the surface, typically 144 natural products or up to 576 chemicals per plate.

Whether the cis–trans assay in yeast will work for a particular receptor is not predictable. Basic receptor research has demonstrated that "adapter proteins" are sometimes required for proper expression. For example, TR, RAR, VDR, ER, and RXR, which all function in yeast, require TR interacting protein (TRIP1) for their ligand activating abilities. However, the yeast *SUG1* gene (suppressor of gal4D lesions) is the homologue of mammalian TRIP1 and is able functionally to replace TRIP1 [42]. In screens to evaluate endocrine modulators, RSP5 was overexpressed with PR and SPT3 with AR. The rationale for this approach is that the overexpression of the human homologues RPF1 and TAF18 in mammalian systems enhanced transcriptional efficiency without altering potency or specificity [53]. It is speculated that *SUG1* and RSP5 act by affecting the turnover of receptors rather than mediating agonist [52]. RXR, VDR, and the steroid receptors are functional in yeast without the additional coexpression of corepressors like TR and the RAR associated corepressor (TRAC), which interact with TR and RAR and do not appear to have a yeast counterpart. Compounds that act as antagonists in mammalian cells function as partial agonists in yeast (tamoxifen

Table 2 *S. cerevisiae* Counterselectable Markers

Marker	Gene product	Selective condition	Host cell requirement
CYH2	ribosomal protein L28 (yeast L29)	cycloheximide	$cyh2^R$
CAN1	arginine permease	canavanine	Arg$^+$
URA3	orotidine-5'-phosphate decarboxylase	5-fluoro-orotic acid	*ura3*
LYS2	α-amino adipate semialdehyde dehydrogenase	α-amino adipate	*lys2*
FPR1	FKBP12, FK-506 binding protein	FK-506 or Rapamycin	*fpr1*, Tor$^+$

for the ER, RU486 for the PR, and Ro 41-5253 for the RARα) potentially because yeast lacks these cofactors [31,42,52]. Therefore in cases where receptors do not function, it may be necessary to coexpress additional factors. Another important feature is that some receptors are covalently modified (i.e., yeast supports the phosphorylation of the glucocorticoid receptor [62]) and therefore the appropriate post-translational modification enzymes must be expressed to obtain ligand-dependent activation of receptors.

The *D. melanogaster* ecdysone receptor (EcR) is an example of a receptor that has not been shown to support ligand mediated transcription in yeast. When expressed in yeast, EcR is a potent transactivator [59]; when it is coexpressed with Usp, it is not ligand inducible. (However, it is not known whether EcR ligands diffuse into the cell.) Yeast extracts with EcR and Usp show specific high-affinity binding of the ecdysone analogue ^{3}H-Ponasterone A [59], indicating that the complex is able to bind the cognate ligand. A chimera between the EcR LBD and the androgen receptor (AR) is transcriptionally silent (AR/EcR), and the "reverse" chimera (EcR with the AR LBD [EcR/AR]) is transcriptionally inducible by testosterone [61]. An interesting observation is that coexpression of this AR/EcR with Usp, which itself is transcriptionally silent, results in reporter gene induction [61].

A screen was designed based on this Usp specific activity to identify small molecule inhibitors of Usp. Usp is an essential gene that functions in many stages of insect development, probably as a "master regulator" [61]. Usp is the homologue of the vertebrate RXR, which is thought to regulate at least eleven different biological pathways [27]. An inhibitor for Usp activity would potentially be lethal to insects and resemble IGR insecticides. The screen employs the yeast canavanine permease, which is the sole means of entry for canavanine, a toxic analogue of arginine, as the reporter gene. An inhibitor of Usp should permit the yeast screening strain to grow in the presence of canavanine. Moreover, this screen is capable of detecting enhanced killing or rescue and therefore can be used to identify agonists as well as antagonists [61].

Yeast-based assays have been used to make a number of important findings. (1) 9-cis RA synergizes with 3,5,3-triiodo-L-thyronine to activate both heterodimeric partners in TRβ:RXRγ, but not in TRβ:RXRα, where RXR is a silent partner (the role all isoforms show in mammalian cells) [42]. (2) When the RXR antagonist LG100754 binds RXR in the heterodimeric RXR/RAR complex (performed using the two-hybrid system), RAR is activated exactly as it would be by its own ligand [63]. (3) CARβ receptor is deactivated by androsanol (also performed using the two-hybrid system), which makes it the first nuclear receptor to be inversely regulated by ligands [27]. (4) Tryptophan and indole compounds are endogenous ligands of the AhR receptor [60].

Future needs for high-throughput screening in this area include the following: (1) Yeast-based steroid receptor screens, i.e., estrogen receptor, progesterone receptor, and testosterone, may be used routinely to ensure the safety of our environment and health [53]. (2) To determine the physiological relevance of receptor isoforms it will be necessary to test all the possible heterodimeric partners, different HREs, and combinations of ligand treatments. For RXR this was calculated to include half a million assays [42]. (3) We need to screen large synthetic chemical and combinatorial libraries that are based on rational drug design, i.e., compounds that are designed to fit into the ligand binding pocket and to target the HATs and HDACs of cofactors [32]. (4) We need to compare the affinity of compounds to receptors of various pest species as opposed to model systems such as *Drosophila*. (5) We need rapidly to evaluate novel receptor targets that may be available from the databases, e.g., the EST database, which was the original source of the pregnane activated receptor [27]. (6) We need to find ligands for the growing family of bHLH-PAS proteins (the dioxin receptor is considered the first member of a large receptor family).

IV. FUNCTIONAL EXPRESSION OF G PROTEIN–COUPLED RECEPTORS IN YEAST

The superfamily of G protein–coupled receptors (GPCRs), characterized by a similarity in structure consisting of seven transmembrane domains, bind a wide variety of ligands that range from small biogenic amines and lipids to large complex proteins [64]. Upon ligand binding a conformational change occurs in the GPCR, leading to activation of heterotrimeric G proteins via a catalytic exchange of GTP for GDP on the α subunit and dissociation of the α subunit from the $\beta\gamma$ complex [65]. The free α subunit and the $\beta\gamma$ complex modulate the activity of a variety of effector proteins, resulting in alterations in second messenger molecules or alterations of cell physiology and/or signal transduction that lead to the cellular response. These effector proteins include adenylyl cyclase, phospholipase Cβ, G protein–coupled Ca^{2+} and K^+ channels, phosphatases, sodium/hydrogen exchangers, and the mitogen activated protein kinase (MAP kinase) signal transduction pathways. First recognized in mammalian cells, GPCR mediated signal transduction pathways have functional homologs in evolutionarily distant organisms like insects, nematodes, plants, and yeast. Recent studies demonstrate that yeast is likely to be a useful model system to study components of the GPCR signaling pathways because of the high level of conservation between the elements of the yeast pheromone response pathway and mammalian GPCR-coupled MAP kinase signaling systems. Advances in GPCR expression in yeast and coupling to the pheromone response pathway suggest that this approach may be particularly useful in examining aspects of structure and function [66]. Furthermore, haploid yeast cells can be altered by the introduction of specific mutations and reporter gene constructs that make them useful as host cells for the development and implementation of sensitive HTS assays designed to identify novel ligands that modulate the activity of the GPCRs [66–68].

A. The Yeast Pheromone Response Pathway

Haploid *S. cerevisiae* cells detect the presence of cells of opposite mating type through binding of peptide mating pheromones to G protein–coupled receptors. Thus, **a** cells secrete **a** factor and express the α factor receptor, Ste2p, while α cells secrete α factor and contain the **a** factor receptor, Ste3p [69,70]. When **a** cells and α cells come into close proximity, mating pheromone is detected by receptors, which initiates the mating process by activating intracellular hetero-trimeric G proteins. Dissociation of the α subunit, Gpa1p, from the complex of β (Ste4p) and γ (Ste18p) subunits allows the βγ complex to activate downstream elements of the pheromone response pathway. Ste20p, a p20 activated kinase (PAK) homolog, stimulates a MAP kinase cascade that consists of the sequential activation of Ste7p (MAP kinase kinase or MEK), Ste11p (MEK kinase), and the MAP kinases Fus3p and Kss1p [71,72]. Upon activation of the pathway, cells undergo a series of changes that prepare the yeast cell to mate with a cell of the opposite mating type. These changes include cell cycle arrest, activation of transcription of pheromone-responsive genes, and formation of mating-related cell structures.

Elements of the yeast pheromone response pathway are remarkably similar to mammalian GPCR signaling systems. This similarity has proved useful for analysis of mammalian GPCRs and G proteins, since yeast GPCRs and G proteins may be functionally replaced with their homologous mammalian counterparts. The yeast system permits analysis of these proteins in isolation, which is not possible using other expression systems. It is predicted that at least 400 GPCR genes may be present in the human genome, and this estimate is as high as 1000 or more if odorant and pheromone receptors are included [73]. Since only about 300 GPCRs have had their cognate ligands identified, a large number of orphan GPCRs for which cognate ligands are not known remain to be characterized. The yeast system can be a valuable tool for the analysis of these receptors, both to study their structure and function and to identify ligands that modulate their activity.

B. GPCR Expression in Yeast

Early studies indicated that heterologous expression of GPCRs in *S. cerevisiae* and other fungal cells resulted in the presence of functional antagonist binding sites in membrane fractions. King et al. reported that the activation of a hetero-logous mammalian β-adrenergic receptor expressed in yeast could be coupled to the pheromone response signal transduction pathway, suggesting that this yeast expression approach could be successfully employed [74]. The coupling was dependent on coexpression of a mammalian Gαs protein in yeast cells lacking the endogenous G protein α subunit, Gpa1p. Binding of the β-agonist, isoproterenol, resulted in activation of the pheromone response pathway, including expression of a pheromone-responsive reporter gene, apparent cell-cycle arrest, and formation of mating specific cell structures.

A number of alterations introduced into the yeast expression system were required to increase its usefulness and flexibility in HTS applications and allow for genetic selections in the presence of agonist [66,75,76]. The terminal cell-cycle arrest response of haploid yeast cells to mating pheromone was eliminated by deletion of the *FAR1* gene, which encodes a negative regulator of G1 cyclins and is thought to serve as the primary interface between the pheromone-response pathway and the cell-cycle regulatory machinery [77,78]. Agonist stimulation of *far1* mutant cells results in activation of the pathway and transcription of pheromone-responsive genes without affecting the cell's ability to grow and divide. A second important modification was introduced to enable yeast cells to grow only in response to an agonist. A pheromone-responsive reporter gene was constructed by placing the gene encoding His3p, an enzyme required for histidine biosynthesis, under the control of the pheromone induced *FUS1* promoter [79]. Hence *his3 far1* mutant yeast cells will grow on media lacking histidine only when agonist is applied to the cells and the GPCR is activated (Fig. 1).

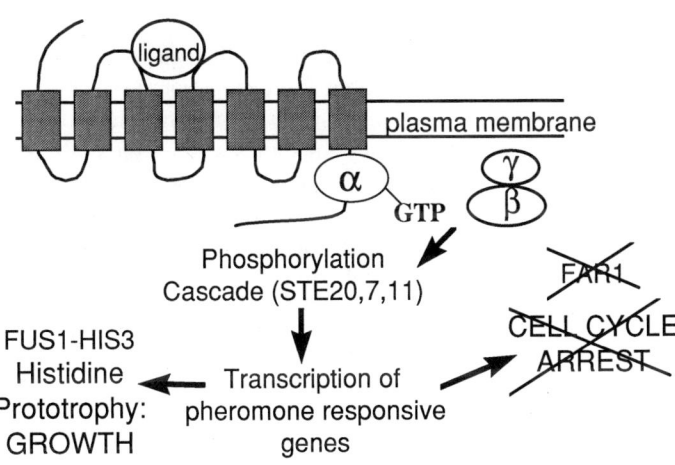

Figure 1   Schematic of the GPCR signaling pathway in engineered yeast cells used for high-throughput screening. α, β, γ: yeast tripartite G protein.

Additional changes have been made to improve the sensitivity of the expression system, including the elimination of desensitization pathways that promote recovery from cell-cycle arrest by reducing the signal transmitted through the pheromone-response pathway. One desensitization pathway, which is induced in response to chronic pheromone stimulation of Gpa1p, allows cells to adapt and continue to grow in the presence of pheromone. This response is mediated by Sst2p (supersensitive), a member of the RGS (regulator of G protein signaling) family of GTPase activating proteins that play an important role in the desensitization of GPCR signaling pathways [80]. Yeast cells lacking Sst2p exhibit pheromone hypersensitivity and are unable to recover from pheromone induced cell-cycle arrest. A second desensitization response, initiated by the pheromone receptors themselves, acts via a poorly understood mechanism to reduce agonist induced signaling [81,82]. In yeast cells optimized for HTS, deletion of the *sst2* and *ste2* genes in *MAT*a cells serves to increase greatly the sensitivity of the yeast cell response to GPCR agonists.

C. HTS Applications

The state of the art in pharmaceutical drug discovery requires mechanism-based screening assays of high selectivity, sensitivity, and throughput. This is achieved by using cloned gene targets in a robust and miniaturizable system with low background (i.e., a high signal-to-noise ratio). Given these criteria, yeast strains that functionally express heterologous GPCRs are ideal for HTS applications. The diversity of GPCRs successfully expressed in yeast so far indicates that the technology will have applicability to a broad range of therapeutic targets. Beyond this, the yeast system has also proven to be flexible with regard to important practical considerations in pharmaceutical drug discovery. For instance, yeast screening assays can be performed on agar plates or in microtiter trays (liquid format), each with specific advantages. Test compounds spotted on yeast cells imbedded in agar will diffuse radially, thus effectively displaying a response over a large concentration gradient. On the other hand, liquid assays can be performed robotically in 96-well or higher formats with fixed test compound concentrations. Another practical consideration is the ability of a screen to test compounds accurately from different sources: organic chemicals dissolved in solvents to natural extracts to synthetic combinatorial libraries. Finally, screens must be designed to identify antagonists vs. agonists at a particular target site. The choice of reporter genes in the yeast GPCR system has broadened its utility for such considerations. For instance, the *HIS3* reporter gene cannot be used in screening natural products, many of which contain histidine. Here, an antibiotic resistance reporter gene (e.g., G418[R]) would be used [66]. To screen for GPCR antagonists, a *CAN1* (canavanine sensitivity) reporter induces a toxic response to an agonist until blocked (rescued) by an antagonist [66].

An example of a successful yeast GPCR assay is the adenosine A_{2a} receptor assay [76]. For this target, with potential uses for agonists in both agriculture and medicine, known purine compounds were difficult and expensive to synthesize. Using an agar plate assay, the A_{2a} receptor expressed in yeast (coupled to the endogenous G_α subunit, Gpa1p) was quickly screened at low cost against a conventional compound library. Of 55,000 compounds tested, 44 hits (0.08%) were retested, of which 12 (0.02%) were positive in a radioligand competition binding secondary assay, indicating that the active compounds bound to the receptor at the known agonist binding site. Among these were nonpurines with submicromolar binding constants, and up to 100 fold selectivity for A_{2a} vs. A_1 receptors. In this screen, the rapidity of screening (<1 week) and its low cost (about 1 cent/sample disposables; <10 cents/sample labor/overhead) contributed to the program's success.

Another successful GPCR yeast assay is the somatostatin subtype 2 receptor (SST$_2$) assay. Here antagonists, with potential uses in both agriculture and medicine, were sought. Somatostatin (SRIF, a 14 amino acid peptide hormone) is inhibitory, causing reduced cAMP levels via interaction with $G_{\alpha i}$. Follow-up assays useful for demonstrating the effect of somatostatin analogs, e.g., inhibition of cAMP accumulation, are difficult to perform because the system must be artificially stimulated before SRIF activity can be detected. Since the yeast assay responds directly to SRIF, agonists can be measured directly. Thus, antagonists were efficiently detected using agar plates containing 10 nM SRIF, where zones of growth inhibition were measured [75,83]. This assay proved to be useful in a conventional analog program, where small, subtype selective peptides were tested individually for agonist and antagonist potency [83]. The yeast assay was also very powerful for screening complex mixtures that required extreme sensitivity to detect activity. A synthetic combinatorial random peptide library containing 160,000 peptides per sample would be virtually impossible to test using conventional assays due to lack of sensitivity. Using the yeast assay, however, a combinatorial library of this kind was successfully screened in a stepwise, iterative fashion. The first round of screening resulted in faint, but discernible, zones of inhibition for antagonists. Each successive round of screening gave incrementally stronger signals, and served to further define the structure of a lead peptide. The final result of these studies was a novel antagonist peptide that showed potent effects in vitro and in vivo [84].

These approaches can be implemented for the increasingly large number of heterologous GPCRs that couple directly to the pheromone-response pathway through the yeast G protein alpha subunit. The coupling of receptors that do not functionally interact with the yeast α subunit will be facilitated by coexpression of cognate mammalian $G\alpha$ proteins. In addition, the $G\alpha$ proteins can be modified by the introduction of mutations that improve functional coupling to the GPCR and $\beta\gamma$ subunit complex. This type of analysis is facilitated by the recently determined crystal structure, which can be used to identify domains of the protein and individual residues that are critical for interaction [85,86].

Agonists that interact with orphan GPCRs can also be identified using assays based on a growth phenotype. Yeast strains expressing orphan GPCRs can be screened with selected agonists, compounds from chemical files and/or combinatorial libraries, to identify surrogate ligands that allow cell growth on selective media. This approach was used to identify an agonist for edg-1, an orphan GPCR thought to be a member of the lysophosphatidic acid (LPA) receptor family [87]. Ligand binding specificity was difficult to demonstrate in cultured mammalian cells due to the ubiquitous presence of LPA receptor subtypes. In this case a yeast expression system was used to demonstrate that phosphatidic acid acts as a high-affinity agonist. This analysis was made possible because of the distinct advantage of expressing a single GPCR subtype in a yeast cell and the ability to couple the receptor to the yeast pheromone-response pathway. As a potentially useful alternative to screening compounds applied to the cell, yeast expression libraries designed for secretion of random small peptides were constructed. The plasmid library is expressed in yeast along with a GPCR to identify cells that express agonists or antagonists of the receptor being expressed. An autocrine loop is established that results in the growth of cells that express an active peptide [88]. Using this scheme, novel agonists and antagonists of the yeast α-mating pheromone receptor [88] and surrogate ligands for the orphan GPCR, FPRL1 were identified [89].

D. GPCR Analysis and Screening with the Yeast Two-Hybrid System

Certain classes of GPCRs, including secretin and growth hormone releasing hormone receptors, possess ligand binding determinants in a large N-terminal extracellular domain that can be used in the yeast two-hybrid system to examine GPCR/ligand interactions. The interaction of the GHRH receptor N-terminal domain with GHRH was evaluated using this system by fusing the complete N-terminus of the human GHRH receptor to one half of the two-hybrid Gal4p protein, and fusing GHRH to the other half [90]. In the two-hybrid system, the expression of a reporter gene that allows growth on selective media occurs only when a protein–protein interaction is formed (see section below). The protein–protein interaction formed between the GHRH receptor domain and GHRH was sufficient to promote growth of yeast cells on selective media, and this interaction was disrupted when specific mutations known to interfere with GHRH binding were introduced into GHRH [90]. This approach may be extended to other members of the secretin class of GPCRs and provides a potentially useful alternative method for investigating receptor–ligand interactions, as well as for high-throughput screen design.

E. Genetic Analysis of GPCRs Expressed in Yeast

The heterologous yeast expression system for GPCRs can also be used to examine structure–function relationships that are difficult to study genetically using other systems, including elucidation of ligand binding sites and GPCR interactions with heterotrimeric G proteins, agonist activation of GPCR activity, and the response to surrogate agonists. Mutations in GPCRs or their signaling pathways can be identified using genetic selections based on the growth phenotype. Constitutively active and dominant-negative mutants of the yeast α factor receptor were identified using genetic approaches that could be employed to identify mutants of heterologous GPCRs [91,92]. In addition, the yeast expression system was used to identify amino acid residues involved in melatonin receptor activation of heterotrimeric G proteins [93]. Similar genetic approaches should be useful for the dissection of the interactions within the heterotrimeric G protein complex, with RGS proteins, and with downstream effector enzymes. Yeast cells also express adenylyl cyclase (CYR1), phospholipase C (PLC1), high-affinity potassium uptake transporter (TRK1), and potassium channel (TOK1). In principle, entirely synthetic signal transduction pathways can be constructed in yeast cells by complementing conditional phenotypes with the corresponding mammalian genes.

V. MACROMOLECULAR INTERACTION TARGETS: YEAST TWO-HYBRID SYSTEMS

A therapeutic target is most often envisioned to be a receptor or an enzyme, where the therapeutic agent constitutes a surrogate agonist, antagonist, or active site inhibitor. However, targets can also be proteins that work in concert with other factors in a complex, in a cascade where other enzymes are the target of enzyme activity, or as integral membrane components transducing a signal. The therapeutic agent can then be viewed as a small molecule that leads directly or indirectly to alterations in protein–protein interactions. Can small molecules that affect interactions of large proteins be detected? Nature offers many examples of such molecules. Benzimidazoles, among other antifungal and antitumor compounds, exert their action by disruption of tubulin protein assembly [94]. Small molecules can also promote association of proteins; for example, taxol and related compounds have this effect on microtubules [95]. Binding of the drugs cyclosporin, FK-506 [96], and Rapamycin [97] to their corresponding immunophilins facilitates the interaction with target proteins.

Yeast two-hybrid technology [98] has evolved into a standard method for detecting protein–protein interactions for gene discovery. It was realized early on that the same methodology could be applied to detect a variety of intracellular interactions and, by extension, for compound discovery. Induction of dimerization [99,100], small molecule–protein interactions [101], and disruption of known protein–protein interactions [102,103] have been demonstrated. In addition, higher order protein interactions [104], as well as protein–RNA [105,106] and protein–DNA [107,108] interactions can be studied with two-hybrid or similar techniques. A recent example of a small molecule inducing protein–protein interactions is relevant to receptor surrogate agonist discovery. In a screen for activators of granulocyte-colony-stimulating factor receptor, a small nonpeptide molecule that presumably mimics the receptor dimerization characteristics of the peptide hormone was detected [109]. Alarcón and Heitman [102] showed that reporter expression induced by hybrid FKBP12 (Fpr1p) and aspartokinase (Hom3p) interactions can be reversed by application of FK-506 to the cells. Young et al. [110] identified small molecule inhibitors of human N-type calcium channels by screening for disruption of hybrid proteins encoding portions of the α1B and the β3 subunits of the channel.

A large number of variations on the two-hybrid theme have emerged in recent years, reviewed in Refs. 111 and 112, and any of these systems could be amenable to compound discovery. One consideration for choosing among these technologies is the characteristic of the compound(s) one desires to detect. If the drug target normally exists in the nucleus, then two-hybrid or one-hybrid screens will be appropriate. But examples exist for nuclear two-hybrid interactions where one might not have expected success. For example, plasma membrane receptors and their cognate soluble ligands have been shown to interact in this system [113]. If there is concern that the yeast nucleus is a poor environment for the particular target interaction, there are a number of techniques that do not depend on transcription for a detectable output. Alternative methods exist for detection, and disruption, of interactions occurring in the cytoplasm [114] or in membranes [115,116]. The latter may be important for discovery of a surrogate ligand for a cell surface receptor where cell permeability of such a compound is not expected, or is not desirable. Alternative protein interaction screens include mammalian and bacterial two-hybrid systems [117–119] and a recent report using fluorescence resonance energy transfer between green fluorescence protein (GFP) fusion proteins [120].

One of the most important considerations for screen design, whether of the two-hybrid type, the conventional enzyme-based type, or other technology, is the method for detection of a hit. Microbial and mammalian cell-based protein interaction screens have been designed using many available reporting systems. The ease of genetic manipulation of yeast allows one to take advantage of a variety of reporting systems as simple as rescue of growth inhibition. Common fluorescence assays, designed for high sensitivity, can also be used in the yeast systems [121,122]. In addition to the reporter itself, the sensitivity of the transcriptional (or other) inducing system should be such that weakly or moderately active compounds will still be detected. See Ref. 123 and the review by Golemis and Brent for a discussion of this topic [124].

One of the most useful reporting systems one can take advantage of is the technique of screening for disruption of interactions by "reverse" two-hybrid techniques [125–127]. These systems rely on the well-known genetic methodology of using counter-selectable markers (Table 2). Using interaction trap and other similar two-hybrid target discovery protocols puts one only a step away from having a workable screen. After validation of the interaction, cells are transformed with a selectable or counter-selectable marker and the screen is ready to run.

The main utility of the yeast two-hybrid technology has been in gene discovery. In the arena of drug screening, this usually means target discovery. But in a broader definition of therapeutics, including gene therapy or transgenic crop generation, protein interaction screens are a direct way of discovering the therapeutic gene.

VI. ANTIFUNGAL AND ANTIBACTERIAL SCREENS

Some of the earliest examples of molecular genetic screen design come from infectious disease research. Most of the current anti-infective agents in clinical use are fermentation products. In contrast, most other therapeutics (with the possible exception of antitumor agents) are based upon chemically synthesized compounds or mammalian hormones. Rediscovery of known antibiotics is a major issue in natural products screening because of the substantial effort required to purify and characterize each active. One indication of the magnitude of this problem is the Kitasato Institute microbial chemistry database, which lists over 16,000 distinct biologically active chemical substances that have been identified from natural products fermentations. This challenge led to the early development of a number of mechanism-based assays to identify selectively rare and novel low-toxicity antibiotics acting on selective targets. Screens have been developed to identify compounds acting on the targets of virtually all antimicrobial agents in clinical use or under clinical evaluation (Table 3), and these approaches have been discussed in a number of excellent reviews [128–134]. This section will therefore only review representative new developments in this field.

Many screens for antibacterial agents have been developed by exploiting the induction properties of antibiotic resistance genes including screens for β-lactam-like, tetracycline-like, and erythromycin-like compounds [135–137]. New screens are developed as additional regulated antibiotic resistance genes are identified. The *vanA* gene cluster was recently used by three groups to design screening assays for cell wall acting antibiotics [138–140]. While these assays have

Table 3 Screen Designs for Detection of Antimicrobial Agents

Anti-infective type	Target	Reported screen design(s)[a,b]
Sulfonamide/trimethoprim	folate metabolism	SS
Quinolone	DNA gyrase	GI
Beta-lactam	bacterial cell wall biosynthesis	GI
Aminoglycoside	bacterial 30S ribosome subunit	R
Tetracycline	bacterial 30S ribosome subunit	GI
Chloramphenicol	bacterial 50S ribosome subunit	GI
Erythromycin/clindamycin (MLS)	bacterial 50S ribosome subunit	GI
Vancomycin (glycopeptide)	bacterial cell wall biosynthesis	GI and PR
Polymyxin B	bacterial cell membrane	GI
Bacitracin	bacterial cell wall biosynthesis	GI
Daptomycin	lipoteichoic acid biosynthesis	PR
Amphotericin B (polyene)	fungal cell membrane	S
Imidazole/triazole	lanosterol 14α-demethylase	SS and GI
Naftifine/tolnaftate	squalene monooxygenase	SS and GI
Echinocandin	$\beta(1-3)$ glucan synthase	E
Benzimidazole carbamate	microtubule assembly	R and SS

[a] SS = supersensitive/resistant pair; GI = gene induction; R = rescue; PR = physiological response; S = sensitive/resistant pair; E = enzyme inhibition.
[b] See Refs. 128–134 for details.

the potential to identify new and useful cell wall acting antibacterial agents, the usefulness of these screens is compromised because *vanA*, unlike many other antibiotic resistance factors, is not induced by a structural feature of glycopeptide antibiotics but is instead induced following damage to the cell wall and plasma membrane. Screens for *vanA* inducers will therefore identify a fairly broad catalogue of compounds including both inhibitors of cell wall synthesis and agents that act directly on the plasma membrane. Additional techniques are needed to identify individual compounds of interest from among a selected group of actives.

Related approaches have been taken to design screens for antifungal agents. Although transcriptionally or translationally regulated antibiotic resistance genes have not been reported in fungi, transcriptionally regulated yeast genes have been exploited in the design of a number of screening systems. Two groups have recently characterized the transcriptional regulation of *ERG3*, a gene encoding an enzyme in the later portion of the ergosterol biosynthesis pathway [141,142]. These groups showed that the transcription of the *ERG3* gene is regulated by cellular ergosterol levels and that mutations and drugs that inhibit the synthesis of ergosterol lead to the up-regulation of this gene. Taking a parallel approach, a reporter fusion incorporating promoter elements from the *ERG11* gene (encoding lanosterol 14α-demethylase) was used to design a screen for antifungal sterol biosynthesis inhibitors [143]. The *ERG11* gene is induced following treatment with sterol biosynthesis inhibitors, and induction specificity by such agents is increased by the introduction of a mutation in *hmg1* (the major locus encoding HMG-CoA reductase) into the genetic background of the screening strain. This assay shows reasonably good specificity for ergosterol biosynthesis inhibitors and detects these compounds with high sensitivity at sublethal concentrations.

Cell wall biosynthesis has for many years been a target of great interest for the discovery of antifungal agents. Many in vitro screening approaches have been described to identify compounds that inhibit fungal cell wall biosynthesis [144]. Some of these screening strategies are general and attempt to identify fungal cell wall biosynthesis agents irrespective of enzyme target. In one recent example of this approach, Zaworski and Gill [145] developed a screen in which cell wall acting agents can be identified using yeast cells expressing a cytoplasm-localized reporter enzyme that is released from cells with damaged cell walls following an osmotic shock.

Another approach is to design screens to identify inhibitors of specific enzymes or isozymes required for cell wall biosynthesis. A number of groups have studied chitin synthase genes from pathogenic fungi such as *Candida albicans* [146–148] and *A. fumigatus* [149–151] to determine the role of specific chitin synthase isozymes in pathogen viability and pathogenicity. Analogous studies have investigated the role of genes encoding glucan synthase subunits in *C. albicans* viability and in vivo drug resistance [152]. Although it is clear that there is one critical enzyme target in glucan biosynthesis (Fks1p), the accumulated evidence suggests that there may not be a single critical chitin synthase target. Therefore, it may be necessary to use a combination of chitin synthase inhibitors to cover all critical targets or to design agents with little or no isozyme specificity. One recent system to find novel isozyme specific chitin synthase inhibitors utilizes genetically engineered deletion mutants of *S. cerevisiae* [153]. A pair of tester strains was developed that expresses a single functional chitin synthase isozyme to support viability. Compounds with an isozyme selective action will inhibit only one of the two strains. In addition, this report exploited the observation that cells expressing Chs3 isozyme activity are calcofluor white sensitive while cells lacking the Chs3 isozyme are calcofluor white resistant. Calcofluor white resistance is also observed following treatment with a Chs3 isozyme inhibitory compound such as nikkomycin, providing a plate assay for such compounds.

While much attention has been paid to chitin synthesis, there have been very few studies targeting screens that exploit chitin degradation. Normal fungal cell growth results from a balance between the synthesis and the degradation of cell wall polymers. In *S. cerevisiae*, the chitin hydrolyzing enzyme chitinase was reported to play an important role in cell separation during growth [103]. Therefore, inhibitors of cell wall degradation as well as synthesis could potentially be useful to control fungal infections. Allosamidin, an insect chitinase inhibitor, was discovered in a screening program for insect growth regulator insecticides [154] and was reported as the inhibitor of *C. albicans* and *N. crassa* chitinases [155,156]. Silverman [157] designed a yeast-based in vivo screen to detect compounds that inhibit a hydrolytic action on the chromogenic substrate, methylumbelliferyl triacetyl chitotriose, but are not toxic to the *S. cerevisiae* cells. The use of this chromogenic substrate makes possible the use of the screen target enzyme as the assay reporter. Allosamidin serves as a high-potency positive control for the assay.

Although this section has focused on in vivo screens, some of the most significant recent progress in the discovery of novel antifungal agents has apparently been made via in vitro screening approaches. Recently, sordarins [158] were discovered in a high throughput in vitro assay for yeast protein synthesis inhibitors. Although the sordarins act on elongation factor 2 (EF-2), which is found in both fungal and mammalian cells, the compounds selectively inhibit protein biosynthesis in fungi. The isolation and characterization of three fungal selective sphingolipid synthesis inhibitors, khafrefungin, rustmycin, and galbonolide B, which act on inositol phosphoceramide synthase, was recently reported [159–161]. The screen used to discover these compounds was not reported, but an in vitro microtiter assay for sphingolipid synthesis was described.

VII. SCREEN IMPLEMENTATION CHALLENGES AND SOLUTIONS

A. Agar-Diffusion Microbial Assays

Liquid or cell-free assays have become very popular in the pharmaceutical arena for the purpose of drug discovery, particularly due to their adaptability to automation, miniaturization, and hands-off data collection and management. However, microbial-based agar diffusion assays provide a great deal of information about a sample that is not possible to obtain with liquid-type assays. The effect of a large range of concentrations of a compound on the test organism can be assessed with a single application, and assay sensitivities in the low nanogram range can be obtained with good reproducibility. In addition, sample activity can be evaluated despite the presence of toxic effects that can mask potential activity when conducted in a liquid assay. Further, contamination from any interfering organism(s) can be readily detected and scoring judgments made accordingly. A disadvantage associated with agar diffusion assays is the need to collect and analyze data manually unless some form of sophisticated image analysis is available. Thus data management for agar-based assays is not as efficient as with liquid-based assays where instrumentation can provide rapid data collection, analysis, and interpretation.

B. High-Density Agar Spotting Techniques for HTS

A variety of sample sources can be used to find potential leads, and the source and characteristics of the samples to be tested will dictate how they are handled.

1. Synthetic Compounds

Synthetic compounds and extracts can be robotically prepared, usually by a central weighing facility. Typically a workstation measures the weight of a sample dispensed into a test tube and calculates the proper solvent volume to add for a desired concentration. Because good solubilization is important for best results, methods to homogenize insoluble samples should be included in any laboratory workstation performing this function. Robotic systems with integrated compound weighing, dissolution, and microplate distribution are particularly useful for this laborious and repetitive task. Typically, master microtiter racks are created, and daughter plates are prepared based upon the individual needs for each screening group. Using automated pipetting devices or liquid handling systems, a specific amount of sample is distributed into microtiter plates having the well density of choice, and samples can be dried by allowing the solvent to evaporate in a fume hood. The dried sample plates can then be distributed to different screening areas without fear of compound spillage. A number of commercially available robotic systems both large and small are available for compound dissolution and storage.

High-throughput agar-based assays can be readily performed with high efficiency. Large bioassay plates containing agar seeded with an appropriate recombinant test organism such as *S. cerevisiae, E. coli,* or filamentous fungi can quickly be prepared [156]. Use of these bioassay plates has the advantage of allowing a large number of samples to be tested with one batch of agar, thus minimizing variation in the test organism seed across plates. Both single and multiple plate sets can be prepared according to the particular target of interest. The large capacity assay plates also have the added benefit of allowing the addition of controls outside of the sample array.

Historically, both small and large bioassay plates have been used in industry to create a matrix of sterile paper disks upon which samples were dispensed, typically not exceeding 20 µL per disk, because greater amounts would produce disk saturation leading to sample running and cross-contamination.

Sample application can be carried out more efficiently in a variety of ways. For example, a 96-well cloning device can be routinely used for spotting agar test plates with small amounts of concentrated sample. This step can be performed either manually or robotically depending upon the needs and financial resources of the laboratory. A large number of samples can quickly and accurately be deposited on an agar surface in up to six 96-well arrays for a total of 576 samples per bioassay plate (Fig. 2). Cloning devices can be purchased in 96 pin and 384-well pin configurations for higher density applications [2,304 samples]. Since

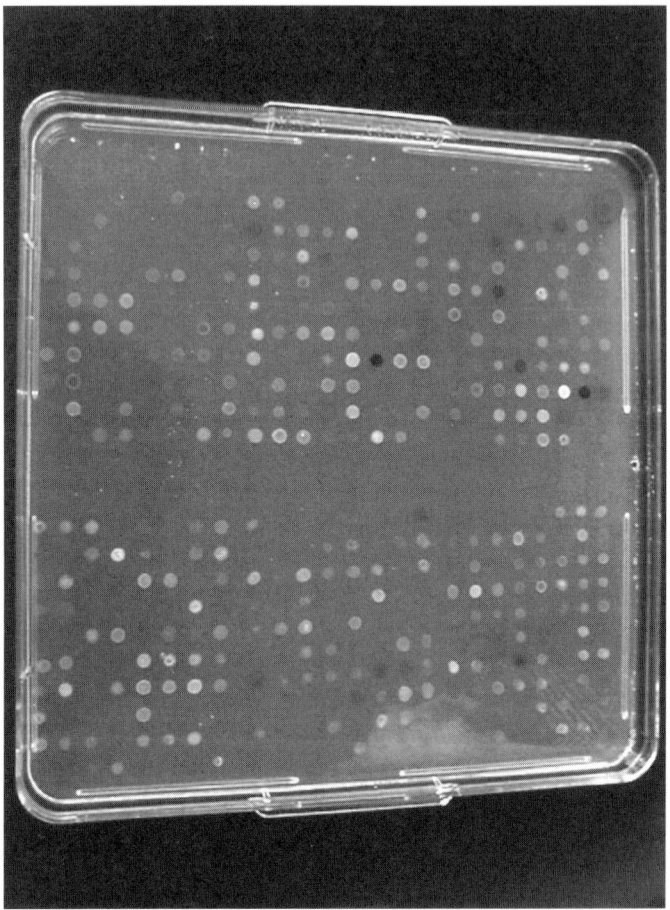

Figure 2 High-density agar spotting of 576 samples using a 96-pin cloning device. (Courtesy of J. C. Walsh, American Cyanamid Company, Princeton, NJ.)

concentrated samples are applied, very small sample volumes (1–10 µL) can be spotted without running on the agar surface. By altering the pin design, the dispense volume can be tailored to meet a variety of screening specifications. Samples applied with a cloning device readily absorb into the agar, thus minimizing cross-mixing of samples. After spotting, the assay plates are incubated as required by each assay protocol and scored accordingly. Caution must be exercised when using high-density arrays, since sample toxicity or robust active responses have the potential to mask activity. Retesting of samples within these areas is necessary to identify which sample is responsible for the response.

2. Natural Products

The basic methodology for conducting microbial-based agar diffusion assays for natural products testing is identical to that for synthetics. Natural products samples, which can potentially contain multiple components in low concentrations, are evaluated using as large a volume of sample as is practical, to maximize the chance of active identification. Wide-bore pipette tips are used for sample distribution to prevent clogging by mycelial fragments and debris that are present in whole broth samples. This is of less concern when testing natural products extracts. In order to maximize the amount of sample applied for testing, agar wells (5 mm) can be bored into an agar surface in an array suitable for high-density testing, and the test wells can be filled manually or using a robotic system [162]. This method allows for significantly larger amounts of sample to be tested than is possible using a standard ¼ inch paper disk. The filled assay plates are then incubated as required and scored according to the criteria established for each assay. Because of the possibility that motile organisms contained in natural products whole broth samples can spread across the agar test surface obscuring the results, appropriate antibiotics can be added to the agar medium. The antibiotic or combination of antibiotics must control contamination without being toxic to the test organism. Minimum inhibitory concentrations must be determined for each antibiotic against the test organism used [163].

C. Alternatives to Agar Diffusion Assays

One of the benefits of using microbial-based assays for novel drug discovery is the flexibility to perform HTS in either agar diffusion assays or liquid assays. Liquid assays are particularly amenable to miniaturization and robotic processing. Advances in miniaturization of labware and the development of new and improved high-density microplate arrays provide an effective means of conducting high-throughput screening. Ultra-high-throughput screening rates of 100,000 compounds per day are now achievable using state-of-the-art microplate robots. Liquid microplate assays incorporating spectrophotometric, fluorescent, or chemiluminescent end points allow rapid quantification and the ability to provide immediate data for analysis and reporting.

D. Criteria for Effective Mechanism-Based Screening

The screen development process is the first step in the discovery process. Mechanism-based assays offer many distinct advantages over conventional ''spray and pray'' methods. Mechanism-based screens should be rapid and inexpensive to perform and should identify compounds that act on the target of interest. Screen sensitivity is extremely important, because limitations are frequently imposed on the quantity of material available for testing. In order to achieve maximize productivity, assays should give clear, robust responses with a minimum of assay interference. Assays that are easy to score offer a greater potential for higher throughput even if scoring is performed without the aid of instrumentation.

E. Use of High-Density Agar Diffusion Assays for Assay Validation

Before implementing a new assay, it is helpful to characterize the assay by observing how it responds when tested against various chemical classes. Libraries consisting of thousands of diverse chemical samples can be maintained exclusively for the purpose of validating new assays. High-density agar diffusion or liquid assays can be quickly performed to characterize a new assay prior to implementation. Using agar assays, validation can be quickly accomplished in either 96- or 384-well format. Utilization of a 384-well format can greatly reduce the sample storage requirements for the collection. Automated liquid handling devices can be used to create high-density storage plates from vials or lower density well formats. Likewise, automated pipetting devices can be used for conducting high-density liquid assays.

F. Adaptability of Assays to Laboratory Automation

Of paramount importance during the development phase of any high-throughput microbial assay is the prerequisite that the assay be adaptable to laboratory automation. Next, screen development and screen implementation personnel must work as a team to optimize the assay to achieve maximum throughput. Laboratory robotics can provide many benefits over manually performed techniques because automated devices are capable of carrying out tasks with a high degree of precision-thus achieving quality of results. Incorporation of modular automated devices in the laboratory allows for maximum application flexibility. Using generic workstations and easy programming techniques, screening applications can be quickly modified as needed. The use of laboratory robotics can minimize the exposure of workers to potentially hazardous materials. Application logging and sample tracking using bar code readers provide an accurate means for data auditing. In situations where common screening applications are carried out in a company across divisions, duplication of laboratory robotics systems can be beneficial due to the potential for corporate standardization.

REFERENCES

1. AG Goodman, A Gilman, TW Rall, AS Nies, P Taylor. Goodman and Gilman's The Pharmacological Basis of Therpeutics. 8th ed. New York: McGraw-Hill, 1990, pp 269–271, 638–639, 814, 1065.
2. EG Sharvelle. The Nature and Uses of Modern Fungicides. Minneapolis: Burgess, 1961, pp 62–63.
3. GT Brooks. Chlorinated insecticides: retrospect and prospect. In: JR Plummer, ed. Pesticide Chemistry in the 20th Century. Washington DC: American Chemical Society, ACS Symposium Series 37, 1977, pp 1–20.
4. R Betina. The Chemistry and biology of antibiotics. In: WTh Nauta, RF Rekker, eds. Pharmacochemistry Library, vol. 5. New York: Elsevier Scientific, 1983, pp 17–23.
5. RF Gaber. Molecular genetics of yeast ion transport. Int Rev Cytol 137:299–353, 1992.
6. RL Nakamura, RF Gaber. Studying ion channels using yeast genetics. Methods Enzymol 293:89–104, 1998.
7. RF Gaber, CA Styles, GR Fink. TRK1 encodes a plasma membrane protein required for high-affinity potassium transport in *Saccharomyces cerevisiae*. Mol Cell Biol 8:2848–2859, 1988.
8. J Ramos, R Alijo, R Haro, A Rodriguez-Navarro. TRK2 is not a low-affinity potassium transporter in *Saccharomyces cerevisiae*. J Bacteriol 176:249–252, 1994.
9. CH Ko, AM Buckley, RF Gaber. TRK2 is required for low affinity $K+$ transport in *Saccharomyces cerevisiae*. Genetics 125:305–312, 1990.
10. H Sentenac, N Bonneaud, M Minet, F Lacroute, JM Salmon, F Gaymard, C Grignon. Cloning and expression in yeast of a plant potassium ion transport system. Science 256:663–665, 1992.
11. JA Anderson, SS Huprikar, LV Kochian, WJ Lucas, RF Gaber. Functional expression of a probable *Arabidopsis thaliana* potassium channel in *Saccharomyces cerevisiae*. Proc Natl Acad Sci USA 89:3736–3740, 1992.
12. SA Goldstein, KW Wang, N Ilan, MH Pausch. Sequence and function of the two P domain potassium channels: implications of an emerging superfamily. J Mol Med 76:13–20, 1998.
13. SA Goldstein, LA Price, DN Rosenthal, MH Pausch. ORK1, a potassium-selective leak channel with two pore domains cloned from *Drosophila melanogaster* by expression in *Saccharomyces cerevisiae*. Proc Natl Acad Sci USA 93:13256–13261, 1996.
14. W Tang, A Ruknudin, WP Yang, SY Shaw, A Knickerbocker, S Kurtz. Functional expression of a vertebrate inwardly rectifying $K+$ channel in yeast. Mol Biol Cell 6:1231–1240, 1995.
15. S Kurtz, G Luo, KM Hahnenberger, C Brooks, O Gecha, K Ingalls, K Numata, M Krystal. Growth impairment resulting from expression of influenza virus M2 protein in *Saccharomyces cerevisiae*: identification of a novel inhibitor of influenza virus. Antimicrob Agents Chemother 39:2204–2209, 1995.

16. KM Hahnenberger, M Krystal, K Esposito, W Tang, S Kurtz. Use of microphysiometry for analysis of heterologous ion channels expressed in yeast. Nat Biotechnol 14:880–883, 1996.

17. KM Hahnenberger, SE Kurtz. A drug screening program for ion channels expressed in yeast. Trends Biotechnol 15:1–4, 1997.

18. RL Nakamura, JA Anderson, RF Gaber. Determination of key structural requirements of a K+ channel pore. J Biol Chem 272:1011–1018, 1997.

19. D Becker, I Dreyer, S Hoth, JD Reid, H Busch, M Lehnen, K Palme, R Hedrich. Changes in voltage activation, Cs+ sensitivity, and ion permeability in H5 mutants of the plant K+ channel KAT1. Proc Natl Acad Sci USA 93:8123–8128, 1996.

20. KA Ketchum, WJ Joiner, AJ Sellers, LK Kaczmarek, SA Goldstein. A new family of outwardly rectifying potassium channel proteins with two pore domains in tandem. Nature 376:690–695, 1995.

21. XL Zhou, B Vaillant, SH Loukin, C Kung, Y Saimi. YKC1 encodes the depolarization-activated K+ channel in the plasma membrane of yeast. FEBS Lett 373:170–176, 1995.

22. JD Reid, W Lukas, R Shafaatian, A Bertl, C Scheurmann-Kettner, HR Guy, RA North. The S. cerevisiae outwardly-rectifying potassium channel (DUK1) identifies a new family of channels with duplicated pore domains. Receptors Channels 4:51–62, 1996.

23. F Lesage, E Guillemare, M Fink, F Duprat, M Lazdunski, G Romey, J Barhanin. A pH-sensitive yeast outward rectifier K+ channel with two pore domains and novel gating properties. J Biol Chem 271:4183–4187, 1996.

24. A Bertl, H Bihler, JD Reid, C Kettner, CL Slayman. Physiological characterization of the yeast plasma membrane outward rectifying K+ channel, DUK1 (TOK1), in situ. J Membr Biol 162:67–80, 1998.

25. R Madrid, H Lichtneberg-Frate. Functional expression of the rat ether a go-go potassium channel in yeast. Folia Microbiol 43:224, 1998.

26. DJ Mangelsdorf, C Thummel, M Beato, P Herrlich, G Schutz, K Umesono, B Blumberg, P Kastner, M Mark, P Chambon. The nuclear receptor superfamily: the second decade. Cell 83:835–839, 1995.

27. B Blumberg, RM Evans. Orphan nuclear receptors—new ligands and new possibilities. Genes Dev 12:3149–3155, 1998.

28. T Perlmann, RM Evans. Nuclear receptors in Sicily: all in the famiglia. Cell 90:391–397, 1997.

29. JC Rowlands, JA Gustafsson. Aryl hydrocarbon receptor-mediated signal transduction. Crit Rev Toxicol 27:109–134, 1997.

30. D Moras, H Gronemeyer. The nuclear receptor ligand-binding domain: structure and function. Curr Opin Cell Biol 10:384–391, 1998.

31. J Torchia, C Glass, MG Rosenfeld. Co-activators and co-repressors in the integration of transcriptional responses. Curr Opin Cell Biol 10:373–383, 1998.

32. BP Klaholz, D Moras. A structural view of ligand binding to the retinoid receptor. Pure Appl Chem 70:41–47, 1998.

33. H Oberlander, DL Silhacek, E Shaaya, I Ishaaya. Current status and future perspectives of the use of insect growth regulators for the control of stored product insects. J Stored Prod Res 33:1–6, 1997.

34. KH Hoffmann, MW Lorenz. Recent advances in hormones in insect pest control. Phytoparasitica 26:323–330, 1998.

35. TG Wilson, J Heinrich. Recombinant BHLH-PAS/JHR polypeptide and its use to screen potential insecticides. Patent application WO9846724, November 17, 1997.

36. M Ashok, C Turner, TG Wilson. Insect juvenile hormone resistance gene homology with the bHLH-PAS family of transcriptional regulators. Proc Natl Acad Sci USA 95:2761–2766, 1998.

37. CS Thummel. From embryogenesis to metamorphosis: the regulation and function of *Drosophila* nuclear receptor superfamily members. Cell 83:871–877, 1995.

38. ST Crews. Control of cell lineage-specific development and transcription by bHLH-PAS proteins. Genes Dev 12:607–620, 1998.

39. DM Duncan, EA Burgess, I Duncan. Control of distal antennal identity and tarsal development in *Drosophila* by spineless-aristapedia, a homolog of the mammalian dioxin receptor. Genes Dev 12:1290–1303, 1998.

40. JE Rutila, V Suri, M Le, WV So, M Rosbash, JC Hall. CYCLE is a second bHLH-PAS clock protein essential for circadian rhythmicity and transcription of *Drosophila* period and timeless. Cell 93:805–814, 1998.

41. CA Meier. Regulation of gene expression by nuclear hormone receptors. J Recept Signal Transduct Res 17:319–335, 1997.

42. TR Butt, PG Walfish. Human nuclear receptor heterodimers: opportunities for detecting targets of transcriptional regulation using yeast. Gene Expr 5:255–268, 1996.

43. N Huang, E vom Baur, JM Garnier, T Lerouge, JL Vonesch, Y Lutz, P Chambon, R Losson. Two distinct nuclear receptor interaction domains in NSD1, a novel SET protein that exhibits characteristics of both corepressors and coactivators. EMBO J 17:3398–3412, 1998.

44. CA Miller, MA Martinat, LE Hyman. Assessment of aryl hydrocarbon receptor complex interactions using pBEVY plasmids: expression vectors with bidirectional promoters for use in *Saccharomyces cerevisiae*. Nucleic Acids Res 26:3577–3583, 1998.

45. P Mak, DP McDonnell, NL Weigel, WT Schrader, BW O'Malley. Expression of functional chicken oviduct progesterone receptors in yeast (*Saccharomyces cerevisiae*). J Biol Chem 264:21613–21618, 1989.

46. SL Dana, PA Hoener, DA Wheeler, CB Lawrence, DP McDonnell. Novel estrogen response elements identified by genetic selection in yeast are differentially responsive to estrogens and antiestrogens in mammalian cells. Mol Endocrinol 8:1193–1207, 1994.

47. N Pocuca, S Ruzdijic, C Demonacos, D Kanazir, M Krstic-Demonacos. Using yeast to study glucocorticoid receptor phosphorylation. J Steroid Biochem Mol Biol 66: 303–318, 1998.

48. K Henry, ML O'Brien, W Clevenger, L Jow, DJ Noonan. Peroxisome proliferator-activated receptor response specificities as defined in yeast and mammalian cell transcription assays. Toxicol Appl Pharmacol 132:317–324, 1995.

49. SL Marcus, KS Miyata, RA Rachubinski, JP Capone. Transactivation by PPAR/RXR heterodimers in yeast is potentiated by exogenous fatty acid via a pathway requiring intact peroxisomes. Gene Expr 4:227–239, 1995.

50. P Mak, HA Fuernkranz, R Ge, SK Karathanasis. Retinoid X receptor homodimers function as transcriptional activators in yeast. Gene 145:129–133, 1994.

51. JW Liu, D Picard. Bioactive steroids as contaminants of the common carbon source galactose. FEMS Microbiol Lett 159:167–171, 1998.

52. A Joyeux, V Cavailles, P Balaguer, JC Nicolas. RIP 140 enhances nuclear receptor-dependent transcription in vivo in yeast. Mol Endocrinol 11:193–202, 1997.

53. KW Gaido, LS Leonard, S Lovell, JC Gould, D Babai, CJ Portier, DP McDonnell. Evaluation of chemicals with endocrine modulating activity in a yeast-based steroid hormone receptor gene transcription assay. Toxicol Appl Pharmacol 143:205–212, 1997.

54. EA Allegretto, RA Heyman. Expression and characterization of retinoid receptors in yeast. In: DB McCormick, JW Suttie, C Wagner, eds. Methods in Enzymology, Volume 282: Vitamins and Coenzymes. New York: Academic Press, 1997, pp 25–32.

55. DR Kirsch. Development of improved cell-based assays and screens in *Saccharomyces* through the combination of molecular and classical genetics. Curr Opin Biotechnol 4:543–552, 1993.

56. DP McDonnell, E Vegeto, MA Gleeson. Nuclear hormone receptors as targets for new drug discovery. Biotechnology (NY) 11:1256–1261, 1993.

57. DX Wen, DP McDonnell. Advances in our understanding of ligand-activated nuclear receptors. Curr Opin Biotechnol 6:582–589, 1995.

58. MJ Garabdeian. Genetic approaches to mammalian nuclear receptor function in yeast. Companion Methods Enzymol 5:138–146, 1993.

59. F de la Cruz, P Mak. *Drosophila* ecdysone receptor functions as a constitutive activator in yeast. J Steroid Biochem Mol Biol 62:353–359, 1997.

60. CA Miller. Expression of the human aryl hydrocarbon receptor complex in yeast. Activation of transcription by indole compounds. J Biol Chem 272:32824–32829, 1997.

61. JN Heinrich, F de la Cruz, DR Kirsch. Screen for ultraspiracle inhibitors. Patent application WO 97/45737, December 4, 1997.

62. N Pocuca, S Ruzdijic, C Demonacos, D Kanazir, M Krstic-Demonacos. Using yeast to study glucocorticoid receptor phosphorylation. J Steroid Biochem Mol Biol 66:303–318, 1998.

63. IG Schulman, C Li, JW Schwabe, RM Evans. The phantom ligand effect: allosteric control of transcription by the retinoid X receptor. Genes Dev 11:299–308, 1997.

64. HG Dohlman, J Thorner, MG Caron, RJ Lefkowitz. Model systems for the study of seven-transmembrane-segment receptors. Ann Rev Biochem 60:653–688, 1991.

65. HR Bourne, DA Sanders, F McCormick. The GTPase superfamily: conserved structure and molecular mechanism. Nature 349:117–127, 1991.

66. MH Pausch, LA Price, EM Kajkowski, J Strnad, F dela Cruz, J Heinrich, BA Ozenberger, JR Hadcock. Heterologous G protein–coupled receptors expressed in *Saccharomyces cerevisiae*: methods for genetic analysis and ligand identification. In: KR Lynch, ed. Identification and Expression of G Protein–Coupled Receptors. New York: John Wiley, 1998, pp 196–212.

67. MH Pausch. G-protein–coupled receptors in *Saccharomyces cerevisiae*: high-

throughput screening assays for drug discovery. Trends Biotechnol 15:487–494, 1997.

68. JR Broach, J Thorner. High-throughput screening for drug discovery. Nature 384: 14–16, 1996.

69. GF Sprague, JW Thorner. In: JR Pringle, EW Jones, JR Broach, eds. The Molecular Biology of the Yeast Saccharomyces: Gene Expression Vol. 2. Cold Spring Harbor: Cold Spring Harbor Laboratory Press, 1992, pp 657–744.

70. J Kurjan. The pheromone response pathway in *Saccharomyces cerevisiae*. Ann Rev Genet 27:147–179, 1993.

71. I Herskowitz. MAP kinase pathways in yeast: for mating and more. Cell 80:187–197, 1995.

72. MC Gustin, J Albertyn, M Alexander, K Davenport. MAP kinase pathways in the yeast *Saccharomyces cerevisiae*. Microbiol Mol Biol Rev 62:1264–1300, 1998.

73. KR Lynch. G protein–coupled receptor informatics and the orphan problem. Methods for genetic analysis and ligand identification. In: KR Lynch, ed. Identification and Expression of G Protein–Coupled Receptors. New York: Wiley-Liss, 1998, pp 54–72.

74. K King, HG Dohlman, J Thorner, MG Caron, RJ Lefkowitz. Control of yeast mating signal transduction by a mammalian beta 2-adrenergic receptor and Gs alpha subunit [erratum appears in Science 251:144, 1991]. Science 250:121–123, 1990.

75. LA Price, EM Kajkowski, JR Hadcock, BA Ozenberger, MH Pausch. Functional coupling of a mammalian somatostatin receptor to the yeast pheromone response pathway. Mol Cell Biol 15:6188–6195, 1995.

76. LA Price, J Strnad, MH Pausch, JR Hadcock. Pharmacological characterization of the rat A2a adenosine receptor functionally coupled to the yeast pheromone response pathway. Mol Pharmacol 50:829–837, 1996.

77. F Chang, I Herskowitz. Identification of a gene necessary for cell cycle arrest by a negative growth factor of yeast: FAR1 is an inhibitor of a G1 cyclin, CLN2. Cell 63:999–1011, 1990.

78. M Peter, A Gartner, J Horecka, G Ammerer, I Herskowitz. FAR1 links the signal transduction pathway to the cell cycle machinery in yeast. Cell 73:747–760, 1993.

79. BJ Stevenson, N Rhodes, B Errede, GFJ Sprague. Constitutive mutants of the protein kinase STE11 activate the yeast pheromone response pathway in the absence of the G protein. Genes Dev 6:1293–1304, 1992.

80. HG Dohlman, J Song, DM Apanovitch, PR DiBello, KM Gillen. Regulation of G protein signalling in yeast. Semin Cell Dev Biol 9:135–141, 1998.

81. JP Hirsch, FR Cross. The pheromone receptors inhibit the pheromone response pathway in *Saccharomyces cerevisiae* by a process that is independent of their associated G alpha protein. Genetics 135:943–953, 1993.

82. A Couve, JP Hirsch. Loss of sustained Fus3p kinase activity and the G1 arrest response in cells expressing an inappropriate pheromone receptor. Mol Cell Biol 16:4478–4485, 1996.

83. RT Bass, BL Buckwalter, BP Patel, MH Pausch, LA Price, J Strnad, JR Hadcock. Identification and characterization of novel somatostatin antagonists [erratum appears in Mol Pharmacol 51(1):170, 1997]. Mol Pharmacol 50:709–715, 1996.

84. WR Baumbach, TA Carrick, MH Pausch, B Bingham, D Carmignac, ICAF

Robinson, R Houghten, CM Eppler, LA Price, JR Zysk. A linear hexapeptide somatostatin antagonist blocks somatostatin activity in vitro and influences growth hromone release in rats. Mol Pharmacol 54:864–873, 1998.

85. MA Wall, DE Coleman, E Lee, JA Iniguez-Lluhi, BA Posner, AG Gilman, SR Sprang. The structure of the G protein heterotrimer Gi alpha 1 beta 1 gamma 2. Cell 83:1047–1058, 1995.

86. DG Lambright, J Sondek, A Bohm, NP Skiba, HE Hamm, PB Sigler. The 2.0 A crystal structure of a heterotrimeric G protein. Nature 379:311–319, 1996.

87. JR Erickson, JJ Wu, JG Goddard, G Tigyi, K Kawanishi, LD Tomei, MC Kiefer. Edg-2/Vzg-1 couples to the yeast pheromone response pathway selectively in response to lysophosphatidic acid. J Biol Chem 273:1506–1510, 1998.

88. JP Manfredi, C Klein, JJ Herrero, DR Byrd, J Trueheart, WT Wiesler, DM Fowlkes, JR Broach. Yeast alpha mating factor structure–activity relationship derived from genetically selected peptide agonists and antagonists of Ste2p. Mol Cell Biol 16: 4700–4709, 1996.

89. C Klein, JL Paul, K Sauve, MM Schmidt, L Arcangeli, J Ransom, J Trueheart, JP Manfredi, JR Broach, AJ Murphy. Identification of surrogate agonists for the human FPRL-1 receptor by autocrine selection in yeast. Nat Biotechnol 16:1334–1337, 1998.

90. EM Kajkowski, LA Price, MH Pausch, KH Young, BA Ozenberger. Investigation of growth hormone releasing hormone receptor structure and activity using yeast expression technologies. J Recept Signal Transduct Res 17:293–303, 1997.

91. JB Konopka, SM Margarit, P Dube. Mutation of Pro-258 in transmembrane domain 6 constitutively activates the G protein–coupled alpha-factor receptor. Proc Natl Acad Sci USA 93:6764–6769, 1996.

92. M Dosil, L Giot, C Davis, JB Konopka. Dominant-negative mutations in the G-protein-coupled alpha-factor receptor map to the extracellular ends of the transmembrane segments. Mol Cell Biol 18:5981–5991, 1998.

93. T Kokkola, MA Watson, J White, S Dowell, SM Foord, JT Laitinen. Mutagenesis of human Mel1a melatonin receptor expressed in yeast reveals domains important for receptor function. Biochem Biophys Res Commun 249:531–536, 1998.

94. LC Davidse. Benzimidazole fungicides: mechanism of action and biological impact. Ann Rev Phytopathol 24–43, 1986.

95. SB Horwitz. Mechanism of action of taxol. Trends Pharmacol Sci 13:134–136, 1992.

96. SL Schreiber, GR Crabtree. The mechanism of action of cyclosporin A and FK506. Immunol Today 13:136–142, 1992.

97. SN Sehgal. Rapamune (Sirolimus, rapamycin): an overview and mechanism of action. Ther Drug Monit 17:660–665, 1995.

98. S Fields, O Song. A novel genetic system to detect protein–protein interactions. Nature 340:245–246, 1989.

99. H Wang, GA Peters, X Zeng, M Tang, W Ip, SA Khan. Yeast two-hybrid system demonstrates that estrogen receptor dimerization is ligand-dependent in vivo. J Biol Chem 270:23322–23329, 1995.

100. P Colas, R Brent. The impact of two-hybrid and related methods on biotechnology. Trends Biotechnol 16:355–363, 1998.

101. EJ Licitra, JO Liu. A three-hybrid system for detecting small ligand–protein receptor interactions. Proc Natl Acad Sci USA 93:12817–12821, 1996.
102. CM Alarcon, J Heitman. FKBP12 physically and functionally interacts with aspartokinase in *Saccharomyces cerevisiae*. Mol Cell Biol 17:5968–5975, 1997.
103. MJ Kuranda, PW Robbins. Chitinase is required for cell separation during growth of *Saccharomyces cerevisiae*. J Biol Chem 266:19758–19767, 1991.
104. CW Xu, AR Mendelsohn, R Brent. Cells that register logical relationships among proteins. Proc Natl Acad Sci USA 94:12473–12478, 1997.
105. U Putz, P Skehel, D Kuhl. A tri-hybrid system for the analysis and detection of RNA–protein interactions. Nucleic Acids Res 24:4838–4840, 1996.
106. DJ Sen Gupta, B Zhang, B Kraemer, P Pochart, S Fields, M Wickens. A three-hybrid system to detect RNA–protein interactions in vivo. Proc Natl Acad Sci USA 93:8496–8501, 1996.
107. MM Wang, RR Reed. Molecular cloning of the olfactory neuronal transcription factor Olf-1 by genetic selection in yeast. Nature 364:121–126, 1993.
108. JJ Li, I Herskowitz. Isolation of ORC6, a component of the yeast origin recognition complex by a one-hybrid system. Science 262:1870–1874, 1993.
109. SS Tian, P Lamb, AG King, SG Miller, L Kessler, JI Luengo, L Averill, RK Johnson, JG Gleason, LM Pelus, SB Dillon, J Rosen. A small, nonpeptidyl mimic of granulocyte-colony-stimulating factor. Science 281:257–259, 1998.
110. K Young, S Lin, L Sun, E Lee, M Modi, S Hellings, M Husbands, B Ozenberger, R Franco. Identification of a calcium channel modulator using a high throughput yeast two-hybrid screen. Nat Biotechnol 16:946–950, 1998.
111. PL Bartel, S Fields. The Yeast Two-Hybrid System. Oxford: Oxford University Press, 1997.
112. RK Brachmann, JD Boeke. Tag games in yeast: the two-hybrid system and beyond. Curr Opin Biotechnol 8:561–568, 1997.
113. BA Ozenberger, KH Young. Functional interaction of ligands and receptors of the hematopoietic superfamily in yeast [erratum appears in Mol Endocrinol 10(8):936, 1996]. Mol Endocrinol 9:1321–1329, 1995.
114. A Aronheim, E Zandi, H Hennemann, SJ Elledge, M Karin. Isolation of an AP-1 repressor by a novel method for detecting protein–protein interactions. Mol Cell Biol 17:3094–3102, 1997.
115. N Johnsson, A Varshavsky. Split ubiquitin as a sensor of protein interactions in vivo. Proc Natl Acad Sci USA 91:10340–10344, 1994.
116. I Stagljar, C Korostensky, N Johnsson, HS Te. A genetic system based on split-ubiquitin for the analysis of interactions between membrane proteins in vivo. Proc Natl Acad Sci USA 95:5187–5192, 1998.
117. G Karimova, J Pidoux, A Ullmann, D Ladant. A bacterial two-hybrid system based on a reconstituted signal transduction pathway. Proc Natl Acad Sci USA 95:5752–5756, 1998.
118. F Rossi, CA Charlton, HM Blau. Monitoring protein–protein interactions in intact eukaryotic cells by beta-galactosidase complementation. Proc Natl Acad Sci USA 94:8405–8410, 1997.
119. MG Kornacker, B Remsburg, R Menzel. Gene activation by the AraC protein can be inhibited by DNA looping between AraC and a LexA repressor that interacts

with AraC: possible applications as a two-hybrid system. Mol Microbiol 30:615–624, 1998.

120. DA De Angelis, G Miesenbock, BV Zemelman, JE Rothman. PRIM: proximity imaging of green fluorescent protein-tagged polypeptides. Proc Natl Acad Sci USA 95:12312–12316, 1998.

121. RS Cormack, K Hahlbrock, IE Somssich. Isolation of putative plant transcriptional coactivators using a modified two-hybrid system incorporating a GFP reporter gene. Plant J 14:685–692, 1998.

122. JM Vieites, F Navarro-Garcia, R Perez-Diaz, J Pla, C Nombela. Expression and in vivo determination of firefly luciferase as gene reporter in *Saccharomyces cerevisiae*. Yeast 10:1321–1327, 1994.

123. J Estojak, R Brent, EA Golemis. Correlation of two-hybrid affinity data with in vitro measurements. Mol Cell Biol 15:5820–5829, 1995.

124. EA Golemis, R Brent. Searching for interacting proteins with the two-hybrid system III. In: PL Bartel, S Fields, eds. The Yeast Two-Hybrid System. Oxford: Oxford University Press, 1997, pp 43–72.

125. HM Shih, PS Goldman, AJ DeMaggio, SM Hollenberg, RH Goodman, MF Hoekstra. A positive genetic selection for disrupting protein–protein interactions: identification of CREB mutations that prevent association with the coactivator CBP. Proc Natl Acad Sci USA 93:13896–13901, 1996.

126. M Vidal, RK Brachmann, A Fattaey, E Harlow, JD Boeke. Reverse two-hybrid and one-hybrid systems to detect dissociation of protein–protein and DNA–protein interactions. Proc Natl Acad Sci USA 93:10315–10320, 1996.

127. CA Leanna, M Hannink. The reverse two-hybrid system: a genetic scheme for selection against specific protein/protein interactions. Nucleic Acids Res 24:3341–3347, 1996.

128. S Omura. Philosophy of new drug discovery. Microbiol Rev 50:259–279, 1986.

129. TD Gootz. Discovery and development of new antimicrobial agents. Clin Microbiol Rev 3:13–31, 1990.

130. RL Monaghan, JS Tkacz. Bioactive microbial products: focus upon mechanism of action. Ann Rev Microbiol 44:271–301, 1990.

131. L Silver, K Bostian. Screening of natural products for antimicrobial agents. Eur J Clin Microbiol Infect Dis 9:455–461, 1990.

132. DR Kirsch, BJ DiDomenico. Mechanism based screening for the discovery of novel antifungals. In: V Gullo, ed. The Discovery of Natural Products with Therapeutic Potential. Boston: Butterworth-Heinemann, 1993, pp 177–221.

133. S Kadam. Mechanism-based screens in the discovery of chemotherapeutic antibacterials. Biotechnology 26:247–266, 1994.

134. RC Goldman, DJ Frost, JO Capobianco, S Kadam, RR Rasmussen, C Abad-Zapatero. Antifungal drug targets: C secreted aspartyl protease and fungal wall beta-glucan synthesis. Infect Agents Dis 4:228–247, 1995.

135. RB Sykes, JS Wells. Screening for beta-lactam antibiotics in nature. J Antibiot (Tokyo) 38:119–121, 1985.

136. DR Kirsch, MH Lai, J McCullough, AM Gillum. The use of beta-galactosidase gene fusions to screen for antibacterial antibiotics. J Antibiot (Tokyo) 44:210–217, 1991.

137. DM Rothstein, M McGlynn, V Bernan, J McGahren, J Zaccardi, N Cekleniak, KP Bertrand. Detection of tetracyclines and efflux pump inhibitors. Antimicrob Agents Chemother 37:1624–1629, 1993.

138. MH Lai, DR Kirsch. Induction signals for vancomycin resistance encoded by the vanA gene cluster in *Enterococcus faecium*. Antimicrob Agents Chemother 40: 1645–1648, 1996.

139. N Mani, P Sancheti, ZD Jiang, C McNaney, M DeCenzo, B Knight, M Stankis, M Kuranda, DM Rothstein, P Sanchet, B Knighti. Screening systems for detecting inhibitors of cell wall transglycosylation in *Enterococcus*. Cell wall transglycosylation inhibitors in *Enterococcus* [erratum appears in J Antibiot (Tokyo) 51(6):607, 1998]. J Antibiot (Tokyo) 51:471–479, 1998.

140. J Grissom-Arnold, WEJ Alborn, TI Nicas, SR Jaskunas. Induction of VanA vancomycin resistance genes in *Enterococcus faecalis*: use of a promoter fusion to evaluate glycopeptide and nonglycopeptide induction signals. Microb Drug Resist 3:53–64, 1997.

141. SJ Smith, JH Crowley, LW Parks. Transcriptional regulation by ergosterol in the yeast *Saccharomyces cerevisiae*. Mol Cell Biol 16:5427–5432, 1996.

142. BA Arthington-Skaggs, DN Crowell, H Yang, SL Sturley, M Bard. Positive and negative regulation of a sterol biosynthetic gene (ERG3) in the post-squalene portion of the yeast ergosterol pathway. FEBS Lett 392:161–165, 1996.

143. MH Lai, DR Kirsch. Enzyme induction screen for ergosterol biosynthesis inhibitors. U.S. patent 5,527,687, June 18, 1996.

144. RL Wood, TK Miller, A Wright, P McCarthy, CS Taft, S Pomponi, CP Selitrennikoff. Characterization and optimization of in vitro assay conditions for (1,3)beta-glucan synthase activity from *Aspergillus fumigatus* and *Candida* for enzyme inhibition screening. J Antibiot (Tokyo) 51:665–675, 1998.

145. PG Zaworski, GS Gill. Use of *Saccharomyces cerevisiae* expressing beta-galactosidase to screen for antimycotic agents directed against yeast cell wall biosynthesis and possible application to pathogenic fungi. Antimicrob Agents Chemother 34: 660–662, 1990.

146. NA Gow, PW Robbins, JW Lester, AJ Brown, WA Fonzi, T Chapman, OS Kinsman. A hyphal-specific chitin synthase gene (CHS2) is not essential for growth, dimorphism, or virulence of *Candida*. Proc Natl Acad Sci USA 91:6216–6220, 1994.

147. CE Bulawa, DW Miller, LK Henry, JM Becker. Attenuated virulence of chitin-deficient mutants of *Candida*. Proc Natl Acad Sci USA 92:10570–10574, 1995.

148. T Mio, T Yabe, M Sudoh, Y Satoh, T Nakajima, M Arisawa, H Yamada-Okabe. Role of three chitin synthase genes in the growth of *Candida*. J Bacteriol 178: 2416–2419, 1996.

149. E Mellado, A Aufauvre-Brown, NA Gow, DW Holden. The *Aspergillus fumigatus* chsC and chsG genes encode class III chitin synthases with different functions. Mol Microbiol 20:667–679, 1996.

150. E Mellado, CA Specht, PW Robbins, DW Holden. Cloning and characterization of chsD, a chitin synthase-like gene of *Aspergillus fumigatus*. FEMS Microbiol Lett 143:69–76, 1996.

151. A Aufauvre-Brown, E Mellado, NR Gow, DW Holden. *Aspergillus fumigatus* chsE:

a gene related to CHS3 of *Saccharomyces cerevisiae* and important for hyphal growth and conidiophore development but not pathogenicity. Fungal Genet Biol 21:141–152, 1997.

152. CM Douglas, JA D'lppolito, GJ Shei, M Meinz, J Onishi, JA Marrinan, W Li, GK Abruzzo, A Flattery, K Bartizal, A Mitchell, MB Kurtz. Identification of the FKS1 gene of *Candida* as the essential target of 1,3-beta-D-glucan synthase inhibitors. Antimicrob Agents Chemother 41:2471–2479, 1997.

153. JP Gaughran, MH Lai, DR Kirsch, SJ Silverman. Nikkomycin Z is a specific inhibitor of *Saccharomyces cerevisiae* chitin synthase isozyme Chs3 in vitro and in vivo. J Bacteriol 176:5857–5860, 1994.

154. S Sakuda, A Isogai, T Makita, S Matsumoto, K Koseki, H Kodama, A Suzuki. Structures of allosamidins, novel insect chitinase inhibitors, produced by actinomycetes. Agric Biol Chem 51:3251–3259, 1987.

155. K Dickinson, V Keer, CA Hitchcock, DJ Adams. Chitinase activity from *Candida* and its inhibition by allosamidin. J Gen Microbiol 135:1417–1421, 1989.

156. R McNab, LA Glover. Inhibition of *Neurospora crassa* cytosolic chitinase by allosamidin. FEMS Microbiol Lett 66:79–82, 1991.

157. SJ Silverman. Screen for inhibitors of chitinase. U.S. Patent 5,561,051, October 1, 1996.

158. JM Dominguez, VA Kelly, OS Kinsman, MS Marriott, dIH Gomez, JJ Martin. Sordarins: a new class of antifungals with selective inhibition of the protein synthesis elongation cycle in yeasts. Antimicrob Agents Chemother 42:2274–2278, 1998.

159. SM Mandala, RA Thornton, M Rosenbach, J Milligan, M Garcia-Calvo, HG Bull, MB Kurtz. Khafrefungin, a novel inhibitor of sphingolipid synthesis. J Biol Chem 272:32709–32714, 1997.

160. SM Mandala, RA Thornton, J Milligan, M Rosenbach, M Garcia-Calvo, HG Bull, G Harris, GK Abruzzo, AM Flattery, CJ Gill, K Bartizal, S Dreikorn, MB Kurtz. Rustmicin, a potent antifungal agent, inhibits sphingolipid synthesis at inositol phosphoceramide synthase. J Biol Chem 273:14942–14949, 1998.

161. GH Harris, A Shafiee, MA Cabello, JE Curotto, O Genilloud, KE Goklen, MB Kurtz, M Rosenbach, PM Salmon, RA Thornton, DL Zink, SM Mandala. Inhibition of fungal sphingolipid biosynthesis by rustmicin, galbonolide B and their new 21-hydroxy analogs. J Antibiot (Tokyo) 51:837–844, 1998.

162. JC Walsh. High throughput, mechanism-based screening techniques for discovering novel agrochemicals. J Biomol Screening 3:175–181, 1998.

163. JC Walsh. Use of robotics for high throughput natural products screening. Proceedings of the International Symposium on Laboratory Automation and Robotics, Boston, 1992, pp 592–605.

164. AJ Salerno, Z He, A Goos-Nilsson, H Ahola, P Mak. Differential transcriptional regulation of the apoAl gene by retinoic acid receptor homo- and heterodimers in yeast. Nucleic Acids Res 24:566–572, 1996.

165. HA Fuernkranz, Y Wang, SK Karathanasis, P Mak. Transcriptional regulation of the apoAl gene by hepatic nuclear factor 4 in yeast. Nucleic Acids Res 22:5665–5671, 1994.

7
Receptor Screens for Small Molecule Agonist and Antagonist Discovery

Ramakrishna Seethala
Bristol-Myers Squibb Company, Princeton, New Jersey

I. INTRODUCTION

At the beginning of twentieth century, J. N. Langley and P. Erlich independently recognized the fundamental features of drug–receptor interaction, i.e., specificity, the basis of cellular recognition and activation and the cellular response [1]. The concept of a receptor includes the key attributes of ligand recognition and signal transduction. The signal transduction process may be mediated through an integral part of the receptor structure or involve receptor interactions with additional nonreceptor proteins [2]. The final proof of the existence of the pharmacological receptors came from the recent advances in biochemistry and molecular biology to purify, sequence, clone, and express receptor proteins. The discovery and development of receptors as drug targets stems from the drug interactions with receptor molecules located in the plasma membranes or in the cytosol of target cells. Several receptors have now been very well characterized. Gene sequences of hundreds of orphan receptors have been identified by homology analysis of genome databases. The ligands of these orphan receptors have to be characterized to determine their functions in order to convert these receptors into new drug targets.

Analysis of the drug targets in current drug therapy showed that there are about 500 molecular targets [3]. Receptors, including cell membrane receptors, nuclear receptors, ion channels, and orphan receptors, represent more than 60%

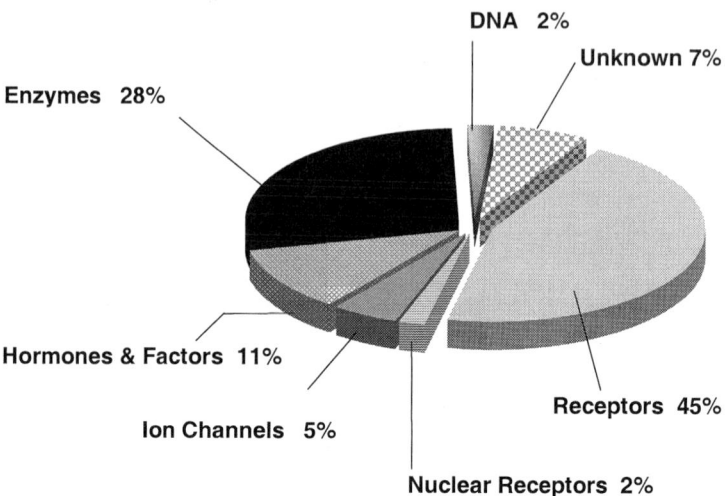

Figure 1 Distribution of drug targets among different types. Receptor targets that include membrane receptors, nuclear receptors, ion channels, and hormones and growth factors represent more than 60% of the total drug targets. (From Ref. 3.)

of the drug discovery targets (Fig. 1). Ligand binding provides a direct approach of in vitro receptor-binding assays. The principle of receptor-binding assays is straightforward. In the conventional ligand–receptor-binding assay, a suitable (high-affinity) radiolabeled ligand is incubated with the chosen receptor preparation. The free radioligand is separated from the receptor-bound ligand, and the total ligand bound to receptor (nonspecific binding plus specific binding) is determined by counting in a scintillation counter. The nonspecific binding is determined in the presence of a large excess (100–1000×) of unlabeled ligand to block the receptor-binding sites of interest and represents the radioligand bound to other receptor sites and to the separation medium such as glass fiber, filter, or assay tubes (Fig. 2A). Specific binding (the ligand bound to the specific receptor-binding sites) is calculated by subtracting the nonspecific binding from the total binding.

Figure 2 (A) Binding of NPY to NPY receptor as a function of ligand concentration. Nonspecific binding was determined in the presence of ×1000 excess of cold ligand. Specific binding was obtained by subtracting nonspecific binding from total binding. (B) Saturation binding of PYY to membranes prepared from CHO-K1 cells and CHO-K1 expressing NPY-Y2R. Inset shows the Scatchard plot from the saturation binding data.

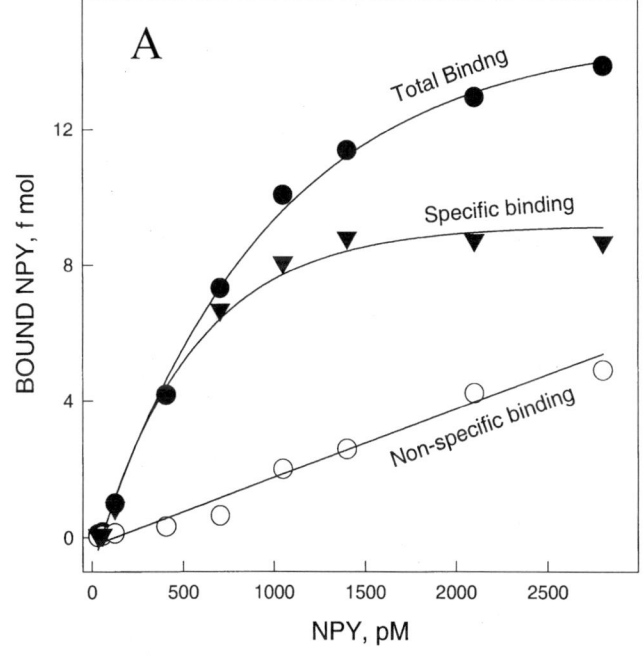

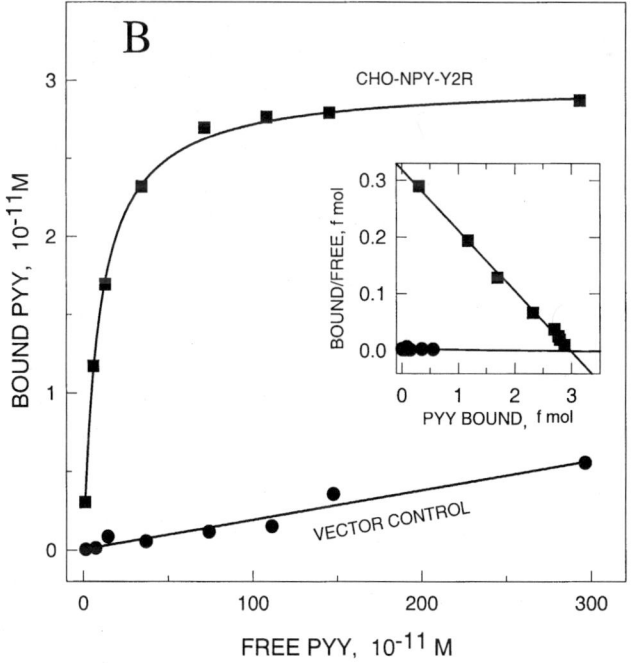

Competition of binding to a specific ligand–receptor-binding site is determined using the radiolabeled ligand at about a concentration of K_d and a predetermined single concentration of various compounds from a compound library. The "hits" (compounds showing the required percentage inhibition) are confirmed by retesting. A full dose–response of the compounds at several concentrations (6–12 concentrations) is determined to obtain the IC_{50} and the K_i for the *hit* compounds. Agonists and antagonists compete in the radioligand-binding to the receptor (they share the ability to bind to a common site on the receptor molecule) but differ in that antagonists are devoid of signal transduction activity. To differentiate a receptor-binding compound as agonist or antagonist, the compound has to be tested in an appropriate signal transduction assay. If the compound activates the signal in a functional assay, it is an agonist, and EC_{50} is determined. Neutral antagonists have no effect on basal receptor activity but inhibit the receptor signal transduction activity generated by a standard agonist for the receptor. Negative antagonists inhibit agonist-independent receptor activity and possess negative intrinsic activity.

II. CLASSIFICATION OF RECEPTOR GROUPS

Receptors are macromolecules that may or may not be a single molecular entity with multiple sites of interactions [4]. The hormone and neurotransmitter receptors present in the plasma membranes are transmembrane glycoproteins. The cytosolic receptors are soluble DNA-binding proteins that belong to a superfamily of nuclear receptors. The primary pharmacological classification of receptors for hormones or neurotransmitters is based on the interaction with synthetic ligands or drugs. Norepinephrine acts on two types of receptors, they are named α- and β-adrenoceptors on the basis of the rank order of potencies for norepinephrine, epinephrine, and other analogs [4,5]. With the discovery of selective α- and β-adrenoceptor antagonists, α-adrenoceptors were differentiated as α1 and α2-adrenoceptors, and based on the agonist interactions β-adrenoceptors were further characterized as β1-, β2-, and β3-adrenoceptors. The structural diversity and multiplicity of a hormone or neurotransmitter receptor subtypes cannot be predicted only on pharmacological data. The current classification criteria recommended by the Committee for Receptor Nomenclature and Drug Classification of the International Union of Pharmacology (IUPHAR) is based on a combination of molecular structure (structural), signal transduction mechanism (transduction), and receptor function (premonitory or operational) [2,6]. The structures and functions of a number of receptors have now been elucidated. Advances in molecular biology also make it possible to determine amino acid sequence of receptors.

Based on structural and transductional characteristics, receptors can be classified into four groups [2,3]: (1) receptors with a single transmembrane segment,

(2) oligomeric receptors with both ligand-binding sites and ion channel complexes, (3) G-protein coupled receptors linked to G-proteins, and (4) nuclear receptors, which are cytosolic, soluble DNA-binding proteins. Receptors can also be classified into superfamilies based on sequence (structure) similarities, which include many receptor proteins that differ pharmacologically but are functionally similar [2], e.g., G-protein coupled receptors (GPCRs), ligand gated-ion channel receptors (LGCRs), voltage-gated-ion channel receptors (VGCRs), tyrosine kinase receptors, tyrosine phosphatase receptors, hematopoietic cytokine receptors, and nuclear receptors. Receptors in each superfamily may be further subclassified into receptor families, and usually they have been named with reference to their endogenous ligands, e.g., neuropeptide Y, endothelin, and epidermal growth factor. Each family of receptors can be classified into subtypes on the basis of relative sequence homologies and functional and signal transduction mechanisms, e.g., neuropeptide Y receptor Y1 (NPYR-Y1), NPYR-Y2, NPYR-Y-Y3, NPYR4, and NPYR-Y5, which show characteristic rank order of potency for antagonist or agonist affinities for each subtype. Different receptor families in a superfamily will have similar structure though there may not be good sequence homology ($\sim 20\%$). Within a receptor family, members of a subfamily are structurally more closely related (~ 50–80%), and each subtype single polypeptide is encoded by distinct gene. In the case of multisubunit oligomeric receptors (LGCR superfamily), each subtype receptor may be made of subunits of different isoforms (isoforms in each family are highly homologous, $> 70\%$ identity). The receptors are basically either membrane or cytoplasmic receptors.

A. Membrane Receptors

All the receptors except the nuclear receptors are membrane receptors and comprise G-protein coupled receptors (GPCRs), ligand gated-ion channel receptors (LGCRs), voltage-gated-ion channel receptors (VGCRs), tyrosine kinase receptors, tyrosine phosphatase receptors, and hematopoietic cytokine receptors.

1. G-Protein Coupled Receptor Superfamily

G-protein coupled receptors (GPCRs) are the largest receptor superfamily with several drugs developed against GPCRs. Members of GPCRs exhibit a common structural motif consisting of seven stretches of hydrophobic amino acid residues that span the membrane and different stretches of amino acids that form extracellular and intracellular loops (Fig. 3). GPCRs are receptors with seven transmembrane spanning regions and transduce the binding of extracellular ligands into intracellular signaling events through GTP-regulatory proteins (G-proteins) [7]. The high-resolution crystal structure of bacteriorhodpsin suggests that the transmembrane (TM) core though consists of polar residues; only a limited number

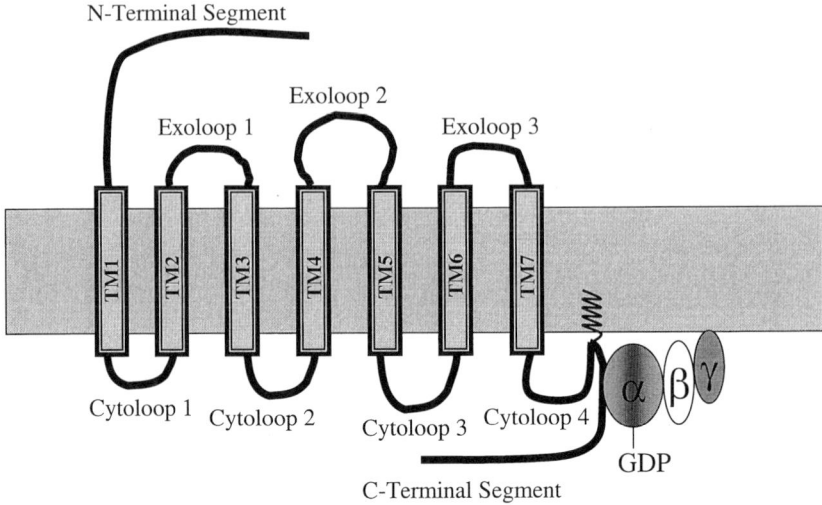

Figure 3 Schematic representation of the general structure of GPCR with trimeric G-protein. GPCRs have seven transmembrane (TM) spanning helices with extracellular N-terminal segment, which is variable in length. The TMs are connected by three extracellular loops (exoloop) and three intracellular loops (cytoloop) and the C-terminal intracellular segment. The ligand receptor interactions are different for different classes of GPCRs. In the inactive state the G-protein is a trimeric complex with GDP bound to the α-subunit.

of water molecules is associated with the TM core. There are extensive hydrogen bonds between residues of the same TM as well as other TMs [8]. Some of GPCRs containing Cys residues in the exoloops often are linked by disulfide, thus constraining the loops and receptor. The GPCRs contain regions involved in ligand binding and another region involved in G-protein coupling. The ligand–receptor interaction involves hydrogen bond, ion pairs, and hydrophobic contacts. G-proteins play important roles in determining the specificity and temporal characteristics of the cellular responses to signals. The GPCRs often have post-translation modifications like N-linked glycosylation on asparagine residue of the extracellularly located N-terminus, palmitoylated on cysteine residues, and phosphorylation on serine/threonine or tyrosine residues. However, the function of these posttranslational modifications in ligand binding or signal transduction is not clear. Phosphorylation/dephosphorylation of sites located intracellularly in the C-terminus may regulate GPCR signaling. Membrane associated G-protein coupled receptor kinases (GRKs) phosphorylate serine/threonine residues of GPCR in active conformation rapidly desensitizing GPCR [9]. The ligand-binding domain (LBD) is a hydrophobic pocket created by transmembranes for receptors with small, nonpeptidc ligands. The ligand-binding site for GPCRs with neuropeptide or peptide hormones as ligands is more complex and consists of multiple extracellular sequences as well as transmembrane regions.

About 1000 distinct mammalian GPCRs have been identified [10]. Additionally, one or two thousand more GPCRs are predicted to be identified within the human genome. Ligands for the orphan GPCRs will have to be discovered. Orphan GPCRs are used as drug targets by screening effectively in melanophores, oocytes, and yeast-based and other formats. GPCRs bind a wide variety of ligands including biogenic amines, neurotransmitters, nucleotides, phospholipids, neuropeptides, growth factors, peptide and nonpeptide hormones, photons, odorants, certain taste ligands, and calcium (Table 1).

2. Receptor Tyrosine Kinase Superfamily

The receptor tyrosine kinase (RTK) superfamily consists of a wide variety of peptide growth factor receptors that possess intrinsic tyrosine kinase activity. RTKs are single polypeptide chain and monomeric in the absence of ligand with the exception of the insulin receptor family. RTKs contain an extracellular ligand-binding domain followed by a transmembrane region and intracellular tyrosine kinase containing catalytic domain (Fig. 4A). Members of the insulin receptor subfamily are disulfide-linked dimers of α and β subunits forming heterotetrameric protein. The insulin receptor contains IRα, which is an extracellular ligand-binding subunit that forms a disulfide bridge with another IRα and with the extracellular region of the IRβ subunit that continues as a transmembrane helix-followed by catalytic tyrosine kinase containing intracellular domain (Fig. 4A). Ligand binding to the extracellular portion of RTK receptors leads to dimerization of monomeric receptors or conformational changes in the heterotetrameric receptor. It results in the autophosphorylation of the cytoplasmic domain by activation of the intrinsic tyrosine kinase catalytic activity that initiates an activating cascade of intracellular pathways (signal transduction) that regulate calcium mobilization, phospholipid and arachidonic acid metabolism, transcriptional regulation, and phosphorylation pathways [11]. Six subfamilies based on structural considerations have been assigned to the RTK superfamily, EGF-receptor, insulin receptor, PDGF-receptor, FGF-receptor, TRK receptor, and EPH/ECK subfamilies. In each subfamily there are several subtypes of receptors.

3. Receptor Protein-Tyrosine Phosphatases

Most members of the receptor protein-tyrosine phosphatases (RPTP) superfamily consist of a large amino-terminal extracellular ligand-binding domain, a single transmembrane spanning domain (~ 25 amino acids), and a large highly conserved carboxyl terminal. The cytoplasmic domain contains two tandem tyrosine

Table 1 Classification of G-Protein Coupled Receptors

Class A	Class B	Class C	Class D	Class E
Rhodopsinlike	Secretinlike	Metabotropic	Fungal	cAMP
Amine	Calcitonin	glutamate/phero-	pheromone	receptors
Adrenoceptors	Corticotropin re-	mone		Dictyostelium
Dopamine	leasing factor	Metabotropic glu-		
Histamine	Gastric inhibitory	tamate		
Serotonin	peptide	Extracellular cal-		
Octopamine	Glucagon	cium-sensing		
Peptide	Growth hormone	Putative pheromone		
Angiotensin	releasing hor-	receptor		
Bombesin	mone			
Bradykinin	Parathyroid hor-			
C5a-anaphylatoxin	mones			
Fmet-leu-phe	PACAP			
Interleukin-8	Secretin			
Chemokine	Vasoactive intesti-			
CCK	nal peptide			
Endothelin	EMR1			
Melanocortin	Latrotoxin			
Neuropeptide Y	Orphan			
Neurotensin				
Opoid				
Somatostatin				
Tachykinin				
Vasopressinlike				
Galanin				
Proteinase activated				
Orexin				
(Rhod)opsin				
Rhodopsin vertebrate				
Rhodopsin Arthropod				
Rhodopsin Mollusc				
Olfactory				
Prostanoid				
Prostaglandin				
Prostacyclin				
Thromboxane				
Nucleotidelike				
Adenosine				
Purinoceptors				
Cannabis				
Platelet activating factor				
Gonadotropin-releasing factor				
Gonadotropin-releasing hormone				
Thyrotropin-releasing hormone				
and secretagogue				
Thyrotropin-releasing hormone				
Growth hormone secretagogue				
Melatonin				
Viral				
Orphan/other				

This information is from http://swift.embl-heidelberg.de/7tm/phylo/phylo.html.

phosphatase domains often separated by an insert (Fig. 4B). The extracellular domains vary among the members of superfamily and may contain either three tandem repeats of fibrinectin-type domains, an immunoglobulin-like (IgL) domain, or an N-terminal carbonic anhydrase-like domain. The members of RPTP include CD45, LAR (a CD45 homologue), LRP, HPTP, RPTP, PTP-P1, and many others [2]. Crystal structures of tyrosine phosphatase PTP-1B, RPTPα, SHP-1, and SHP-2 have been determined [12].

4. Cytokine Receptor Superfamily

Cytokines are small proteins ($Mr = 20-30$ kDa) that exhibit profound and often lineage specific effects on the formation and maturation of hematopoietic cells through cell surface cytokine receptors [13]. The cytokine receptor superfamily members are single transmembrane proteins that contain a cytokine receptor homology region (200–250 amino acids) of two fibronectin III (FNIII) domains in the extracellular domain. The cytokine receptor superfamily has four subgroups, IL-1 cytokine, Class I cytokine and Class II cytokine receptors, and tumor necrosis factor (TNF) families (Table 2). The cytokine ligands induce biological responses such as differentiation, proliferation, and cell death.

IL-1α and IL-1β polypeptides are the ligands for the two types of IL-1 receptors, type I IL-1 and type II IL-1 receptors, that contain immunoglobulin domains. Type I IL-1 is a larger receptor and is found on T cells and fibroblasts. Type II IL-1 is a smaller receptor and is present on B cells, monocytes, neutrophils, and bone marrow cells. Class I cytokine receptors consist of hematopoietic cytokine receptors and are characterized by the presence of one or two conserved 200 amino acid extracellular domains that contain two FN-III modules. A second region is characterized by a conserved cysteine motif (four conserved cysteines and one tryptophan) in the N-terminal FN III domain and a common Trp-Ser-X-Trp-Ser (WSXWS) sequence (the cytokine binding site) that is located in the C-terminal FNIII domain proximal to the transmembrane domain [2,13,14]. The Class I cytokine receptor family has been subdivided into subfamilies. In the GH-R family, cytokine binding to a single receptor-binding subunit promotes the formation of a functional high-affinity receptor dimer through the conserved cysteines. In the other three subfamilies, cytokine binding does not form dimers, and these contain specialized ligand-binding subunits with a short cytoplasmic domain (α-chain) that cannot transduce an intracellular signal on its own. After ligand binding to this α-chain subunit, it associates with a signaling chain (gp130, gp140, or IL-2γc). Class II cytokine receptors consist of interferon receptors in which ligand binding to the receptor induces dimerization of the receptor and activation of receptor-associated JAK kinases [15]. The TNF receptor family consists of a single transmembrane receptor protein with a considerable homology in an extracellular domain and a short intracellular domain with less sequence homology. Ligand binding to the receptor induces the formation of a trimer of receptors or more complex oligomers.

5. Ligand-Gated Ion Channel Superfamily

The ligand-gated ion channel superfamily consists of the acetylcholine receptor (AchR) family (AchR muscle and neuronal subtypes) and the serotinin 5HT3, $GABA_A$, glycine, purinergic, P2x, and ionotropic glutamate receptors. The ligand-gated ion channels are composed of multiple subunits of integral membrane protein homo-oligomers or hetero-oligomers. The subunits are arranged in a ring, and the central axis forms the ion channel. Interaction of the ligand with the receptor channel directly mediates rapid changes in the ionic permeability of the intrinsic ion channel component of the receptor, allowing the selective movement of ions down their electrochemical gradients. An essential feature of the receptor channel is the gate, which controls the flow of ions through the channel and is located at some distance from the ligand binding sites. The ion channel screening strategies have been reviewed in Chapter 10.

6. Voltage-Gated Ion Channel Superfamily

The voltage-gated ion channel superfamily consists of voltage-gated sodium, calcium, and potassium channels. Subtypes of the Na^+ and Ca^{2+} channels are all large polypeptides, termed the α or $\alpha 1$ polypeptides ($\sim$ 250 kDa), containing four homologous repeating domains. Each domain contains six hydrophobic transmembrane spans (S1–S6), and S4 contains a large number of basic residues. The S4 segment is conserved, and this sequence functions as the channel's voltage sensor. The α-polypeptide folds four domains into a transmembrane array

---

Figure 4 Schematic illustration of the general structure of protein tyrosine kinase receptors (PTKR) and protein phosphatases. (A) Members of PTKR superfamily are a single ploypeptide chain consisting of single transmembrane receptors with an extracellular LBD, a transmembrane region, and an intracellular catalytic (tyrosine kinase) domain, with the exception of the insulin receptor family. The insulin receptor family consists of two subunits, an extracellular α-subunit that contains LBD and a β-subunit that has a short extracellular segment followed by transmembrane and intracellular catalytic domain regions. Two α-subunits are connected by a disulfide bridge and are also connected to the β-subunit by a disulfide link forming a heterotetramer of α- and β-chains. The extracellular segments in different subfamilies are composed of cysteine-rich domains, immunoglobulinlike domains (PDGF-family). (B) Protein-tyrosine phosphatase receptors consist of a single polypeptide chain with a variable extracellular domain (with IgG-like domain or carbonic anhydrase domain, or fibronectin domains), a single TM region, and a large highly conserved cytoplasminc domain containing two tandem PTPase domains.

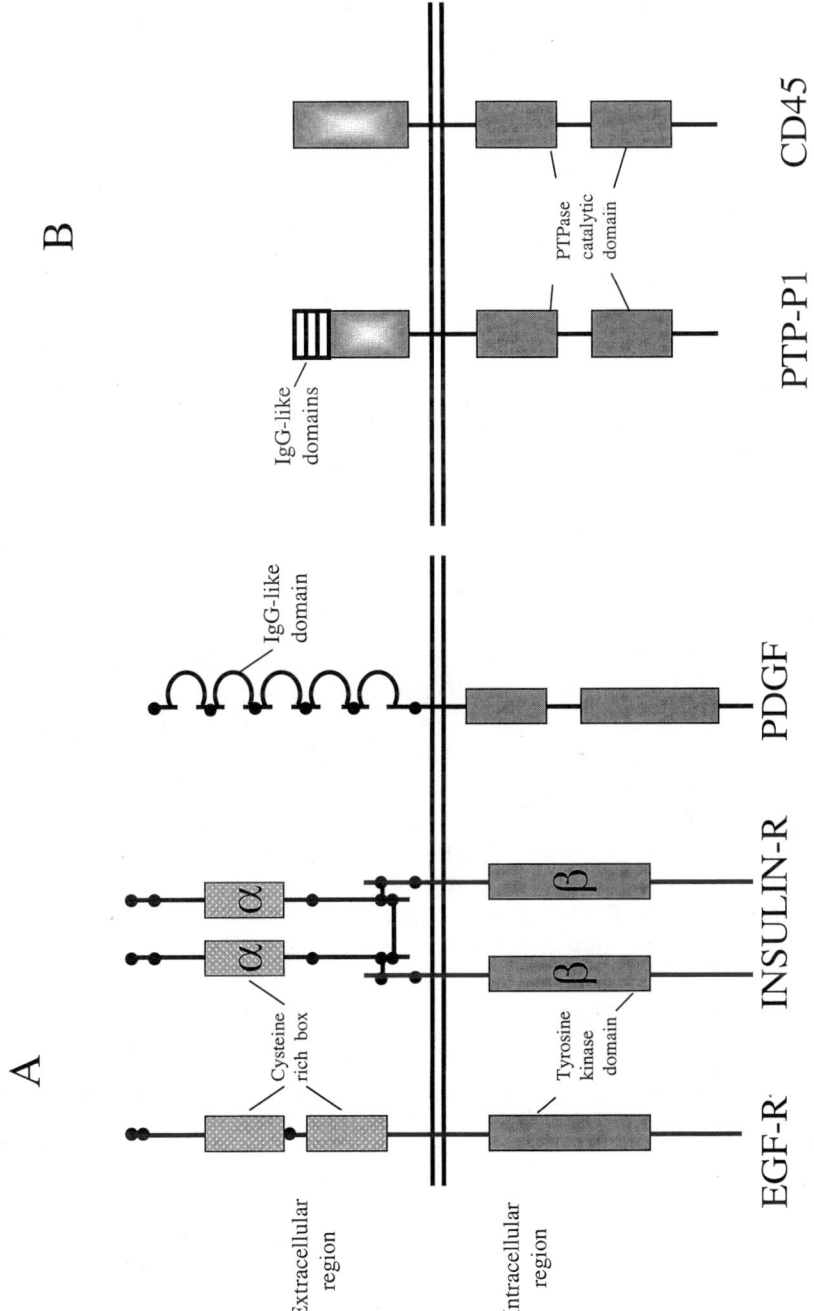

Table 2 Cytokine Receptor Superfamily

---

1. IL-1 cytokine receptor family
   Type I IL-1 receptor
   Type II IL-1 receptor
2. Class I cytokine receptor family
   i. Growth hormone receptor family
      Erythropoietin (EPO) receptor, growth hormone (GH-R) receptor, Prolactin
      (PRL-R) receptor, thrombopoietin (TPO) receptor, G-CSF-R, leptin receptor
   ii. IL-2 or γC receptor family
      IL-2, IL-4, IL-7, IL-9, IL-13, and IL-15 receptor
   iii. IL-3 receptor family
      IL-3, IL-5, GM-CSF receptors
   iv. IL-6 receptor family
      IL-6, IL-11, CNTF, OSM, and LIF receptors
3. Class II Cytokine receptor family
   Interferon (IFN)α/β, IFNγ, IL-10, and tissue factor receptors
4. Tumor necrosis factor receptor (TNFR) family, TNFR-1, TNFR-2, TNFR-RP, NGF,
   CD27, CD30, and CD40 receptors

---

that surrounds a central water-filled pore. Functional voltage-gated K^+ channels are also tetrameric structures of homologous or heterologous α-subunits. Each subunit contains six hydrophobic segments (S1–S6). The functional voltage-gated channel is governed by interconvertible states (closed, open, and inactivated) of the channel. An inactivated state is a closed state that cannot react to form open channels upon depolarization of the membrane. Channels at rest are distributed between resting and inactivated conformations. Changes in electrical potential exert force on the charges in the S4 segment and cause a conformational change to an activated ion-conducting state. Conversion from the open to the inactivated state is a time-dependent process that varies for channel subtype [2].

7. Orphan Receptors

Homology screening approaches have led to the identification of an increasing number of orphan GPCRs. The orphan receptors thus exhibit the structural characteristics shared by all members of the superfamily, but on the basis of their primary sequence they do not belong to any of the receptor subfamilies [16]. These orphan receptors are expected to bind to yet unidentified and undescribed novel ligands and may prove to be important receptors for drug targets. Orphan receptors have been used as molecular targets as in the case of the growth hormone secretogogue receptor (GHSR) as drug targets [17]. Recently, a 28-amino-acid natural peptide, ghrelin, was discovered as a specific natural ligand for this orphan GPCR, GHSR [18]. Another example is characterizing an orphan GPCR localized in the hypothalamus as an orexin receptor after the identification of its natural ligand orexin [19].

B. Intracellular Receptors

The intracellular hormone receptors, unlike the membrane receptors, are located either in the cytoplasm or in the nucleus of the cell.

1. Nuclear Receptor Superfamily

The nuclear receptor (NR) superfamily is composed of steroid receptors including androgen, estrogen, glucocorticoid, mineralcorticoid and progesterone receptor, nonsteroidal receptors like thyroid hormone, vitamin D and retinoic acid receptors, ecdysone receptors, which are found in insects, and orphan NR receptors (Table 3), which constitute the majority of the members of the NR superfamily [20]. The NRs are transcription factors that regulate the development and metabolism through control of gene expression. All nuclear receptors are modular proteins that contain one DNA-binding domain (DBD) and one ligand-binding domain (LBD) (Fig. 5). The N-terminal domain (A/B) contains a cell-specific and promoter-specific transactivation domain termed AF-1 that functions autonomously and in a ligand-independent manner. It is the least conserved domain (< 15% sequence homology) across the superfamily. The centrally located DBD (C domain) has a highly conserved sequence (> 50% sequence homology). It contains conserved cysteine residues that form two zinc fingers that are involved in DNA recognition and binding in concert with residues in N-terminal (A/B) and hinge (D) regions. The DNA-binding domain also contains nuclear localization sequences. DBD is also important in the dimerization of the receptor. The C-terminal LBD though is similar in length between receptors; the sequence is highly variable across the family (sequence homology 15–60%). LBD is important for transactivation, hetero- and homodimerization, and is involved in binding of heat shock proteins to inactive receptor. In a classical steroid receptor function, ligand binds to LBD of cytosolic complex containing the receptor and chaperone proteins changing protein conformation in the C-terminus dissociating the corepressor proteins from the transcription complex. The liganded receptor is translocated to the cell nucleus, where the activated receptor protein binds directly to specific DNA sequences of hormone response elements contained in the enhancer and promoter regions of target genes, resulting in transcriptional activation [21]. The first structure of LBD of a nuclear receptor determined was that of unliganded RXRα [22]. Since then, LBDs of several nuclear receptors have been crystallized with and without agonist or antagonists, and the structure of the binding pockets of ligands, agonists, and antagonists has been determined.

Table 3 Nuclear Receptor Superfamily

1. Steroid receptors
     Glucocorticoid receptor (GR)
     Mineralcorticoid receptor (MR)
     Androgen receptor (AR)
     Estrogen receptor (ERα,β)
     Progesterone receptor
2. RXR heterodimeric receptors
     Vitamin D receptor
     Thyroid receptor (TRα,β)
     Retinoic acid receptor (RARα,β,γ)
     Ecdysone receptor (EcR)
3. Orphan receptors
   Orphan receptors (dimeric)
     PPARα,β,γ
     RXRα,β,γ
     HNF-4
     COUP-TF
     TR2α,β
     GCNF
     REV-ERB
     LXR
     PXR
     BXR
     FXR
   Orphan receptors (monomeric)
     SF-1
     ERRα,β
     NGF1—Bα,β,γ
     ROR α,β,γ
     TLX

NRs in the absence of ligand binding are associated with other proteins either in the cytoplasm or in the nucleus. In the absence of ligand, steroid receptors form quaternary complexes with chaperones like heat shock proteins that prevent their interaction with DNA. RXR-heterodimeric receptors bind to their cognate DNA-response elements in the absence of ligand and cause transcriptional repression. In absence of ligand, NR is often associated with corepressor proteins inhibiting basal transcription of target genes. On ligand binding, steroid receptors dissociate from the chaperone proteins and associate with hormone response elements on the DNA and interact with coactivation proteins. Ligand binding changes the conformation of RXR-heterodimeric receptors and activates or

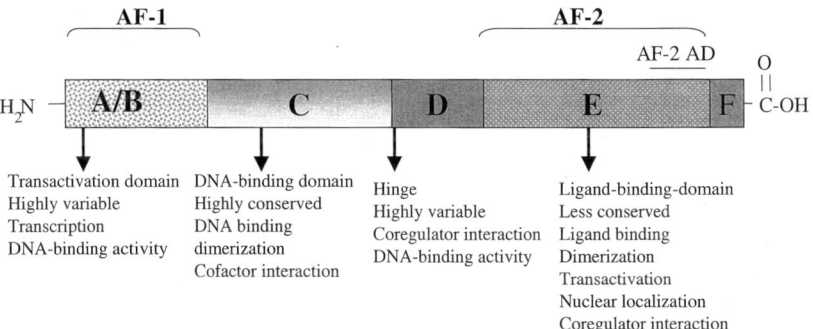

Figure 5 Schematic illustration of a nuclear hormone receptor. The highly conserved DBD (C) is flanked by less well conserved N-terminal and C-terminal regions. Ligand binds to the LBD (E), which is moderately conserved. Dimerization functions are located in the C and E regions. Ligand-dependent (AF2) and independent (AF1) transactivation functions are located in the receptor A/B and E domains, respectively. DNA binding is regulated by residues in the DBD (C) in cooperation with residues in the N-terminal (A/B) region and hinge region (D).

represses transcription of the genes by binding to various coactivators or corepressors.

Orphan Nuclear Receptors. Molecular cloning has identified several orphan receptors sharing similar structure to the nuclear receptors (Table 3). However, the ligands for these orphan receptors are not known at present. Some of these receptors have been recently matched with physiological ligands, e.g., 9-cis retinoic acid was found to bind and activate three of the nuclear receptors classified originally as orphan receptors and later named as RARs [2]. Similarly, though the physiological ligands of many orphan nuclear receptors have not been identified at present, they provided potentially important drug targets as in the case of orphan receptors like peroxisome proliferator activation receptor-γ (PPARγ), which plays a role in adipogenesis and the discovery of thiazolidienediones in the treatment of noninsulin dependent diabetes and regulation of cholesterol metabolism by steroidogenic receptor (SF-1), liver L receptor α (LXRα), and farnesoid L receptor (FXR).

III. RECEPTOR BINDING

A. Basic Considerations

In developing a ligand–receptor-binding assay, a high affinity ligand is needed, and the receptor membranes should have a reasonable distribution of the receptor.

The assay conditions have to be optimized to reduce the nonspecific binding and increase the specific binding. Traditionally, the receptor-binding assay consists of binding with a high-affinity ligand, and the binding of other compounds is determined in a competition binding assay.

1. Radioligand

Radioligand preferably should be a selective, high-affinity ligand for the receptor to decrease NSB and receptor crossover. The radioligand should be chemically and radiochemically pure. A ligand that interacts with high affinity and specificity for a receptor, does not guarantee that after radiolabeling it will retain the same affinity and provide a suitable radioligand. Chiral radioligands are preferred, since the less-active enantiomer may interfere or complicate analysis. It is preferable to have high specific activity for the radioligand (e.g., with $^{125}I \sim 2200$ Ci mmol) wherever possible. Small molecule agonists labeled with the bulky ^{125}I may sometimes change the binding characteristics of the ligand, and in these cases ligand may be labeled with ^{3}H (29–87 Ci/mmol). The fmols of radioligand bound to the receptor can be calculated from the radioactivity assuming 80, 70, and 50% efficiency of counting for ^{125}I, ^{35}S, and ^{3}H, respectively.

For ^{125}I (2200 Ci/matom): $2200 \times 2.2 \times 0.8 = 3,872$ cpm/fmol.

For ^{3}H (29 Ci/matom): $58 \times 2.2 \times 0.5 = 66$ cpm/fmol (assuming two ^{3}H atoms per mol).

For ^{35}S (1500 Ci/matom): $1500 \times 2.2 \times 0.7 = 2,310$ cpm/fmol.

For ^{125}I-ligand, the affinity for the receptor should be in the sub-nM range, and the typical expression level of the receptor should be about 25,000 receptors per cell, which will correspond to about 100 fmol receptor/mg of protein. Assuming under optimal assay conditions with 10 μg of membrane per assay and 10% occupancy of the receptor, a few thousand CPM of ligand will be detected. Similarly, for ^{3}H ligand, the affinity should be in the nM range, and a much higher expression of receptor (about 250,000 receptors per cell) is required, which corresponds to about 1000 fmol of receptor/mg protein. About 50 μg of membrane per assay has to be used to detect a few thousand CPM of ligand bound. For ^{35}S ligand, the affinity should be in the nM range and may be used with 10 μg of membrane per assay from recombinant cells expressing about 100,000 receptors per cell.

In nonradioactive ligand–receptor-binding assays such as fluorescence-based assays, a high affinity ligand is labeled with a fluorophore, which forms a receptor-fluorescent ligand complex. Fluorescein-labeled ligands have been used to bind GPCR receptors [23–29]. Also, a naturally occurring high-affinity fluorescence ligand fluormone ES1 or fluorescein-labeled estrogen (ES2) was used for a fluorescence polarization estrogen receptor-binding assay [30].

2. Assay Conditions

For a ligand–receptor-binding assay, the optimal assay buffer has to be determined. Normally, an isotonic or hypotonic buffer may be used. The pH optimum for the binding reaction should be determined. The ligand stability also may depend on pH. The requirements of monovalent cations such as Na^+ or K^+, divalent cations such as Mg^{2+} or Mn^{2+}, sulfhydryl reagents such as dithiothreitol (DTT), and other cofactors should be determined. When peptide ligands are used, protease inhibitors and BSA are often included in the buffer to stabilize the peptide, and the peptide stock solutions are stored in siliconized polypropylene tubes to reduce the ligand binding to the tubes. BSA reduces the nonspecific binding of the peptide ligand and also reduces the protease activity and increases the stability of membranes or cells. BSA increases the NSB of hydrophobic and lipidlike ligands by binding, and with these ligands BSA can be replaced with human γ-globulin or gelatin in the buffer. The optimal temperature (4°, 20°, 25°, 30°, 37°C) for the binding reaction should be determined. Normally, for HTS, ambient temperature will be advantageous. If there is a large variation of binding activity with temperature for a receptor-binding assay, it is advisable to use a defined temperature at which optimal binding activity is observed as the room temperature fluctuates. When whole cells are used for binding, to minimize internalization of the ligand, the binding reaction is carried out at 4°C. The receptor concentration used in the assay ideally will utilize about 10% of the radioligand added in the assay. However, receptor concentrations that utilize 50% of the radioligand may just be acceptable. In general, the concentration of radioligand used in a receptor binding assay is at about the K_d, except that with a very high-affinity, high-specific radioactive ligand, a lower than K_d concentration is used, and when low-specific radioactive ligand of very high affinity is used, a higher than K_d concentration is employed in the binding reaction.

3. Analysis of Binding

Ligand–receptor binding in the simplest system is analyzed here (for other complex systems consult Refs. 31–33). In a simplest case, the binding of labeled ligand L* to receptor R is a simple bimolecular association reaction given by the equation

$$R + L^* \underset{k_{-1}}{\overset{k_1}{\rightleftharpoons}} RL^* \tag{1}$$

where k_1 is the association rate constant (on-rate) of the ligand–receptor interaction, k_{-1} is the dissociation constant (off-rate) of the ligand–receptor complex, and R and L^* are the free concentrations of receptor and ligand. The rate of change in the receptor–ligand complex with time is given by the difference between the rate of formation ($k_1 \cdot R \cdot L^*$) and breakdown ($k_{-1} \cdot RL^*$) of the ligand–receptor complexes.

$$\frac{d(RL)}{dt} = k_1 \cdot R \cdot L^* - k_{-1} \cdot RL^* \qquad (2)$$

At binding equilibrium, the rates of association and dissociation of the ligand–receptor complex will be equal. Thus at equilibrium, the rate of change of concentration of the complex becomes zero.
Then

$$d(RL^*)/dt = 0 \qquad \text{and} \qquad k_1 \cdot R \cdot L^* = k_{-1} \cdot RL^* \qquad (3)$$

Thus

$$RL^* = \frac{k_1}{k_{-1}} R \cdot L^* = K \cdot R \cdot L^* \qquad (4a)$$

and

$$K = \frac{k_1}{k_{-1}} \qquad (4b)$$

where K is the affinity or association constant of the binding reaction and is the ratio of the association to the dissociation rate constant. The dissociation constant is the inverse of the association constant.

$$K_d = \frac{1}{K} = \frac{k_{-1}}{k_1} \qquad (5)$$

At equilibrium, the concentration of the ligand–receptor complex is given by the Langmuir isotherm:

$$RL^* = \frac{R_t \cdot K \cdot L^*}{1 + K \cdot L^*} = \frac{R_t \cdot L^*}{K_d + L^*} \qquad (6)$$

Where R_t is total of the binding sites. A plot of occupancy RL^*/R_t against $\log_{10}$ free ligand concentration ($\log_{10}L^*$) generates a sigmoidal curve.

A plot of free ligand concentration against the bound ligand gives the concentration dependence of the equilibrium binding of a labeled ligand to a receptor. A typical saturation curve shows at low concentration a linear dependence of RL^* on free ligand; as ligand increases, the slope of the curve decreases, and eventually RL^* reaches saturation at higher ligand concentrations.

A linear transformation of the equation $RL^* = R_t \cdot K \cdot L^*/(1 + K \cdot L^*)$ gives $RL^*/L^* = R_t \cdot K - RL^* \cdot K$. The plot RL^*/L^* (bound/free) vs. RL^* (bound) is called the Scatchard plot. When making a Scatchard plot, specific binding and free ligand cpm are used for RL^*/L^* ratio on the axis. A more rigorous alternative is to express specific binding in sites/cell or fmol/mg protein and free ligand concentration as nM providing correct units for the slope. Extrapolation of the Scatchard plot to the x axis gives an estimate of R_t (B_{max}), and the slope gives an estimate of K or $1/K_d$ (Figure 2B). A Scatchard plot for reversible binding to a single population of receptors possessing a single affinity for ligand gives a linear relation. The plot may deviate from linearity when there is positive cooperativity between binding sites in an oligomeric structure (convex upward curve) or negative cooperativity within an oligomer due to multiple binding sites with different affinities for the ligand (concave upward). These deviations can also result from many artifacts. Another data transformation of saturation binding data that is frequently employed to determine whether ligand–receptor interactions occur via a bimolecular reaction that obeys the mass action law is the Hill plot. In the Hill plot, log $(LR^*/(R_t - LR^*))$ is plotted against log L^*. The Hill coefficient (n_H slope factor) of 1 suggests that ligand is binding to a single species of receptor via a simple reversible bimolecular reaction, and if n_H is > 1, this suggests positive cooperativity; $n_H < 1$ suggests negative cooperativity or heterogeneity of the binding sites. Since Hill plots are usually not completely linear, the data gathered over the range of 30–70% occupancy is considered in calculating the n_H value.

4. Quantitation of the Potency of Competing Agents

In the competition-binding assay, the ability of various concentrations of the test compound (6–12 concentrations differing by log, ½ log or increments of double) in competing a single fixed concentration of ligand (1–2 $\times$ K_d) at a particular receptor concentration is determined. The potency of compounds is calculated as the concentration of competitor that effectively competes for 50% of the specific ligand binding (IC_{50}) and is calculated from the competition-binding curve (Fig. 6). The IC_{50} values of competitors can be obtained by indirect Hill plots, logit-log analysis, computer-based nonlinear regression analysis, or visual inspection. The relationship used in the indirect Hill plots is

$$\log B_1/(B - B_1) = n \log[I] + n \log IC_{50} \qquad (7)$$

where B is the amount of binding in the absence of competitor, B_1 is the amount of binding in the presence of competitor I, and [I] is the concentration of competitor. When percent control specific activity is plotted on the y axis against $\log_{10}$ competitor concentration on the x axis, the IC_{50} is given by the competitor concentration corresponding to 50% control specific binding, and the slope gives the apparent Hill coefficient. A slope of -1 suggests that the radioligand and the competitor interact with a single receptor population. The IC_{50} determined in the competition-binding studies depends on the radioligand concentration [L], and

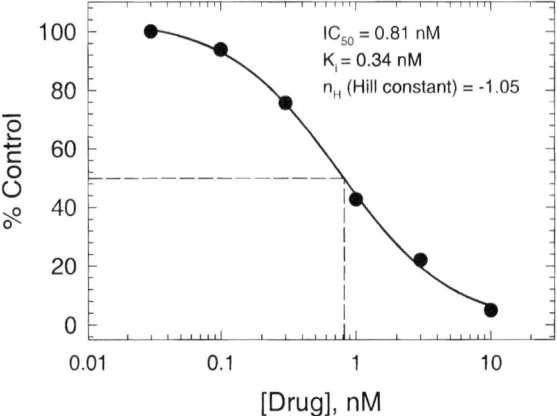

Figure 6 Competition of ligand receptor binding by drug compounds. IC_{50}, the concentration of compound required to inhibit binding activity by 50%, is determined by plotting the percentage control activity against log concentration of the compound. IC_{50} determined for the compound is 0.81 nM. The data is fitted with nonlinear regression curve fit analysis-sigmoidal fit.

with the Cheng and Pursoff equation we can calculate the K_i value for the competitor:

$$IC_{50} = K_i \frac{1 + [L]}{K_d} \qquad \text{or} \qquad K_i = \frac{IC_{50}}{1 + [L]/K_d} \tag{8}$$

Where K_i is equilibrium dissociation constant for the competitor I, and K_d is the equilibrium dissociation constant for the radioligand L. Thus when [L] is at K_d, $IC_{50} = 2 \times K_i$, and when [L] is present at trace concentrations ($[L] \ll K_d$), $IC_{50} = K_i$. In the presence of an increasing concentration of competitive antagonists, fewer and fewer receptors are available for occupancy by the ligand. Thus competitive antagonists suppress agonist-mediated responses by blocking access of the agonist to its specific receptor. The dose–response relationship of an agonist is shifted to the right in the presence of increasing concentrations of an antagonist, giving a series of parallel curves if the antagonist interacts in a truly competitive and fully reversible fashion (Fig. 7A). K_d for the competitive antagonist can be determined by Schilid equation,

$$\frac{[A']}{[A]} - 1 = \frac{[B]}{K_{d_B}} \tag{9}$$

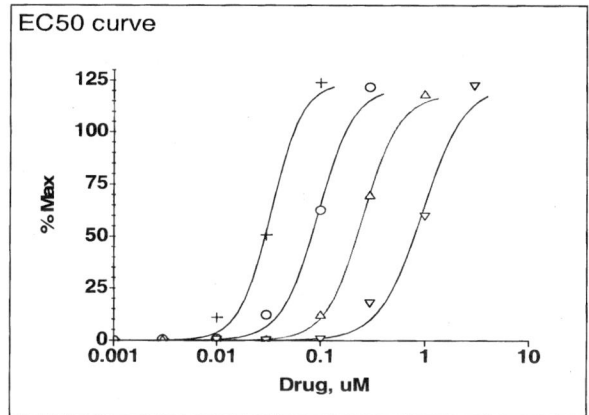

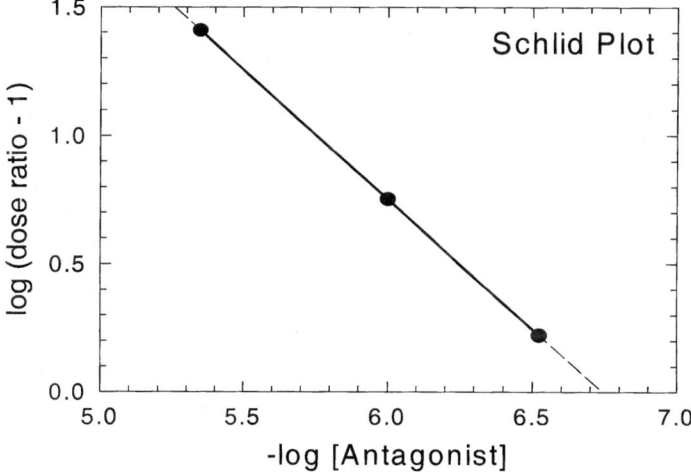

Figure 7 Determination of the K_d for the receptor interaction with a competitive antago-nist. (A) Dose–response relationship of antagonist in the absence (control) or presence of increasing concentration of a competitive antagonist. (B) Schlid plot of the data in panel A to calculate Kd_b. Competitive antagonists suppress agonist-mediated responses by blocking access of the agonist to its specific site on the receptor.

[A']/[A] is the ratio of agonist concentrations that elicits an equal response in the presence [A'] or absence [A] of antagonist and denoted as X. Equation 9 can be rewritten by substituting X for [A']/[A] and taking logarithms:

$$\log(X - 1) = \log[B] - \log K_{d_B} \qquad (10)$$

A plot of log $(x - 1)$ against log antagonist permits the determination of K_{d_B} from the x intercept (Fig. 7B).

B. Receptor Preparations

The source of receptor may be whole tissue membranes, cultured cells, cell membranes, solubilized receptors, or soluble receptors. Some of these methods of receptor preparation are detailed below.

1. Whole Tissue Membranes

Plasma membrane fractions are prepared from a tissue of interest. The tissue is homogenized in cold buffer either in a Potter–Elvehjem homogenizer (soft tissues) or in a Polytron homogenizer (tough tissues) and filtered through two layers of cheesecloth. The filtered tissue homogenate is centrifuged at 600 × g for 10 min, and the supernatant is centrifuged at 100,000 × g for 1 h at 4°C. The supernatant is discarded, the membrane pellet washed twice by suspending in buffer, and centrifugation, to remove endogenous receptor effectors such as nucleotides, ions, and proteolytic enzymes. Protein in the washed membrane pellets is determined, aliquoted into cryotubes, and frozen immediately and stored at −80°C. Prior to the availability of the cloned receptor targets, receptors obtained from animal tissue homogenates have been utilized for screening, which had the disadvantages of heterogeneity, nonhuman pharmacology, and lower receptor expression levels. Membrane fraction may be a preferred approach for receptor binding assays to increase specific activity of low-capacity receptors and decrease NSB.

2. Cultured Cells

Human receptors are preferred as drug targets because the receptors from nonhuman sources may have different pharmacological properties. Mammalian cells over-expressing human or mammalian receptors are preferred for receptor assays as the receptors can be expressed at very high levels (10–1000 K receptors/cell) suitable for automated HTS assays. Most of the cDNAs of human receptors are intellectual property (IP) and cannot be used freely without paying large sums of money. This limits the use of the recombinant receptors. Alternately, clonal cell lines of tumor origin which have higher expression of the receptors than the wild-type cells or receptor bearing natural cell lines as adherent or in suspension cultures provide a more physiological system.

Cultured cells either as adherent or suspension cells can be used for receptor binding and functional activity. Cells grown to near confluency in flasks are plated to the required cell density in 96-well microplates and grown for 1 or 2 days. The growth medium in the plate is removed and incubated with medium without fetal calf serum for 2 to 4 h. Cells are washed with PBS and used for receptor binding. The disadvantages include internalization of radioligand using whole cells, and in these cases the binding reaction can be performed at 5°C. When using suspension cells, cells grown in flasks are dissociated with a dissociation medium (GIBCO), taken into a medium with low fetal calf serum ($< 1\%$), and plated into assay-ready plates (microplates with the compounds) at the cell density optimized for the assay and incubated with labeled ligand and compound for binding assay or with compound for functional assay and processed.

3. Cell Membranes

Cultured cells over-expressing the receptor are harvested by scrapping or treating with dissociation buffer, and the cell suspension is centrifuged. The cell pellet is suspended in hypotonic buffer for 30 min followed by 1 to 3 freeze-and-thaw cycles to break the cells, or they are homogenized in a Potter–Elvehjem or Dounce homogenizer or Polytron homogenizer. The plasma membrane fraction is prepared from the cell homogenate as described for whole-tissue membrane preparation. The membrane fraction is aliquoted into cryotubes and stored at $-80°C$. For the best pharmacology of the receptors, cell membranes are preferred to whole cells, as the ligand can be internalized in cells and may have higher nonspecific binding.

4. Solubilized Receptors

Solubilized receptors can be obtained by treating the membranes with appropriate detergent followed by centrifugation [34]. Receptors with single transmembrane spanning domains can be solubilized best with nonionic detergents like Triton X-100, n-octylglucoside, or Nonidet P-40. These detergents break weak protein–protein interactions and remove endogenous lipids. The critical micellar concentration for these detergents is relatively low (0.3 mM) and form 90 kDa micelles, making removal of the detergent difficult. Solubilization of ion channel receptors like nAChR, $GABA_AR$, and glyR can be achieved with sodium cholate or deoxycholate. As (the micellar weight is relatively low (1–4 kDa), can be removed by dialysis. Some GPCRs can be solubilized using zwitterionic detergents like CHAPS, CHAPSO, digitonin, and digitonin–cholate mixtures. GPCRs solubilized with Triton X-100 or sodium cholate loose receptor activity. Binding ligand to the receptor prior to solubilization can increase retention of the receptor activity. The maximum solubilized receptor activity under optimal conditions does not exceed 35%. The solubilized receptors are utilized mostly in the purification of the receptor protein by one-step affinity chromatography; they are used to determine the protein sequence and in structure studies. Simplicity of solubilized receptor preparation is an advantage. The disadvantages are assay difficulties and conformational uncertainty. Usually, the assays using the solubilized receptor are tedious and not practical for multiple-well plate assays and hence are not widely used.

5. Soluble Receptors

The receptors of the nuclear receptor superfamily are cytoplasmic receptors and can be purified by one-step affinity chromatography if engineered with an appropriate fusion protein such as glutathione-S-transferase (GST), maltose binding protein (MBP), or histidine$_{(6-12)}$ (His$_{6-12}$) tag. Ligand binding domains (LBDs) of cytoplasmic receptors have been expressed in *E. coli*, and receptor activity has been shown in the bacterial extract and in the purified receptor preparations. Receptors can also be expressed in insect cells using bacculo virus expression systems such as GST, MBP, and His$_{6-12}$ fusion proteins with a signal sequence to enable them to be produced as recombinant soluble extracellular receptors. LBD expressed as fusion protein is purified by appropriate affinity chromatography depending on the protein tag. The purified protein is concentrated to concentrations of 1 mg/mL or higher in storage buffers containing required cofactors, aliquoted and stored at $-80°C$ until use.

6. Expression of Human and Mammalian Receptors

Before the availability of cloned receptors, drug-screening programs used mammalian tissue homogenates, which have the disadvantages of heterogeneity of the receptors, low levels of receptor expression, and nonhuman pharmacology. With the advances in recombinant DNA technology, now the human receptor subtypes can be cloned into appropriate mammalian cell lines along with signal transduction proteins to have functional receptor proteins. Stable expression of a single human receptor subtype in a cell line permits selective pharmacological screening and to develop highly selective drugs that discriminate between various receptor subtypes. These recombinant receptor proteins generally have the same ligand binding characteristics as that of the native receptor. If a recombinant receptor expressed in a cell line is linked to the signal transduction mechanism, the functional activity can also be monitored in addition to the receptor binding activity, which enables determination of the binding affinity of lead compounds and rapid characterization of agonist and antagonist activity. Although human receptors expressed in human or mammalian cells show essentially the same binding properties as the native receptor, the receptor pharmacology is shown to be host-cell-specific, making pharamcological characterization difficult in heterologous systems.

Several receptor studies suggested that ligand binding and signal transduction properties when expressed in heterologous systems depend upon the host cell and the expression level of the receptor (receptor density). When a receptor is expressed in different cells, with each type of cell having few specific functional proteins, the binding characteristics may be somewhat similar, but the functional coupling may be different, resulting in different functional properties. As far as possible it will be advantageous to find a null cell for that receptor, which will not complicate the receptor pharmacology of the subtype being expressed. Although a high level of membrane expression of GPCRs in insect cells using the bacculo virus system is achieved, showing appropriate receptor pharmacology for antagonists, agonists showed low-affinity binding (uncoupled) and lacked high-affinity binding (functionally coupled) [2]. When the receptors are expressed in yeast, they showed low-affinity binding and lacked high-affinity binding due to a lack of functional coupling, and thus they are not comparable to that expressed in mammalian cells. Coexpression of the recombinant receptor and the appropriate G-proteins combined with mutation of the endogenous yeast G-protein α-subunit were able to produce functional coupling (detailed in Chap. 6). Construction of chimeric receptors has helped in the identification of functional domain and cytoplasmic domains involved in the signal transduction of many receptors. The critical amino acids involved in ligand binding and signal transduction have been identified by point mutations. These molecular approaches together with molecular modeling approaches will identify the structural determinants of drug binding and allosteric sites on the receptor for drug interactions.

In drug discovery, the receptor of choice is a human receptor subtype from appropriate tissue. If there are no IP issues, the human receptor subtype of interest is expressed in human cells closely resembling the tissue target. Due to IP issues with many human receptors, heterologous expression of the human receptors is not always possible. To avoid infringement on patents of those receptors and to get around this problem, instead of human receptors, receptors from other animal species or chimeric receptors wherever possible may be expressed. However, caution should be exercised when using other nonhuman receptor subtypes, as compounds may be species specific, compounds effective in one species may not be effective on human receptor subtypes in appropriate human cells. To obtain a cell line with the best ligand binding and signal transduction properties, the receptor is usually expressed in several human and mammalian cell lines and screened for optimal binding activity and functional activity. The best receptor expressing cell line that is more representative of the native human receptor pharmacology, and the expression is reproducible from passage to passage, should be selected for use. When a cell line is selected, several clones of the recombinant cells will have to be screened for best expression and stability of the expression.

Some important characteristics to look for in selecting a cell line are discussed in Chap. 3.

All the members of a superfamily of nuclear receptors (cytoplasmic receptors) consist of functional domains of transactivation, DNA-binding, and ligand binding [21]. These individual domains have been expressed in bacteria as fusion proteins of GST, MBP, or $(His)_6$ and shown to be active. This enables the production of large amounts of purified active LBD of the receptor to study the high-resolution receptor structure by physical methods such as x-ray crystallography, NMR, and electron microscopy, which will help in understanding the ligand binding sites and precisely define drug–receptor interactions [35]. LBDs have been expressed in *E. coli* as fusion proteins with a protein tag, often with a protease-sensitive peptide sequence in between. The crude cell extracts have shown good LBD activity and may be used without further purification for screening of binding assays. *E. coli* with expressed LBD is grown in the presence of agonists or antagonists to increase the stability of the receptor, purified by affinity chromatography; the protein tag is removed with appropriate protease treatment, and the pure LBD with or without agonist/antagonist is used for crystallization to determine the structure and binding pockets (regions) of the receptor.

C. Ligand-Receptor Binding Assays

Most of the earlier receptor drug discovery used radioligand-binding assays in which receptor-bound ligand is separated from free ligand. To increase the throughput and to be able to automate the receptor-binding assays, homogeneous assays are being explored.

1. General Separation Methods for the Bound from Free Ligand

a. Radioligand Assays. Filtration and centrifugation are the common methods of separation when cell suspension or membrane preparations are used with radioligand. If solubilized receptor is used, the methods of separation are gel-filtration, precipitation by poyethyleneglycol, or charcoal adsorption. In the case of adherent cells, after incubation of cells with radioligand in a microtiter plate, the unbound radioligand is removed and the plate is washed ($\times$5) with buffer, the cells are extracted with 0.5 M NaOH, and scintillant is added and plate is counted in a scintillation counter.

FILTRATION ASSAY. The filtration assay is the most common assay format used with receptor membranes or whole cells. The binding reaction is performed in a 100 µL volume in a 1.2 mL tube in 96-well racks (Marsh tubes in racks). The reaction contents are filtered rapidly onto a 96-well GF/C, GF/B, or GF/F plate (pretreated with 0.1% polyehyleneimine to reduce NSB) in a cell harvester (Packard) or a 96-well plate filtration unit from another manufacturer, and washed (4 × 1 mL buffer). The washed plate is air-dried; the bottom of the plate is sealed, and scintillant is added and counted in a scintillation plate counter. The filtration assay only gives medium throughput and cannot be used in the high-density plate format as the filtration devices are not yet available to increase the throughput for HTS.

CENTRIFUGATION ASSAY. The binding reaction is done in a 100 µL volume in a 96-well microtiter plate. The reaction is terminated by adding 50 µL reaction to 1 mL washing buffer in a 1.2 mL tube 96-rack. The tube rack is centrifuged immediately, and the supernatant is aspirated. If the radiolabel is ^{125}I, the tube is placed in a 15 × 75 mm tube and counted in a gamma counter. If the radiolabel is ^{3}H, 100 µL Soluene 350 is added to each tube and solubilized overnight by agitation on a shaker. Scintillant is added; the contents are mixed, placed in plastic scintillation vials, and counted. The centrifugation assay is labor intensive, low throughput, and not adaptable to HTS.

GEL-FILTRATION ASSAY. The gel-filtration method is used for receptor binding assays for soluble receptors, and detergent solubilized receptors. The LBD of NRs can be expressed in *E. coli* as a fusion protein with GST, myelln basic protein, or His_6 tag. The NR LBD purified by affinity chromatography can be used for binding with radioligand by the gel-filtration method. After incubation of LBD with ^{3}H-radioligand (at a concentration of K_d) in a microtiter plate in a 100 µL volume, 50 µL of the reaction mixture is applied on the top of 1.0 mL Sephadex G-25 in a 96-well gel block (Edge Biosystems) (precentrifuged to remove the excess of liquid) placed on top of a clean 96-well collection plate and briefly centrifuged. The unbound radioactive ligand stays in the gel, and the receptor bound radiolabel is excluded from the gel. To the filtered samples, Microscint-40 (Packard) is added, mixed for 1 h, and the radioactivity bound to the LBD is measured in a TopCount (Packard) or MicroBeta (Wallac). Because of the centrifugation and scintillant addition steps, this assay is only medium throughput; it generates radioactive waste, and the per assay cost is very high.

PRECIPITATION BY POLYETHYLENE GLYCOL. This method is used for the assay of binding to soluble receptors. The binding reaction is done in 100 µL volume in a 1.2 mL tube in a 96-rack. At the end of incubation, 7.5 µL of 3.3% (w/v) bovine γ-globulin followed by 42.5 µL of 36% (w/v) poyethyleneglycol (6000–8000) are added. The contents are mixed thoroughly, incubated for 30–60 min on ice, and centrifuged at 10,000 × g for 10 min. The supernatant is removed, the pellet is dissolved in the buffer, and scintillant is added and counted in a counter. Alternatively, the protein precipitate is filtered onto a GF/C filter plate in a harvester and washed in 5 × 1 mL buffer with 7.5% PEG. Scintillant is added and counted for radioactivity. This is a low-throughput assay and is not useful for HTS.

CHARCOAL ADSORPTION.   This method is used for the assay of binding to soluble receptors. The binding reaction is done in a 100 μL volume in a 1.2 mL tube in a 96-rack. At the end of incubation, 300 μL of charcoal suspension in buffer is added, mixed vigorously, incubated on ice for 30 min, and centrifuged. The free ligand binds to charcoal, and the receptor-bound ligand stays in solution. About 100 μL of supernatant from each is transferred into another microtiter plate. Scintillant is added, mixed, and counted. This is a low-throughput assay and is not useful for HTS.

ADHERENT CELL ASSAY.   The cells grown fresh in a 96-well plate are washed with phosphate buffered saline (PBS), and cells are incubated with reaction components in a total volume of 100 μL. After incubation, the reaction mix is removed, washed (5 × 300 μL), and extracted with 50 μL of 0.5 M NaOH. To the solubilized cell extract scintillant is added, mixed, and counted. The assay has low throughput and is not adaptable for HTS.

b. Heterogeneous Binding Assays with Nonradioligands.   Receptor binding assays have also been developed with nonradioligands using ELISA and TRF assays.

ELISA ASSAY.   Nonradioactive solid-phase ELISA has been used in HTS as a receptor binding assay. A sensitive HTS assay used for type 1 interleukin-1 receptor binding consisted of immobilizing the ligand, IL-1Ra, in the wells of a 96-well plate, incubating with receptor sIL-1R with or without compound, washing the plate with PBS buffer, incubating with antibody to the extracellular domain of IL-1R, washing 4× in buffer, and developing with *o*-phenylenediamine [36]. The signal increased with increasing binding of receptor to the immobilized ligand. Similarly, a sensitive solid-phase ELISA was developed for screening biotin-labeled PDGF (PDGF BB), by coating PDGFR onto a 96-well plate, followed by incubation with PDGF BB, washing excess unbound biotin-PDGF BB, further incubation with neutravidin-horseradish peroxidase, and the ligand bound is quantitated by developing color [37]. The signal is proportional to the amount of receptor binding.

TIME-RESOLVED FLUOROMETRIC (TRF) ASSAY.   A TRF assay has been reported for human galanin receptor, subtype 1 (hGalR1) with adherent monolayer cells or cell membranes using europium-labeled galanin (Eu^{3+}-galanin) [38]. Eu^{3+}-galanin has the same affinity as that of unlabeled galanin to galanin receptor. The recombinant cells expressing galanin receptor are incubated with Eu^{3+}-galanin, the reaction is terminated by washing 5× in cold buffer, DELFIA enhancement solution is added, and the amount of ligand bound to the receptors is measured by time-resolved fluorometry. The TRF assay is a sensitive nonradioactive assay and gives appropriate pharmacology to that previously described for hGalR1. This TRF assay can be used for HTS for finding small molecule or peptide agonists and antagonists for a wide variety of receptors.

2. Homogeneous (Nonseparation) Receptor–Ligand Binding Assays

The traditional radioligand binding assays involve separation of free ligand from receptor-bound ligand by filtration or centrifugation when membranes are used and with solubilized receptor protein gel filtration, precipitation, and charcoal adsorption assays. All these radioligand binding assays involve several steps, generating large volumes of radioactive liquid waste, and are very tedious and labor intensive for screening. These assays are at best low-to-medium throughput assays. In the homogeneous assay formats, the free ligand (unbound) need not be separated from the bound ligand, unlike the heterogeneous assays. Some of the homogeneous assays that have been used for receptor-binding assays include assays based on the scintillation proximity assay (SPA), FlashPlate, fluorescence polarization (FP), fluorescence confocal microscopy (FCS), TR-FRET, Alpha-Screen, chemiluminescence, and others.

a. Radioactive Homogeneous Binding Assays

SPA SCREEN. SPA has been successfully applied to receptor-binding assays by immobilizing receptors directly to SPA beads by a number of coupling methods (SPA principles are detailed in Chap. 4). When the radioligand binds to the receptor immobilized on the SPA bead, it will be in close proximity to stimulate the bead to emit light, whereas the unbound radioligand is too distant from the bead to transfer energy and goes undetected and need not be separated from the receptor-bound radioligand (Fig. 8). The SPA method is generally applicable to ^{3}H- and ^{125}I-labeled ligands, and several labeled ligands are available for a large spectrum of receptors from vendors. When labeled natural ligands are not available, high-affinity agonist or antagonist is radiolabeled by custom radiolabel synthesis groups within or outside the organization.

Two generic beads, wheat-germ agglutinin (WGA)-PVT SPA beads and polylysine(PL)-YS SPA beads, have been used for receptor assays. The WGA-PVT bead (density 1.05 g/cm^3, capacity 10–30 mg membrane protein per mg bead, binding to N-acetyl-b-D-glucosamine residue in membranes) has been widely used for receptor assays. With the PL-YS bead (capacity 10 mg per mg bead), the interaction is a nonspecific electrostatic interaction between the positively charged PL-YS bead and negative charges of membranes. If the recombinant receptor is expressed with GST or $(His)_{6-12}$ tags, the glutathione SPA bead or the copper SPA bead can be used, respectively. Neuropeptide Y (NPY) receptor is a GPCR, and several subtypes based on the pharmacology of binding have been identified. In addition to heterologous expression of these receptors in mammalian cells, some cell lines have been characterized of selectively expressing a

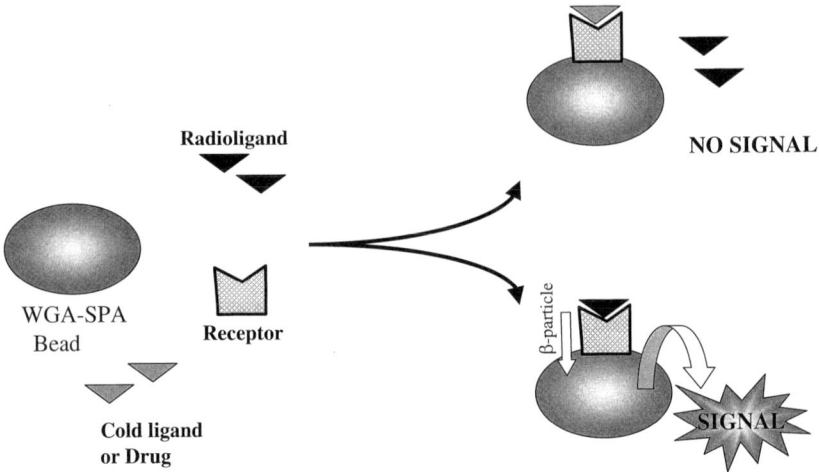

Figure 8 Schematic presentation of an SPA receptor ligand binding assay. A membrane receptor through the N-acetyl b-D-glucosamine residue in the membranes binds to wheat germ agglutinin (WGA) coated on the SPA bead. When the radiolabeled ligand binds to the receptor attached to the SPA bead, it will be in close proximity to the scintillant on the bead, generating photons, which can be measured in a TopCount or MicroBeta plate counter. When a compound competes with radiolabeled ligand in binding to the receptor, the radiolabel is displaced by the competitor and will have a reduced signal. The radiolabel, which is not bound to the SPA bead, will not be in close proximity of the scintillant and will not give any signal.

particular subtype of NPY receptor, such as SMS-KAN cells expressing NPY-Y_2 receptors and SKNMC cells expressing NPY-Y_1 receptor. ^{125}I-NPY binding to membranes by filtration assay is compared with SPA assay using the WGA-PVT SPA bead and NPY agonists (NPY analogs) for competing the binding to NPY-Y_2 receptor (Fig. 9). Though the signal-to-noise ratio is higher in a filtration assay than in an SPA assay, the profiles of competition of ^{125}I-NPY binding by NPY analogs were similar, the IC_{50}s obtained were similar, and the rank order of potency for the agonists was the same in both assays. This suggests that the SPA assay can substitute for the conventional radioligand-binding assay; as it is a homogeneous assay, the throughput can be increased, and the assay can also be automated.

Solubilized receptor approaches involving antibodies or biotinylation have also been used. Biotinylation of receptor can be achieved by chemical methods [39] or the enzymatic method [40,41]. In the enzymatic biotinylation, a unique proprietary biotinylation sequence (Avidity, Denver, CO) is engineered at the

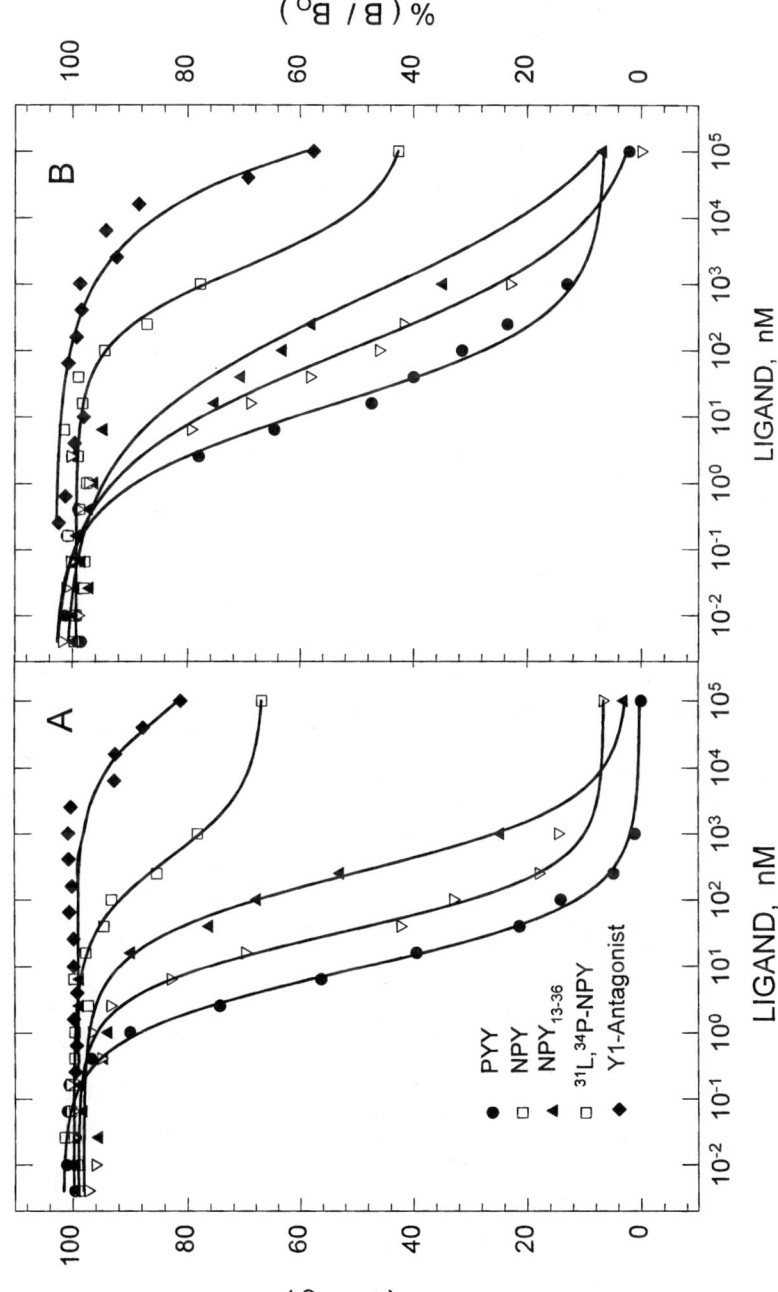

Figure 9 Comparison of NPY receptor competition binding assays by (A) filtration and (B) SPA methods. The IC_{50} values obtained using NPY-Y2 R membranes for various agonists and antagonists were similar in both methods, and the rank order potency of the compounds was the same in both methods, suggesting that the SPA method can substitute for the traditional filter binding assay.

end of a receptor that can be biotinylated with biotin ligase in the purified preparation or by coexpression of biotin ligase in the cell. With biotinylated receptor, streptavidin SPA beads can be used to capture the receptor, and the radioligand bound to the receptor due to close proximity will stimulate the scintillant on the bead to emit a light signal. Purified PPARγ-LBD is biotinylated nonenzymatically and assayed with ^{3}H-agonist and the SPA-streptavidin bead [39]. A receptor-binding assay by SPA also can be done with LBD of a NR expressed with a His$_6$ tag using a SPA-copper bead and ^{3}H-ligand. LBD-His$_6$ is incubated with radioligand (at about K_d concentration) and an SPA-copper bead, and the microplate is counted in a plate counter. These SPA receptor-binding assays are robust homogeneous assays adaptable to HTS.

It is important to confirm that there is little or no nonspecific interaction of the radioligand with the bead. Nonspecific interactions can be reduced by the addition of BSA or detergent or by variation of the ionic strength of the buffer or antagonist. There are three possible formats of addition of the SPA bead: addition of a bead precoupled to the receptor membranes (or receptors), addition of beads at the time of radioligand addition, or delayed addition of the bead after radioligand and membranes or receptors have attained equilibrium. The optimum bead-to-membrane ratio has to be determined at a fixed concentration of radioligand, usually at or around K_d, to obtain highest specific binding. Optimal concentration of the ligand is determined to maximize the signal-to-noise ratio.

FLASHPLATE ASSAYS. The FlashPlate is a 96-well polystyrene microplate with plastic-scintillant-coated wells. The FlashPlate is precoated with polyethylene imine (0.1%) for 24 h at 4°C, washed with buffer, and coated with recombinant receptor (5 μg protein/well) for another 24 h [42,43]. The nonspecific binding sites are blocked by treatment with 1% BSA, and the plates can be stored for 6 weeks or longer at 4°C. The binding assay is done in the receptor-coated FlashPlate by adding the radioligand, competitor, and buffer, incubating, and counting the plate in a TopCount. Washing with PBS or assay buffer after incubation can reduce background. Although the assay can be a homogeneous assay, the plates have to be coated with receptor, which may not be uniform batch to batch. A new generic WGA coated FlashPlate is now available that can be used for receptor-binding assays with receptor membranes and ^{125}I-ligands.

b. Nonradioactive Homogeneous Receptor-Binding Assays. The homogeneous radioactive methods reduce the radioactive waste generated compared to conventional radioactive receptor-binding assays. However, these assays still use environmentally sensitive radioisotopes and thus create handling problems. Some homogeneous nonradioactive methods that obviate this problem are described below.

FLUORESCENCE POLARIZATION (FP) RECEPTOR-BINDING ASSAY. The throughput for receptor-radioligand-binding assays is generally low except for

SPA assays. The FP is a very robust homogeneous assay platform that has been gaining prominence in the receptor-binding assay field (the principles of FP are described in Chap. 4). Though the utility of this technique for cell-surface receptors was demonstrated with fluorescein-labeled ligands by analytical methods [23–27], the FP assay was not used for screening compounds in HTS of membrane receptors until recently. With new advances in FP measurement with sensitive fluorescence plate readers (LJL's Analyst and Acquyest, PolarStar from BMG, Polarion from Tecan, Victor2V from Wallac) that can measure in 96-, 384-, or 1536-well plates, it is now possible to extend FP to membrane receptors. Indeed, recently, FP receptor ligand-binding assays have been demonstrated for GPCRs, e.g., vasopressin V1a, δ-, κ-, and μ- opioid receptors, β1-adrenocepter, 5-HT3, neurotensin, and melanocortin-3,4- and 5 receptors [28,29]. For successful application of the FP method to HTS receptor-binding assays, the membrane receptor has to be expressed in a high copy number ($\sim$ 100,000 per cell or 0.5–2.0 pmol per mg membranes) in each cell. In addition, the fluorescent ligand has to be very high affinity for the receptor ($\sim$ 1 nM), and a substantial amount of tracer ligand ($>$ 20%) has to bind to the receptor to be able to see changes in FP values.

The FP method has been successfully used for nuclear receptors [30]. Nuclear receptors are a superfamily of ligand-induced transcription factors that are regulated by binding of lipid-soluble ligands. NR-LBD can be recombinantly expressed as a fusion protein and purified by affinity chromatography retaining the ligand-binding activity. FP assays amenable for HTS using purified NR protein and a fluorescent ligand have been developed for estrogen receptors ERα and ERβ using full-length recombinant receptors [30] and Fluormone® ES1 (a naturally occurring intrinsically fluorescent nonsteroid estrogen) or Fluormone® ES2 (fluorescein-labeled estrogen) as ligand. When using fluorescein derivatives, the sample is excited with polarized light at λ 495 nm, and the emission at λ 530 nm parallel and perpendicular to the plane of excitation is measured. The ligand is incubated with the receptor protein for 1 h, and the FP signal is measured in a plate reader. The kinetics of binding can also be measured easily. The assay produced a robust signal of 200 milli P. The ERα-FP competition assay gave the same rank order of potency (estradiol $>$ rolaxifen $>$ tamoxfen $\gg$ testosterone) as was observed with the radioactive estradiol gel-filtration assay (Fig. 10). The FP ligand-binding assay can be extended to other NRs using fluorescein, Bodipy, or rhodamine-labeled agonists.

FLUOROGENIC ASSAY. 1-Anilinonaphthalene-8-sulfonic acid (ANS) is extensively used as a fluorescent probe that interacts with hydrophobic pockets of proteins and is used for the identification of competitive inhibitors of fatty acid binding protein [44]. ANS is nonfluorescent in aqueous solutions but becomes appreciably fluorescent when bound to hydrophobic pockets of proteins and other molecules. ANS binding to NR-LBD resulted in an increase in fluorescence, and

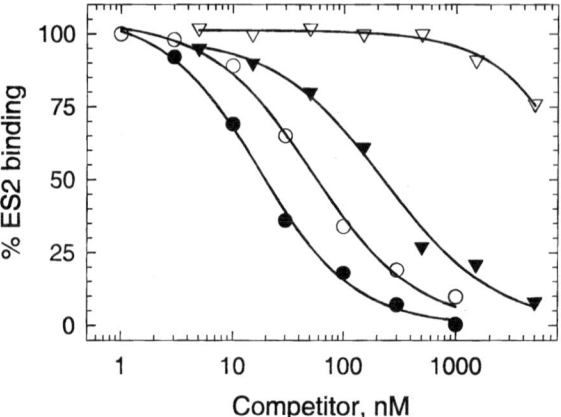

Figure 10   Fluorescence polarization competition binding assay for estrogen receptor α. Flurmone ES2 (estrogen labeled with fluorescein) was used as the ligand. The results showed that the rank order of potency of Estradiol (●) > Raloxifen (○) > Tamoxifen (▼) ≫ Testosterone (▽).

the bound ANS is competed with agonists and antagonists (Fig. 11). The ANS fluorogenic assay is a homogeneous assay adaptable to HTS. The disadvantages of this assay are that it requires relatively large amount of protein and that ANS binds to hydrophobic pockets not necessarily in the active binding site. Even when the screening of a compound library is done with this assay, the hits have to be confirmed by another suitable method.

FMAT RECEPTOR–LIGAND-BINDING ASSAYS.   FMAT is fluorescence-based, homogeneous cell and bead based nonradioactive assay (described in Chaps. 3 and 4). FMAT can be used for G-protein coupled receptors and other membrane receptors on intact whole cell with peptide/protein ligands. Ligand labeled with *Cy5* dye is coated on the bead surface and incubated with cells in a 96-, 384-, or 1536-well plate. The binding of fluorescent ligand to the whole cell receptor on the plate is measured in mm^2 without interference from background fluorescence. Different size beads can be labeled with different *CY5*-dye-labeled ligands for different receptors, and the cells expressing these receptors are incubated with these beads. The fluorescence associated with each receptor-bound fluorescent ligand can be measured, thus enabling a screen for more than one receptor binding (multiplexing). Other receptors tested by FMAT include substance P, Neuropeptide Y, galanin, neurokinin A, bradykinin, somatostatin, angiotensin, and nuclear receptors [45].

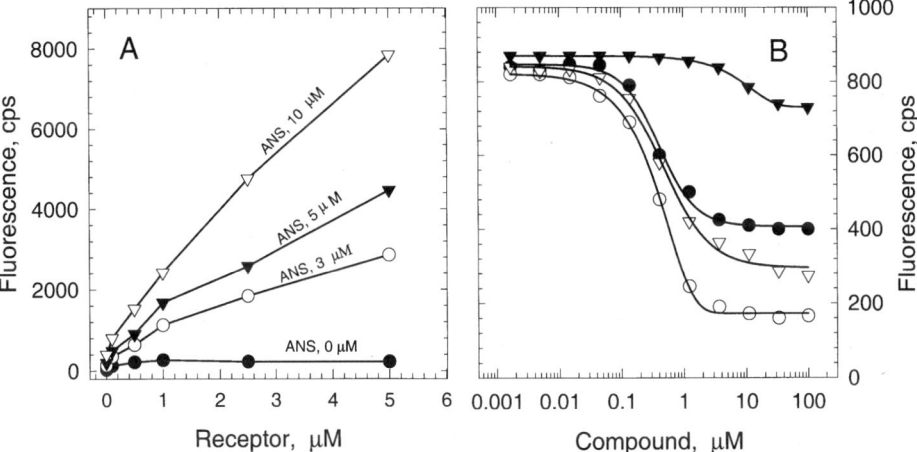

Figure 11 A fluoregenic assay using anilinonaphthalene sulfonic acid (ANS) for an orphan nuclear receptor. ANS in aqueous solution is nonfluorescent, and when it binds to protein it yields fluorescence. (A) Dose-dependent response to receptor concentration. Fluorescence signal increased with increasing ANS and receptor concentrations. (B) Competition binding of ANS by agonists. The known agonists competed ANS binding to the receptor, but some of them competed only partially, whereas they competed fully in a gel-filtration binding assay with a ³H-agonist, suggesting that ANS also binds to other hydrophobic pockets in addition to the agonist binding pocket.

HTRF SCREEN.   To generate a HTRF signal for a membrane receptor with a peptide ligand, a semidirect labeling method can be used. The ligand can be labeled with europium cryptate [(Eu)K], and a monoclonal antibody against a nonbinding region of the receptor is labeled with XL665. In the competition-binding assay the signal is reduced with agonists and antagonists binding to the receptor. A HTRF competition-binding assay for epidermal growth factor (EGF) receptor using EGF- [(Eu)K], anti-EGFR antibody labeled with XL665, and A431 cell membranes containing EGFR was described [46]. In another HTRF receptor–ligand assay, recombinant human interleukin 2 (IL-2) and a monoclonal antibody (nonneutralizing antibody) against the human IL-2 receptor α-chain were labeled with europium chelate and Cy5, respectively [47]. When the Eu^{3+}-labeled ligand is incubated with crude hIL-2Rα membranes from recombinant bacculo virus-infected Sf9 insect cells, a ligand–receptor complex is formed allowing FRET to occur between Eu^{3+} and Cy5 upon excitation of the donor, and the FRET signal is measured. The FRET assay showed that it could be used for measuring the binding kinetics and for HTS.

VESICLELIKE PARTICLE TECHNOLOGY. Evotec's new vesiclelike particle (VLiP) technology can be used for GPCR binding assays. Specific tags attached at the C-terminus of GPCR interact with Gag with high affinity and result in cotransport of GPCR with Gag to the cell membrane and incorporation into the membrane. When Gag concentration is high enough, VLiPs start budding and are released from cells, with each VLiP containing up to 100 functional GPCRs. Binding of a fluorescently labeled ligand to GPCR-embedded VLiPs can be directly monitored in a confocal fluorescence plate reader as a homogeneous assay. TAMARA-labeled endothelin-1 (ET-1) binding to VLiPs containing human ET_AR has been measured by fluorescence intensity distribution analysis in a confocal fluorescence plate reader [48]. This assay when compared to CHO membranes containing ET_AR assay is more sensitive, and the profile was identical.

VLiP-based orphan receptor homogeneous assays were also developed, wherein the VLiPs loaded with orphan receptor when binds a putative agonist or antagonist, the GPCR undergoes conformational change exposing thiol groups that can interact with a specific dye, and the signal can be measured [48].

AMPLIFIED LUMINESCENT PROXIMITY HOMOGENEOUS ASSAY SCREEN (ALPHASCREEN™). ALPHAScreen™ is a homogeneous assay (described in Chap. 4) and has been applied for a TNFα competition receptor-binding assay using streptavidin donor beads, biotin-TNFα, and soluble subunits of TNFR1 and anti-sTNFR1 antibodies conjugated to acceptor beads. Using this assay, IC_{50}s measured for TNFα and TNFβ were 6.6 nM and 780 nM, respectively [49].

ELECTROCHEMILUMINESCENCE (ECL) ASSAY. ECL technology was developed by IGEN International as ORIGEN technology (described in Chap. 4). An ECL binding assay for granulocyte colony stimulating factor receptor (GCSFR) has been described using ruthenylated GCSF (Ru-GCSF) and paramagnetic beads coated with antiGCSFR antibody or beads coated with sheep anti-mouse IgG complexed with anti-mouse GCSFR [50]. IgG-precoated beads complexed with anti-mouse GCSFR gave a better signal than beads coated only with anti-mouse GCSFR. AntiGCSFR antibody-coated IgG beads were incubated with test compound for 30 min, Ru-GCSF was added and incubated for 1 h, ORIGEN assay buffer containing tripropylamine was added, and the signal was determined in the ORIGEN analyzer. The magnetic beads with antibody to receptor were captured on the surface by a magnet; receptor bound Ru-GCSF, TPA was introduced into the flow cell, and voltage was applied. Both TPA and Ru^{2+} were oxidized. TPA by losing a proton becomes a reducing agent and transfers an electron to Ru^{3+} and is promoted to an excited state Ru^{2+} and decays to a ground state Ru^{2+} producing a photon, which is measured in ORIGEN analyzer. In the assay, the magnet retains the beads and TPA washes the free ligand, reducing the background; no separation steps are required. ORIGEN M-8 with 8 modules can be used for HTS.

D. Receptor Binding for Different Classes of Receptors

1. GPCRs

Different ligand binding modes may exist for different subfamilies of GPCRs. For example, receptors for small molecules (amines, nucleotides, eicosanolds, and lipid moieties), peptide hormone (PTH, GLP-1, glucagon, calcitonin, vasoactive intestinal peptide [VIP]) receptors, protease activated (thrombin) receptors, glycoprotein hormone (LH, FSH, hCG, TSH) receptors, and neurotransmitter (Ca^{2+}, glutamate, GABA) receptors have different binding sites. [^{125}I]-Ligands have been used in filtration assays with cell or tissue membrane preparation of receptors. Fluorescent peptide ligands have been used in receptor-binding assays by the FP method [28,29].

The small molecule ligands (biogenic amines, nucleotides, eicosanoids, and lipid moieties) bind primarily in the TM core and exoloops by several distinct mechanisms. Small peptide ligands (N-formyl met-leu-phe and GnRH peptide) bind in the TM core and exoloops of the receptor. The C-terminal region of angiotensin II enters the TM core of the receptor, and the N-terminal amino acids seem to pair with exoloops 2 and 3.

Glucagon, calcitonin and VIP are 30–40 amino acid peptide hormones, and their receptors have a 116–147 amino acid N-terminal segment. Though the N-terminal segment is primarily responsible, the exoloops also are required for high-affinity binding of the ligand.

Protease ligands such as thrombin bind and cleave the N-terminal segment of the receptor. The new N-terminal segment acts as a tethered ligand and interacts with exoloops to generate a signal. The released peptide Met-Arg binds to platelets and stimulates aggregation. The new NH_2 terminus generated acts as a tethered ligand, which in turn activates the receptor. The thrombin receptor activating peptide (TRAP), and other peptides containing a similar sequence as the tethered ligand, mimic its action. Activation of the human platelet thrombin receptor can occur without prior cleavage of the receptor by peptides mimicking TRAP sequence SFLLRR or SFFLRR (human and rodent respectively). The binding assay consists of competition of binding of ^{3}H-SFLLRR, a hexamer TRAP to human platelet thrombin receptor in a filtration-binding assay.

A two-step interaction was proposed for PTH, wherein the receptor first forms a transient PTH-N-terminal segment of the receptor; next this complex interacts with the membrane domain of the receptor to generate a signal.

Glycoprotein hormones (TSH, FSH, LH, hCG) bind to the very long N-terminal segment ($\sim$ 350 amino acids), which in turn interacts with exoloops to generate a signal. Primarily, [^{125}I]ligands have been used in receptor binding assays by the filtration method (described in Sec. III.C.1.a). Wherever possible, receptor binding with fluorescent peptides by FP assay have been used in a homogeneous format.

2. Nuclear Receptors

The ligand binding to the cytoplasmic receptors can be assayed in whole cells or with the soluble receptor. When employing native or recombinant mammalian cells expressing the nuclear receptor, the ligand binding to the receptor can be assayed using radiolabeled ligand (^{3}H-ligand) by filtration assay (see Sec. III. C.1.a). When using cell-free extracts of LBD expressed in *E. coli* or purified LBD, the radiolabeled ligand to the receptor can be assayed by a gel-filtration assay, which separates the bound ligand from free ligand (see Sec. III.C.1.c). With LBD expressed as fusion protein with His$_6$ or GST, radiolabeled ligand, and copper-SPA bead or GST-bead, respectively, a HTS SPA assay can be devised for ligand–receptor-binding (see above, ''SPA Screen''). FP ligand–receptor-binding assays with appropriate fluorescent ligand (or agonist or antagonist) have been used in HTS (see above, ''Fluorescence Polarization'').

IV. FUNCTIONAL (SIGNAL TRANSDUCTION) ASSAYS

Receptor-binding screens can identify molecules interacting with receptors at binding sites or allosteric sites. To identify the lead compounds from the binding screen that are of interest, to subject them further to in vivo testing, they have to be successful in the cell-based functional assays. Upon ligand binding to receptor, a series of signaling pathways is activated, leading to downstream intracellular interactions. Different classes of receptors may be coupled to different signal transduction pathways, however; there may be some common functions in different classes of receptors. Some functional screens are described in Chap. 8. An attempt will be made to discuss many of the known functional assays.

A. G-Protein Coupled Receptor Superfamily

Agonist binding to GPCRs may result in some combination of the following cellular responses: stimulation or inhibition of adenyl cyclase (AC), activation of phospholipase C (PLC) and generation of inositol triphosphate (IP$_3$), activation of protein kinase C (PKC) and an increase in intracellular Ca^{2+}, activation of phospholipase A$_2$ (PLA$_2$) and the generation of the arachidonic acid as second messenger, activate phospholipase D (PLD), act on MAPK pathway through ras signaling cascade and JNK pathway through rac 1/Cdc42 dependent biochemical route, and activation of inwardly rectifying K$^+$ channels or voltage-dependent N-type as well as P/Q type Ca^{2+} channels (Fig. 12).

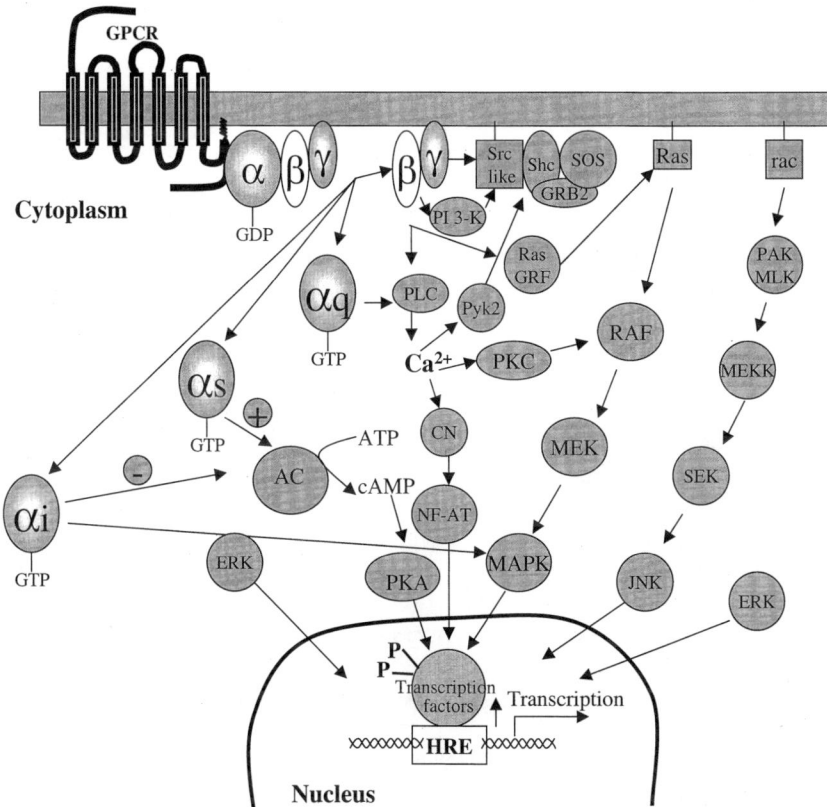

Figure 12 Signal transduction mechanisms involved in G-protein coupled receptors (GPCRs). GPCRs of each subfamily may be coupled to different G-proteins and may activate certain pathways. The pathway connecting GPCRs to low molecular weight GTPases and kinase cascades is not yet fully known.

GPCRs have 7 transmembrane (TM) helices; the intracellular loops that connect these helices form the G-protein binding domain. Binding of ligand to GPCR causes changes in 3 and 6 TM helices, which effects conformation of G-protein-interacting intracellular loops of the receptor and activates the G-protein binding site, which are previously masked. Binding of agonist to a stimulatory receptor induces coupling of heterotrimeric G-protein, and induces GDP release from GDP-$G\alpha_s$ protein (Fig. 13). GTP binding to $G\alpha_s$ protein leads to the dissociation of $G\alpha_s$-GTP subunit from the $G\beta\gamma$ complex and activation of downstream effectors by both $G\alpha$-GTP and free $G\beta\gamma$ subunits. Sixteen α subunits have been

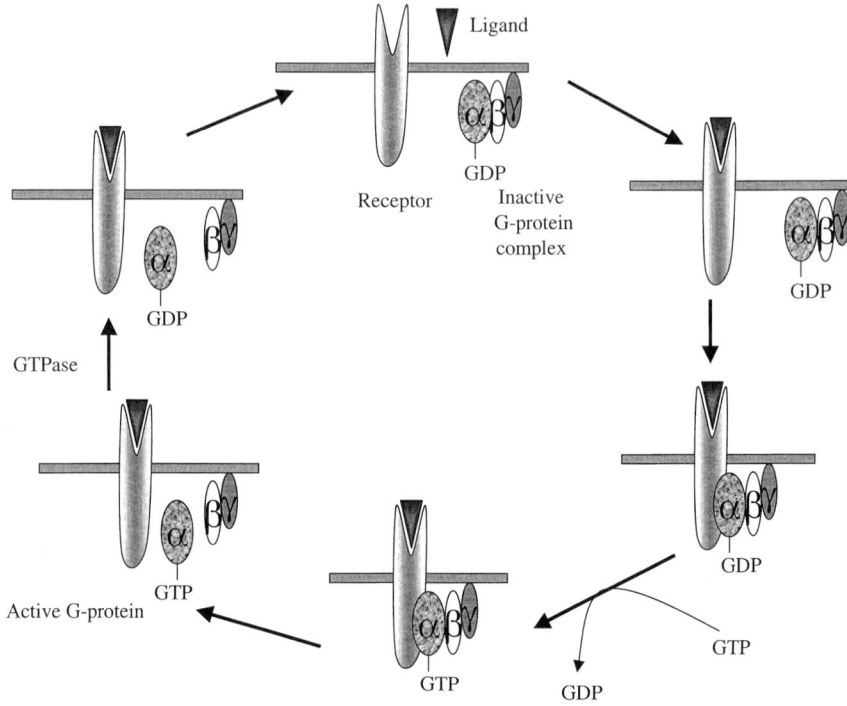

Figure 13 G-protein regulatory cycle showing that the activated receptor interacts with heterotrimeric G-protein complex and induces GDP release from the G-protein. The G-protein activation leads to GTP-binding to the Gα-subunit. The βγ subunits of heterotrimeric G-proteins enhance receptor interaction with the α subunit. Free Gβγ formed after Gα-GTP dissociation from Gβγ is an activator of many effectors. The Gα-GTP after GTPase action forms Gα-GDP, which on reassociation with Gβγ forms inactive G-protein complex.

cloned, which are divided into four families α_s, α_i, α_q, α_{12}, and α_{13}; eleven β subunits and five γ subunits have been identified [10]. In the GTP-bound active conformation, a new surface is formed on Gα* subunits, and they interact with effectors 20–100-fold higher affinity than in their GDP-bound state. Gα$_s$* activates AC, Gα$_i$* inhibits AC, effects ion channels, phospholipases and phosphodiesterases, Gα$_t$* activates photoreceptor cGMP phosphodiesterase, and Gα$_q$* activates phospholipase C-β [18,51]. GTPase intrinsic to Gα-subunit hydrolyzes GTP from Gα-GTP to GDP, deactivating the G-protein. G-protein deactivation is rate limiting for turnoff of the cellular responses.

1. GTPγS Binding Assay

GPCRs are 7TM-spanning domains with 3-extracellular and 4-intracellular loops. In the resting state (receptor unoccupied), the G-heterotrimeric protein is in an inactive state with GDP bound to Gα, and in this state there is low (basal) rate of GDP-GTP exchange. Binding of agonist to GPCR induces coupling of heterotrimeric G-protein to the receptor and induces GDP release from GDP-$Gα_s$ protein. This allows GTP binding to $Gα_s$ protein to a great extent. Using [^{35}S]GTPγS, a nonhydrolyzable GTP analogue, its binding to agonist-induced receptor is seen as an increase in radioactivity associated with the receptor.

[^{35}S]GTPγS binding to agonist-activated GPCR membranes can be measured by incubation of receptor membranes with [^{35}S]GTPγS in the presence of 10 μM GDP (to reduce the basal binding) followed by filtration onto a filter plate using a cell harvester to separate the reaction mixture. SPA assay can be used with wheat germ agglutinin coated SPA bead, which captures the GPCR, and [^{35}S]GTPγS bound to the receptor will be in close proximity to scintillant on the bead and give a signal. In FlashPlate technology, a membrane fraction is incubated with [^{35}S]GTPγS in wells of microplate and centrifuged after incubation; the supernatant is discarded and the plate is counted in a microplate counter [52,53]. This assay has a low signal-to-noise ratio of 1.2 to 3. Nevertheless, the data correlated with the ligand binding assay and functional assays suggest that this assay can be used as a functional assay for G-protein activation. The results obtained by conventional filtration, FlashPlate technology, and SPA assay correlated very well, suggesting that SPA and FlashPlate technology provide rapid HTS methods for measuring [^{35}S]GTPγS binding.

This [^{35}S]GTPγS binding assay can be used for orphan GPCRs where ligand and signaling pathways are not known, and for other GPCRs for which signal transduction mechanisms are not well characterized.

2. Adenyl Cyclase Assay

AC is regulated by the G-proteins $Gα_s$, $Gα_i$ and $Gα_q$ and Ca^{2+}-calmodulin [54]. A number of techniques are available to measure AC activity, the simplest assay being measurement of the product of the reaction, cAMP. The other methods involve incubation of membranes with ^{3}H-adenine or [$α$-^{32}P]-ATP, isolating ^{3}H-cAMP or [$α$-^{32}P]-cAMP, respectively, by chromatography and quantitation by counting the cAMP fraction [55,56]. However, these methods are tedious and the throughput is low, so that these methods are not suitable for HTS.

a. Intracellular Cyclic AMP Assays. The classical cAMP assay procedures are involved in isolating cAMP fraction (by chromatography) and determination of cAMP levels by competition for ^{3}H-cAMP binding to the regulatory subunit of cAMP-dependent protein kinase [57], competition of radio labeled cAMP standard by radioimmunoassay, ELISA assay, or SPA assay [58]. Recent cAMP assay kits developed based on rdioimmunoassays or ELISA assays are able to determine cAMP levels in cell supernatants or extracts in microtiter plate format in multisteps. Further advances in cAMP assay development has made it possible now to determine cAMP levels in cells following drug treatment in fewer steps as a homogeneous assay. The homogeneous cAMP kits include Biotrak™ SPA based cAMP assay (Amersham), FlashPlate™ adenyl cyclase assay kit (NEN), HEFP™ cAMP assay kit (LJL Biosystems), and ALPHAScreen cAMP assay (BioSignal, Inc.) (Table 4).

A one-step Biotrak cAMP screen assay is described here. Cells expressing a GPCR of interest are plated in 96- or 384-well plates overnight, the medium is removed, the cells are treated with compound in buffer for 60 min, and the cells are lysed with lysis buffer; a solution containing ^{125}I-cAMP tracer, cAMP antiserum, and SPA anti-rabbit serum is added and incubated overnight. Different standard amounts of cAMP in assay buffer containing lysis buffer served as standard curve. On each plate, six wells without compound serve as blanks (B_0), and six wells with standard agonist (at a concentration that gives the maximum effect) serve as B_{max}. The plates are counted (each well for 2 min) in the TopCount (Packard) or Microbeta (Wallac). Plotting B/B_0 as a function of log cAMP concentration generates a standard curve. From the standard curve the amount of cAMP in the unknown samples is calculated, converted to percentage of maximal response (% max response) (Fig. 14) and plotted against log agonist concentration, the curve is fitted by a sigmoidal equation, and EC_{50} is determined. For accurate calculation of the EC_{50}, the cAMP levels % max response have to be plotted against the compound concentration in the radioimmunoassay, SPA assay, FP assay, or APHA screen assay. Instead, if the raw radioactive counts or fluorescence counts are plotted, the EC_{50} curve shifts to the left, resulting in a 10- to 100-fold lowering of the EC_{50} concentration (10- to 100-fold higher efficacy). The competition curve is a semi log curve, and hence the first 50% competition of counts is not equivalent to a 50% change of cAMP levels. This artificial EC_{50} will not correlate well with other receptor parameters and will pick up low active, poor agonists as good agonists. So caution must be exercised in processing the data. The assay can also be performed as a two-step assay with removal of the medium after treating the cells with compound. In cell suspension assay cells are added to the wells of a microplate, and compound is added and incubated for 60 min. Lysis buffer is added and to the dissolved cell extract, SPA-bead, antibody, and cAMP tracer are added and incubated over night. The microplate is counted to determine the amount of cAMP in the samples. The assay can also be miniaturized to 384-well plates to increase throughput. The conventional cAMP assays because of low throughput were used at best as secondary assays to confirm the lead compounds. With the development of the homogeneous cellular

Table 4 Cyclic AMP Assays

Assay	Rad/nonrad	Assay type	Reagents	Readout	Provider
ELISA assays	Nonradioactive	Heterogeneous	cAMP-ab, 2°-ab-coated plate, cAMP conjugated with HRP or alk. phos.	Colorimetric, fluorescence plate reader	Sigma and several other vendors
Radioimmuno assay	Radioactive	Heterogeneous	[³H] or [¹²⁵I]-cAMP, cAMP-ab coated bead	Radioactive counting γ-counter or β-plate counter	Amersham, NEN
FLISA assay	Nonradioactive	Homogeneous	biotin cAMP, cAMP-ab coated bead, Cy5-streptavidin	CY5 Fluorescence based FMAT reader	PE Biosystems
Fluorescence Polarization assay	Nonradioactive	Homogeneous	Fluorescent cAMP, cAMP-ab	FP-plate reader	LJL
SPA assay	Radioactive	Homogeneous	[¹²⁵I]cAMP, cAMP-ab, 2°-ab-SPA bead	Radioactive counting β-plate counter (TopCount or MicroBeta)	Amersham
Electrochemiluminescence assay	Nonradioactive	Homogenous	cAMP-ab coated magnetic bead, ruthenyl-cAMP	Bead-based, chemiluminescence ECLM8 reader	IGEN
AlphaScreen	Nonradioactive	Homogeneous	Biotin-cAMP, streptavidin donor bead, cAMP-ab-acceptor bead	Bead-based, fluorescence reader AlphaQuest	BioSignal
FlashPlate	Radioactive	Homogeneous	cAMP-ab coated Flash-Plate, [¹²⁵I]cAMP	Radioactive counting β-plate counter	NEN

Classical cAMP assays involved isolation of cAMP in acid extracts from cells (by first preparing cell extracts) and quantitation by ELISA assays. Now, several homogeneous and heterogeneous cAMP kits are available based on different technologies from different vendors. Some of those are given here.

A. cAMP standard curve

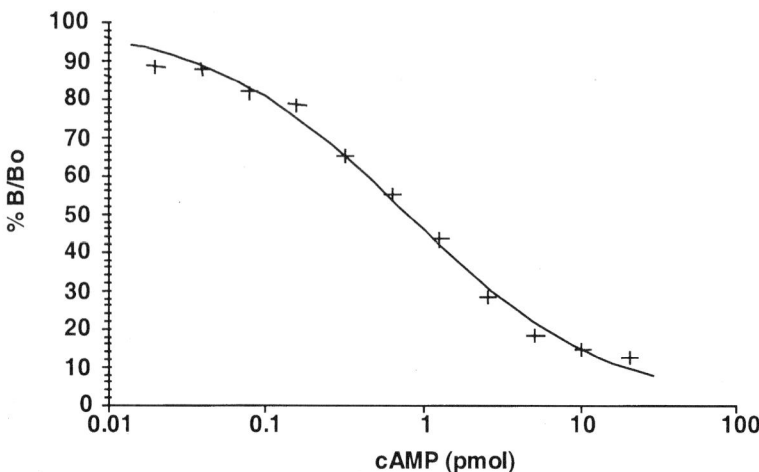

B. EC50 curve

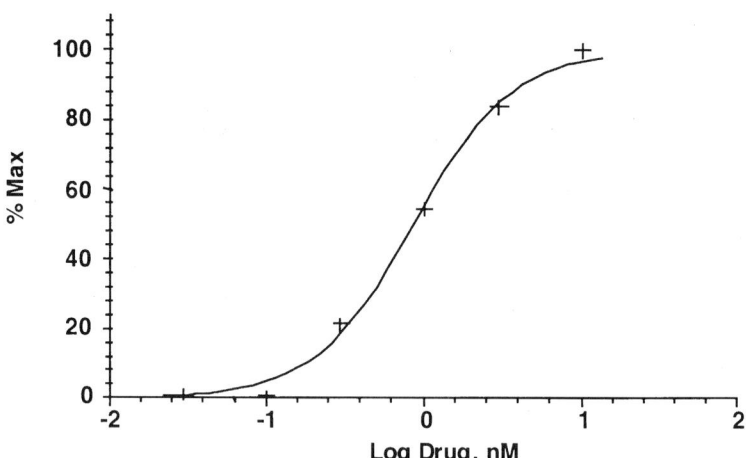

Figure 14 SPA cAMP assay for screening signal transduction via adenyl cyclase. (A) Cyclic AMP standard curve. (B) Activation of adenyl cyclase by agonist of GPCR. CHO-K1 cells expressing GPCR, when treated with agonist activates $G\alpha_s$ and adenyl cyclase and consequently increases intracellular cAMP. There is a good dose–response of agonist in stimulating cAMP levels.

cAMP assays (Table 4), it is possible to use this assay as a primary HTS screen for finding lead compounds from screening compound libraries.

When an agonist binds to receptors coupled to $G\alpha_i$ activation results in the inhibition of AC, which cannot easily be detected in normal cells because the basal cAMP levels are in the lower range of detection. To detect the agonist activity (inhibition of AC activity), first AC activity is stimulated in the cells by forskolin, isoproterenol, or suitable agonist treatment that increases cAMP levels to detection levels and then the agonist induced AC inhibition can be measured. NPY, and galanin receptors are coupled to $G\alpha_i$ receptors, and cells expressing these receptors are treated with forskolin to stimulate AC before the ligand induced AC inhibition is measured [59].

A reporter (luciferase) based cAMP assay was described [60] wherein CHO cells stably expressing human β2-AR were transiently transfected with reporter plasmids containing luciferase gene under transcriptional control of 6 or 12 cAMP response elements (CREs). In these cells, stimulation of β-AR with 20 μM forskolin resulted in a 35-fold induction of luciferase activity. This CRE-directed luciferase reporter gene assay has been used for functional assay of GPCRs that involve activation of AC such as dopamine D1, D2 receptors, adenosine receptor, and calcitonin receptor [60–62] and for GPCR agonists that inhibit AC activity as in the case of 5HT-receptor [61]. The EC_{50} values obtained for agonist activation are the same and for a stimulatory GPCR are about 10-fold higher using a cell-based inhibitory GPCR CRE-directed luciferase reporter gene assay, compared to cAMP accumulation assays, and the rank order of potency of drugs is preserved between these assays [61]. Also, a 100-fold amplification of the signal was found with a reporter assay compared with a direct cAMP accumulation assay, suggesting that the reporter assay offers greater sensitivity. This reporter gene assay for cAMP is a rapid nonradioactive assay, more sensitive and adaptable for HTS. Thus this assay is an excellent alternative to traditional methods of cAMP measurement in the functional evaluation of agonists and antagonists for GPCRs that involve cAMP in their signal pathways.

3. Melanophore-Based Receptor Functional Activity

A bioassay monitoring receptor-mediated pigment dispersion in an immortalized *Xenopus laevis* melanophore cell line has been used as a functional assay for GPCR and RPTKs [63]. The amphibian melanophores contain melanosomes that are filled with melanin pigment and are regulated by the hormones α-melanocyte stimulating hormone (α-MSH) and melatonin. The GPCRs and RPTKs when expressed in the melanophores can use the endogenous cell signaling system within the melanophore to mediate cell darkening or lightening due to pigment dispersion or aggregation, respectively (Fig. 15). In cells expressing G_s-linked receptors, ligand binding activates AC and activates pigment dispersion and cell darkening similar to α-MSH [63–66]. RPTKs that are functionally linked to activate protein kinase C (PKC), such as bombesin and PDGF, also stimulate pigment dispersion [66]. The pigment dispersion can be read in a microtiter plate reader as an increase in absorbance. In melanophores expressing $G\alpha_i$-linked receptors, stimulation with an agonist causes inhibition of AC and causes pigment aggregation and cell lightening and is measured as a decrease in absorbance as in the case of melatonin. The melanophore assay was also extended to erythropoietin receptor, a cytokine receptor [63].

Thus a rapid microtiter-based functional assay for screening compounds that interact with GPCRs, RPTKs, and cytokine receptors has been developed either by transient or stable expression of the receptors in a melanophore cell line derived from *X. laevis* [63,66]. Melanophores do not appear to express many mammalian receptors and may represent null cells for human membrane receptors; hence they can be transiently or stably transfected with the receptor cDNA of interest and can be assayed rapidly. Transiently or stably expressed receptors in amphibian melanophores are plated in a 96-well microtiter plate (20,000/well) in Leibovitz medium (L-15) and incubated at 26°C. For those receptors that activate pigment dispersion, the cells are preincubated with 1 nM melatonin for 2 h in L-15 medium at room temperature to allow complete aggregation of the pigment, and the absorbance is read before the addition of drugs (A_i). Drugs are added to the wells and A_{620} is measured after 30 min (A_f).

For receptors that activate pigment aggregation, the cells are preincubated for 2 h in light at room temperature to allow complete dispersion of the pigment, and the absorbance is read. Various concentrations of ligand or drug are added, and the pigment aggregation is measured after 30 min by reading A_{620nm} in a microtiter plate reader. NPY1 R is a GPCR which is coupled to $G\alpha_q$, and ligand stimulation inhibits AC. Melanophores expressing hNPY1R transiently showed ligand-dependent aggregation which was abolished by a nonpeptide antagonist (Fig. 16A–C). A stable melanophore cell line expressing the hNPY1R also showed ligand-dependent pigment aggregation and correlated very well with ligand binding to the receptor (Fig. 16D). The receptor expressing melanophores can be reused after washing out the test compounds and allowing reequilibration with L-15 medium overnight, and these cells responded to ligand and drugs. The correlation between ligand binding and melanophore pigment dispersion or aggregation and other functional signals suggests that the melanophore bioassay can be used to screen for ligands to GPCRs, RPTKs, and cytokine receptors. This is a rapid homogeneous functional bioassay and can be used to distinguish antagonists and agonists. Some of the disadvantages of this system include inconsistent transfection in melanophores with cDNA of the receptor of interest, very slow growth of melanophores (long doubling time, > 2 days), and loss of pigment after a few passages requiring periodic replacement with new cells.

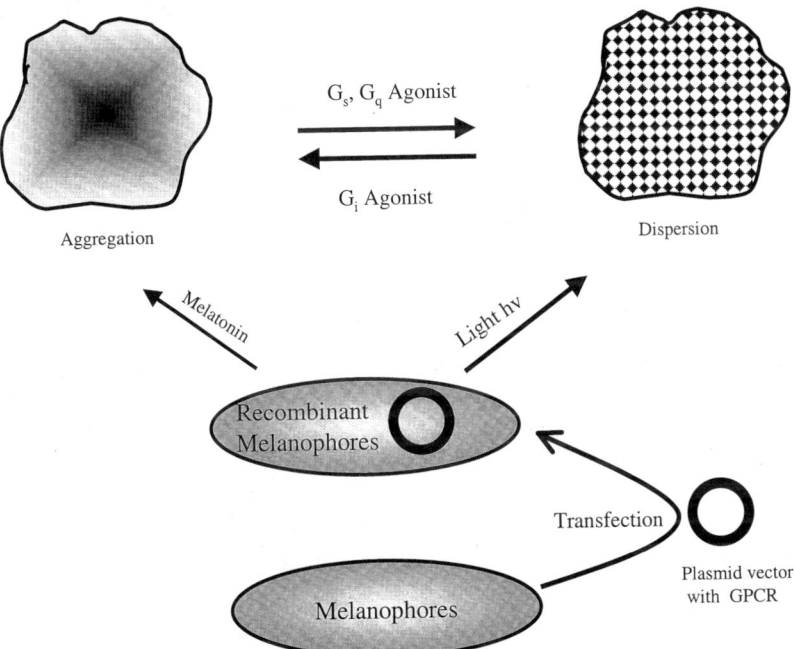

Figure 15 Schematic illustration of melanophore activation by GPCRs. Melanophores are transfected with a plasmid vector containing cDNA encoding GPCR. Activation of Gi-coupled GPCR in melanophores inhibits adenyl cyclase and consequently phosphorylation, resulting in aggregation of melanosomes. Activation of Gs- or Gq-coupled GPCRs in melanophores activate adenyl cyclase and activate either PKA or PKC, increasing phosphorylation, which results in dispersion of melanosomes. Melanophores can be brought to the aggregated state by activating the endogenous melatonin receptor with melatonin or can be brought into the dispersed state by exposing them to light.

4. Inositol Triphosphate Measurements

Following activation of GPCR by agonist binding, G-protein-mediated signal transduction activates phospholipase C (PLC) to hydrolyze the lipid precursor phosphatidylinositol 4,5-bisphosphate to give second messengers diacyl glycerol (DAG) and inositol (1,4,5)-triphosphate (IP$_3$). DAG activates protein kinase C (PKC) and regulates cellular functions. IP$_3$ is a second messenger that controls many cellular processes by generating internal Ca^{2+} signals. IP$_3$ binds to an IP$_3$ receptor (that resembles the Ca^{2+} mobilizing ryanodine receptors of muscle) to mobilize stored Ca^{2+} and to promote an influx of external Ca^{2+}. A radioreceptor IP$_3$ assay (a competitive ligand binding assay) was developed by NEN Life Science, in which the sample containing IP$_3$ competes a fixed amount of tracer (radioactive IP$_3$) for a fixed number of receptor binding sites. The IP$_3$ is extracted from the cells with trichloroacetic acid (TCA), and the TCA is removed by extraction with TCTFE-trioctylamine. IP$_3$ in the aqueous phase is assayed by incubation with [^{3}H]-IP$_3$ tracer and IP$_3$ receptor membranes. The contents are centrifuged, the supernatants are removed, and the radioactivity associated with membranes is counted.

Conventional methods of measuring phosphatidylinositol (PI) turnover involved a low-throughput method of treating cells grown in a 6- or 12-well plate with compound, separating IPs from inositol in the cell extract by ion exchange chromatography in mini columns, and counting the eluted IPs after mixing with scintillant in a 5 mL vial. This method was labor intensive and low throughput and at best could be used for confirmation of the lead compounds. Agonist-induced PI turnover can be measured by measuring inositol phosphate (IPs) accumulation in cells expressing the selected GPCR in a 96-well format [67,68]. Approximately 10,000–50,000 cells/well are plated in a 96-well plate and incubated overnight with [^{3}H]-inositol to label the intracellular pool; the medium is removed and incubated with agonist for 60 min. Cells are washed with PBS, and cold 5% TCA is added and incubated for 30 min. The TCA extract is then added to AG1-X8 resin (Bio-Rad) in a 96-well fiberglass multiscreen filter plate, washed (3×) with 5 mM myoinositol and eluted with 50 μL of 1 M ammonium formate–0.1 M formic acid into an opaque 96-well plate. Scintillant is added, the contents are mixed, and the radioactivity is counted in a TopCount (Packard) or Microbeta (Wallac) plate counter. The throughput by this method has been increased substantially from the conventional assay to medium throughput so it can be used as a secondary assay.

A simple nonradioactive liposome lysis assay for quantification of IP$_3$ using specific IP$_3$ monoclonal antibody has been described [69]. The assay is specific for IP$_3$. The assay is homogeneous with simple addition of reagents to a microtiter plate and mixing and reading in a fluorescence plate reader. The levels of IP$_3$ in CHO-IR stimulated with insulin measured by a liposome lysis assay are comparable with that determined by the conventional radioactive assay. The sensitivity of this liposome assay is low, and it has to be improved for use for quantification of IP$_3$ in small samples [69].

5. Phospholipase C Assay

PLC is a key enzyme in the signal transduction of many cell-mediated responses. At least three families of PLC enzymes, β, δ, and γ, are present. The β and δ are regulated by the GPCR pathway and the receptor-mediated tyrosine kinase pathway, respectively. PLCs hydrolyze phosphoinositide lipids giving rise to second messengers IP$_3$ and DAG. DAG is a potent activator of protein kinase C. IP$_3$ acts as a second messenger in stimulating the intracellular release of Ca^{2+}

stores from the endoplasmic reticulum through its specific interaction with IP_3 receptor. FlashPlate (NEN) is coated with phospholipid substrate, [1-^{3}H-inositol]-phosphatidylinositol 4,5-bisphosphate (PIP_2) [70]. Control cells or cells expressing the receptor are added in the wells of the FlashPlate and incubated with agonist or test compounds at 37°C. The radiolabeled substrate bound to the plate is hydrolyzed in the assay with enzyme activity releasing radioactivity, which can be monitored over a period of 2 h by counting in a TopCount or MicroBeta plate counter. This is a negative assay wherein the SPA signal is reduced with enzyme activity. The assay is homogeneous and can be adapted to HTS.

6. PKC Assay

At least 12 isotypes of PKCs have been characterized and can be grouped into three groups [71]: (1) the conventional PKC isotypes, α, β1, βII, and γ, are Ca^{2+} dependent and activated by DAG and phosphatidylserine (PS); (2) novel PKCs, δ, ϵ, η, and θ are Ca^{2+} independent and regulated by DAG and PS; (3) atypical PKCs, ζ and ι/λ, are Ca^{2+} independent and do not require DAG but are regulated by PS. PKC is a single polypeptide with a C-terminal catalytic domain and an N-terminal regulatory domain. These isoforms differ in structure, substrate requirement, expression, and localization. The phosphorylation of proteins on serine and threonine by PKC is essential for the regulation of several biological functions. PKC activity is altered in different disease states. Activation of GPCRs coupled to $G\alpha_q$ stimulates PLC, which produces the second messengers IP_3 and DAG. DAG is a potent activator of PKC. There are few in vitro PKC-specific pharmacological agents (over other kinases) and fewer have good selectivity for individual PKC isoforms. LY333531 is a PKCβ-specific inhibitor that shows potential for therapeutic use in cardiovascular disease and cancer [71].

The conventional assays utilize the transfer of radiolabeled phosphate from [^{32}P or ^{33}P]-[ATP] to a peptide (specific phosphorylation sequence) or protein followed by separation of radiolabel ATP from the phosphorylated peptide or protein. A homogeneous SPA assay has been developed in which a biotinylated peptide substrate is incubated with ^{33}P-ATP, PKC, and streptavidin-coated SPA bead in the presence of Ca^{2+}, PS, and DAG [72]. The ^{33}P-phosphorylated biotinylated peptide binds by affinity to streptavidin SPA bead and comes in close proximity to the scintillant coated on the SPA bead emitting light. The signal is measured in a Topcount or MicroBeta plate counter. A homogeneous FlashPlate assay for PKC has been described. Recently, FP assays have been reported [30,73] in which a selective ser/thr containing peptide is phosphorylated by incubation with ATP and PKC in the presence of Ca^{2+}, PS, and DAG. Reaction is stopped with the addition of EDTA, fluorescent phosphopeptide (tracer), and a specific monoclonal anti-posphoserine antibody, further incubated and read in a FP plate reader. The phosphorylated peptide competes with the binding of tracer to antibody, and thus activity is inversely proportional to the FP signal. Several recombinantly expressed PKC isoforms are available from PanVera, and FP-PKC assay kits are commercially available from PanVera and LJL Biosystems and other vendors. A new method of measurement of different kinase activation in single mammalian cells has been described [74]. In this technique, fluorescent substrate peptides for different protein kinases are microinjected into mammalian cells. The final cellular concentration of different fluorescent peptide substrates in the cell was 0.01–1 µM using a laser–micropipette system. Kinase substrate loaded cells when treated with pharmacological or physiological stimuli and subjected to capillary electrophoresis to analyze different products formed thus accessing the kinase activities. This technique simultaneously measures several enzymes within the same cell [74].

7. Phospholipase A_2 (PLA$_2$) Assay

Two classes of PLA$_2$ based on their cellular localization have been described, namely cytosolic (cPLA$_2$) and secretary (sPLA$_2$). PLA$_2$ hydrolyzes sn-2-acyl bond of sn-3-phosphoglycerides giving rise to equimolar amounts of free fatty acid and lysophospholipid. A homogeneous PLA$_2$ SPA assay is available from Amersham. This is a signal-decrease assay format wherein the SPA substrate is radiolabeled phosphatidyl-ethanolamine biotinylated on the ethanolamine head group binds to streptavidin-coated SPA beads generates SPA signal due to the proximity of the radiolabel to the scintillant of the SPA bead [75]. PLA$_2$ activity hydrolyzes the substrate removing the radiolabel group from the substrate attached to the SPA bead, which results in a decrease in SPA signal proportional to the enzyme activity. Thus the activity of enzyme is inversely proportional to the SPA signal.

8. Ca^{2+} Assay

Agonist activation of GPCRs, which can couple to the $G\alpha_{q/11}$ class of G-proteins, often increases transient cellular Ca^{2+} concentrations. Although calcium concentration of mammalian cells is very high (1 mM), the cytosolic Ca^{2+} is low (10–100 nM) and stimulation of a receptor can increase it to 400–1000 nM Ca^{2+} and activate calcium responsive events. Ca^{2+} can be measured by fluorescence-based dyes or bioluminescence methods.

   a. Fluorescence-Based Calcium Assays.   The fluorescent calcium dye fura-2 is a dual wavelength ratiometric dye that exhibits a shift in fluorescence intensity; the excitation peak shifts between 335 nm for the bound calcium and 362 nm for the free calcium form of the dye. Ratiometric calcium dyes are better for quantitative measurements of intracellular calcium but are not well suited for HTS.

   Fluo-3, fluo-4, and calcium green-1 are single (visible) wavelength calcium dyes essentially nonfluorescent unless bound to Ca^{2+} and undergo ~ 100-fold

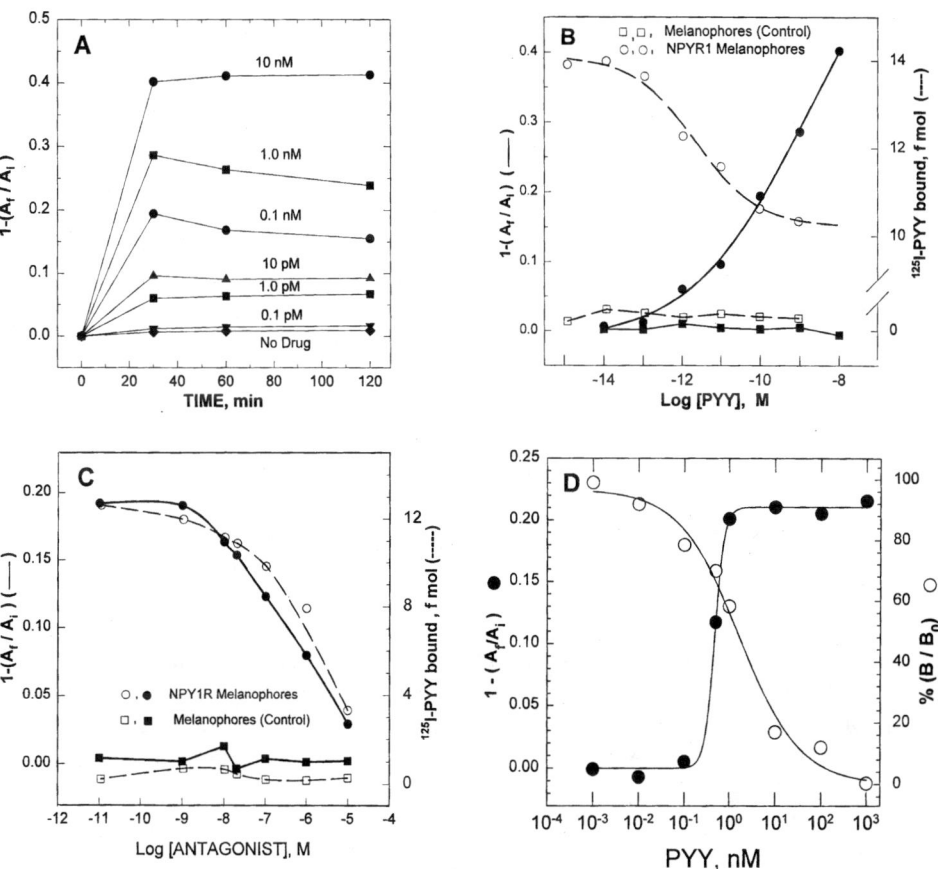

Figure 16 Functional bioassay for hNPY1R in Xenopus melanophores. (A) PYY-induced pigment aggregation in a ligand-controlled manner in NPY-Y1 melanophores. The effect of PYY on aggregation reached a maximum within 30 min and remained unchanged thereafter up to 120 min. Mock transfected melanophores (control) did not respond to PYY. (B) Dose–response of PYY on pigment aggregation and ^{125}I-PYY binding in melanophores expressing NPY-Y1R transiently. (C) The effect of antagonist on pigment aggregation induced by PYY and ^{125}I-PYY binding. The aggregation induced by 0.1 nM PYY was abolished by the NPY-Y1-specific antagonist in a dose-dependent manner with an IC_{50} of 0.4 μM. Antagonist also competed ^{125}I-PYY bound with an IC_{50} of 0.4 μM. Both curves are superimposable, suggesting that the compound is a true antagonist of hNPY1R. (D) Effect of PYY on aggregation and ^{125}I-PYY binding to melanophore stables expressing NPY-Y1R.

increase in fluorescence upon binding Ca^{2+}. These dyes are used in the majority of intracellular calcium determinations. Some receptor classes and/or cell types give better response to fluo-3, while others respond to calcium green-1. Both dyes excite at 488 nm and the emission is 500–560 nm.

The Fluorescent Imaging Plate Reader (FLIPR) (Molecular Devices, Sunnyvale, CA) was developed to perform high-throughput quantitative optical screening for cell-based fluorescent assays [76,77]. The FLIPR can be used for measurements of intracellular calcium, intracellular pH, intracellular sodium, and membrane potential. The FLIPR measures fluorescence signals in all the wells of a 96- or 384-well plate simultaneously by imaging, with kinetic updates in the subsecond range. Ca^{2+} transients can be measured both in adherent cells and in suspension cells. Typically, recombinant cells or natural cell lines expressing the receptor of interest are grown as monlayers in 96- or 384-well plates. The medium is removed and buffer is added. Cells are loaded for 1 h at 37°C with Ca^{2+}-sensitive fluorescent dyes calcium green-1, fluo-4, or fluo-3 AM and extra cellular dye is washed out. With cells in suspension, plates have to be centrifuged for washing out medium or dye. A plate is read in the FLIPR for blank readings of the cells. Agonistic compound is added to each well simultaneously in the FLIPR, and the dye released is measured, which is proportional to transient increases in the intracellular Ca^{2+}. The FLIPR reads in all the wells of a plate simultaneously by imaging, which enables the study of real-time kinetics. Typically, a plate is read every 1 sec for the first 2 min and every 6 sec later up to 10 min. The Ca^{2+} transients depend on the cell type and are generally completed in 10 min. The Ca^{2+} spikes are predictable for each type of receptor and cell and hence for calculation of Ca^{2+} response the counts between certain time periods are measured and blank is subtracted. The antagonist activity can be measured after loading cells with the dye and incubating with antagonist followed by the addition of standard agonist. The antagonist decreases the Ca^{2+} transient due to the standard agonist (Fig. 17, top). The data in ASCI form is imported to an appropriate database, and the percentage of maximal response with a standard agonist is calculated and plotted against agonist concentration to obtain EC_{50} or plotted against antagonist concentration to obtain IC_{50} (Fig. 17, bottom). Real-time kinetic data gives additional pharmacological information for ranking relative potencies of drugs and gives information on the kinetics of the drug–receptor interaction. Measurement of functional response provides data on affinity, efficacy, and function of each drug, and also full agonists, partial agonists, and antagonists can be distinguished within a single assay.

b. Luminescence-Based Calcium Assays. The intracellular $[Ca^{2+}]_i$ measurements have been mostly done with fluorescence-based assays. Luminescence-based assays are being developed for HTS with the advances in imaging instrument technology. Intracellular $[Ca^{2+}]_i$ also can be measured using bioluminescent

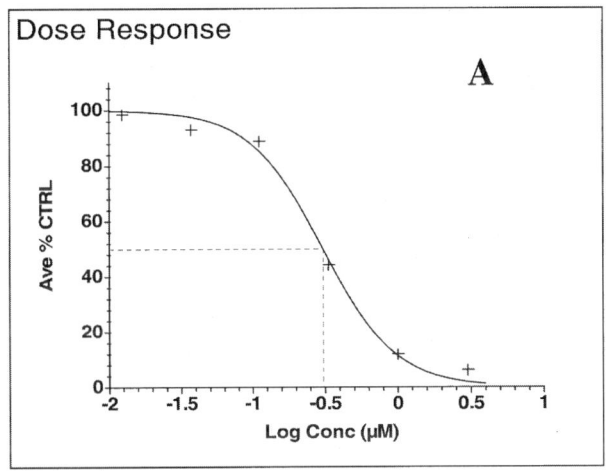

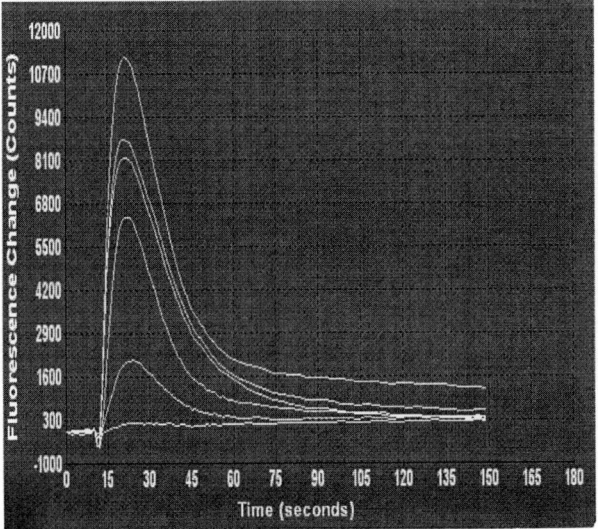

Figure 17 Measurement of intracellular Ca^{2+} using a fluorescent imaging plate reader (FLIPR). Cells from a natural cell line expressing the desired GPCR are loaded with calcium sensing dye Fluo-3 incubated with various concentrations of antagonist compound; agonist is added, and the transient Ca^{2+} is measured at 3 sec intervals for 3 min. The transient Ca^{2+} levels are measured as counts either under the curve or maximum minus minimum peak height. (Top) Dose–response curve of an antagonist on a Ca^{2+} signal. The IC_{50} obtained for the antagonist compound was 33-nM. (Bottom) The transient Ca^{2+} response traces are given. The Ca^{2+} curves line up well and decrease with increasing concentration of antagonist.

aequorin in a luminometer [78,79]. Aequorin is produced in the jellyfish *Aequorea victoria*. It is a photoprotein composed of the apoaequorin protein bound to the prosthetic group coelenterazine and molecular oxygen. It has three EF-calcium binding sites, and when calcium binds to these sites coelenterazine is converted to coelenteramide, which results in the emission of blue light (470 nm). The signal in mammalian cells occurs within 30 s, and the intensity of the aequorin flash is proportional to the Ca^{2+} concentration. Ca^{2+} measurements by aequorin assay have been validated for many GPCRs and calcium channels and the dose–responses are similar to the values obtained with fluorescent dye methods. HEK293 cells stably expressing apoaequorin are cotransfected with serotinin receptors 5HT2a and 5HT2c; when treated with ligand 5HT dose-dependently stimulated the luminescence of aequorin.

Calcium-regulated reporter assays are being developed by constructing transcription factors (NFAT, CREB) that become activated when there is a rapid rise in $[Ca^{2+}]_i$. The calcium-regulated promoter is fused to reporter gene luciferase, β-lactamase, or β-galactosidase. These transcription factors on activation bind to unique promoter and stimulate transcription of the reporter protein. CHO-K1 cells transfected with CCK1 receptor, a $G\alpha_q$-coupled GPCR along with NFAT-luciferase reporter gene, produced a dose–response signal curve with CCK-8 that was similar to that obtained using a fluorescent dye [78].

9. Microphysiometer Assays

When a ligand binds to a GPCR for which signal transduction like AC stimulation is poorly coupled may also couple to other signal pathways. Microphysiometry detects receptor activation and other physiological changes in live cells by monitoring the activity of energy metabolism. The Cytosensor microphysiometer measures the cellular acidification rate as a reliable index of the integrated functional response to receptor activation [80]. Activation of a receptor is followed by an increase in acidification rate within a few minutes that can be measured in a microphysiometer. Microphysiometry requires functional coupling to ligand/receptor binding, but it is not necessary to have the details of the mechanism of that coupling. This may have special use for functional assay for orphan receptor for which knowledge about function and transduction is assumed and a natural ligand is missing.

Cells expressing the recombinant receptor in suspension culture in a 96-well plate are stimulated with the addition of agonist in the medium. The acidification rate is measured every 2 min by picking a 4 μL sample into an LAPS array system by pumping through eight fluid channels. The LAPS chip is micromachined to form four sensor sites in each of eight parallel fluid channels. Adherent cells are attached to the underside of a coverslip that forms the ceiling of the fluid channels. The throughput for the Cytosensor microphysiometer is low

($\sim$ 104 assays/day) and cannot be used for HTS; nevertheless it is useful for optimizing cell lines with transfected receptor, for agonists and antagonists in a variety of receptor assays.

10. Phosphoinositide 3-Kinase (PI 3-Kinase)

PI 3-kinases are a subfamily of lipid kinases that catalyze the addition of phosphate at the 3-position of the inositol ring of phosphoinositides, phosphatidyl inositol (Ptdlns), Ptdlns(4)P, and Ptdlns(4,5)P$_2$. Only about 0.25% of the total inositol containing lipids is phosphorylated at the 3-position, and these lipids are hypothesized to regulate cell functions, 5% phosphorylated at the 4-position and 5% at the 4- and 5-positions [81]. Nine members of the PI 3-kinases are found so far, which can be grouped into three classes according to the substrate molecules. PI3-kinase is involved in many different cellular functions including growth, differentiation, apoptosis, and cytoskeleton rearrangement in response to a variety of different signals [81]. Wortamannin and LY294002 are two powerful PI3-kinase inhibitors and have been utilized in determining the role of PI3-kinase in cellular functions in cells over expressing the activated form of PI3-kinase [82]. A homogeneous FlashPlate assay has been described for PI3-kinase. PI3-kinase is incubated with [γ-^{33}P]ATP in a FlashPlate coated with PI3-kinase substrate, phosphatidylinositol 4,5-bisphosphate (PIP$_2$) at 30°C for 60 min. Following incubation, the reaction mixture is aspirated and washed twice with PBS to decrease background. Due to the PI3-kinase reaction, ^{33}PO$_4$ is transferred to PIP$_2$ immobilized on the surface of the FlashPlate forming PI-3,4,5-P$_3$. The bound radioactivity is determined by counting in a TopCount or MicroBeta plate counter. The FlashPlate assay is comparable to the conventional solution phase assay, and the IC$_{50}$ values obtained for Wortamannin and LY294002 (1.6 nM and 0.44 μM respectively), which agreed with the values obtained in solution phase assay and with the literature values reported [83].

B. RTK Receptor Family

Ligand binding to RTK family receptors lead to dimerization of monomeric receptors or conformational changes in heterotetrameric receptors resulting in autophosphorylation of specific tyrosines in the cytoplasmic domain (Fig. 18). Tyrosine autophosphorylation stimulates the intrinsic receptor tyrosine kinase activity and/or generates recruitment sites for downstream signaling proteins containing phosphotyrosine-recognition domains such as the Src homology 2 (SH2) or the phospho-tyrosine-binding (PTB) domain [84]. Efficient phosphorylation of the substrates by RTK also requires association of the substrate to the activated RTK. The PTB domain of insulin receptor substrate 1 (IRS1) binds at pTYR972 of activated IR, and the pleckstrin homology domain targets IRS-1 to

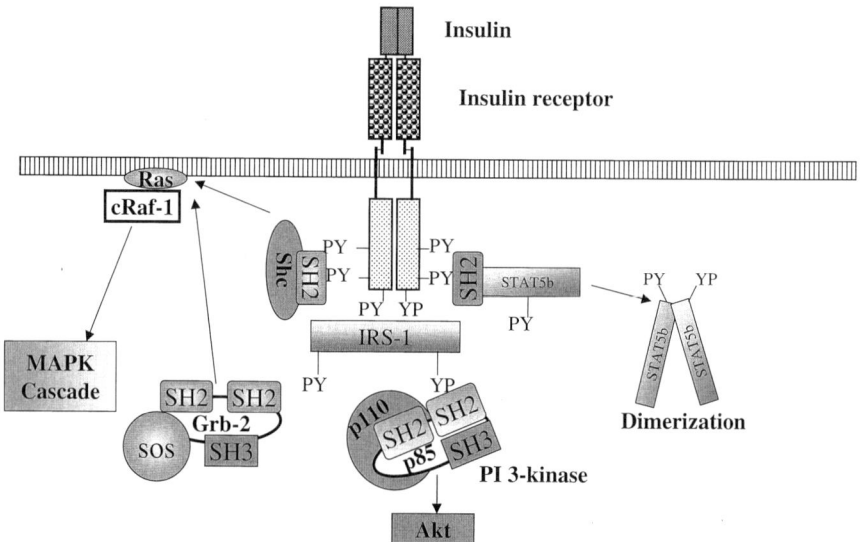

Figure 18 Schematic illustration of signaling cascades for the cellular responses to insulin–insulin receptor (IR) interactions. Insulin binding to its receptor activates tyrosine kinase activity resulting in autophosphorylation followed by binding of IR substrate (IRS), Shc, and STAT5b to specific PY residues of the receptor through SH2 domains. IRS-1 is a docking protein that induces various signaling pathways like MAPK through the Grb-2-SOS pathway, PI3 kinase, and Akt pathway. STAT5b binds to PY residues of IR and gets phosphorylated, followed by dimerization and translocation into the nucleus, where it induces gene expression.

the plasma membrane enabling IR kinase to phosphorylate several tyrosines in IRS-1 [85].

1. PTK Assays

Signal transduction of RTK family receptors can be monitored by assaying for PTK activity. PTK activation has been used as the primary screen for various receptors in this family. Several different PTK assays including both radioactive methods and nonradioactive methods have been used for screening (Table 5). In the conventional radioactive PO_4 transfer protein kinase assay, the radioactive phosphate from $[\gamma\text{-}^{32}P]$ or $[\gamma\text{-}^{33}P]$-ATP is incorporated into tyrosine of a protein/peptide substrate and is measured by binding the protein/peptide to phosphocellulose (P-81) filter discs or precipitating with TCA and filtration [86,87].

Table 5 Common Tyrosine Protein Kinase Assays

Assay type	Properties	Special reagents required	Comments
Filtration assay	Radioactive, heterogeneous	$[\gamma^{-33}P]$-ATP, biotin peptide, streptavidin, or phosphocellulose filter plates	Radioactive waste, laborious, heterogeneous, medium throughput not suited for HTS
ELISA assay	Nonradioactive, heterogeneous	PY antibody, secondary antibody, substrate coated on microtiter plate	Laborious, heterogeneous, medium throughput not suited for HTS
TRF-ELISA assay	Nonradioactive, heterogeneous	Eu-PY antibody	Laborious, heterogeneous, medium throughput not suited for HTS
SPA assay	Radioactive, homogeneous	$[\gamma^{-33}P]$-ATP, biotin peptide, streptavidin SPA bead	Radioactive waste, homogeneous, suitable for HTS
FlashPlate Assay	Radioactive, homogeneous	$[\gamma^{-33}P]$-ATP, biotin peptide coated on to streptavidin FlashPlate	Radioactive waste, requires precoating of the peptide to microtiter plate, homogeneous, suitable for HTS
Fluorescence polarization assay	Nonradioactive, homogeneous	Fluorescent phosphopeptide, PY antibody	Homogeneous, tracer is fluorescent labeled, suitable for HTS and uHTS
HTRF/Lance assays	Nonradioactive, homogeneous	Biotin peptide, APC–streptavidin, Eu-PY antibody	Homogeneous, two labeled reagents are used, CY5 or XL665 labeled generic acceptors available, suitable for HTS
Electrochemilumnescence	Nonradioactive, homogeneous	Biotin peptide, streptavidin magnetic bead, ruthenyl PY antibody	Homogeneous, two labeled reagents are used, suitable for HTS with multichannel reader ECLM8
Alpha Screen	Nonradioactive, homogeneous	Biotin peptide, streptavidin donor bead, PY-antibody receptor bead	Homogeneous, two labeled beads are used, suitable for HTS

A normal PTK reaction needs kinase, Mg^{2+}-ATP, and a kinase substrate. In the methods here, other special reagents required for that method are given.

Biotin-labeled peptides have been used as kinase substrates, and the ^{32}P-phosphorylated peptides are captured on high-capacity streptavidin-coated membranes [88]. In the solid-phase radioactive phosphate transfer assay the substrate (peptide/protein) is bound to a ScintiStrip microplate (Wallac); or a FlashPlate (Dupont NEN) coated with scintillant is used for HTS [89,90]. Alternatively, nonisotopic ELISA-type assays have been used in which a microtiter plate is coated with the peptide substrate, kinase reactants are incubated, excess reagents are washed, further incubated with a PY antibody, washed, incubated with a secondary antibody conjugated to alkaline phosphatase or horseradish peroxidase, washed, finally incubated with the color or fluorescence developing reagents and read in a plate reader [91]. A sensitive ELISA microtiter plate assay using europium cryptate (EuK) labeled PY antibody has been described [92]. A HTRF PTK assay has been described in which biotinyl peptide is phosphorylated by PTK and immunocomplexed to EuK-PY antibody [93,94]. The biotin peptide binds to XL665 (allophycocyanin, APC) labeled streptavidin. The two flurophores are brought into close proximity and the energy from donor Eu is transferred to acceptor APC and the signal is read in a time-resolved fluorometer. In the SPA-based PTK assay, biotinyl peptide is ^{33}P-phosphorylated by PTK and binds to streptavidin-coated SPA beads [95]. In the PTK assay based on the electrochemi-luminescence method, the phosphorylated biotinyl peptide complexes to Ru^{2+}-PY antibody and binds to streptavidin-coated magnetic beads brought to the electrode by a magnet where due to the redox reaction it produces light and is measured in an Origen analyzer [86].

Recently, a nonradioactive immunological PTK assay based on fluorescence polarization (FP) was described [97,98]. In this direct FP-PTK assay, a fluorescenylated peptide substrate is incubated with the kinase, ATP, and PY antibody. The phosphorylated peptide product is immunocomplexed with the PY antibody, resulting in an increase in the polarization signal. A FP competition immunoassay in which the kinase phosphorylates a peptide or protein and the phosphorylated peptide or protein will compete with fluorescent tyrosine phosphorylated peptide tracer that forms an immunocomplex with PY antibody as in a typical immunoassay was described [98,99]. The advantages of these FP-PTK assays over the other kinase assays include the use of inexpensive nonisotopic substrate; it is a homogeneous assay with no separation, precipitation, or washing steps, and it can be miniaturized for a 384-well microtiter plate to increase throughput. The simplicity and speed and homogeneous nature of this method make it ideal for HTS in a small molecule drug discovery program.

2. Mitogen Activate Protein Kinase (MAPK)

MAPKs are serine/threonine kinases that are activated by phosphorylation by MAPK kinase (MAPKK) on threonine and tyrosine, and MAPKK in turn is activated by MAPK kinase kinase (MAPKKK). MAPKs play prominent role in mediating intracellular responses to various physiological stimuli including growth factors, cytokines, and stress conditions. The MAPKs can be divided into three subgroups: p38MAPK (p38^{MAPK}), stress-activated protein kinase/c-Jun N-terminal kinase (SAPK/JNK), and extracellular-regulated protein kinases (ERKs). MAPK pathways regulate the activities of a variety of transcription factors and other cellular proteins involved in gene expression.

As MAPK is activated by phosphorylation, quantitation of phosphorylation of MAPK is a measure of MAPK activity. SDS-PAGE followed by immunoblotting for phosphorylated MAPK (electrophoretic mobility shift due to phosphorylation) measures MAPK activity under different responses. This is a low-throughput assay, and a higher throughput cell-ELISA in a 96-well plate has been reported [100]. The cells grown in 96-well plates are treated with stress reagents or growth factors; the cells are washed and fixed on the plate, treated with primary antibody, washed, treated with biotinylated secondary antibody, washed, horseradish peroxidase coupled streptavidin, washed, and color developed and read in a plate reader. The transient phosphorylation due to H_2O_2 or PDGF treatment is in agreement with the SDS-PAGE results. Inhibitors (pyridinyl imidazole compounds) of p38^{MAPK} have been shown to bind to the ATP binding site and inhibit p38^{MAPK} kinase activity. A filtration binding assay using a radiolabeled pyridinyl imadazole compound that binds to p38^{MAPK} has been developed [101]. Small molecules competing in this binding assay have been found to inhibit p38^{MAPK} activity. Using biotinylated peptide substrate, homogeneous SPA bead [102], SPA FlashPlate [103], and AlphaScreeen [104], MAPK assays have been developed. A high-throughput SPA for the Raf/MEK/ERK kinase cascade has been described [105]. When purified components are incubated together, cRaf-1 phosphorylates and activates MEK1. Activated MEK1 phosphorylates and activates ERK2. Activated ERK2 phosphorylates the biotin peptide. The assay detects inhibitors of cRaf-1, MEK1, and ERK2, and the specific target of inhibition has to be determined in further specificity assays.

3. Dimerization Assay

Dimerization of growth factor, cytokine, or erythropoietin receptor is a prerequisite for all downstream effects. A high-throughput dimerization assay for erythropoietin receptor was described [106]. The extracellular domain of the erythropoietin receptor (rEpoR) was expressed in *E. coli*. A modified version of this protein with protein kinase A substrate site (^{33}P-rEpoR) incorporated into rEpoR was also expressed in *E. coli* and phosphorylated in vitro using PKA and [γ-^{33}P]ATP. The dimerization assay consists of coating rEpoR to a 96-well high-binding plate, the addition of ^{33}P-rEpoR along with or without erythropoietin mimetic peptide (EMP-1), which induces dimerization, and incubation for 16 h at 4°C. After washing the plate, scintillant was added and counted in the TopCount. In the absence of EMP-1, no dimerization was seen, and EMP-1 induced dimerization was competed by cold rEpoR. This dimerization assay can be used to identify compounds that promote dimerization. Based on this assay, a dimerization assay can be developed for other receptors.

A protein-fragment complementation assay (PCA) was developed for quantitative characterization of protein–protein interactions in vivo in mammalian cells based on the murine enzyme dihydrofolate reductase (DHFR) [107] (Fig. 19A). Fragments of DHFR F[1,2] correspond to 1–105, and F[3] corresponds to 106–186 amino acids of murine DHFR and are fused to a fragment containing extracellular and transmembrane-domain of erythropoietin receptor (EpoR) via a 5-amino acid linker resulting in EpoR (1-270-F[1,2] and EpoR (1-270)-F[3]. These constructs when cotransfected into CHO DUKX-B11 cells were treated for 30–60 min with Epo or with a peptide agonist EMP1 and DHRF substrate fluorescein-conjugated methotrexate (fMTX). The cells were washed to remove the unreacted fMTX and reincubated for 30 min in select medium to allow for the efflux of unbound fMTX. When EpoR(1-270-F[1,2] and EpoR (1-270)-F[3] dimerize, induced by ligand or agonist, DHRF enzyme activity results from association-folding of two fragments. The enzyme activity can be measured in a flow cytometer, or a fluorescence reader. EP showed saturable binding isotherms with K_ds of 164 pM and 168 nM respectively. This assay can be used for HT-drug screening and quantitative analysis of induction or disruption of protein–protein interactions.

An in vivo large-scale library-versus-library screening strategy for identifying optimally interacting pairs of heterodimerizing polypeptides has been reported [108]. The murine DHFR (mDHFR) was genetically designed to two complementary fragments. Two leucine zipper libraries were fused semirandomly at the positions adjacent to the hydrophobic core to either of the mDHFR and cotransformed to *E. coli*. When the library peptides interact, the interaction reconstitutes DHFR enzyme activity allowing bacterial growth. The two libraries formed heterodimers with varying stability. To isolate polypeptides that interact with greater stability, stringency of selection was increased by using more weakly associating mDHFR fragments. Multiple passages and selection of the pooled, selected colonies in liquid culture produced the best heterodimer(s) that can be used for further in vivo strategies.

In another approach for monitoring protein–protein interactions, chimeric proteins composed of proteins of interest were fused to nonfunctional complementing β-galactosidase (β-gal) Δα, Δω peptides [109] (Fig. 19B). When the proteins interact to form a complex, the β-gal Δα, Δω peptides associate producing functional activity of the enzyme. Thus the protein interaction can be monitored by assaying for β-gal activity in the lysates by chemiluminescence detec-

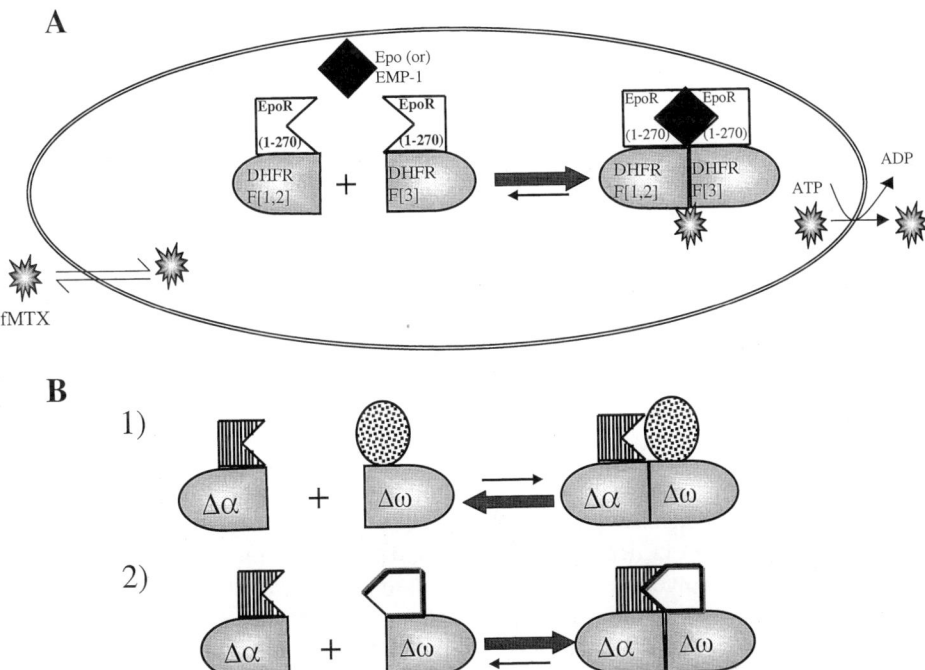

Figure 19 (A) Schematic representation of a murine dihydrofolate reductase (DHFR) protein complementation assay. Erythropoietin receptor (Epo R) extracellular and transmembrane domains [EpoR(1-270)] are fused to each of the two complementary fragments of DHFR, F[1,2] or F[3], and stably cotransfected in CHO cells. Contransfectants grown in the presence of 2 nM Epo in nucleotide-free medium are selected and incubated with DHFR inhibitor fluorescein-conjugated methotrexate (fMTX). In the presence of Epo or peptide agonist EMP1, dimerization of EpoR(1-270)-F[1,2] and EpoR(1-270)-F[3] reconstitutes DHFR to which fMTX is bound and can be detected in FACS or fluorescence spectroscopy. (B) Schematic representation of β-galactosidase protein complementation assay in intact eukaryotic cells. (1) When the β-gal mutants Δα and Δω are fused to proteins that do not dimerize, the association to active β-gal is not favored. (2) When the Δα and Δω mutants are fused to proteins-that can dimerize, the formation of active β-gal results. β-Gal activity can be measured by chemiluminescence.

tion. This method can be used for the assessment of specific protein dimerization interactions and for screening compounds that effect dimerization.

4. Reporter Gene Assay

Upon stimulation of insulin receptor (IR), insulin receptor kinase is activated by autophosphorylation. Major substrates for IR kinase include IRS-1, Shc and signal transducer and activator of transcription-5 (STAT5), which interact via the Src-homology-2 (SH2) domain with phosphorylated tyrosine residues of IRβ [110]. STAT5 after phosphorylation by IR dimerizes and translocates to the nucleus resulting in gene induction. A cellular assay has been developed with a luciferase gene reporter construct under control of a STAT5-inducible promoter, which showed insulin-mediated induction of STAT5-dependent luciferase activity. This cellular bioassay monitors IR kinase activity in a simple luciferase readout system in insulin-responsive cells like fibroblasts and is adaptable to HTS for insulin-mimetic compounds and IR kinase antagonists.

5. SH2 Interactions

Tyrosine phosphorylation regulated by tyrosine protein tyrosine kinases and protein tyrosine phosphatases plays an important role in cellular regulation, growth, and proliferation. A critical step in the tyrosine phosphorylation signal transduction pathways involves molecular recognition in protein–protein interactions via SH2 domains. SH2 domains are approximately 100 amino acid noncatalytic motifs that specifically bind to phosphotyrosine. SH2 domains and their associated catalytic and noncatalytic proteins constitute critical signal transduction targets for drug discovery, for example, SH2 domains of Src, Grb2, Shc, P85/PI3K, Gap for cancer targets, Hck for immune disease, Syk for allergy and asthma, STATs for inflammatory disease targets.

A SPA-based assay for antagonists of binding interaction between [^{125}I]-diphosphorylated peptide corresponding to the human T-cell receptor ζ-1 immunoglobulin receptor family tyrosine-based activation motif (ITAM) with the tandem SH2 domain of the ZAP-70 protein tyrosine kinase has been reported [111]. The ZAP-SH2 domain expressed in *E. coli* was purified and biotinylated. The assay consists of incubation of biotinylated ZAP-SH2 with streptavidin SPA bead and [^{125}I]-ζ-1 ITAM in the presence or absence of a test compound and measuring radioactivity in a TopCount. A FP assay was developed by PanVera with Src-SH2 domain fused with GST and interacting fluorescent phosphorylated peptide.

6. Receptor Internalization Assays

Membrane receptors upon ligand activation are recruited to clathrin-coated pits, internalized to an early-endosomal compartment. From endosome a fraction of the receptor is either recycled back to the plasma membrane and majority of the receptor is degraded in lysosome. EGF-induced activation and phosphorylation of EGFR is followed by down-regulation of EGFR. The phosphorylated ligand-bound EGFR is recruited to clathrin-coated pits and the activated receptor is internalized and transported to early endosome. To measure the internalization of EGFR, the cells are exposed to biotin-EGF, the cells are washed, the membrane receptor associated biotin-EGF is removed by acid wash, the cells are fixed and incubated with horseradish peroxidase conjugated streptavidin, and the internalized EGFR is determined by measuring the absorbance at 490 nm [112]. The internalization of EGFR in HER14 cells was time and EGF-dose dependent. The internalization of EGFR was inhibited by H_2O_2 in a dose-dependent manner. This assay is adaptable for HTS.

Many GPCRs undergo ligand-dependent homologous desensitization accompanied by aggregation of the receptor followed by internalization. When a ligand binds to GPCR, a specific GPCR-kinase (GRK) phosphorylates GPCR and the phosphorylated GPCR binds to arrestin, which leads to uncoupling of GPCR from G-protein, resulting in loss of sensitivity to ligand (desensitization). The uncoupled GPCR aggregates in clathrin-coated pits and internalized to an early-endosomal compartment. From endosome the GPCR is either recycled back to the plasma membrane or degraded in lysosome. Internalization of PTH-receptor or β2-adrenergic receptor (β2 AR) stably expressed in HEK 293 cells as green fluorescent protein (GFP) fusion conjugates in presence of their respective ligands was studied using ArrayScan™ (cellomics, Pittsburgh, PA) using an algorithm capable of measuring internalization [113]. ArrayScan II has a unique optical path that allows rapid automated scans through the bottom of clear-bottom microplates by high content screening. ArrayScan software identifies and measures individual features and structures within each cell in a field of cells and is capable of analysis of several hundreds of cells in parallel. The internalization of GFP-labeled receptor is dependent on the ligand concentration and time of incubation. Before treatment of the cells with ligand, the GFP-PTHR was present on the cell periphery with little intracellular staining. Rapid internalization occurred within minutes and was dependent on the PTH concentration. Similarly, with GFP-β2 AR isoproterenol promoted a concentration- and time-dependent internalization. High content screening is a homogeneous format that can be used for the study of intracellular events, and subcellular localization.

C. Receptor Protein-Tyrosine Phosphatases

The mechanism by which the receptor-phosphatases (PTPases) initiate transmembrane signaling in response to external ligands is not well understood. The physiological ligands for RPTPs have not been identified. PTP-specific inhibitors are being sought for understanding the role played by PTPs in signaling pathways.

Vanadium in a proper oxidation state has a similarity to phosphate and complexes within the PTP catalytic site and is an effective PTP Inhibitor. CD45 is essential for T-cell receptors to couple to second messenger pathways, IL-2 production, and antigen specific proliferative response in response to specific antigen. CD45 activates TCR-associated tyrosine kinase (src-family protein kinases) by dephosphorylating the regulatory phosphotyrosine residues. Thus the main function of RPTPs is to dephosphorylate regulatory phosphotyrosine residues. Antagonists to RPTPs have been shown to increase tyrosine phosphorylation levels of many cellular proteins and increase the intracellular Ca^{2+} levels and T-cell activation.

PTPase activity can be assayed essentially with methods that are the reverse of the protein kinase assay in that a phosphorylated peptide/protein serves as substrate and dephosphorylated product or phosphate released is measured. PTPase activity can be determined by estimating the inorganic phosphate released, disappearance of the substrate using SPA, FP, HTRF, dephosphorylation as in the case of protein kinase can be employed for finding inhibitors. PTPase assays using hydrolysis of p-nitrophenyl phosphate, 4-methyl-7-hydroxycoumarinyl phosphate are also commonly used at pH > 7 which is above the optimal pH for PTPases. A sensitive fluorogenic and chromogenic assay using 3,6-fluorescein diphosphate (FDP) has been described for CD45, protein tyrosine phosphatase-1B at pHs < 7 [114]. The substrate FDP has no absorption > 330 nm and when hydrolyzed by PTPase forms fluorescein monophosphate which absorbs maximally at 445 nm and further hydrolysis of FMP produces fluorescein which absorbs maximally at 490 nm. The assay is a very simple, homogeneous, and sensitive method that can be used for HTS. Ligand-induced dimerization plays an important role in the regulation of tyrosine kinase receptors by autophosphorylation and activation of tyrosine kinase activity. However, ligand-induced dimerization also plays an important role in the RPTPs but CD45 is negatively regulated. The phosphatase domain forms a symmetrical dimer in which the catalytic site of one molecule is blocked by specific contacts with a wedge from another [115].

D. Cytokine Receptors

Cytokine receptors are associated at the intracellular domain of the receptor with soluble cytoplasmic tyrosine kinases called JAKs [116]. Upon binding a ligand, a cytokine receptor will dimerize and result in the activation of receptor-associated JAK1 and JAK2 by transphosphorylation (Fig. 20). The activated JAKs phosphorylate tyrosine in the endodomains of the receptor distal to the JAK binding domain. These receptor phosphotyrosines and the 4–5 carboxy proximal amino acids constitute the SH2 binding/recognition domain for SH2 domains of STATs. Appropriate STATs are recruited to the phosohorylated receptor forming a complex with receptor. At the receptor, JAKs phosphorylate STATs on a conserved tyrosine. Activated (phosphorylated) STATs are released from the receptor and dimerize through the interaction of the SH2 domain of one STAT with

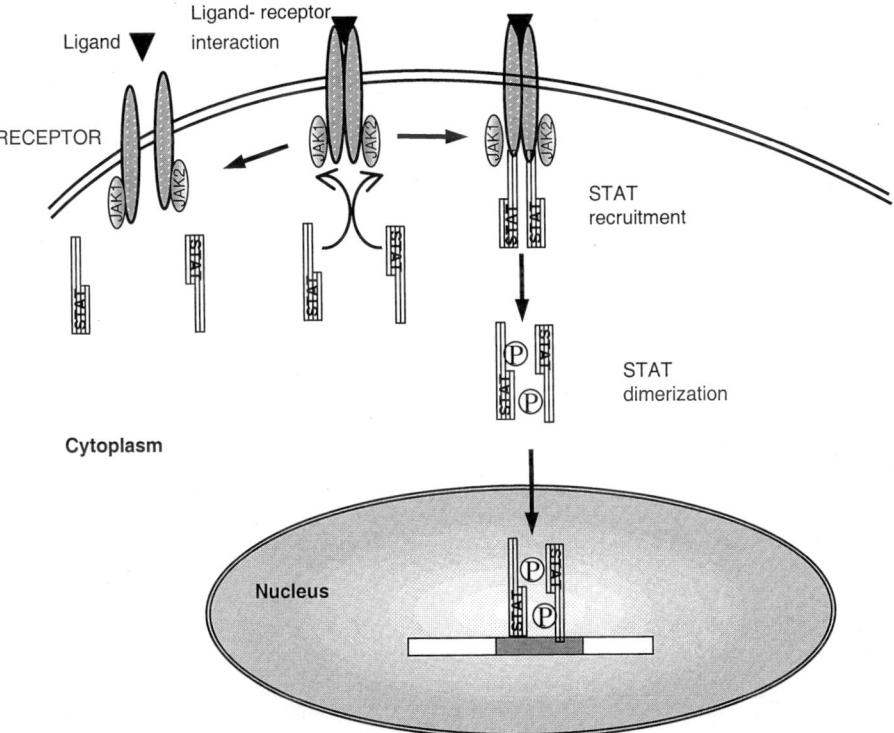

Figure 20 Schematic illustration of cytokine signal-transduction pathways. Upon ligand binding, the cytokine receptor dimerizes and activates Jak kinases associated with the receptor, which phosphorylated tyrosine residues of the receptor. STAT proteins are recruited to the receptor and bind through SH2 domain where they are phosphorylated on tyrosine by JAKs. Activated STATs are released and dimerize by interacting with SH2 one with the PY of the other and vice versa. The STAT dimer translocated into the nucleus where it binds to the enhancer and activates transcription.

phosphotyrosine of the other STAT. These dimers are translocated to the nucleus where they bind to members of IFN-gamma activation site (GAS) family of enhancers and activate transcription of the target genes [15,116]. The polypeptide ligands that activate the JAK-STAT pathway bind to PTK-receptors, non-PTK receptors, and GPCRs. The dimerization assay is described above in Sec. IV.B.3.

E. Ion Channels

HTS for ion-channel function requires sensitive assays that report ion channel activity in living cells. Electrophysiology is the gold standard functional assay that gives information on the state-dependence inhibition. The patch clamp assay, which involves clamping either voltage or current across a cell membrane, permits a detailed characterization of ion-channel gating, permeability, and drug metabolism. This assay is a low-throughput assay (100–150 assay points/week) and cannot be used for HTS. Because of the wealth of information obtained by the patch clamp technique, it is used as a secondary assay to provide unequivocal biophysical data regarding the activity and action of hits from primary screens [117–120]. Fully automated patch clamp assays are being developed that will increase the throughput. Radioactive flux assays measuring ion flux through the channel can be used for K^+, Na^+, and Cl^- channels. This end-point assay throughput is medium and is amenable to automation.

Optical readout assays using fluorescent probes to measure ion-channel-dependent changes, either intracellular ion concentration or membrane potential, are desirable for miniaturization and automation. These assays can report the activities of all pharmacologically active functional sites of the target. Intracellular Ca^{2+} can be measured using fluorescent indicators in a FLIPR (Molecular Devices, Sunnyvale, CA), which allows kinetic measurements. The membrane potential can be measured with commonly available lipophilic, negatively charged BIS-Oxonol dyes, e.g., DiBAC4 [117,120]. The FRET-based membrane potential assay using voltage sensor dyes was developed by Aurora Biosciences (San Diego, CA). This improved membrane-potential sensor is based on FRET between voltage-sensing oxanol dyes and voltage-insensitive donor fluorophores associated with cell membranes. The FRET assay has been shown to retain the oxonol probes reporting real kinetics of the membrane potential [117,120].

F. Nuclear Receptors

Members of the NR superfamily regulate gene expression by binding to cis-active elements in target genes and either activating or repressing transcription. Ligand binding to NRs modifies the DNA-binding and transcriptional properties of these receptors resulting in the activation or repression of target genes. Ligand binding induces conformational change in NRs and promotes association with diverse nuclear proteins that may function as coactivators of transcription through a conserved sequence motif present in the coactivators (Fig. 21). The coactivator proteins that associate with NRs in a ligand-dependent manner varied from 2 for estrogen receptor (ERAP) to 12 for vitamin D receptor (DRIPS). In addition to the receptor binding assays, transcription, DNA binding, and coactivator binding assays are in use for functional screens of NRs.

1. Transcription Assays

The ligands for NR, on entering the cells, bind to their cognate receptors with high affinity and induce a conformational change that activates the receptor. In some cases, the receptor may dissociate from heat shock proteins and conformational changes that allow the receptor to bind with other proteins involved in transcriptional regulation. Activated receptors bind to specific DNA sequences called hormone response elements (HRE) and increase transcription of the linked downstream gene [121]. High-throughput transcription assays have been developed in which the NR is expressed in a cell line under a certain selection of antibiotic, and in the same cell a plasmid-containing reporter gene like luciferase under the control of a promoter containing appropriate HRE is expressed under selection of a second antibiotic. Incubation of the cell with a ligand activates the expressed NR receptor, dimerizes and translocated into nucleus where it binds to the HREs in the reporter plasmid and induces the expression of the reporter gene (Fig. 21). The agonist-induced expression of the reporter gene results in the increased production of the reporter gene-product, which can be quantitatively assayed. The common reporters used for this purpose include firefly luciferase

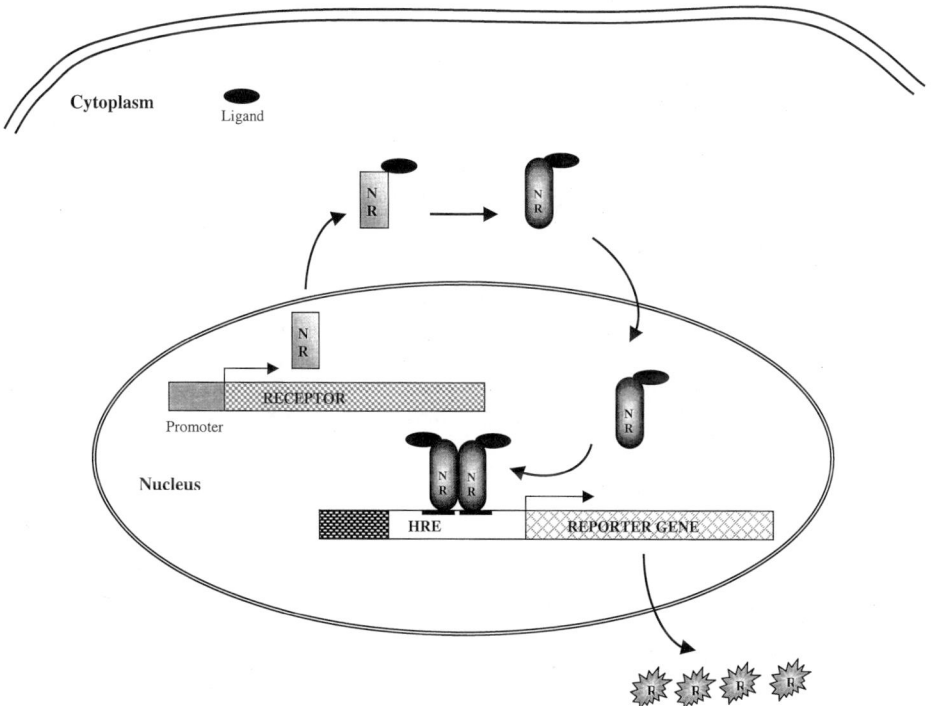

Figure 21 Schematic illustration of the regulation of gene expression by a nuclear receptor (NR). Upon binding of ligand, NR undergoes conformational changes and is translocated into the nucleus, where it binds to HRE as a dimer and activates gene expression.

enzyme, secreted alkaline phosphatase (SEAP), β-lactamase, and β-galactosidase. Luciferase activity can be assayed as a homogeneous assay by the LucLite™ kit from Packard, the LucScreen™ kit from Tropix, the SteadyGlo™ kit from Promega, or other kits. Similarly, SEAP enzyme kits such as attophos, and for β-galactosidase, fluorescence detection kits, are available. β-Lactamase is a proprietary reporter assay technology from Aurora Bioscience and the assay can be automated. The reporter assays are described in Chapter 4.

2. Binding of NR to NR Response Element

When ligand binds to the NR, the complex translocates into the nucleus where it acts as a transcription factor binding to NR responsive elements NRRE in the DNA and modulates cellular functions. About 50 base pair double stranded DNA to which fluorescein attached is used as substrate for binding to NR in FP assay. Binding of an increasing concentration of estrogen receptor to 1 nM fluorescein-labeled estrogen response element increased the signal from 60 to 260 milliP with a K_d of 4.5 nM for estrogen receptor. This is a homogeneous assay that can be miniaturized to high-density plates.

NRs interact with transcriptional coactivators regulating transcription. The conserved sequence motif LXXLL (where L is leucine and X is any amino acid) of coactivators is a signature sequence that facilitates the interaction of different proteins with NRs [122,123]. The interaction between NRs and coactivators are traditionally measured in vitro by a pull-down assay using ^{35}S-coactivator protein or by a nonradioactive pull-down assay with GST-coupled coactivators [124]. These pull-down assays are not amenable to screening large number of agonists and antagonists. A homogeneous HTS FP assay was described wherein rhodamine-labeled LXXLL peptide is incubated with GST-fused ligand domain of NR and FP signal is read [125]. Compounds that possess agonist activity will increase the FP signal. This homogeneous FP assay can be used for screening agonists for orphan NRs in HTS mode for which the coactivators are not known.

V. CONCLUSIONS

More than half of the drug discovery targets comprise receptors including membrane receptors, soluble receptors, and ion-channel receptors. Receptors have been grouped into superfamilies based on sequence and structural similarities. Several subtypes and isoforms have been identified by molecular biology approaches, which define the structure and function of receptor families. The receptor subfamilies within a superfamily may have different ligand binding and signal transduction properties. Receptor-subtype-specific drugs have been identified by HTS using human receptor subtypes expressed in human or mammalian cells. Drugs with therapeutic activity coming from HTS will give information of molecular specificity of a subtype of receptor that can effectively be used for subtype-specific targets in discovery efforts. Multiple receptor subtypes with unique pharmacological characteristics have been discovered that have specific tissue distribution. These cloned and expressed human receptor subtypes provide molecular targets for the development of tissue-specific novel subtype-selective drugs that will be devoid of the side effects associated with nonselective drugs.

With the completion of the sequencing of the human genome, new gene sequences are being discovered. Several orphan G-protein coupled receptors, nuclear receptors, and single transmembrane tyrosine kinase receptors have been identified in the human genome whose functions are unknown. These orphan receptors are useful for drug discovery only if their functions are defined, but elucidation of function by genetic methods is not an easy task. Some of these orphan receptors are of immediate interest as potential drug discovery targets, as they represent novel receptor subtypes of subfamilies that have members that are therapeutically important drug targets. With the advances in new technologies and experimental strategies and an increase in throughput by miniaturization it is possible to identify ligands that activate or inhibit the orphan receptors. Several sensitive biophysical methods have been developed that are capable of monitoring molecular changes within a single cell. New biochemical approaches and HT assay systems for receptors that modulate gene expression can be used in identifying the ligands binding to orphan receptors. The current challenges will be to determine the molecular functions of the orphan receptors so that the usefulness of the receptor in a disease target can be determined. The advances in genomics and proteomics will present opportunities for the development of better drugs targeted to specific receptor subtypes and new drugs for diseases that were not possible before. The signal transduction pathways are very complex, and these pathways for each receptor subtype are not well defined. The progress in assay technologies and instrumentation will permit rapid measurements of signal events in the single cell, and it may be possible to elucidate the signal transduction mechanisms with more precision, enabling the development of specific drugs for specific receptor targets.

ACKNOWLEDGMENTS

I thank my colleagues Zhengping Ma, Rajasree Golla, and Dorothy Slusarchyk for their help in the preparation of some of the illustrations.

REFERENCES

1. DJ Triggle. Pharmacological receptors: a century of progress. Highlights in Receptor Res 1–5, 1984.

2. JM Herz, WJ Thomsen, GG Yarbrough. Molecular approaches to receptors as targets for drug discovery. J Receptor Signal Trans Res 17:671–776, 1997.

3. J Drews. Drug discovery: a historical perspective. Science 287:1960–1964.

4. PPA Humphrey. The characterization and classification of receptors. J Receptor Signal Trans Res 18:345–362, 1998.

5. RP Ahlquist. A study of the adrenergic receptors. Am J Physiol 153:586–600, 1948.

6. D Hoyer, PPA Humphrey. Nomenclature and classification of transmitter receptors: an integrated approach. J Receptor Signal Trans Res 17:551–568, 1997.

7. HE Hamm. The many faces of G protein signaling. J Biol Chem 273:669–672, 1988.

8. TH Li, M Grossman, I Ji. G protein-coupled receptors. I. Diversity of receptor–ligand interactions. J Biol Chem 273:17299–17302, 1988.

9. RJ Lefkowitz. G-Protein coupled receptors. III. New roles for receptor kinases and β-arrestins in receptor signaling and desensitization. J Biol Chem 273:18677–18680, 1998.

10. JS Gutkind. The pathways connecting G-protein coupled receptors to the nucleus through divergent mitogen-activated protein kinase cascades. J Biol Chem 273:1839–1842, 1998.

11. JP Whitehead, SF Clark, B Urso, DE James. Signaling through the insulin receptor. Cur Opin Cell Biol 12:222–228, 2000.

12. P Hof, S Pluskey, S Dhe-paganon, MJ Eck, SE Shoelson. Crystal structure of the tyrosine phosphatase SHP-2. Cell 92:441–450, 1998.

13. JA Wells, AM deVos. Hematopoietic receptor complexes. Ann Rev Biochem 65:609–634, 1966.

14. RC Skoda. Specificity of signaling by hematopoietic cytokine receptors: instructive versus permissive effects. J Receptor Signal Trans Res 19:741–772, 1999.

15. MH Heim. The Jak-Stat pathway: cytokine signaling from the receptor to the nucleus. J Receptor Signal Trans Res 19:75–120, 1999.

16. O Civelli, HP Nothacker, A Bourson, A Ardati, F Monsma, R Reinscheid. Orphan receptors and their natural ligands. J Receptor Signal Trans Res 17:545–550, 1997.

17. M Ankersen, TH Hansen, I Ahnfelt-Ronne, AM Kappelgaard. Growth hormone secretagogues: recent advances and applications. Drug Disc Today 4:497–506, 1999.

18. M Kojima, H Hosoda, Y Date, M Nakazato, H Matsuo, K Kangawa. Ghrelin is a growth-hormone-releasing acylated peptide from stomach. Nature 402:656–660, 1999.

19. T Sakurai, A Amemiya, M Ishii, I Matsuzaki, RM Chemelli, H Tanaka, SC Williams, JA Richardson, GP Koalowski, S Wilson, JR Arch, RE Buckingham, AC Haynes, SA Carr, RS Annan, DE McNulty, WS Liu, JA Terrett, NA Elshoubagy, DJ Bergsma, M Yanagisawa. Orexins and orexin receptors: a family of hypothalamic neuropeptides and g-protein coupled receptors that regulate feeding behavior. Cell 92:1, 1998.

20. RV Weatherman, RJ Fletterick, TS Scanlan. Nuclear-receptor ligands and ligand-binding domains. Ann Rev Biochem 68:559–581, 1999.

21. CA Meier. Regulation of gene expression by nuclear hormone receptors. J Receptor Signal Trans Res 17:319–335, 1997.

22. W Bourguet et al. Crystal structure of the ligand-binding domain of the human nuclear receptor RXR-α. Nature 375:377–382, 1995.

23. GG Chicchi, MA Cascieri, MP Graziano, T Calahan, MR Tota. Fluorescein-Trp25-Exendin-4, a biologically active fluorescent probe for the human GLP-1 receptor. Peptides 18:319–321, 1997.

24. P Tairi, R Hovius, H Pick, H Blasey, A Bernard, A Surprenant, K Lundstrom, H Vogel. Ligand binding to the serotonin-5-HT3 receptor studied with a novel fluorescent ligand. Biochemistry 37:15850–15864, 1998.

25. G Turcatti, H vogel, A Chollet. Probing the binding domain of the NK2 receptor with fluorescent ligands: evidence that heptapeptide agonists bind differently. Biochemistry 34:3972–3980, 1995.

26. MR Tota, S Daniel, A Sirotina, KE Mazina, TM Fong, J Longmore, GD Strader. Characterization of a fluorescent substance P analog. Biochemistry 33:13079–13086, 1995.

27. MR Tota, L Xu, A Sirotina, CD Strader, MP Graziano. Interaction of [fluorescein-Trp25]glucagon with the human glucagon receptor expressed in *Drosophila schneider* 2 cells. J Biol Chem 270:26466–26472, 1995.

28. M Allen, J Reeves, G Mellor. High throughput fluorescence polarization: a homogeneous alternative to radioligand binding for cell-surface receptors. J Biomol Screen 5:63–69, 2000.

29. P Banks, M Gosselin, L Prystay. Fluorescence polarization assays for high throughput screening of G protein coupled receptors. J Biomol Screen 5:159–167, 2000.

30. GJ Parker, TL Law, FJ Lenoch, RE Bolger. Development of high throughput screening assays using fluorescence polarization: nuclear receptor–ligand binding and kinase/phosphatase assays. J Biomol Screen 5:77–88, 2000.

31. HE Rosenthal. Graphical method for the determination and presentation of binding parameters in a complex system. Anal Biochem 20:525–532, 1967.

32. EC Hulme, NJM Birdsall. Strategy and tactics in receptor binding studies. In: EC Hulme, ed. *Receptor Ligand Interaction—A Practical Approach*. IRL Press at Oxford University Press, New York 1990, pp 63–176.

33. LE Limbird. Cell surface receptors: a short course on theory and methods. Martinus Nijhoff, Boston, 1986.

34. T Haga, H Kaga, EC Hulme. Solubilization, purification and molecular characterization of receptors: principles and strategy. In: EC Hulme, ed. *Receptor Biochemistry—A practical approach*. IRL Press at Oxford University Press, New York, 1990, pp 1–50.

35. M Schapira, BM Raaka, HH Samuels, R Abagyan. Rational discovery of novel nuclear hormone receptor antagonists. Proc Natl Acad Sci USA 97:1008–1013, 2000.

36. E Sarubbi, SD Yanofsky, RW Barrett, M Denaro. A cell-free nonisotopic, high-throughput assay for inhibitors of Type-1 interleukin-1 receptor. Anal Biochem 237:70–75, 1996.

37. CW Mahoney. High-throughput nonradioisotopic determination of binding of platelet-derived growth factor to platelet-derived growth factor receptor b-extracellular domain using biotinylated ligand with enzyme-linked immunosorbent assay. Anal Biochem 276:106–108, 1999.

38. J Liu, M Gallagher, RA Horlick, AK Robbins, ML Webb. A time-resolved fluorometric assay for galanin receptors. J Biomol Screen 3:199–206, 1998.
39. J Nichols, DJ Parks, TG Consler, SG Blanchard. Development of a scintillation proximity assay for peroxisome proliferator-activated receptor γ ligand binding domain. Anal Biochem 257:112–119, 1998.
40. PJ Schatz. Use of peptide libraries to map the substrate specificity of a peptide-modifying enzyme: a 13 residue consensus peptide specifies biotinylation in *Escherichia coli*. Biotechnology 11:1138–11431, 1993.
41. S Duffy, KL Tsao, DS Waugh. Site-specific, enzymatic biotinylation of recombinant proteins in *Spodoptera frugiperda* cells using biotin acceptor peptides. Analytical Biochem 262:122–128, 1998.
42. V Breu, B Butscha, M Clozel. A simple high throughput FlashPlate assay to quantify ET-antagonists in plasma. FlashPlate® File #8, 1998.
43. R Bosse, R Garlick, B Brown, L Menard. Development of non separation binding and functional assays for G-protein-coupled receptors for high throughput screening: pharmacological characterization of the immobilized CCR5 receptor on FlashPlate. J Biomol Screen 3:285–292, 1998.
44. E Kurian, WR Kirk, FG Prendergast. Affinity of fatty acid for rRat intestinal fatty acid binding protein: further examination. Biochem 35:3865–3874, 1996.
45. S Miraglia, EE Swartzman, J Mellentin-Michelotti, L Evangelista, C Smith, I Gunawan, K Lohman, EM Goldberg, B Manian, P-M Yuan. Homogeneous cell and bead assays for high throughput screening using flurometric microvolume assay technology. J Biomol Screen 4:193–204, 1999.
46. AJ Kolb, JW Burke, G Mathis. Homogeneous, time-resolved fluorescence method for drug discovery. In: JP Devlin, ed. *High Throughput Screening*. New York: Marcel Dekker, 1997, pp 345–360.
47. K Stenroos, P Hurskainen, S Eriksson, I Hemmila, K Blomberg, C Lindqvist. Homogeneous time-resolved IL-2IL-2Rα assay using fluorescence resonance energy transfer. Cytokine 10:495–499, 1998.
48. EVOTEC Technical Note. VLiPs: GPCRs in a nut-cell.
49. ALPHAScreen™ Technical Note ATN-009, 2000.
50. S Gopalakrishnan, U Warrior, D Burns, DR Groebe. Evaluation of electrochemiluminescent technology for inhibitors of granulocyte colony-stimulating factor receptor binding. J Biomol Screen 5:369–375, 2000.
51. S Watson, A Arkinstall. *The G-Protein Linked Receptor Facts Book*. Academic Press, San Diego, 1994.
52. J Watson, JV Selkirk, AM Brown. Development of FlashPlate technology to measure [^{35}S]GTPγS binding to Chinese Hamster ovary cell membranes expressing the cloned human 5-HT1b receptor. J Biomol Screen 3:101–105, 1998.
53. DS Gembitsky, S Lovas, RF Murphy. Development of a high throughput functional assay for structure-activity studies of neurokinin A anlog. J Biomol Screen 3:183–188, 1998.
54. AG Gilman. G-proteins and regulation of adenyl cyclase. JAMA 262:1819–1825, 1989.
55. G Krishna, B Weiss, BB Brodie. A simple sensitive method for the assay of adenyl cyclase. J Pharmcol Exp Ther 163:379–385, 1968.

56. J Ramachandran. A new simple method of separation of adenosine 3′, 5′-cyclic monophosphate from other nucleotides and its use in the assay of adenyl cyclase. Anal Biochem 43:227–239, 1971.

57. SPH Alexander. The measurement of cyclic AMP levels in biological preparations. Meth Mol Biol 41:79–89, 1995.

58. JK Horton, PM Baxendale. Mass measurements of cyclic AMP formation by radio-immunoassay, enzyme immunoassay and scintillation proximity assay. Meth Mol Biol 41:91–112, 1995.

59. PM Rose, PB Fernandes, JS Lynch, ST Frazier, SM Fisher, K Krishna, B Kienzle, R Seethala. Cloning and functional expression of a cDNA encoding a human type 2 neuropeptide Y receptor. J Biol Chem 270:22661–22664, 1995.

60. SS Vansal, DR Feller. An efficient cyclic AMP assay for the functional evaluation of β-adrenergic receptor ligands. J Receptor Signal Trans Res 19:853–863, 1999.

61. SE George, P Bungay, LH Naylor. Evaluation of CRE-directed luciferase reporter gene assay as an alternative to measuring cAMP accumulation. J Biomol Screen 2:235–240, 1997.

62. A Himmler, C Stratowa, AP Czemilofsky. Functional testing of human dopamine D1 and D5 receptors expressed in a stable cAMP responsive luciferase reporter cell lines. J Recept Res 13:79–94, 1993.

63. MD Carrithers, LA Marotti, A Yoshimura, MJ Lerner. A melanophore-based screening assay for erythropoietin receptors. J Biomol Screen 4:9–14, 1999.

64. MN Potenza, GF Graminski, MR Lerner. A method or evaluating the effects of ligands upon Gs protein coupled receptors using a recombinant melanophore-based bioassay. Anal Biochem 206:315–322, 1992.

65. ME Nuttall, JC Lee, PR Murdock, AM Badger, F Wang, JT Laydon, GA Hofman, GR Pettman, JA Lee, A Parihar, BC Van Wagenen, J Fox, M Gowen, RK Johnson, MR Mattern. Amphibian mealnophore technology as a functional screen for antagonists of G-protein coupled 7-transmembrane receptors. J Biomol Screen 4:269–277, 1999.

66. GF Graminiski, MN Lerner. A rapid bioassay for platelet-derived growth factor β-receptor function. Nature Biotechnol 12:1008–1011, 1999.

67. Y Tian, L-H Wu, F-Z Chung. High throughput 96-well plate assay for receptor-mediated phosphatidylinositol turnover. J Biomol Screen 2:91–97, 1997.

68. M Chengalvala, B Kostek, DE Frail. A multi-well filtration assay for quantitation of inositol phosphates in biological samples. J Biochem Biophys Methods 38:163–170, 1999.

69. Y Nishio, S Nagata, M Umeda, R Shirai, T Yokogawa, S Ihara, Y Fukui. Quantification of phosphatidylinositol 3,4,5-triphosphate by liposome lysis assay with specific monoclonal antibodies. Anal Biochem 285:270–273, 2000.

70. TR Mullinax, G Henrich, P Kasila, DG Ahern, EA Wenske, C Hou, D Argentieri, ME Bembenek. Monitoring inositol-specific phospholipase C activity using a phospholipid FlashPlate®. J Biomol Screen 4:151–155, 1999.

71. KJ Way, E Chou, GL King. Identification of PKC-isoform-specific biological actions using pharmacological approaches. Trends Pharm Sci 21:181–187, 2000.

72. B Schnurr, C Schachtele. Use of FlashPlate for automated kinase assays. FlashPlate File #6, 1998.

73. JJ Wu, DR Yarwood, Q Pham, MA Sills. Identification of a high affinity anti-phosphoserine antibody for the development of homogeneous fluorescence polarization assay of protein kinase C. J Biomol Screen 5:23–30, 2000.

74. GD Meredith, CE Sims, JS Soughayer, NL Allbritton. Measurement of kinase activation in single mammalian cells. Nat Biotech 18:309–312, 2000.

75. Scintillation Proximity Assay Technical Bulletin.

76. KS Schroeder, BD Neagle. FLIPR: a new instrument for accurate, high throughput optical screening. J Biomol Screen 1:75–80, 1996.

77. TR Miller, DG Witte, LM Ireland, CH Kang, JM Roch, JN Masters, TA Esbenshade, AA Hancock. Analysis of apparent noncompetitive responses to competitive H1-histamine receptor antagonists in fluorescent imaging plate reader based calcium assays. J Biomol Screen 4:249–258, 1999.

78. LC Mattheakis, LD Ohler. Seeing the light: calcium imaging in cells for drug discovery. Drug Disc Today Supplement 1:15–19, 2000.

79. MD Ungrin, LMR Singh, R Stocco, DE Sas, M Abramovitz. An automated aequorin luminescence-based functional calcium assay for G-protein coupled receptors. Anal Biochem 272:34–42, 1999.

80. ML Alajoki, GT Baxter, WR Bemiss, D Balu, LJ Bousse, SDH Chan, TD Dawes, KM Hahnenberger, JM Hilton, P Lam, RJ McReynolds, DN Modlin, JC Owicki, JW Parce, D Redington, K Stevenson, HG Wada, J Williams. High-performance microphysiometry in drug discovery. In: JP Devlin, ed, *High Throughput Screening*. New York: Marcel Dekker, 1997, pp 427–442.

81. LE Rameh, LC Cantley. The role of phosphoinositide 3-kinase lipid products in cell function. J Biol Chem 274:8347–8350, 1999.

82. BT Burgering, PJ Coffer. Nature 376:599–602, 1995.

83. B Maswoswe, H Xie, P Kasila, L-A Yeh. A homogeneous phosphoinositide 3-kinase assay on phospholipid FlashPlate platforms. NEN Technical Bulletin, 2000.

84. SR Hubbard, M Mohammadi, J Schlessinger. Autoregulatory mechanisms in protein-tyrosine kinases. J Biol Chem 273: 11987–11900, 1998.

85. L Yenush, KJ Makati, J Smith-Hall, O Ishibashi, MG Myers Jr, MF White. J Biol Chem 271:24300–243006, 1996.

86. JJ Witt, R Roskoski Jr. Anal Biochem 66:253–258, 1975.

87. JE Casnellie. Assay of protein kinases using peptides with basic residues for phosphocellulose binding. Meth Enzymol 200:115–120, 1991.

88. A Tereba. High density protein kinase assay with subattomole sensitivity. J Biomolec Screening 3:29–33, 1998.

89. AF Braunwalder, DR Yarwood, T Hall, M Missbach, KE Lipson, MA Sills. A solid-phase assay for the determination of protein tyrosine kinase activity of c-src using microtitration plates. Anal Biochem 234:23–26, 1996.

90. GR Nakayama, MP Nova, Z Parandoosh. A scintillating microplate assay for the assessment of protein kinase activity. J Biomolec Screening 3:43–48, 1998.

91. JS Cleaveland, PA Kiener, DJ Hammond, B Schacter. A microtiter-based assay for the detection of protein tyrosine kinase activity. Anal Biochem 190:249–253, 1990.

92. AF Braunwalder, DR Yarwood, MA Sills, KE Lipson. Measurement of the protein tyrosine kinase activity of c-src using time-resolved fluorometry of europium chelates. Anal Biochem 238:159–164, 1996.

93. AJ Kolb, PV Kaplita, DJ Hayes, Y-W Park, C Pernell, JS Major, G Mathis. Tyrosine kinase assays adapted to homogeneous time-resolved fluorescence. Drug Disc Today 3:233–242, 1998.
94. P Ollikka, I Hemmilia, P Kivela, M Vaisala, K Blomberg. Fourth Annual Conference of the Soc Biomol Screening, 21–24 September, Baltimore, MD, 1998.
95. Y-W Park, RT Cummings, L Wu, S Zheng, PM Cameron, A Woods, DM Zaller, AI Marcy, JD Hermes. Homogeneous proximity tyrosine kinase assays: scintillation proximity assay versus homogeneous time resolved fluorescence. Anal Biochem 269:94–104, 1999.
96. R. Williams. Electrochemoluminescence: a new assay technology. IVD Technology 28–31, 1995.
97. R Seethala, R Menzel. A homogeneous, fluorescence polarization assay for src-family tyrosine kinases. Anal Biochem 253:210–218, 1997.
98. R Seethala. Fluorescence polarization competition immunoassay for tyrosine kinases. Methods 22:61–70, 2000.
99. R Seethala, R Menzel. A fluorescence polarization competition immunoassay for tyrosine kinases. Anal Biochem 255:257–262, 1998.
100. R Wit, J Boonstra, AJ Verkleij, JA Post. Large scale screening assay for the phosphorylation of mitogen-activated protein-kinase in cells. J Biomol Screen 4:277–284, 1999.
101. U Warrior, XG Chiou, MP Sheets, RJ Sciottti, JM Parry, RL Simmer, BW Surber, DJ Burns, BA Beutel, KW Mollison, SW Djuric, JM Trevillyan. Development of a p38 kinase binding assay for high throughput screening. J Biomol Screen 3:129–135, 1998.
102. Amersham Pharmacia Biotech home page on World Wide Web URL: http://www.apbiotech.com/.
103. NEN Life Science Products home page on World Wide Web URL: http://www.nen.com/.
104. MAP kinase assay. BioSignal, A Packard Biosience Company. AlphaScreen Technical Note AN007-Asc, 2000.
105. OB McDonald, WJ Chen, B Ellis, C Hoffman, L Overton, M Rink, A Smith, CJ Marshall, ER Wood. A scintillation proximity assay for the raf/MEK/ERK kinase cascade: high-throughput screening and identification of selective enzyme inhibitors. Anal Biochem 268:318–329, 1999.
106. DE Biazzo, H Motamedi, DF Mark, SA Qureshi. A high-throughput assay to identify compounds that can induce dimerization of erythropoietin receptor. Anal Biochem 278:39–45, 2000.
107. I Remy, SW Michnick. Clonal selection and in vivo quantitation of protein protein interactions with protein-fragment complementation assays. Proc Natl Acad Sci USA 96:5394–5399, 1999.
108. JN Pelletier, KM Arndt, A Pluckthun, SW Michnick. An in vivo library versus-library selection of optimized protein-protein interactions. Nature Biotech 17:683–690, 1999.
109. F Rossi, CA Charlton, HM Blau. Monitoring protein-protein interactions in intact eukaryotic cells by β-galactosidase complementation. Proc Natl Acad Sci USA 94:8405–8410, 1997.

110. P Storz, H Doppler, J Horn-Muller, B Groner, K Pfizenmaier, G Muller. A cellular reporter assay to monitor insulin receptor kinase activity based on STAT-5 dependent luciferase gene expression. Anal Biochem 276:97–104, 1999.

111. MP Sheets, UP Warrior, H Yoon, KW Mollison, SW Djuric, JM Trevillyan. A high-capacity scintillation proximity assay for the discovery and evaluation of ZAP-70 tandem, SH3 domain antagonists. J Biomol Screen 3:139–144, 1998.

112. BR Conway, LK Minor, JZ Xu, JW Gunnet, R DeBiasio, MR D'Andrea, R Rubin, R DeBiasio, K Giuliano, L Zhou, KT Dermarest. Quantification of G-protein coupled receptor internalization using G-protein coupled receptor-green fluorescent protein conjugates with the ArrayScan™ high-content screening system. J Biomol Screen 4:75–86, 1999.

113. R deWit, CMJ Hendrix, J Boonstr, AJ Verkleij, JA Post. Large-scale screening assay to measure epidermal growth factor internalization. J Biomol Screen 5:133–139, 2000.

114. Z Huang, Q Wang, HD Ly, A Govindarajan, J Schigetz, R Zamboni, S Desmarais, C Ramachandran. 3,6-Fluorescein diphosphate: a sensitive fluorogenic and chromogenic substrate for protein tyrosine phosphatases. J Biomol Screen 4:327–333, 1999.

115. R Majeti, AM Bilwes, JP Noel, T Hunter, A Weiss. Dimerization-induced inhibition of receptor protein tyrosine phosphatase function through an inhibitory wedge. Science 279:88–91, 1998.

116. C Schindler, I Strehlow. Cytokines and STAT signaling. Adv Pharmacol 47:113–174, 2000.

117. JE Gonzalez, K Oades, Y Leychkis, A Harootunian, PA Negulescu. Cell-based assays and instrumentation for screening ion-channel targets. Drug Disc Today 4:431–439, 1999.

118. J Denyer, J Woeley, B Cox, G Allenby, M Banks. HTS approaches to voltage-gated ion channel drug discovery. Drug Disc Today 3:323–332, 1998.

119. JJ Clare, SN Tate, M Nobbs, MA Romanos. Voltage-gated sodium channels as therapeutic targets. Drug Disc Today 5:492–520, 2000.

120. AW Mulvaney, Cl Spenser, S Culliford, JJ Borg, SG Davies, RZ Kozlowski. Cardiac chloride channels: physiology, pharmacology and approaches for identifying novel modulators of activity. Drug Disc Today 5:506–505, 2000.

121. P Lamb, J Rosen. Drug discovery using receptors that modulate gene expression. J Receptor Signal Trans 17:531–543, 1997.

122. DM Heery, E Kalkhoven, S Hoare, MG Parker. A signature motif in transcriptional co-activators mediates binding to nuclear receptors. Nature 387:733–736, 1997.

123. XF Ding, CM Anderson, H Ma, RM Uht, PJ Kushner, MR Stallcup. Nuclear receptor-binding sites of coactivator glucocoticoid receptor interacting protein 1 (GRIP1) and steroid receptor coactivator 1 (SRC-1): multiple motifs with different binding specificities. Mol Endocrinol 12:302–313, 1998.

124. O Bakker, HC vanBeeren, T Emrich, HJ Holtke, WM Wiersinga. Interaction between nuclear hormone receptors and coactivators analyzed using a nonradioactive "Pull-Down" assay. Anal Biochem 276:105–106, 1999.

125. K Lustig, P Baeuerle, H Beckmann, J-L Chen, B Shan. Nuclear hormone receptor drug screens. Patent WO 99/27365, 1999.

8
Functional Assay Screens

Maria L. Webb, Robert A. Horlick, and Bassam Damaj
Pharmacopeia, Inc., Princeton, New Jersey

Kirk McMillan
Exelixis, South San Francisco, California

I. INTRODUCTION

The rapid pace of gene discovery in the 1990s has accelerated the innovation and application of related discovery sciences: medicinal chemistry and high-throughput screening (HTS). To understand a gene product's function and definitively ascertain its therapeutic potential, more targets are screened against ever-expanding collections of small molecules than before. It is now commonplace to describe compound collections in the millions and the throughput of a screen in hundreds of thousands per week. The numbers are staggering, but the process of target discovery, lead discovery, lead optimization, and development of a drug candidate demands testing numerous and diverse molecules against newly identified gene products to ascertain their biological, and potentially pathological, roles.

As the pace of screening accelerates, there is an increasing need for functional assays on living cells. However, some of the most important kinds of these cellular assays, including those for receptor and ion-channel activity, have been difficult to adapt to high-throughput applications. Because of the molecular diversity, complexity, and extra- or intracellular locations of target molecules, the multiple signaling pathways, and the downstream intracellular interactions, the approaches one must employ to the discovery of activating or inhibiting molecules has also evolved. Traditional approaches often involve using radioactive or fluorescent ligands, in the presence and absence of the test molecule, to monitor the inhibition or activation of the target receptor or enzyme, respectively. These approaches most often utilize a recombinant human molecule expressed in a null cell or expressed cell-free. The signals or "readouts" are robust and specific, and because of the nature of the assay, they have well-defined mechanisms of action. These types of screens have been the basis of the pharmaceutical industry's approach to HTS. However, some targets, such as cytokines or growth factor receptors, involve protein–protein interactions and complex or undefined signal pathways. These targets are often of great therapeutic interest but have not yielded quality leads through standard cell-free screening approaches. The reasons for this are many, including the relative lack of high-affinity small molecules in our present-day compound collections, but we have speculated that alternate approaches may yet yield desirable small molecule leads. Indeed, cell-based functional assays that recapitulate the endogenous cellular cascade have been successful in several screens run at Pharmacopeia. The readouts in such assays often involve the measurement of a downstream cell function such as chemotaxis, or de novo expression of a cell surface molecule or of a particular cytokine. Typically, specificity must be carefully controlled by counter-screening against related, but distinct, target molecules, and the mechanism of action of putative leads meticulously evaluated.

Advances in molecular cloning and expression, fluorescent labeling, measurement of ion fluxes, transcription-based reporters, fluorescent antibodies, imaging, and miniaturization techniques have led to the development of several cell-based approaches to HTS. This chapter will review some recent advances and applications in functional approaches to HTS. In addition, we will discuss several recent innovations in recombinant technologies that underlie many advances in signal generation, signal amplification, and specificity critical to functional screening.

II. APPLICATION OF RECOMBINANT TECHNOLOGY TO CELL-BASED HTS

A. Episomal vs. Integrated Expression of Cloned Genes

The accelerated pace of drug screening has led to an increased need for rapid generation of stable cell lines expressing suitable concentrations of the cloned, recombinant targets for the screening efforts. Classical methods for generating stable cell lines have generally relied on the integration of the recombinant gene into random locations within the genome of the host cell. The integrating gene is subjected to unpredictable rearrangements [1,2], and expression is strongly influenced by flanking chromosomal sequences and variations in gene copy number [3]. Following selection, only 5–30% of the resulting clonally derived cell lines functionally express the recombinant protein of interest [4,5]. Methods that use lox-cre, locus control regions, or IRES (internal ribosome entry sites) to circumvent these problems have been described, but these solutions still require integration of the DNA and subsequent isolation of clonally derived populations of cells for reliable long-term expression [5–7].

More recently, episomal vectors have been used for rapidly generating stable cell lines. These vectors do not integrate into the host cell chromosome and are therefore not subject to variability of expression due to position effects that are typical of integrating constructs. The need for the clonal isolation of expressing cells is therefore obviated, and the entire population of transfected cells can be pooled and used in approximately two weeks. The transfection efficiency of these replicating vectors can be several orders of magnitude higher than with integrating vectors [8–10] contributing to the speed and efficiency with which the stable cell lines can be made.

Often, in order to obtain sufficiently high levels of recombinant protein, efforts are taken to amplify the copy number of transfected recombinant gene. These methods, such as the use of methotrexate to select for coamplification of a recombinant gene with a dihydrofolate reductase selectable marker, are generally cumbersome and can require many months to accomplish [11]. Analysis of cell lines generated using episomal vectors reveals that the steady-state levels of recombinant RNA present often equals or exceeds the concentration of GAPDH, an abundant housekeeping gene product that typically represents approximately 0.8–3.6% of the poly(A^+) species present in the cell (Fig. 1) [10,12]. Furthermore, EBV-based vectors are generally maintained at 2–50 copies per cell, but most often at between 5 to 10 copies per cell (Fig. 2) [13–16]. Since this class of vector can therefore be considered already amplified, and because the level of RNA is so elevated, additional gene amplification is generally not necessary. The nature of the factors that govern episomal copy number are not yet understood, but it is thought that the particular combination of host cell background and episomal vector are key determinants. Cell lines transfected with the identical episomal construct but using very different transfection conditions will nonetheless stabilize at approximately the same number of gene copies and steady-state RNA per cell [10].

B. Use of Single vs. Multiple Episomes

Among targets for drug discovery, the need to express two or more recombinant proteins is frequently encountered. For example, numerous receptors, transporters, and ion channels are composed of multiple subunits that must be present in stoichiometric quantities for functional activity. In addition, signal transduction cascades contain many potential targets of interest for drug intervention. Traditionally, integrating vectors have been used in order to express multiple gene products. The cointegration and concomitant coexpression with desired stoichiometry of two or more genes is a rare, labor-intensive event to obtain and exploit

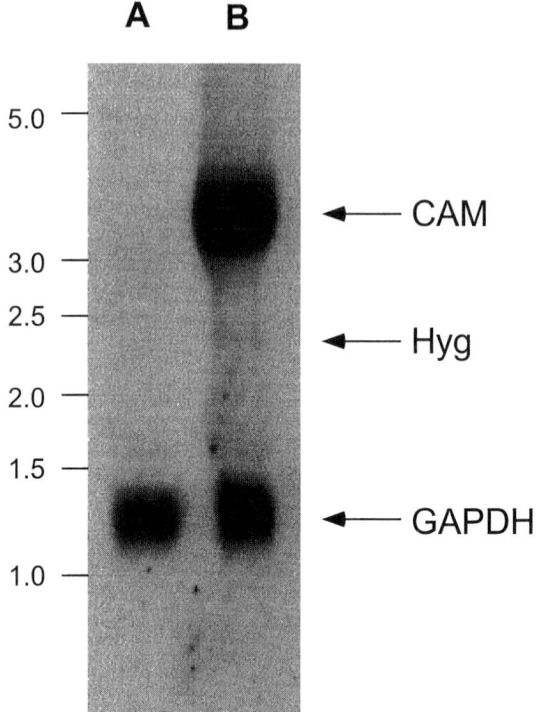

Figure 1 Northern blot analysis. Concentration of steady-state RNA of CAM, a recombinant cell adhesion molecule, is compared to the housekeeping gene, GAPDH (glyceraldehyde-3-phosphate-dehydrogenase). RNA from the hygromycin resistance marker (Hyg) is also visible. CAM, transcribed from a CMV promoter, and hyg, transcribed from a weak herpes simplex virus thymidine kinase (tk) promoter, are both contained on the episomal vector. Numbers to the left of the figure correspond to transcript length (in kilobases). Lanes A and B represent RNA isolated from untransfected and transfected 293EBNA cells (Invitrogen), respectively. Based on densitometer scans of several different exposures, the RNAs in this figure are represented at approximately the following ratios: 2:1: <0.1 for CAM:GAPDH:Hyg, respectively.

[17]. The use of a single episome for the high-level production of multiple genes has also proven somewhat difficult. When using the strong cytomegalovirus (CMV) immediate early promoter twice on one construct, we have encountered the phenomenon of promoter occlusion [18] in which transcription from one of the two promoters has been significantly dampened (R.A.H., unpublished observations). Furthermore, cloning multiple expression cassettes into one episome produces exceptionally large constructs that are cumbersome to generate and re-

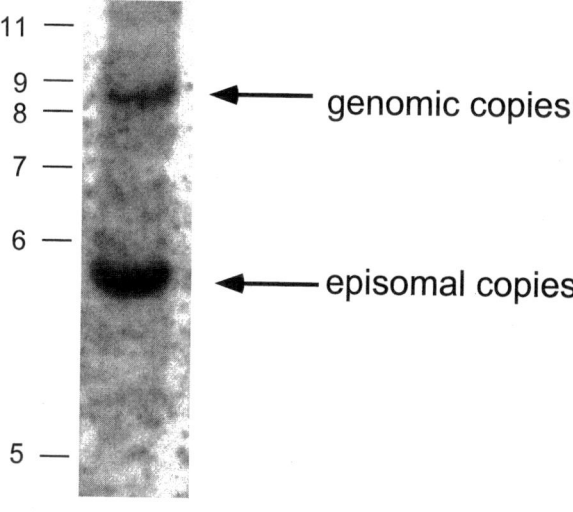

genomic copy #   3

episomal copy # 7-8

Figure 2 Genomic Southern blot analysis. The lane contains DNA isolated from 293EBNA cells 8 weeks after transfection with pE3hyg containing the coding sequences of the rat $G_{i\alpha2}$ gene. The probe used, $G_{i\alpha2}$, illuminates 2 bands, one at 8.5 kb representing the genomic $G_{i\alpha2}$ copies endogenous to 293EBNA cells, and the other at 5.8 kb representing the episomal copies. The parental cell is hypotriploid with a modal chromosome number of 64 [15,16]. Therefore the genomic band is most likely derived from a triploid chromosome and represents three copies. Scanning densitometer measurements of several exposures indicate that the episomal band is represented at seven or eight copies per cell.

quires a separate construct for each combination of genes to be employed. Therefore, in order to accelerate the development of assays requiring simultaneous expression of multiple recombinant targets, we have used multiple independently replicating episomes (Fig. 3). We have found that the addition of a second, third, or fourth episomal construct does not affect expression, copy number, or steady-state RNA concentrations from episomes already contained within the cell (R.A.H., unpublished observations). By using multiple episomes, each containing a single recombinant gene of interest, one can transfect any combination of vectors at will in a procedure that we term ''combinatorial transfections.''

The choice of an appropriate cell line that contains characteristics desirable or even essential to a HTS format is an important part of the assay development process. Considerations such as the lineage from which the cell line was derived,

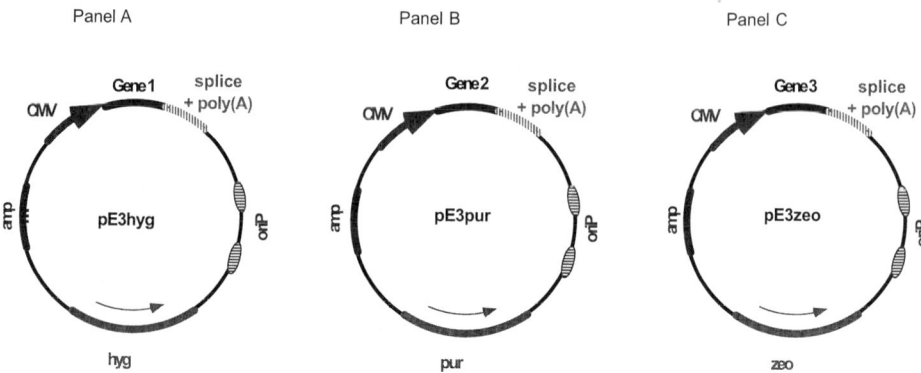

Figure 3 Schematic representation of episomal vectors described in this work. Vectors exist as a set of nearly identical constructs that differ only in selectable marker. Construct pE3hyg contains the hygromycin antibiotic resistance coding sequence, pE3pur confers puromycin resistance, etc. Resistance markers are abbreviated as follows: Zeo, zeocin; bla, blasticidin; oua, ouabain; and gpt, xanthine-guanine phosphoribosyl transferase. In cases where multiple episomes are used, pE3hyg contains "gene 1," pE3pur contains "gene 2," pE3zeo contains "gene 3," etc.

adherent properties, absence of interfering receptor subtypes or protein isoforms, or the presence of necessary ancillary protein subunit or signal transduction components are all fundamental. Various cell lines such as HEK293 (293EBNA), CV1, the somatic cell hybrid line D98/raji [19], the osteosarcoma line 143 [19], and numerous EBV⁺ or EBNA1⁺ lymphoid cells that already express EBNA1 can be readily obtained from Invitrogen or the American Type Cell Culture [16]. In the event that a cell type not already expressing EBNA1 is preferred, one need only simultaneously introduce into the cell line of choice two episomes, one encoding EBNA1 and a second encoding the desired recombinant gene. In this case, it is important that EBNA1 be transcribed from a strong promoter so that sufficient concentrations of EBNA1 will be reached rapidly enough to enable the cell to maintain both episomes before they are lost from the cell population or integrated into the host chromosome. Commercially available vectors generally do not contain a recognizable promoter driving expression of EBNA1, and we have observed instability of expression when using these constructs. The use of multiple episomes permits even further flexibility in the choice and use of cell lines for drug discovery. For instance, T-antigen can be introduced via one of the episomal complements in order to immortalize a desired cell type, or the coding sequence for the macrophage scavenger receptor can be introduced in order to confer a strongly adherent phenotype to cells in culture [20]. In all cases, we find that cell lines are stable and ready for use in high-throughput assays in 2 to 3 weeks.

Lastly, one can exploit the effects of signal transduction cascades on activation or repression of gene transcription by using appropriate response elements cloned upstream of reporter genes such as luciferase to generate functional high-throughput drug screening assays. The consensus sequences of many different types of response elements have been described in detail. Synthetic oligonucleotides comprising these sequences can be concatenated and cloned adjacent to a minimal promoter-reporter gene expression cassette to produce convenient readouts sensitive to changes in transcription factor activity. Alternatively, promoter regions from cellular genes containing appropriate response elements can be used in place of synthetic constructs. The signal transduction responsive expression cassette can be cloned onto an episomal vector and used combinatorially with any of a number of receptors to provide stable cell lines rapidly. Rate-limiting proteins in the signal transduction cascade can also be added to increase the magnitude of the signal (R.A.H., unpublished observations).

III. FUNCTIONAL SCREENING BY USING CALCIUM MOBILIZATION

A. General Model for Calcium Mobilization

Activation of cells by numerous and varied extracellular stimuli modulates the intracellular levels of several ions, including sodium, potassium, hydrogen, and calcium. Calcium is used as an index of cellular activation. For example, activation of leukocytes by chemokines leads to a rapid and transient increase in the intracellular level of free calcium that is dependent to varying degrees on both influx of calcium from outside the cell and release of calcium from intracellular stores [21]. Because changes in the intracellular calcium concentration constitute an early and transient signaling event following cell activation, it has been difficult to monitor. However, recent advances in detection capabilities and automation have improved the measurement of calcium mobilization and have fostered the use of cell-based screening for calcium changes.

The fundamental feature of calcium as a signaling element is its maintenance in an extreme state of disequilibrium. Indeed, mammalian cells, whose total concentration of calcium is about 1–2 mM, maintain their intracellular concentration of free cytoplasmic calcium in the range of 0.1–0.2 μM despite a large unfavorable electrochemical gradient. This sets the stage for the requisite large, rapid, and transient change in the concentration of free cytoplasmic calcium that characterizes the activation of most cells. Intracellular calcium is distributed among several pools that include free, lipid or protein bound, and calcium that is accumulated in various storage compartments. For purposes of the application discussed here, one can assume that the free cytoplasmic calcium is regulated by exchanges between intracellular storage pools and with the extracellular milieu, which typically contains about 1–2 mM of calcium [22].

The mechanism responsible for the release of calcium from intracellular storage pools in nonmuscle cells has been linked to the phospholipase C-mediated generation of inositol triphosphate. Cell stimulation is also accompanied by increases of extracellular calcium through channels that differ from other (voltage-dependent) calcium channels. These specialized channels are known as calcium release-activated calcium channels or CRAC [23]. The nature of the linkage between the depletion of intracellular stores and the activation of CRAC is still conjectural. Cells sense that calcium storage pools have been depleted and respond by allowing calcium to flow in from the external milieu [23]. Thus, there are at least three parameters of calcium mobilization that can be monitored as indices of activation by chemokines and other cell stimuli factors: (1) changes in the concentration of free cytoplasmic calcium, (2) discharge of calcium from the intracellular storage pools, and (3) influx of calcium [19]. Below, we will concentrate on the ways that measurement of free cytoplasmic calcium, as an indication of activation by chemotactic factors, is employed for high-throughput functional screening for small molecule antagonists or agonists.

B. Methods for Measuring Intracellular Calcium

We will review direct methods of intracellular calcium measurement, i.e., those in which some form of calcium sensor is incorporated into a cell. The response of the sensor is compared with responses observed in the presence of calcium standards, so that the detection signal can be converted into an intracellular calcium concentration. To date, three such direct methods have been used successfully. The first method is that of phosphoproteins. These substances are obtained from luminescent organs of coelenterates and emit light in a reaction that is catalyzed by calcium. Since the first use of this method in the mid-1960s [24], two other major techniques have been developed. One of them employs calcium-sensitive microelectrodes. These are similar to the glass microelectrodes used for conventional electrical recording from cells, but with the important addition of a calcium "sensor" dissolved in an organic phase in the electrode tip. The sensor makes the electrode specifically permeable to calcium, so the electrode potential (after correction for the membrane potential) varies with the intracellular calcium concentration and hence can be used to measure it. The last method uses calcium-sensitive dyes (metallochromic indicators). This method has a long history, but the first dyes to be used, such as murexide, were too insensitive to be of much use for intracellular calcium measurement. During the mid-1970s, several groups demonstrated that the "azo" dyes (arsenazo III has been the most used dye in this family) could be used to measure intracellular calcium in a variety of preparations [25]. More recently, other suitable dyes have been described, and some of these undergo a change in fluorescence (i.e., re-emission of photons, following absorbance of photons of shorter wavelength) on binding calcium. Fluorescence has proven to be one of the most efficient and sensitive methods to measure changes in intracellular levels of free cytoplasmic calcium.

C. Fluorescent Calcium Indicators

The commonly available fluorescent indicators for calcium fall into two operational classes: single-wavelength intensity-modulating dyes and dual-wavelength ratiometric dyes, which are referred to as single-wavelength (SW) indicators and ratiometric indicators, respectively. For SW indicators, changes in calcium concentrations bring about changes in the intensity of their fluorescence excitation and emission spectra, whereas the spectral maxima remain essentially unchanged. Ratiometric indicators exhibit not only intensity changes with fluctuating calcium concentration but the calcium-free and calcium-bound forms of the indicator have distinct spectra, the maxima of which are located at different wavelengths. The two ratiometric dyes most commonly used are Fura-2 and Indo-1 [26]. The two most commonly used SW indicators, Fluo-3 and calcium-green, incorporate fluorescein chromophores and are therefore excited at wavelengths typical for fluorescein. They both exhibit the largest intensity changes in their transition from calcium-free to calcium-bound forms (40- to 1000-fold [27,28]). This change can be an advantage because, compared with other SW indicators, similar changes in calcium concentrations result in larger changes in brightness for fluo-3 and calcium-green-2. Because fluorescence quantum efficiency can range only from 0 to 1, the large intensity difference between calcium-bound and calcium-free forms implies that the unbound forms of the two indicators must be only very weakly fluorescent. This would result in a low resting fluorescence background in the cells, thus providing another advantage to the use of these dyes.

D. Imaging Systems

The use of commercially available fluorescent imaging systems has become routine in HTS. One commercially proven fluorescent imaging plate reader currently available for calcium-based HTS is the FLIPR, developed by Molecular Devices [29]. The salient features of FLIPR are that it detects fluorescent signals rapidly, in real time, and with precision. Using a combination of standard and nonstandard integration of optics, fluidics, and temperature control, FLIPR is fit for homogeneous, kinetic, cell-based fluorometric assays such as the measurement of intracellular calcium, membrane potential, and intracellular pH. The key advantage of the system is that it simultaneously stimulates and reads all 96/384 wells of a microplate within 1 to 2 seconds. Although under ideal conditions FLIPR can be used for both adherent and nonadherent cell lines, we have found that its use for HTS is practical with adherent cell lines only (B.D., unpublished observations).

FLIPR is powered by an argon laser that is used to excite a fluorescent indicator dye. The emitted light is detected using a proprietary optical scheme. The use of an argon laser as a source of light limits the use of fluorescent calcium dyes that are currently available to only two: Fluo-3/AM and calcium-green. A cooled CCD camera is used as an integrating detector, accumulating the fluorescent signal during the period in which it is exposed to the image, making it extremely sensitive. A data point is taken from each of the 96 wells within one second. Sensitivity is further enhanced by proprietary cell-layer isolation optics that permit signal discrimination on cell monolayers. This eliminates the undesirable extracellular background fluorescence found in most fluorescent assays. Conventional optical detection based on viewing living cells at the bottom of a microplate well also detects fluorescence from the fluid above the cells, leading to a high level of background fluorescence that tends to obscure the cellular signal. Other artifacts caused by temperature fluctuations and fluid mixing can also lead to inaccurate results.

The key advantage of the FLIPR system is the ability to perform calcium mobilization assays in a high-throughput fashion. There are several disadvantages to the current format, however. Among these are (1) the substantial base price, which precludes its access to many investigators, (2) the argon laser light source, which limits the choice of dyes to just two fluorescent dyes, (3) the present optical system, which functions best with adherent cell lines, and (4) that the argon laser needs special installation for extensive cooling, thus requiring its own designated space.

IV. TRANSCRIPTION-BASED SCREENING WITH REPORTER GENES

Alteration in gene expression levels is an indicator of cellular activation. Concentrations of a particular mRNA in a cell can be measured, but such methods are not suitable for HTS. The methodology for measuring changes in gene expression for HTS was facilitated by the advent of both synthetically constructed and naturally derived inducible promoters and response elements. These cis-acting DNA regions respond to intracellular signals, such as changes in levels of cAMP, by increasing transcription of covalently linked, downstream reporter genes. Changes in transcription rates from a particular promoter serve as an index of activation of a cellular pathway. Reporter gene technology has evolved over the past five years to the point that there are now many vectors with different combinations of response elements, promoters, and reporter genes commercially available. This section will highlight the use of several inducible response elements and promoters, and reporter genes, and discuss the issues that underlie the best choices for HTS.

A. Inducible Promoters and Response Elements

Inducible promoters and response elements are critical tools for the use of reporter gene technology, as they provide the ultimate on-switch that activates transcription of the reporter gene. Promoter activity is controlled or modulated by regulatory regions of the DNA that are usually, but not always, found in the 5' end of the gene. These regulatory regions encode the response elements that act as sensors of activation of a cell signal pathway. Ideally, for screening purposes, the promoter should be strong but controlled by the desired response element in a way that maintains baseline transcription at low levels in the absence of the cell activation. The cytomegalovirus promoter is a strong promoter, but constitutive expression of the reporter gene often leads to high background. The herpes virus thymidine kinase, SV40, and growth hormone promoters have been used in many constructs.

The response elements used in HTS depend to a large extent on what signal pathway is activated. Genes that are transcriptionally regulated by cAMP contain one or more copies of the octameric sequence TGACGTCA [30,31] or cAMP response element (cre). Binding of the agonist isoproterenol to β-adrenergic receptors causes an increase in the cAMP levels, which activates protein kinase A (PKA). Activated PKA translocates to the nucleus and phosphorylates the transcription factor known as cAMP response element binding protein (CREB), which in turn binds to cre elements in the DNA. A similar but distinct heptameric sequence, TGACTCA, encodes a response element activated by tumor-promoting phorbol esters, such as 12-O-tetradecanoylphrobol 13-acetate (TPA). TPA activates protein kinase C and induces similar changes in cell morphology and gene expression via the intracellular signaling molecule diacylglycerol [32]. The TPA response element (tre) binds the Jun homodimer or the Jun/Fos heterodimer [33,34]. More recently, a response element identified in the interleukin-2 gene promoter that binds the nuclear factor of activated T cells (NFAT) has been used to drive reporter genes [35–37]. NFAT is a member of a family of transcription factors. NFAT is a multicomponent transcription factor found in the nucleus of activated T cells. The immunosuppressant drugs cyclosporin and FK506 inhibit NFAT activity by inhibiting dephosphorylation of the NFAT component proteins by calcineurin and preventing nuclear translocation [38,39]. Although each of these response elements generally monitors different intracellular signals, providing the potential to couple the activation of different pathways to a reporter gene, the best studied is cre, and much work is needed to understand and utilize additional cis-acting elements to induce gene transcription for HTS.

B. Reporter Genes

The most critical considerations for HTS are that the gene product be easily detected and stable in the assay system. Detection is limited by the constitutive expression of a gene raising the background; thus many reporter genes for use in mammalian systems encode for nonmammalian enzymes. This allows for robust signal generation over low background. For example, parathyroid hormone (PTH) binds to the extracellular PTH receptor and activates the adenylyl cyclase pathway in osteoblast-like UMR-106 cells. Transfection of UMR-106 cells with a firefly luciferase reporter gene under the control of cre led to a 40-fold increase in luciferase activity [40]. The most commonly used reporter genes are luciferase, choramphenicol acetyltransferase (CAT), secreted alkaline phosphatase (SEAP), β-galactosidase (β-Gal), β-glucuronidase (β-Gus), and green fluorescent protein (GFP). Most of these reporters have been reviewed in detail recently [41,42], and only luciferase will be discussed below.

Luciferase from *Photimus pyralis* (firefly) has become the most widely used reporter as it offers great sensitivity over an essentially null background in mammalian cells [43,44]. Commercially available reagents allow for longer assay kinetics enabling the luminescent reaction to be measured as a glow reaction. The dynamic measurement range is ~7 orders of magnitude, and the ability to conduct plate imaging is especially attractive as assay miniaturization occurs and screens are conducted in 384-well and 1536-well formats [45]. Several mutant luciferases are available and emit light over a wavelength range of 548–612 nm [46]. This allows the investigator to use two different constructs if an internal control is desired. An early limitation was the requirement for cell lysis; lysis is no longer essential with a naturally secreted luciferase from *Vardula hilendorfii* [47].

V. CELL-BASED SCREENS

A. Cell-Based vs. Mechanism-Based Screens

Cell-based screening, in which an intact cell is used to measure a particular response, had been a primary screening strategy prior to the broad use of recombinant technologies. Recombinant technology has made reagents readily available for use in assays as components of the cell and thus facilitated the focus on mechanism-based or target-based screening. This allowed for a focused approach targeting specific biological molecules. One could define the potency of a new drug at a molecular interaction site and reduce nonspecific side effects. A disadvantage of mechanism-based screening, however, is that a cell pathway or cell function may be known but the precise molecular target may be unclear. In addition, target-based screening does not take into account cell permeability, and while this is not disadvantageous for cell-surface targets, this issue is substantive

Human THP-1 Monocytes (25,000 cells/well) in 100 µl/well

↓ Incubate overnight at 37 °C/5%CO$_2$

TNF-α Induction/Compound Treatment Microtiter Plate

↓ Induce with 1 µg/mL lipopolysaccharide (*E. coli*), 1%DMSO and test compounds.

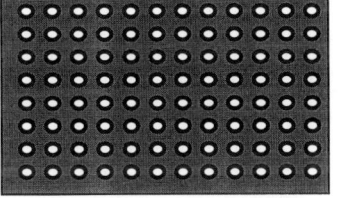

Incubate overnight at 37 °C/5%CO$_2$

Transfer 50 µL of culture medium to an ELISA plate coated with a monoclonal antihuman TNF-α antibody.

TNF-α ELISA Microtiter Plate

↓ Determine TNF-α production by ELISA using a biotinylated rabbit IgG secondary antibody and horseradish peroxidase-antirabbit IgG conjugate.

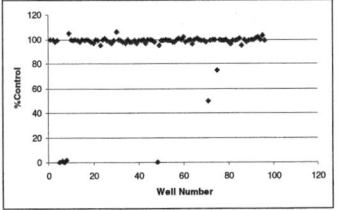

Normalized Data

↓ Normalize data to % of control TNF-α expression by comparison to unstimulated cells.

Figure 4 Schematic representation of cell-based assay for TNF-α. Culture media of THP-1 cells.

for intracellular targets. Cell-based approaches broaden the target horizon and assure the identification of compounds with desirable cell permeability. The two largest disadvantages with cell-based screening are the absence of a defined mechanism of action and the need for more numerous specificity controls. In the following section, we will discuss a cell-based screen and compare it to a mechanism-based screen.

B. Case Study: Screen for TNF-α Expression Inhibitors

An example of a cell-based expression screen run at Pharmacopeia is that for inhibitors of tumor necrosis factor-α (TNF-α) production. In a search to identify inhibitors of TNF-α expression, we adapted an assay to a high-throughput screen that measured TNF-α concentration in the culture media from stimulated THP-1 human monocytic cells (Fig. 4). Compounds were screened at a final concentra-

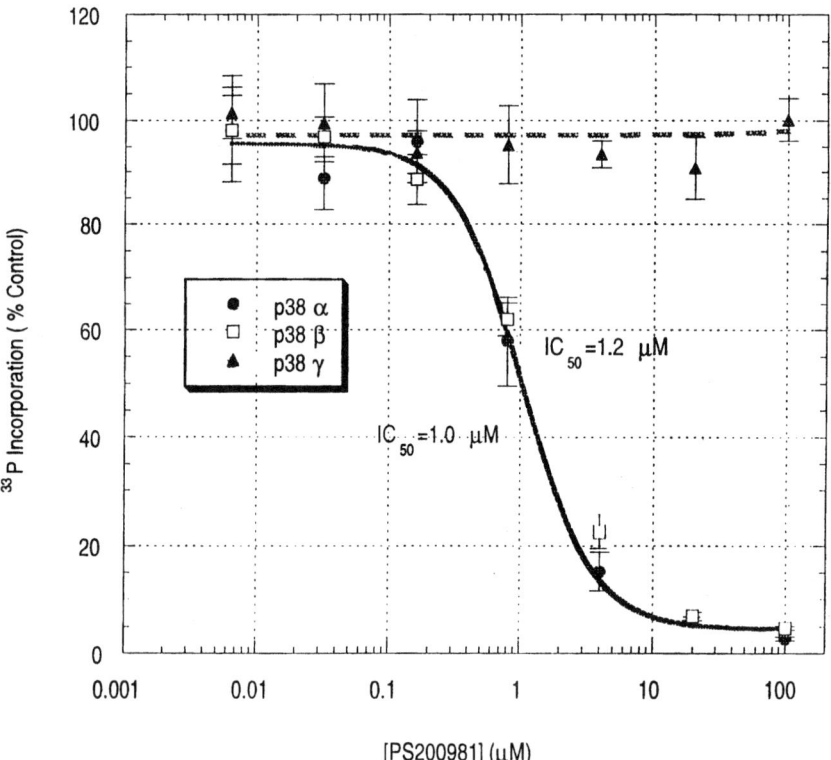

Figure 5 Inhibition curves showing effect of PS200981 on p38α,β,γ kinases.

Table 1 Kinase Selectivity of PS200981

Kinase	PS200981 (IC$_{50}$ in µM)
P38α	1
P38β	1.3
P38γ	>50
ERK-2	>50
PKA	>50
Src pp60	>50
Jak3	>50

tion of 1–5 µM in a multiplexed format. Thus 20 compounds per well were evaluated in the primary screen. Using these conditions, cytotoxicity and nonspecific effects were not observed, and 100,000 compounds were screened per day. Screening of approximately 2 million compounds led to the identification of a series of compounds from a combinatorial library that demonstrated a clear synthon preference and structure–activity relationship. Compound PS200981 inhibited TNF-α expression in THP-1 cells with an IC50 of 0.25 µM.

Recent work had shown that the p38 kinase regulated production of TNF-α and other proinflammatory cytokines [48–50]. p38 kinases are members of the mitogen-activated protein kinase family that includes the extracellular signal-regulated kinases (ERKs) and c-jun N-terminal kinases (JNKs). MAP kinases transduce extracellular stimuli in the regulation of such diverse cell functions as proliferation and stress response. The role of p38 kinase in cytokine regulation was confirmed by scientists at SmithKline Beecham using a class of bicyclic imidazoles that are specific for the p38 kinase [51]. These observations prompted us to examine the effect of PS200981 on p38 kinase as a potential target in the cell-based assay of TNF-α production. PS200981 inhibited p38 but not other protein kinases examined (Fig. 5, Table 1) with the exception of the highly related isoform p38β. Thus a novel molecular class of p38 kinases inhibitors was identified through a cell-based screen for TNF-α production.

VI. CONCLUSIONS

Advances in molecular cloning and expression, fluorescent labeling, measurement of ion fluxes, transcription-based reporters, fluorescent antibodies, imaging, and miniaturization techniques have led to the development of cell-based high-throughput screens. Targets such as cytokines, growth factors, ion channels, and GPCRs have been interrogated using cell-based screening with good success.

The issues in these screening approaches are whether the readouts are amenable to HTS and the specificity of the response. Improved technologies and tools, and attention to utilizing counter-screens against related, but distinct, target molecules, have made cell-based screening a viable alternative for lead identification.

REFERENCES

1. DM Robins, S Ripley, AS Henderson, R Axel. Transforming DNA integrates into the host chromosome. Cell 23:29–39, 1981.
2. DM Robins, R Axel, AS Henderson. Chromosome structure and DNA sequence alterations associated with mutation of transformed genes. J Mol Appl Genet. 1: 191–203, 1981.
3. E Lacy, S Roberts, EP Evans, MD Burtenshaw, FD Costantini. A foreign beta-globin gene in transgenic mice: integration at abnormal chromosomal positions and expression in inappropriate tissues. Cell 34:343–358, 1983.
4. V Gurtu, G Yan, G Zhang, IRES Bicistronic expression vectors for efficient creation of stable mammalian cell lines. Biochem Biophys Res Comm 229:295–298, 1996.
5. S Rees, J Coote, J Stables, S Goodson, S Harris, MG Lee. Bicistronic vector for the creation of stable mammalian cell lines that predisposes all antibiotic-resistant cells to express recombinant protein. Biotechniques 20:102–110, 1996.
6. S Fukushige, B Sauer. Genomic targeting with a positive-selection lox integration vector allows highly reproducible gene expression in mammalian cells. Proc Natl Acad Sci USA 89:7905–7909, 1992.
7. J Yu, JH Bock, JL Slightom, B Villeponteau. A 5′ beta-globin matrix-attachment region and the polyoma enhancer together confer position-independent transcription. Gene 139:139–145, 1994.
8. RF Margolskee. Epstein-Barr virus based expression vectors. Curr Top Microbiol Immunol 158:67–95, 1992.
9. A Jalanko, A Kallio, I Ulmanen. Comparison of mammalian cell expression vectors with and without an EBV-replicon. Arch Virol 103:157–166, 1988.
10. RA Horlick, K Sperle, LA Breth, CC Reid, ES Shen, AK Robbins, GM Cooke, BL Largent. Rapid generation of stable cell lines expressing corticotropin-releasing hormone receptor for drug discovery. Protein Expr Purif 9:301–308, 1997.
11. CS Simonsen, M Walter, AD Levinson. Expression of the plasmid-encoded type I dihydrofolate reductase gene in cultured mammalian cells: a novel selectable marker. Nucleic Acids Res 16:2235–2246, 1988.
12. M Piechaczyk, JM Blanchard, L Marty, C Dani, F Panabieres, S El Sabouty, P Fort, P Jeanteur. Post-transcriptional regulation of glyceraldehyde-3-phosphate-dehydrogenase gene expression in rat tissues. Nucleic Acids Res 12:6951–6963, 1984.
13. JL Yates, N Warren, B Sugden. Stable replication of plasmids derived from Epstein-Barr virus in various mammalian cells. Nature 313:812–815, 1985.
14. PB Belt, H Groeneveld, WJ Teubel, P van de Putte, C Backendorf. Construction and properties of an Epstein-Barr-virus-derived cDNA expression vector for human cells. Gene 84:407–417, 1989.

15. FL Graham, J Smiley, WC Russell, R Nairn. Characteristics of a human cell line transformed by DNA from human adenovirus type 5. J Gen Virol 36:59–74, 1977.
16. R Hay, J Caputo, TR Chen, M Macy, P McClintock, Y Reid. ATCC Catalog of Cell Lines and Hybridomas, 7th ed. Rockville, MD: American Type Culture Collection, 1992, p 148.
17. MI Cockett, R Ochalski, K Benwell, R Franco, J Wardwell-Swanson. Simultaneous expression of multi-subunit proteins in mammalian cells using a convenient set of mammalian cell expression vectors. Biotechniques 23:402–407, 1997.
18. T Kadesch, P Berg. Effects of the position of the simian virus 40 enhancer on expression of multiple transcription units in a single plasmid. Mol Cell Biol 6:2593–2601, 1986.
19. D Reisman, J Yates, B Sugden. A putative origin of replication of plasmids derived from Epstein-Barr virus is composed of two cis-acting components. Mol Cell Biol 5:1822–1832, 1985.
20. AK Robbins, RA Horlick. Macrophage scavenger receptor confers an adherent phenotype to cells in culture. Biotechniques 25:240–244, 1998.
21. BB Damaj et al. Diverging signal transduction pathways activated by interleukin 8 (IL-8) and related chemokines in human neutrophils. IL-8 and Gro-alpha differentially stimulate calcium influx through IL-8 receptors A and B. J Biol Chem 271:20540–20544, 1996.
22. SR McColl, PH Naccache. Calcium mobilization assays. Methods Enzymol 288:301–309, 1997.
23. R Penner, C Fasolato, M Hoth. Calcium influx and its control by calcium release. Curr Opin Neurobiol 3:368–374, 1993.
24. EB Ridgway, CC Ashley. Calcium transients in single muscle fibres. Biochem Biophys Res Commun 39:229–234, 1967.
25. NC Kendrick. Purification of arsenazo III, a Ca^{2+}-sensitive dye. Anal Biochem 76:487–501, 1976.
26. G Grynkiewicz, M Poenie, RY Tsien. A new generation of Ca^{2+} indicators with greatly improved fluorescence properties. J Biol Chem 260(6):3440–3450, 1985.
27. A Minta, JP Kao, RY Tsien. Fluorescent indicators for cytosolic calcium based on rhodamine and fluorescein chromophores. J Biol Chem 264(14):8171–8178, 1989.
28. RP Haugland. Handbook of Fluorescent Probes and Research Chemicals. Eugene, Oregon: Molecular Probes, 1991.
29. KS Schroeder, BD Neagle. FLIPR: a new instrument for accurate, high throughput optical screening. J Biomolecular Screening 1:75–80, 1996.
30. MR Montminy, KA Sevarino, JA Wagner, G Mandel, RH Goodman. Identification of a c-AMP-responsive element within the rat somatostatin gene. Proc Natl Acad Sci USA 83:6682–6686, 1986.
31. PJ Deutsch, JP Hoeffler, JL Jameson, JF Habener. Cyclic AMP and phrobol ester-stimulated transcription mediated by similar DNA elements that bind proteins. Proc Natl Acad Sci USA 85:7922–7926, 1988.
32. A Hata, Y Akita, K Suzuki, S Ohno. Functional divergence of protein kinase C (PKC) family members. J Biol Chem 268:9122–9129, 1993.
33. P Angel, M Imagawa, R Chiu, B Stein, RJ Imbra, HJ Rahmsdorf, C Jonat, P Herrlich,

M Karin. Phorbol ester-inducible genes contain a common cis element recognized by a TPA-modulated trans-acting factor. Cell 49:729–739, 1987.

34. T Curran, BR Franza Jr. Fos and jun: the AP-1 connection. Cell 55:395–397, 1988.

35. J Karttunen, N Shastri. Measurement of ligand-induced activation in single viable T cells using the lacZ reporter gene. Proc Natl Acad Sci 88:3972–3976, 1991.

36. J Navarro, C Punzon, JL Jimenez, E Fernandez-Cruz, A Pizarro, M Fresno, MA Munoz-Fernandez. Inhibition of phosphodiesterase type IV suppresses human immunodeficiency virus type 1 replication and cytokine production in primary T cells: involvement of NF-kB and NFAT. J Virol 72:4712–4729, 1998.

37. M Saxena, S Williams, J Gilman, T Mustelin. Negative regulation of T cell antigen receptor signal transduction by hematopoietic tyrosine phosphatase. J Biol Chem 273:15340–15344, 1998.

38. J Park, NR Yaseen, PG Hogan, A Rao, S Sharma. Phosphorylation of the transcription factor NFATp inhibits its DNA binding activity in cyclosporin A-treated human B and T cells. J Biol Chem 270:20653, 1995.

39. KT-Y Shaw, AM Ho, A Raghavan, J Kim, J Jain, J Park, S Sharma, A Rao, PG Hogan. Immunosuppressive drugs prevent a rapid dephosphorylation of transcription factor NFAT1 in stimulated immune cells. Proc Natl Acad Sci USA 92:11205–11209, 1995.

40. B Fluhmann, U Zimmermann, R Muff, G Bilbe, JA Fischer, W Born. Parathyroid hormone responses of cyclic AMP-, serum- and phrobol ester-responsive genes in osteoblast-like UMR-106 cells. Mol Cell Endocrinol 139:89–98, 1998.

41. W Scheirer. Reporter gene assay applications. High Throughput Screening 401–412, 1997.

42. CM Suto, DM Ignar. Selection of an optimal reporter gene for cell-based high throughput screening assays. J Biomolecular Screening 2:7–9, 1997.

43. JR deWet, KV Wood, DR Helinski, M DeLuca. Cloning of firefly luciferase cDNA and the expression of active luciferase in *Escherichia coli*. Proc Natl Acad Sci USA 82:7870–7873, 1985.

44. TO Baldwin. Luciferase: the structure is known, but the mystery remains. Structure 4:223–228, 1996.

45. S Inouye, Y Ohmiya, Y Toya, FI Tsuji. Imaging of luciferase secretion from transformed Chinese hamster ovary cells. Proc Natl Acad Sci USA 89:9584–9587, 1992.

46. N Kajiyama, E Nakano. Isolation and characterization of mutants of firefly luciferase which produce different colors of light. Prot Eng 4:691–693, 1991.

47. EM Thompson, S Nagata, FI Tsuji. *Vargula hilgendorfii* luciferase: a secreted reporter enzyme for monitoring gene expression in mammalian cells. Gene 96:257–262, 1990.

48. JC Lee, PR Young. Role of CSBP/p38/RK stress response kinase in LPS and cytokine signaling mechanisms. J Leukocyte Biol 59:152–157, 1996.

49. K Schulze-Ostioff, D Ferrari, K Riehemann, S Wesselborg. Regulation of NF-kB activation by MAP kinase cascades. Immunobiology 198:35–49, 1997.

50. L New, J Han. The p38 MAP kinase pathway and its biological function. Trends Cardiovasc Med 8:220–228, 1998.

51. DE Griswold, PR Young. Pharmacology of cytokine suppressive anti-inflammatory drug binding protein (CSBP), a novel stress-induced kinase. Pharmacol Commun 7:323–329, 1996.

9
Enzyme Screens

Thomas D. Y. Chung
DuPont Pharmaceuticals Company, Wilmington, Delaware

Dennis J. Murphy
Hercules Incorporated, Wilmington, Delaware

I. INTRODUCTION

The targeting of specific enzymes of metabolic pathways, lipid metabolism, signal transduction, protein processing, and inflammatory cascades that are shown to be involved in disease processes is a proven method used in drug discovery. Many marketed medicines such as Rheumatrex™ (methotrexate), Mevacor™ (lovostatin), Capoten™ (captopril), Zovirax™ (acyclovir), Cipro™ (ciprofloxacin), fluorouracil, Retrovir™ (zidovudine-AZT), clavulinic acid, Proscar™ and Propecia™ (finesteride), and disulfuram have resulted from this strategy. There are also examples of up-regulation of key regulatory enzymes or the use of specific enzymes themselves as therapeutics (e.g., streptokinase as an antithrombolytic).

The identification of enzymes as potential targets for drug discovery has been facilitated by the increasing breadth and depth of available public and commercial DNA sequences (e.g., *Expressed Sequence Tags*), transcriptional fingerprints, and protein sequence and protein structure databases. In some cases, a new enzyme is found to be related by both sequence and structure to another enzyme or family of enzymes already established as valid drug targets, such as the ADAMS family of proteases [1,2]. Sometimes high-resolution crystal structures of both free and inhibitor bound enzymes are available, and homology modeling of the putative target against the known reference structure affords early guidance for the design of inhibitors and an understanding of the structure–activity relationships (SAR) of inhibitors. Furthermore, the ability to genetically modify, deregulate, or knock out the target enzyme gene in bacteria, yeast, and small transgenic animal models has facilitated the experimental validation of the enzyme as a target, once identified by sequence comparisons.

As an increasing number of enzymes are identified, cloned, expressed, and validated, pharmaceutical companies are challenged to mount drug discovery campaigns rapidly against these enzyme targets. Rational design approaches have proven their merit where much structural data on target enzymes and several "privileged structural" classes of known inhibitors existed. Comprehensive high-throughput screening (HTS) of large chemical collections, including novel random and targeted combinatorial libraries, is a proven complementary method for rapidly identifying novel compounds as starting points for chemical syntheses and drug design programs, or to provide potential inhibitors to aid in target validation.

General assay development, HTS hit selection processes, assay statistics, and other operational issues common to most HTS practitioners are covered elsewhere in this book. This chapter instead provides guidance on the advantages and disadvantages, the strategies and tactics, and the principles and practices used by several practitioners of HTS as applied to enzyme screens. It also highlights some of the exceptions from the "ideal" kinetics that make assay development and execution a real technical challenge. The definition restricts itself to screens where an enzyme is assayed in a relatively isolated in vitro context, rather than as part of a cell-based pathway or reporter screen. An HTS must be appropriately designed (strategy) and properly configured (tactics) to maximize both efficiency and effectiveness. An improperly designed HTS may fail to detect the desired classes of inhibitors even if it is trivial to execute with automation, consume too much reagent, take too long to complete, or cost too much to execute. A successful enzyme HTS requires skills and knowledge in chemistry, enzymology, physics, automation, and engineering, which often best result from cross-functional, interdepartmental teams. Finally, properly designed and configured high-throughput enzyme assays are often robust, precise, and accurate enough to provide IC_{50} determinations for lead optimization activities that follow.

II. PROS AND CONS OF ENZYME SCREENS

Enzyme-based screens offer advantages due to the measurement of a clearly defined enzyme activity. The development of SAR is more tractable as off-target activities are not confounding (e.g., in whole cell screens the inhibitor must traverse membranes, so permeability and active transport confound potency SAR). Often there is considerable knowledge of active sites or the relevant contact surfaces. In vitro systems allow independent manipulation of conditions to tune the "sensitivity" of an assay to detect desired classes of inhibitors. These choices define the downstream follow-up of potential inhibitors. Also an in vitro system permits adjustment of the fluxes in a multienzyme pathway to allow detection of cumulative weak inhibition at several loci and to screen an entire pathway (*vide infra*). Additionally, cell-free systems allow the ability to prepare a single lot of enzyme and store aliquots for the entire screen, rather than relying on living cells grown in tissue culture or obtained in relatively small batches from whole tissue.

However, in vitro enzyme screens are inherently artificial systems that can only approximate the relevant true physiological state. There is the possibility that key regulatory and feedback mechanisms are missing that affect binding to the target. Also they require some level of isolation and purification of the enzyme from some recombinant or natural source, which always raises the possibility that critical component(s) may be purified away or that contaminants may add undesired activities or mask desired ones. Finally, while new pharmacophores may be found from the increased accessibility of the target in vitro, they may prove to be intractable with regard to permeability properties on the target organism. Historically, this has been the case in antibacterial drug discovery.

III. PRINCIPLES, STRATEGIES, AND TACTICS FOR ENZYME SCREENS

When an HTS scientist accepts a research-level enzyme assay for implementation into an HTS campaign, there are several guiding principles that should be considered before deciding on final configuration, format, and implementation. There should be as thorough a knowledge of the relevant biological, physical, and enzymological properties of the enzyme target as possible [3].

First, the biological properties of the enzyme would ideally be known. Its role in metabolism, coupling to other enzymes, site of expression, occurrence and distribution in species and tissues, intracellular location, genes, precursors, existence of isozymes, and the effects of its deficiency must be considered in evaluating whether the enzyme is a viable, accessible target for the ultimate treatment modality. All these factors weigh in evaluating the risk of assaying the enzyme in an in vitro versus an in situ environment. In the test tube an enzyme's environment is very different from its natural environment in the cell, and these differences may affect its behavior and hence the ability to find inhibitors that affect the relevant reactions. An intracellular enzyme that "cross-talks" with many other enzymes (e.g., as part of a serial pathway, at a branch point, or part of an amplification cascade) may be difficult to inhibit specifically without undesirable effects on related pathways (e.g., MAP kinases).

Historically, biochemists have studied soluble cytoplasmic enzymes or circulating proenzymes and approximated the physiological Na^+, K^+, and Cl^- concentrations and the pH of the cytosol or serum. However, many target enzymes are membrane bound. An enzyme that is loosely associated with the outer or inner leaflet of a membrane, or rapidly cycles between cytosol and the inner leaflet, will affect the configuration of the final assay. Diacyl glycerol kinase is a good example: the rate-limiting step was in part the transbilayer diffusion of diacyl glycerol from the outer leaflet to the inner cytoplasmic leaflet [4]. Some integral membrane enzymes require membrane components, in-plane membrane dimerization, or association with membrane-bound coproteins for proper activity. These would clearly need to be taken into consideration in assay design. If an enzyme is post-translationally modified or modified in vivo upon external signaling, and these modifications affect the enzyme activity, stability, or localization of the enzyme, it may be difficult to supply biologically relevant enzyme from recombinant sources, and "natural" sources may be the only alternative.

Second, what relevant physical properties are known about the target enzyme, such as homogeneity, isozymes, molecular weight, isoelectric point, and stability to various conditions of storage and assay? The gene and amino acid sequences, tertiary and quaternary structures, prosthetic groups, cofactors, and essential catalytic and structural residues are relevant to the physical and enzymatic properties of the target enzyme. A clear understanding of the level of purity, cofactors, or associated proteins should be obtained. The enzyme must be of sufficient purity that it reflects the relevant biochemistry. If the enzyme preparation is contaminated with other enzyme activities that interfere with the primary signal generation, divert substrate to other paths, or modify potential inhibitors or substrate, the observed signals will be misleading. Nonenzyme protein contaminants may also be problematic if they are capable of low-affinity, high-capacity binding of potential inhibitors. This can be assessed by obtaining reference values of K_i or IC_{50}s of inhibitors with known protein binding at several concentrations of serum proteins (e.g., HSA or BSA). There is of course the potential difficulty of losing relevant accessory protein factors, cofactors, coenzymes, or essential trace components during an attempted purification [5]. Also structural cellular components may be necessary for activity such as a lipid containing membrane, e.g., 3-hydroxybutyrate dehydrogenase requires lipid for activation [6]. In some cases purification actually alters properties of enzyme due to partial solubilization or proteolytic modification, e.g., lactate dehydrogenase [7]. In some cases the enzymatic activity being monitored is completely lost at some step of purification, and systematic reconstitution experiments need to be done. In other cases the effect may be a more subtle loss of specific activity or a change in mechanism without the relevant cofactor, and these could be missed if the specific activity is not carefully monitored during purification. Such was the case for calcium-calmodulin where Ca^{+2} was identified as the essential cofactor only after arduous "add-back" experiments of chromatographic washes. Reconstitution of a multi-subunit enzyme can be difficult, if the purification alters the relative abundance or stoichiometry of their subunits [8]. For topoisomerases and polymerases, specific protein factors increase the processivity of the polymerase [9–12], and thus an imprudent choice of the enzyme preparation could bias a screen to detect elongation versus initiation inhibitors.

The crystallinity of an enzyme is no guarantee of purity as it is not uncommon that the first crystals obtained are often only 50% pure. However, even in utilizing ''pure'' enzymes to characterize the kinetics of an enzyme, one must be alert to additional reactions that may indicate low levels of contaminating activity when high concentrations of enzyme are used. Practically for HTS, what is desired is some measure of ''kinetic purity or competence'' of the enzyme before HTS commences. Generally the k_{cat}/K_m or turnover number must be reasonable as compared to the literature. Recombinant proteins can often be expressed as fusions with peptide and nonpeptide tags, which can serve as the basis for ensuring a selective and specific detection. For example, in vivo biotin-tagged recombinant topoisomerase expressed into crude nuclear extracts was sufficient for a generation of a selective signal [13] using strepavidin-scintillation proximity beads (SPA).

Third, the enzymatic properties should be known in some detail: the reaction sequence, coenzyme or prosthetic group involvement, stereochemical, substrate, and inhibitor affinities and specificities, reversibility of inhibitors, the nature of chemical conversion, the number of active sites, and the mechanism of action. One ideally begins with a relatively detailed knowledge of the enzyme kinetics, since this is the foundation for understanding how an enzyme works in chemical terms and in designing inhibitors. Kinetic characterization, such as specific activity, allostery, pH, temperature, and ionic and solvent effects must be in hand from the literature, therapeutic group collaborators, or internal studies. The key strategic point is to determine the kinetically predominant form(s) of the enzyme under the HTS conditions. This will largely dictate the kinds and classes of inhibitors identifiable by the enzyme screen. Asked another way, are the conditions chosen to maximize the chance of finding desired inhibitors? In order to make these choices rationally, correct values of the relevant kinetic constants such as k_{cat}, K_ms, and pK_as of all relevant substrates should be obtained as well as the pH, ionic strength, solvent, and temperature profiles for the overall reaction.

Before a screen can be sensibly configured, basic kinetic behavior and steady-state parameters for the isolated enzyme preparation need to be obtained. One must ascertain whether the enzyme turns over substrate and functions in a catalytic cycle, simply reacts stoichiometrically, or progresses to a dead-end complex. With some multistep, multifunctional systems one substrate may indeed turn over while another does not. For example, bacterial RNA polymerases elongate progressively by adding nucleotide triphosphates to the 3' end of the growing transcript. However, in the absence of termination sequences, the RNA polymerase complex does not release from the template DNA, so the transcription reaction does not reinitiate on new DNA templates but is "trapped" on the first template. Additional nucleotide triphosphate turns over but DNA template does not [14,15].

The HTS scientist must make rational guesses at the forms of all free and substrate bound forms of the enzyme in assay design. As a specific example, many protein kinases have been screened using radiolabeled ATP at total ATP concentrations at or below K_m, and most of the inhibitors found were competitive to the ATP binding site and were often ATP analogs. However, recently available nonradioactive electrochemiluminescent detection chemistry allows high "physiological" ATP concentrations in the HTS assay [16]. Screening in this format opens the possibility of uncovering novel classes of protein kinase inhibitors.

IV. PRACTICAL ENZYMOLOGY AND BASIC KINETIC CHARACTERIZATION

A. Measurement of Initial Velocity

The progress curves of most well-behaved enzyme reactions are of the general form of Figure 1. Substrate depletion and product formation are of course mirror images of each other [43], and the curves can sometimes be modeled by a first-order kinetic decay. However, the fall-off in velocity with time is most often not first order, as will be described below. Therefore they do not often fit to equations of standard homogeneous chemical reactions [18]. The usual approach adopted is to study the enzyme under initial velocity conditions where these various effects have not yet had time to operate and the conditions are accurately known. This "initial rate" is the tangent to the progress curve passing through the origin. Experimentally, one usually finds that the curves are practically straight lines as long as the amount of signal change does not exceed 10–20% of the total during this period (inset Fig. 1 and Fig. 3, curve a).

It would be kinetically rigorous to determine progress curves from continuous monitoring of a sample. Nevertheless, for most HTS applications, the mandate of rapid data accumulation and large numbers of compounds to test usually makes a single point estimation of an initial rate at a fixed time point after initiation the preferred format. Therefore, it is critical during the initial HTS assay development that full progress curves are obtained over the range of substrate and enzyme concentrations for which kinetic parameters will be estimated. Only then can one be certain that initial velocity is correctly estimated from the linear region of the progress curve. Figure 2 illustrates the point for three progress curves that show increasing curvature at higher activity levels (by increasing enzyme, substrate, pH, or coenzyme). If a longer end point, t_2, is used instead of t_1 there would be a severe underestimation of the initial velocity for the faster progress curve. One might think, therefore, if the end point were taken for the

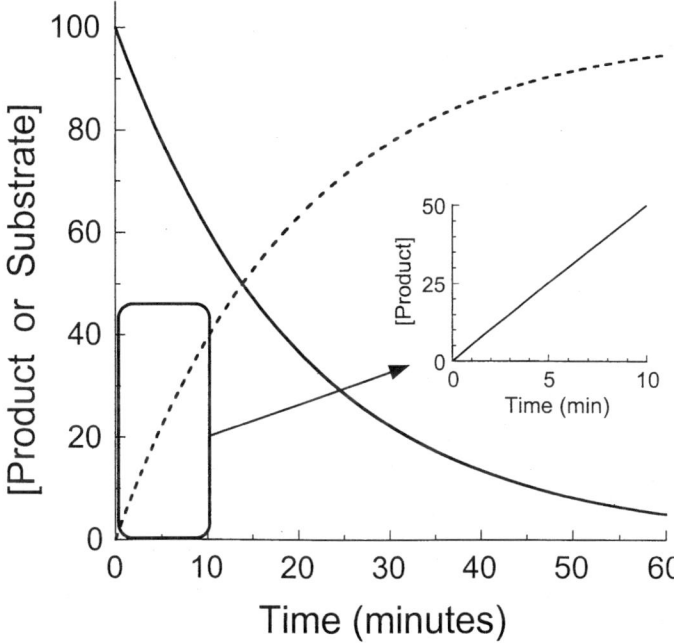

Figure 1 Typical example of progress curve showing loss of substrate (——) and forma-tion of product (----). Inset displays linearity of initial rate period.

linear portion of the fastest progress curve, that all slower progress curves would give accurate initial velocity estimates. However, progress curves are often not first order for many reasons such as a lag period (Fig. 3, curve b) or initial bursts (Fig. 3, curve c). If an enzyme is unstable or a component of the assay is inhibitory or inactivates the enzyme during turnover, the progress curve will show a time-dependent loss of enzyme activity, eventually flattening out completely (Fig. 3, curve d). Occasionally, factors can cause an increase in velocity during the first part of a reaction yielding S-shaped or autocatalytic curves (Fig. 3, e). It is not easy to decide whether initial velocity, maximal velocity, or the slope at some time is the best measure of enzyme activity for these complex progress curves. In these cases, it is best to determine the cause of the lag or acceleration and then modify the test procedure to eliminate it. For example, D-aminoacid oxidase requires a flavin coenzyme that combines slowly, so the amount of active enzyme increases for a short time until the enzyme–coenzyme complex can be formed [19]. This was eliminated by preincubation of enzyme with coenzyme and then starting the reaction with the addition of substrate. This is consistent with the concept that one must know the forms of the enzyme to make sense of what one

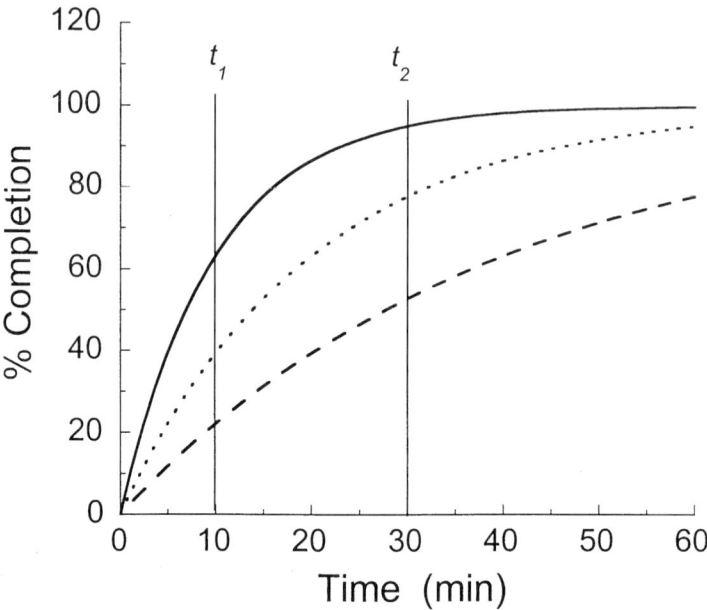

Figure 2 Effect of choice of time points t_1 and t_2 from progress curves to estimate initial rate from a single point determination at 4 (—), 2 (----), and 1 (····) units of enzyme activity.

is screening against. Another case is where a concentrated stock solution of enzyme may be partially aggregated and needs time to dissociate to a fully active form upon (not instantaneous) dilution into the assay (Fig. 3, curve b). This can be resolved by preincubating the enzyme after dilution into reaction buffer and then adding substrate after a period determined to eliminate the lag phase (Fig. 3, curve a). Enzymes can also be reversibly complexed with an inhibitor (e.g., a metal complex) that can dissociate upon dilution. In this case activity *decreases* with increasing $[E]_{tot}$. The main factors that determine the initial velocity of a reaction are enzyme and substrate concentrations, pH, temperature, and the presence of activators or inhibitors in the assay. All of these must be carefully controlled and maintained uniformly within an assay plate, across all plates, and for the screening period of an HTS campaign.

With a soluble, single enzyme it is straightforward to adjust the enzyme concentration to achieve a linear reaction over a convenient time window. This is necessary whether running a continuous or a discontinuous (end point) assay. To follow a linear reaction it is best, if possible, to take initial and final time points. This is because two points determine the line and thus the slope, and

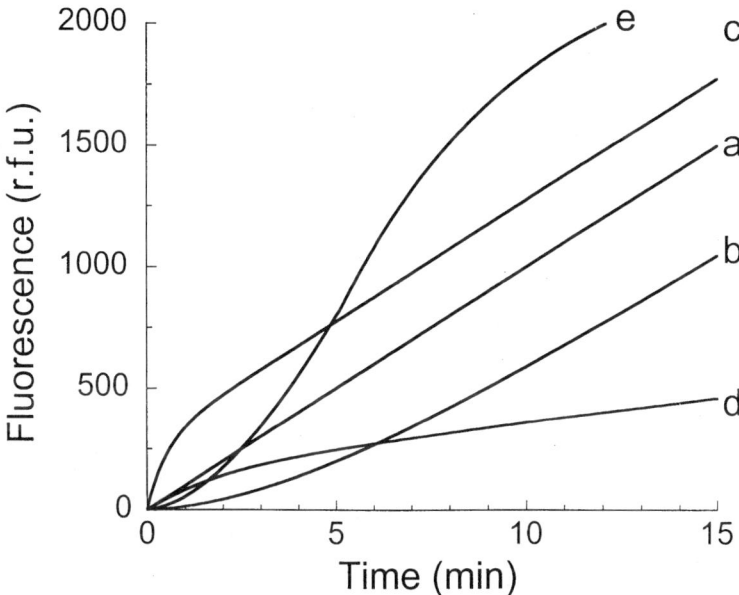

Figure 3 Nonideal progress curves from typical enzyme assays: normal linear (a), lag (b), initial burst (c), unstable enzyme (d), and autocatalytic (e).

therefore interfering compounds can often be weeded out because they would affect only the intercept and not the slope. This requires that the compound shifts the initial and final signal by the same absolute amount, which is not necessarily the case with fluorescence detection. It is also of value to examine the initial time point to see if there is a dramatic effect of the compound on the assay signal when there is minimal product formed.

B. Linearity of Initial Velocity with Enzyme Concentration

For the great majority of actual cases studied, the initial rate of enzyme reaction increases linearly with increasing enzyme concentration, as in Figure 4 (curve a) [20]. Indeed, this is necessary for modeling the effect of different levels of inhibition of an enzyme on an assay. However, this relationship must be experimentally verified, since departures from linearity are warning signs to look for artifacts in the assay system and many of them can be rectified experimentally. Some of the more common examples are

Displaced linear curve (Fig. 4, curve b): a toxic impurity poisons enzyme until free E overcomes the level of toxic impurity.

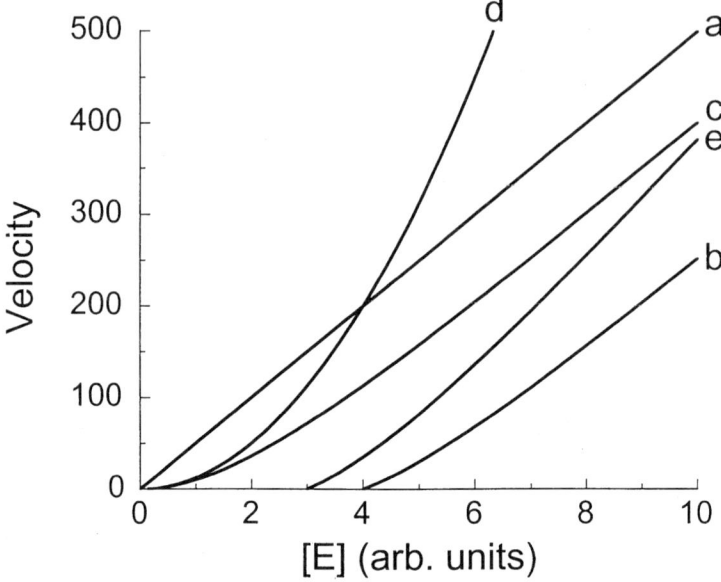

Figure 4 Examples of initial rate versus enzyme concentration plots with upwards curvature. Normal (a), toxic impurity (b), activator (c), active aggregated subunits (d), and complex subunit aggregation and activation (e).

Upward curvatures: a reversible coenzyme or activator is part of the enzyme preparation. v is proportional to the square of $[E]_{tot}$, until the enzyme is saturated with coenzyme or activator at high $[E]_{tot}$, whereupon a linear rate dependence obtains (Fig. 4, curve c). This behavior is observed with proteinases known to need some activator, such as in benzoyl-arginineamide hydrolysis by ficin [20]. Suboptimal amounts of thiols in the enzyme preparation give upward curvature with or without cyanide activator. When thiogylcollate is added, the curvature disappears (e.g., Fig. 4, curve a). Another example is phosphofructokinase, which is active only in its aggregated form [20], so velocity increases as some power of the concentration of the enzyme (Fig. 4, curve d). AMP binds to the aggregated active form, whereas ATP favors the dissociated form [21], so as ATP is hydrolyzed the enzyme becomes increasingly active. However, the behavior is more complicated with ATP, as ATP binds to both aggregated and dissociated forms of the enzyme, acting both as an activator of activity and a modulator of aggregation state. Its profile is a displaced activation curve with a different linear terminal rate at high enzyme concentration (Fig. 4, curve e).

Downward curvatures: are a more common occurrence. They can result

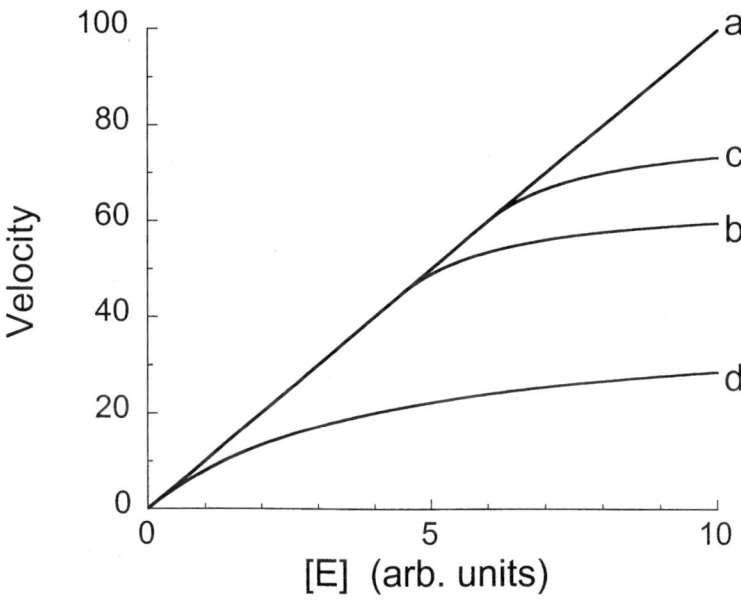

Figure 5 Examples of initial rate versus enzyme concentration plots with downwards curvature. Normal (a), limiting coupling enzyme or assay component (b), addition of more limiting enzyme or assay component (c), and time-dependent inactivation or presence of inhibitor in enzyme preparation (d).

from exceeding the linear range of a method, rather than true decrease in enzyme activity. If a test method depends upon a second enzyme or limiting reagent (post-reaction detection or coupled enzyme assays), it may become limiting as the concentration of the first enzyme increases (Fig. 5, curve b). In this case, increasing the concentration of the second enzyme or reagent will extend the linear portion of the curve, and the plateau will be observed at a higher signal level (Fig. 5, curve c). In the case of optical detection methods, deviations from the Beer–Lambert law are pronounced when absorbance measurements are greater than 2.0, since the transmitted light that the photometer measures is now only 1% incident light. Instrumental noise is often a limiting factor at these low light levels. With fluorescence, inner filter effects can occur at high concentrations of a fluorescent substrate or product, and these can actually lead to *decreasing* signal with increasing concentration (see Fig. 6, curve b). This can be diagnosed by checking the proportionality of the fluorescence of a sample after a large dilution (e.g., a 20-fold dilution yielding only 5-fold less fluorescence). These effects can also be corrected if the absorption of the sample is known at the excitation and

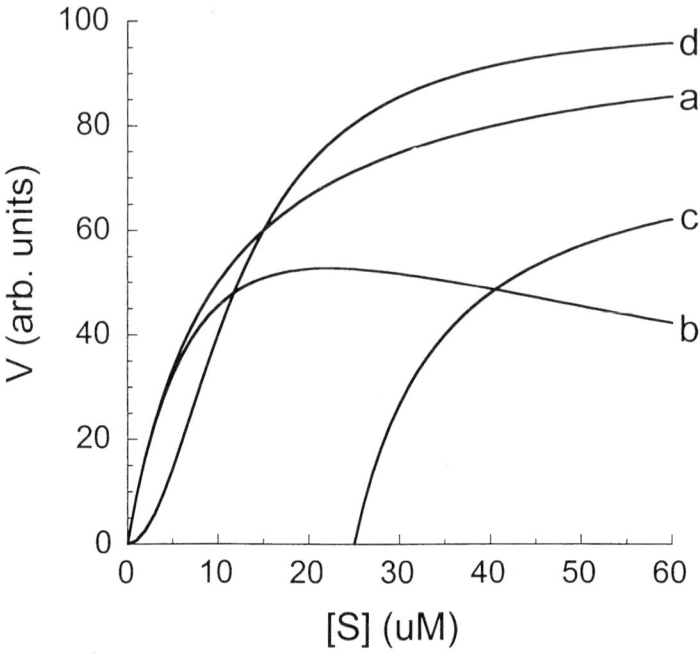

Figure 6 Ideal versus nonideal V versus S behavior. Normal hyperbolic kinetics (a), solubility or method limitations (b), interfacial partitioning (c), and positive cooperativity (d).

emission wavelengths used in the fluorescence measurement [22]. These effects are distinct from the simple systematic underestimation resulting from a poor choice of an end point estimation of the initial rate from a nonlinear region of the progress curve (as in Fig. 2).

In a multienzyme system, if a very high concentration of ''pure enzyme'' is used and this enzyme has an appreciable affinity for a requisite coenzyme, the other enzyme could be inhibited, slowing down the overall rate.

If one of the assay components introduces a low level of reversibly complexed inhibitor whose concentration also increases as enzyme concentration increases (Fig. 5, curve d) premature plateauing occurs. Dialysis or further purification of the enzyme or assay components will remove this artifact (Fig. 5, curve a). If however the enzyme is aggregated, but only the dissociated form of the enzyme is active, dialysis would not remove this inhibition, e.g., reversible aggregation. The enzyme could be unstable with time, and the effect is more pronounced at higher enzyme concentration [20] (Fig. 5, curve d).

Trivial reasons, such as change in actual pH or ionic strength, or water activity as very high enzyme concentrations are used, especially if the concentrated enzyme stock has high salt or glycerol.

In summary, the concentration of enzyme used must be within its linear range, and the assay method must also be within its linear range for detection.

C. Enzyme Stability

The stability of an enzyme target must be determined for all conditions that the enzyme will be exposed to during the development, validation, and execution of an HTS. Obviously, all assays must operate over a region of pH, temperature, ionic strength, and solvents where the enzyme is reasonably stable during the determination of initial rates. Enzymes often denature at surfaces, so it is important to avoid froth during additions of assay components, and to be cognizant of the surface-to-volume of various microplate well geometries, especially when detergents of any sort are used to solubilize a component of the reaction mixture. The stability or solvent tolerance of an enzyme must be checked over the concentration range that the enzyme will be exposed to during assay setup. Often, to avoid a prolonged mixing step for an assay, the enzyme may be exposed to a higher than final concentration of solvent during preincubation of compound with enzyme.

If linear progress curves are not obtained over the 10–20% conversion range, the stability of the enzyme must be verified from its preincubation under the final conditions of the assay without substrate. The stability of the enzyme under the conditions of storage of the "working dilution" of enzyme on the robotic platform must also be determined, since this will control the throughput, the reagent waste, and the length of unattended operation and size of a batch run of screening plates. It is important to determine stability of the enzyme under conditions chosen for long-term storage (e.g., days, weeks, or months), especially if a screen is to run over an extended period, or if a screen is resurrected for a second campaign after a chemical collection has grown significantly. This stability should be ideally monitored by measuring activity and physical integrity (e.g., SDS-PAGE). It is worth noting that while it is empirically possible to preserve the total signal by adding a larger volume of stock, a significant specific activity change is a warning sign for potential artifacts due to a changing preparation.

One may also need to tune the reaction conditions in the case of an enzyme that inactivates rapidly (<10 min, e.g., the 5-lipoxegenases [23] and cyclooxygenases [24]). Here one must add sufficient concentrations of enzyme to achieve detectable signal before turnover-dependent inactivation.

During storage of enzyme for HTS it is most preferable to have one large single or combined lot that is aliquotted in sufficient quantity for each batch run, rapidly frozen down as a concentrated stock to be stored in deep freeze ($-20°C$ to $-80°C$). The freeze–thaw stability of an enzyme stock should be checked early on over several cycles and storage periods. In general, freezing and thawing should be as fast as possible, with gentle mixing to avoid the formation of pH and concentrations gradient in situ. In some cases, enzymes do not freeze well as solutions, so stabilizers such as BSA, gelatin, glycerol, sucrose, or cyclodextrins can be added, or the enzyme can be stored as a salt pellet (e.g., ammonium sulfate) [25,26].

D. Substrate Effects on Initial Rates and Choice of Substrate

Knowledge of the K_m of the substrate of an enzyme is necessary to set the concentrations desired in an assay. To find both competitive and noncompetitive inhibitors, one does not want to be too far above K_m, subject to having enough substrate to give adequate rate and final signal. With soluble substrates it is often desirable to manipulate the assay conditions to bias the hits from screening to a particular binding site. If one does not want to miss competitive inhibitors, set the substrate concentration less than 3 times K_m, so that the $1 + [S]/K_m$ term does not get too large in the relationship for competitive inhibitors:

$$IC_{50} = K_i\left(1 + \frac{[S]}{K_m}\right)$$

If, on the other hand, one wishes to exclude competitive inhibitors (e.g., molecules that bind at the ATP site of kinases) then one should set $[S] > 10 \times K_m$.

While most treatises of Michaelis–Menten kinetics use single substrate reactions, the vast majority of enzyme-catalyzed reactions involve more than one substrate. Figure 6 (curve a) is a typical rectangular hyperbolic plot of v versus $[S]$, which is used to determine the K_m of substrate. $[S]$ is estimated by $[S]_{tot}$, since usually $[S] \gg [E]$. However, other methods such as a Dixon plot must be used to determine the K_m (or K_i) of very high affinity substrates (or inhibitors), since the low range of concentrations used are often of the same magnitude as $[E]_{tot}$ [27]. It is worthwhile to point out that the assumptions of steady-state of all enzyme–substrate intermediates used to derive Michaelis–Menton kinetics fit experimental fact but do not prove mechanism. In fact, the Langmuir isotherm and chain reaction mechanism will give the same equations. Normalized concentration curves are rectangular hyperbolas, and the semilog plot looks like a classic IC_{50} [28] or Langmuir absorption isotherm [29]. For enzymes with more than one substrate the apparent K_m of each substrate will vary, depending upon the saturation level of the other substrates. An estimate of one should be made at saturating levels of the other. However, if these concentrations are not achievable, the apparent K_m must be estimated for the conditions of the assay with some knowledge of the mechanism (random, ordered, or ping-pong) [30].

Nevertheless, the v vs. [S] plot can fall off sooner than predicted by the rectangular hyperbola. This occurs for several reasons that must be kept in mind and dealt with experimentally.

Substrate insolubility can account for a premature drop-off. If the solubility of a substrate is not very great in comparison to its K_m, the v vs. [S] plot plateaus earlier (Fig. 6, curve b). The drop-off in rate can be more abrupt than shown if the substrate solubility isotherm is very steep. This was observed with liver carboxyesterase acting on the related homologs ethyl butyrate and tributyrin. However, the opposite effect can be observed with amphiphilic substrates. For optical measurements of rate, the apparent drop-off can be due to optical artifacts (inner-filter effects, breakdown of Beer–Lambert relationship), as mentioned above. An enzyme that exhibits product inhibition would also yield a plot much like curve b in Figure 6. The presence of inhibitors or low levels of contaminating product in the substrate preparation that can form abortive or dead-end complexes may also give a premature plateau in the Michaelis plot.

Pancreatic lipase [31,32] does not act on dissolved substrates but is only active when adsorbed at an ester–water interface, as apparently the active center is only exposed therein. In this case a hydrophobic water insoluble substrate in an emulsion gives a typical Michaelis curve (Fig. 6, curve a). However, a more soluble substrate such as methyl butyrate gives no activity until some of it partitions into the interface phase after the aqueous phase is saturated, which results in a Michaelis curve with a "lag" (Fig. 6, curve c).

The presence of more than one enzyme in a partially purified enzyme preparation that acts on the same substrate would give deviations from classical hyperbolic kinetics, and if this is observed, additional purification of the enzyme may be necessary.

Finally, multimeric enzymes that bind substrate cooperatively, display complex nonhyperbolic kinetics that can be modeled by the modification of the classic Michaelis–Menton equation:

$$V = \frac{V_{max}[S]^n}{K' + [S]^n}$$

For positive cooperativity, binding of substrate to one active site increases the affinity of the remaining site(s), resulting in a positive Hill coefficient, n, and K', which is related to K_m but also contains terms related to the effect of substrate occupancy. For a Hill coefficient of 2, the v vs. [S] plot shows an initial lag and then a much more rapid increase in rate, followed by a much more rapid plateau (Fig. 6, curve d). Negative cooperativity behaves similarly to having inhibitors in the substrate, but with much more pronounced downwards initial curvature.

E. Catalytic Efficiency of the Substrate

For many in vitro HTS targets, the natural substrate is not available in sufficient quantities, and often a surrogate substrate or modified substrate is used. Often the choice of substrate determines the technical feasibility of an assay as well as the detection methodology of choice. Therefore some work must be done to assure that this artificial substrate recapitulates most of the significant interactions of the real substrate with the enzyme. For large substrates, k_{cat}/K_m gives a measure of whether the substrate mimics the physiological one. For small active sites, where recognition involves only active site interactions, small model substrates are often reasonable mimics of the natural substrate. The substrate with the lowest K_m is not necessarily the best for an assay; using it at subsaturating levels may not yield enough product for detection. If a substrate is very poor in terms of both binding and catalytic rate, one must question if critical enzyme substrate interactions are missing for binding, and that the kinetic mechanism or critical chemical residues are different for the studied substrate. The concern is that the measured inhibition by compounds may be working in a completely different way from what was expected. With a very poor substrate, it would be difficult to characterize the inhibition class of a compound, as well as obtain detectable products levels.

F. Effects of pH, Temperature, Ionic Strength, and Solvent

Figure 7 gives a typical behavior of an enzyme as a function of pH. The apparent pH optimum for the activity of an enzyme can be due to reversible effects on the velocity, an effect on the affinity of a substrate, or an effect on enzyme stability [33,34]. These can be distinguished experimentally. Stability can be measured by preincubation of the enzyme at different pHs before assaying activity after standardizing the pH to that in the assay. This effect would be time dependent and nonreversible and can be eliminated by short incubation times. A reversible ionization that affects catalytic groups but does not affect enzyme stability would give the same activity after readjustment to the optimal pH.

Effects of pH on substrate affinity can be eliminated by using a sufficiently high concentration of substrate to saturate the enzyme at all pHs used. But this does require a check of the K_m as a function of pH to ensure that at the assay pH the substrate is still several fold K_m. At high [S], all enzyme is complexed as ES, so changes in the state of ionization of the free enzyme or free S will not affect V_{max}, only K_m.

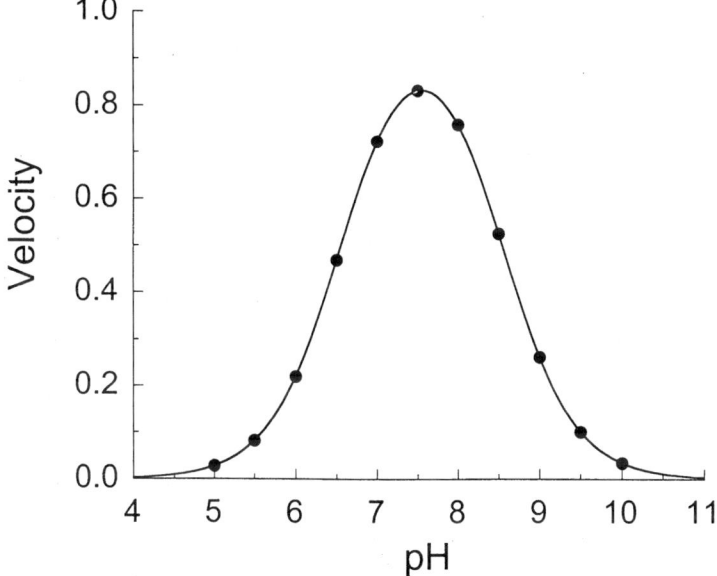

Figure 7 pH dependence of enzyme activity with acidic and basic titratable groups affecting velocity.

The pH effects on velocity [54] are more complex and depend upon the ionizing groups, their proximity to the active center, and the breakdown path and rates of each ionized species. Velocity measurements at saturating substrate (V_{max} conditions) yield pK_as of the ES forms. Velocity measurements at subsaturating substrate (V_{max}/K_m conditions) yield pK_as of both free E and S.

Inhibition of enzymes may be greatly affected by pH in both the reversible and irreversible case if an ionizable group is involved in the binding or inactivation. The pH chosen for a screen can also have direct effects on the apparent inhibitory potency of compounds being screened. In the case of stromelysin, the inhibitory potencies of peptide phosphonamidates were highest at lower pH (pH 6), so a screen run at the typical "physiological" pH 7.4 would have missed this class of compound [35].

Temperature can affect the stability of the enzyme, V_{max}, K_m, or the pK_a of functional groups. It can also affect activators or different enzymes in the assay. While this might suggest that the effects of temperature are extremely complex, actually they can be readily distinguished experimentally. Stability effects can be studied by preincubation at temperatures prior to assaying. Effects on K_m and V_{max} can be determined by usual methods. Effect on the pK_as of the components can be separately determined. In a multienzyme system, each enzyme can be studied separately.

The rate of inactivation of enzymes in solution increases rapidly with the temperature. In nearly all cases, inactivation becomes virtually instantaneous at temperatures well below 100°C, with most below 70°C (excepting thermophilic enzymes). The inactivation of enzymes by heat is nearly always due to denaturation of the enzyme protein. Unfolding followed by aggregation leads to irreversible inactivation [36]. However, inactivation rates vary considerably with conditions of solution, e.g., pH, protein concentration, protective action of substrate, inhibitors, and other substrates. A heat-coagulated *O*-methyl transferase can also be restored to full activity refolding from 8 M urea or 6 M guanidine [37].

While cooling enzymes usually stabilizes them, there have been several reports of cold labile enzymes, such as mutant *E. coli* inorganic pyrophosphatases [38], a rabbit muscle skeletal AMP deaminase [39], and *E. coli* glutaminase B [40]. In each of these cases, there is an isomerization, oligomeric association, or prosthetic group dissociation favored in the cold.

V. ASSAY VALIDATION CHECKS

In addition to the standard assay validation criteria, such as signal reproducibility and stability, the ability to detect known hits on several days, etc., there are several additional checks that must be considered for enzymatic assays.

During the initial assay development it is also critical to obtain the ''plateau'' or limit value for the measured signal, and then to determine if that value makes physical sense with the expected equilibrium position for product conversion. For instance, if the progress curve at plateau indicates that less than 1% of total substrate is consumed, there is clearly a more complicated set of kinetic processes going on. The initial rates of conversion should be consistent with the literature values and mechanism for the class of enzyme, if known. Additionally, at the point of plateau, addition of additional substrate should yield additional reaction, unless the enzyme is tied up in a dead-end complex or the system has severe product inhibition. Product inhibition can be directly tested by addition of the known product to the assay.

One component of assay validation is to reproduce literature K_is or IC_{50}s (if conditions are identical for assays) and to check effects of the order of addition of inhibitor versus substrate. If no compounds are available, the best that can be done is to model an inhibition curve by using enzyme concentrations determined to yield 90%, 50%, 10% activity of the enzyme used in the standard assay. However, information about the onset and duration of inhibition (biphasic progress curves) is lost by this method. If the time chosen to estimate the initial rate is too short, slow-onset inhibitors might be lost since the apparent activity will be underestimated.

VI. EXAMPLES OF CHALLENGING SYSTEMS

A. Membrane Utilizing Enzymes

Soluble enzymes, which act on membranes, are particularly challenging for developing and interpreting HTS assays and results. The best studied enzymes in this class are the phospholipases, in particular phospholipase A2, because of its ability to release arachidonic acid, which is metabolized to cytoactive prostaglandins and leukotrienes [41–45]. This enzyme has proven to be problematic because dissociation of the enzyme from the membrane surface, which separates enzyme from substrate, functionally inhibited it [46–50]. Nonspecific membrane active agents (e.g., amphiphiles) often produce this behavior [51–60]. They are useless as drug leads, however, since they do not bind to the enzyme and could not be used in a clinical application. Jain and collaborators have devised an ingenious assay system for the low molecular weight phospholipases A2 that overcomes this problem [46,61,62]. These enzymes have a cationic surface that adsorbs to the membrane. Jain took advantage of this feature by utilizing an anionic phospholipid as the substrate (e.g., phosphatidyl methanol derivatives). Combination of the cationic enzyme with the anionic membrane leads to a functionally irreversibly bound enzyme ($K_d < 10^{-15}$); typical amphiphiles are unable to dislodge the enzyme. Once bound, only compounds that actually bind to the enzyme inhibit the reaction. In cases where it is not possible to bind the enzyme to the membrane irreversibly, probes for simple dissociation of the enzyme from the membrane need to be used (e.g., HDNS [60,61], dansylated hexadecylphosphoethanolamine) to weed out artifacts that desorb the enzyme [63–65]. An alternative and nearly definitive method is to utilize protection of the active site histidine from alkylating agents [52,66]. If a compound is binding at the active site, then the rate of inactivation is slowed.

B. Multitarget or Balanced Pathway Screening

A novel strategy was used in finding inhibitors of the biosynthesis of cell wall components [67]. Because all of the murein biosynthetic enzymes are valid targets, there was no a priori way of determining which was best. The pathway was reconstituted in vitro to assay simultaneously six of the soluble expressed enzymes (MurA-MurF). At subsaturating substrate (near K_m), the concentration of each enzyme was optimized empirically to equalize the fluxes of substrates through each enzyme, so that an inhibitor of any one of the enzymes in the pathway is equally likely to be detected. The assay required the preparation of only the first substrate, since the product of each enzyme serves as the substrate for each subsequent downstream enzyme. The kinetics observed was consistent with a sequential buildup of each intermediate to a concentration sufficient to support the next catalyzed reaction in the pathway. This methodology increases the efficiency of screening several targets simultaneously, but more importantly, it may uncover inhibitory compounds that are more effective at inhibiting a pathway by modest simultaneous inhibition of several enzymes of the pathway than by having a potent inhibitor of a single enzyme. Resistance to such inhibitors would be expected to be negligible, as mutations at several target genes would have to occur concurrently in a single generation. Technically, this assay required the careful monitoring of at least two products.

C. Coupled Systems, Unfavorable Equilibrium

If there is an unfavorable equilibrium for a particular reaction of interest, coupling to a favorable reaction can drive the overall reaction to completion [68]. Ideally the second reaction generates products that also yield a usable detection signal (e.g., an irreversible formation of a gas, or colored precipitate, or strong fluorophore). In coupled systems one must make sure that the overall rate of the downstream coupling reactions is greater than the enzyme reaction of interest, i.e., the reaction of interest is the rate-determining step. This can be determined empirically by increasing the amount of coupling enzymes at a fixed concentration of target enzyme and primary substrates, until the observed rate plateaus (as in Fig. 5, curves b and c). At that point any increased relative amount of coupling enzyme will not increase the rate. The fold increase of coupling enzymes and substrates that can be achieved above this plateau break point will give the HTS scientist a guide to the insensitivity of the coupling system to potential inhibitors. For example, if one arranges the coupling system to proceed at a rate 100 times faster than the rate of product formation from the primary enzyme of interest, the coupling system can be inhibited 99% by an inhibitor but still not affect the apparent rate of the overall reaction. This will give confidence that any inhibition will be due to direct inhibition of the primary enzyme.

For a coupled assay, this increase of downstream rate cannot always be practically achieved because enzymes or substrates of the coupling system, or their solubility, stability, or availability, may limit enzymatic reaction. In such cases, the assay must be set up with as high a ratio of secondary to primary enzyme as is practical, with an awareness of the effect of inhibitors changing the rate-determining step.

Note also that for a coupled system, if the downstream enzymes and substrates are partially rate limiting, there will be a lag in the progress curve until the system reaches steady state. The lag phase will be more pronounced as the coupling enzymes limit a greater portion of the rate.

VII. PREFERRED AUTOMATION COMPATIBLE HTS FORMATS, SIGNAL GENERATION, AND DETECTION

In a general sense enzyme assays have an inherent advantage over simple binding assays that make them easier to automate. Enzymes catalytically turn over multiple molecules of substrate into product. This inherent amplification improves the sensitivity of enzyme assays over stoichiometric binding assays. Chromogenic and fluorogenic enzyme assays that use appropriately labeled substrates are very amenable to HTS, as they are often homogeneous mix-read or mix-stop-read, requiring very few operations. There are multiple examples of assays and commercially available labeled substrates suitable for many fluorescent and colorimetric-based signal detection and generation technologies. Colorimetric, chemiluminescent, fluorescence resonance energy transfer (FRET), homogeneous time-resolved fluorescence (HTRF), and fluorescence polarization (FP) assays have been described. There are FRET [69,70] and colorimetric [71] assays for HCV protease. FRET assays have been reported for metalloproteases [72,73], trypsin [74], and aspartyl proteases [75]. Colorimetric assays exist for HIV protease [76], which replaces an earlier radiometric assay [77]. Fluorogenic [78] and colorimetric [79] assays exist for beta-glucan synthase. FP assays have been configured for tyrosine kinases [80–82] and proteases [83]. A homogeneous chemiluminescent assay for telomerase has been reported [84]. There are also post-reaction capture methods to detect reaction products or remaining substrate for stopped reactions. Homogeneous HTS detection of ligand induced c-*fos* mRNA expression by hybridization capture with branch DNA oligonucleotides [85] or biotin-linked oligonucleotides. These homogeneous post-reaction analysis methods sometimes are preferred despite an increased automation burden. Often labeled primary substrates are difficult to make, the enzyme is fastidious in its acceptance of any modified substrates, or labeling introduces additional artifacts.

Automation of a filtration assay is cumbersome [86] and likely will not miniaturize well. However, even classical radioactive TCA assays can be converted into homogeneous scintillation proximity-based (SPA™ beads or Flashplates™) (e.g., HCV helicase [87]) or nonradioactive fluorescence-based formats, as discussed above. Even where homogeneous assays are not possible, there are many sensitive nonradioactive post-reaction technologies based on solid phase capture of product and signal amplification [88] such as classical enzyme-linked signal amplification (ELISA) [89,90], lanthanide chelate-labeled substrate, detection antibody [91], or hybridization capture [85].

As a rule of thumb, the fewer steps and manipulations the better for steady signal and automation friendliness. A homogeneous nonseparation method is preferable. If separation is necessary, ELISA washing is easier to automate than filtration. Nonradioactive separation is still preferable to nonseparation radioactivity, and nonradioactive is preferable to radioactive. Robocon Incorporated (Vienna, Austria) has automated centrifugation for HTS with a "smart" centrifuge, but it is slow and has limited throughput. However, recently a 96-array of miniaturized zonal centrifuge rotors [97] has been reported. Even where "automation" is possible, the pragmatic throughput needs must be considered. For instance, a centrifugation assay may be made that takes 10 min/plate on the robot, but only 2 min of manual manipulation. One has to judge whether the turnkey automation really increases throughput or whether there are alternative formats that can be implemented. In our experience, a significant proportion of HTS assays are critically limited by the stability of key reagents in short supply. Therefore in such cases a fully integrated robotics system with a long cycle time for a critical reagent addition made it impossible to achieve serial treatment of each plate. In addition, there is considerable "dead volume" in many robotics stations. For these, a batch process utilizing addition of the critical enzyme to as many plates as possible in as short a time is preferable, using 96- or 384-head pipetting stations with low dead volume reservoirs.

Minimization of pipetting steps is preferable. For most enzyme assays, reactions can be started by addition of substrate or enzyme. The common practice is usually to add enzyme to compounds to be tested, then add substrate to initiate reaction after some preincubation period. This in theory allows interrogation of slow-binding inhibitors. In HTS assays, additional mixing steps increase the time taken to make additions. Additionally, if crossover contamination is to be minimized, tips have to be washed or disposed of and then new tips reseated on a pipetting head, creating an increased waste stream and manipulation time. A practical solution for many robotic or workstation systems is forcefully to pipette a relatively large fraction of total volumes of an assay (20–50 µL) in each step, wherein the addition step mixes the contents in the well of a microtiter plate very efficiently. This also allows the use of a more accurate "to deliver" rather than "to contain" dispensing mode on most multichannel 96-well pipetting heads.

VIII. SCREENING FOR NEW ENZYME ACTIVITIES

While most of this review has focused on HTS of enzymes in search of *inhibitors*, HTS for enzyme *activity* is also being pursued on an industrial scale. The primary goal is the discovery of new biocatalysts [93], but it should not be long before pharmaceutical screening for particular activities is undertaken to circumvent the problem of DNA sequence with no assignable function. Screening for proteases, kinases, or phosphatases, which have major roles in signal transduction, would have the ability to find novel targets that have no DNA or primary sequence homology to known classes of enzymes.

High-throughput screening for enzyme activity has been successfully done by the technique of expression cloning, where genes from thermophilic [94] or hyperthermophilic organisms [95,96] are subcloned into a garden-variety expression system. Screening searches for activity against a particular substrate or panel of substrates. The assays must be robust enough to work in the presence of the host cell contents, and thus substrates that generate strongly absorbent and fluorescent products are typically used. The other area in which HTS is utilized in enzyme screening is in directed evolution in which a library of mutant enzymes is assayed for changes in activity/stability from wild type [97–99]. As is the case with functional screening, each colony (ideally) contains a different clone, and an independent measure of activity for each clone is desired.

IX. CONCLUSIONS

The transition of a research bench assay to a sustainable automation-compatible HTS that still preserves the essential features of the research assay is the goal of all HTS labs. Enzyme HTS is a proven and preferred methodology for drug discovery. Assay development is straightforward for ideal enzymes and detection methods. When, however, deviations from simple kinetic behavior are encountered, the HTS scientist must carefully assess these behaviors to identify their causes and find creative means to adapt HTS assays to accommodate them. This review details some of the more common deviations from ideal behavior and their experimental resolution. Additionally, the process of appropriate assay strategy, design, and configuration have been exemplified to yield more efficient and effective HTS campaigns. Finally, these principles are useful in the discovery of new inhibitors and new enzymes.

ACKNOWLEDGMENT

The authors would like to thank Dr. Robert A. Copeland for a critical reading of the manuscript and helpful discussions on text and figures.

REFERENCES

1. L Killar, J White, R Black, J Peschon. Adamalysins. A family of metzincins including TNF-alpha converting enzyme (TACE). Ann NY Acad Sci 878:442–452, 1999.
2. AL Stone, M Kroeger, QX Sang. Structure–function analysis of the ADAM family of disintegrin-like and metalloproteinase-containing proteins. J Protein Chem 18: 447–465, 1999.
3. M Dixon, EC Webb. Enzymes. The study of an enzyme. In: CJR Thorne, KF Tipton, eds. Enzymes. 3d ed. New York: Academic Press, 1979, pp 12–13.

4. P Badola, CR Sanders. *Escherichia coli* diacylglycerol kinase is an evolutionarily optimized membrane enzyme and catalyses direct phophoryl transfer. J Biol Chem 272:24176–24182, 1997.

5. M Dixon, EC Webb. Enzyme isolation. In: CJR Thorne, KF Tipton, eds. Enzymes. 3d ed. New York: Academic Press, 1979, p 23.

6. S Fleischer, HG Bock, P Gazzoti. In: GF Azzone, JE Klingenberg, E Quagliarello, N Siliprandi, eds. Membrane Proteins in Transport and Phosphorylation. Amsterdam: North-Holland, p 124, 1974.

7. KD Bush, JA Lumpkin. Structural damage to lactate dehydrogenase during copper iminodiacetic acid metal affinity chromatography. Biotechnol Prog 14:943–950, 1998.

8. LL Rokosz, SJ O'Keefe, JN Parsons, PM Cameron, JJ Burbaum. Reconstitution of active human calcineurin from recombinant subunits expressed in bacteria. Protein Expr Purif 6:655–664, 1995.

9. NT Le Gac, G Villani, PE Boehmer. Herpes simplex virus type-1 single-strand DNA-binding protein (ICP8) enhances the ability of the viral DNA helicase-primase to unwind cisplatin-modified DNA. J Biol Chem 273:13801–13807, 1998.

10. PA Der Garabedian, G Mirambeau, JJ Vermeersch. Mg^{2+}, Asp-, and Glu-effects in the processive and distributive DNA relaxation catalyzed by a eukaryotic topoisomerase. Biochemistry 30:9940–9947, 1991.

11. M Fairman, G Prelich, T Tsurimoto, B Stillman. Identification of cellular components required for SV40 DNA replication in vitro. Biochim Biophys Acta 951:382–387, 1988.

12. Hohn KT, Grosse F. Processivity of the DNA polymerase alpha-primase complex from calf thymus. Biochemistry 26:2870–2878, 1987.

13. CG Lerner, AYC Saiki. Scintillation proximity assay for human DNA topoisomerase I using recombinant biotinyl-fusion protein produced in baculovirus-infected insect cells. Anal Biochem 240:185–196, 1996.

14. PH von Hippel. An integrated model of the transcription complex in elongation, termination, and editing. Science 281:660–665, 1998.

15. C Condon, C Squires, CL Squires. Control of rRNA transcription in *Escherichia coli*. Microbiol Rev 59:623–45, 1995.

16. C Lehel, S Daniel-Issakani, M Brasseur, B Strulovici. A chemiluminescent microtiter plate assay for sensitive detection of protein kinase activity. Anal Biochem 244: 340–346, 1997.

17. RA Copeland. Steady state kinetics of single substrate enzyme reactions. In: Enzymes, A Practical Introduction to Structure, Mechanism, and Data Analysis. New York: VCH, 1996, p 94.

18. M Dixon, EC Webb. Enzyme techniques. In: Enzymes. 3^d ed. New York: Academic Press, 1979, p 7.

19. M Dixon, P Kenworthy. D-aspartate oxidase of kidney. Biochim Biophys Acta 146: 54–76, 1967.

20. M Dixon, EC Webb. Enzyme kinetics. In: Enzymes. 3^d ed. New York: Academic Press, 1979, pp 47–51.

21. EC Hulme. The isotope-exchange reactions of ox heart phosphofructokinase. Biochem J 122:181–187, 1971.

22. RA Copeland. Experimental measures of enzyme activity. In: Enzymes, A Practical Introduction to Structure, Mechanism, and Data Analysis. New York: VCH, 1996, p 143.

23. MD Percival, D Denis, D Riendeau, MJ Gresser. Investigation of the mechanism of non-turnover-dependent inactivation of purified human 5-lipoxygenase. Inactivation by H_2O_2 and inhibition by metal ions. Eur J Biochem 210:109–117, 1992.

24. JK Gierse, CM Koboldt, MC Walker, K Seibert, PC Isakson. Kinetic basis for selective inhibition of cyclo-oxygenases. Biochem J 339:607–614, 1999.

25. M Dixon, EC Webb. Effects of temperature. In: Enzymes. 3^d ed. New York: Academic Press, 1979, pp 164–166.

26. RA Copeland. Enzyme stability. In: Enzymes, A Practical Introduction to Structure, Mechanism, and Data Analysis. New York: VCH, 1996, pp 183–185.

27. M Dixon. The graphical determination of Km and Ki. Biochem J 129:197–202, 1972.

28. RA Copeland. Dose–response curves of enzyme inhibition. In: Enzymes, A Practical Introduction to Structure, Mechanism, and Data Analysis. New York: VCH, 1996, pp 204–209.

29. W Kortlandt, HJ Endeman, JO Hoeke. A three-parameter Langmuir-type model for fitting standard curves of sandwich enzyme immunoassays with special attention to the alpha-fetoprotein. Anal Biochem 162:5–10, 1987.

30. RA Copeland. Enzymes with multiple substrates. In: Enzymes, A Practical Introduction to Structure, Mechanism, and Data Analysis. New York: VCH, 1996, pp. 263–268.

31. SJ O'Connor, BM Sutton. Interfacial interactions between proteins and mammalian lipases. Adv Colloid Interface Sci 28:1–34, 1987.

32. C Chapus, M Semeriva, C Bovier-Lapierre, P Desnuelle. Mechanism of pancreatic lipase action. 1. Interfacial activation of pancreatic lipase. Biochemistry 15:4980–4987, 1976.

33. M Dixon, EC Webb. Enzymes. 3^d ed. New York: Academic Press, 1979, pp 148–153.

34. M Dixon, EC Webb. Enzymes. 3^d ed. New York: Academic Press, 1979, p 138.

35. M Izquierdo-Martin, RL Stein. Mechanistic studies on the inhibition of stromelysin by a peptide phosphonamidate. Bioorg Med Chem 1:19–26, 1993.

36. J Backmann, G Schafer, L Wyns, H Bonisch. Thermodynamics and kinetics of unfolding of the thermostable trimeric adenylate kinase from the archaeon *Sulfolobus acidocaldarius*. J Mol Biol 284:817–833, 1998.

37. D Bhattacharyya, TK Hazra, WD Behnke, PL Chong, A Kurosky, JC Lee, S Mitra. Reversible folding of Ada protein (O6-methylguanine-DNA methyltransferase) of *Escherichia coli*. Biochemistry 37:1722–1730, 1998.

38. IS Velichko, SE Volk, VYu Dudarenkov, NN Magretova, VYa Chernyak, A Goldman, BS Cooperman, R Lahti, AA Baykov. Cold lability of the mutant forms of *Escherichia coli* inorganic pyrophosphatase. FEBS Lett 359:20–22, 1995.

39. M Ranieri-Raggi, A Raggi. pH-dependent cold lability of rabbit skeletal muscle AMP deaminase. Biochim Biophys Acta 742:623–629, 1983.

40. S Prusiner, JN Davis, ER Stadtman. Regulation of glutaminase B in *Escherichia coli*. I. Purification, properties, and cold lability. J Biol Chem 251:3447–3456, 1976.

41. J Balsinde, MA Balboa, PA Insel, EA Dennis. Regulation and inhibition of phospholipase A2. Ann Rev Pharmacol 39:175–189, 1999.

42. F Bartoli, HK Lin, F Ghomashchi, MH Gelb, MK Jain, R Apitz-Castro. Tight binding inhibitors of 85-kDa phospholipase A2 but not 14-kDa phospholipase A2 inhibit release of free arachidonate in thrombin-stimulated human platelets. J Biol Chem 269:15625–15630, 1994.

43. M Murakami, S Shimbara, T Kambe, H Kuwata, MW Winstead, JA Tischfield, I Kudo. The functions of five distinct mammalian phospholipase A2s in regulating arachidonic acid release. Type IIa and type V secretory phospholipase A2s are functionally redundant and act in concert with cytosolic phospholipase A2. J Biol Chem 273:14411–14423, 1998.

44. JV Bonventre, Z Huang, MR Taheri, E O'Leary, E Li, MA Moskowitz, A Sapirstein. Reduced fertility and postischaemic brain injury in mice deficient in cytosolic phospholipase A2. Nature 390:622–625, 1997.

45. N Uozumi, K Kume, T Nagase, N Nakatani, A Ishii, F Tashiro, Y Komagata, K Maki, L Ikuta, Y Ouchi, J Miyazaki, T Shimizu. Role of cytosolic phospholipase A2 in allergic response and parturition. Nature 390:618–622, 1997.

46. OG Berg, J Rogers, BZ Yu, J Yao, LS Romsted, MK Jain. Thermodynamic and kinetic basis of interfacial activation: resolution of binding and allosteric effects on pancreatic phospholipase A2 at zwitterionic interfaces. Biochemistry 36:14512–14530, 1997.

47. MK Jain, MH Gelb, J Rogers, OG Berg. Kinetic basis for interfacial catalysis by phospholipase A2. Methods Enzymol 249:567–614, 1995.

48. MH Gelb, MK Jain, AM Hanel, OG Berg. Interfacial enzymology of glycerolipid hydrolases: lessons from secreted phospholipases A2. Ann Rev Biochem 64:653–688, 1995.

49. MK Jain, MH Gelb. Phospholipase A2-catalyzed hydrolysis of vesicles: uses of interfacial catalysis in the scooting mode. Methods Enzymol 197:112–125, 1991.

50. MK Jain, OG Berg. The kinetics of interfacial catalysis by phospholipase A2 and regulation of interfacial activation: hopping versus scooting. Biochim Biophys Acta 1002:127–156, 1989.

51. MK Jain, W Yuan, MH Gelb. Competitive inhibition of phospholipase A2 in vesicles. Biochemistry 28:4135–4139, 1989.

52. MK Jain, BZ Yu, JM Rogers, AE Smith, ET Boger, RL Ostrander, AL Rheingold. Specific competitive inhibitor of secreted phospholipase A2 from berries of *Schinus terebinthifolius*. Phytochemistry 39:537–547, 1995.

53. MH Gelb, MK Jain, OG Berg. Inhibition of phospholipase A2. FASEB J 8:916–924, 1994.

54. MK Jain, BZ Yu, OG Berg. Relationship of interfacial equilibria to interfacial activation of phospholipase A2. Biochemistry 32:11319–11329, 1993.

55. MK Jain, J Rogers, HS Hendrickson, OG Berg. The chemical step is not rate-limiting during the hydrolysis by phospholipase A2 of mixed micelles of phospholipid and detergent. Biochemistry 32:8360–8367, 1993.

56. T Honger, K Jorgensen, D Stokes, RL Biltonen, OG Mouritsen. Phospholipase A2 activity and physical properties of lipid-bilayer substrates. Methods Enzymol 286:168–190, 1997.

57. MK Jain, J Rogers. Substrate specificity for interfacial catalysis by phospholipase A2 in the scooting mode. Biochim Biophys Acta 1003:91–97, 1989.

58. MK Jain, DV Jahagirdar. Action of phospholipase A2 on bilayers. Effect of inhibitors. Biochim Biophys Acta 814:319–326, 1985.

59. F Ghomashchi, BZ Yu, ED Mihelich, MK Jain, MH Gelb. Kinetic characterization of phospholipase A2 modified by manoalogue. Biochemistry 30:9559–9569, 1991.

60. MK Jain, WL Vaz. Dehydration of the lipid-protein microinterface on binding of phospholipase A2 to lipid bilayers. Biochim Biophys Acta 905:1–8, 1987.

61. OG Berg, BZ Yu, J Rogers, MK Jain. Interfacial catalysis by phospholipase A2: determination of the interfacial kinetic rate constants. Biochemistry 30:7283–7297, 1991.

62. MK Jain, J Rogers, DV Jahagirdar, JF Marecek, F Ramirez. Kinetics of interfacial catalysis by phospholipase A2 in intravesicle scooting mode, and heterofusion of anionic and zwitterionic vesicles. Biochim Biophys Acta 860:435–447, 1986.

63. Y Cajal, MK Jain. Synergism between mellitin and phospholipase A2 from bee venom: apparent activation by intervesicle exchange of phospholipids. Biochemistry 36:3882–3893, 1997.

64. MK Jain, BP Maliwal. Spectroscopic properties of the states of pig pancreatic phospholipase A2 at interfaces and their possible molecular origin. Biochemistry 32: 11838–11846, 1993.

65. T Bayburt, BZ Yu, HK Lin, J Browning, MK Jain, MH Gelb. Human nonpancreatic secreted phospholipase A2: interfacial parameters, substrate specificities, and competitive inhibitors. Biochemistry 32:573–582, 1993.

66. MK Jain, BZ Yu, J Rogers, GN Ranadive, OG Berg. Interfacial catalysis by phospholipase A2: dissociation constants for calcium, substrate, products, and competitive inhibitors. Biochemistry 30:7306–7317, 1991.

67. KK Wong, DW Kuo, RM Chabin, C Fournier, LD Gegnas, ST Waddell, F Marsilio, B Leiting, DL Pompliano. Engineering a cell-free murein biosynthetic pathway: combinatorial enzymology in drug discovery. J Am Chem Soc 120:13527–13528, 1998.

68. E Valero, R Varon, F Garcia-Carmona. Kinetic study of an enzymic cycling system coupled to an enzymic step: determination of alkaline phosphatase activity. Biochem J 309:181–185, 1995.

69. M Taliani, E Bianchi, F Narjes, M Fossatelli, A Urbani, C Steinkuhler, R DeFrancesco, A Pessi. A continuous assay of hepatitis C virus protease based on resonance energy transfer depsipeptide substrates. Anal Biochem 240(1):60–67, 1996.

70. N Kakiuchi, S Nishikawa, M Hattori, K Shimotohno. A high throughput assay of the hepatitis C virus nonstructural protein 3 serine proteinase. J Virol Methods 80(1): 77–84, 1999.

71. R Zhang, BM Beyer, J Kurkin, R Ingram, FG Njoroge, WT Windsor, BA Malcolm. A continuous spectrophotometric assay for the hepatitis C virus serine protease. Anal Biochem 270(2):268–275, 1999.

72. DM Bickett, MD Green, C Wagner, JT Roth, J Berman, GM McGeehan. A high throughput fluorogenic substrate for stromelysin (MMP-3). Ann NY Acad Sci 732: 351–355, 1994.

73. DM Bickett, MD Green, J Berman, M Dezube, AS Howe, PJ Brown, JT Roth, GM

McGeehan. A high throughput fluorogenic substrate for interstitial collagenase (MMP-1) and gelatinase (MMP-9). Anal Biochem 212(1):58–64, 1993.

74. S Grahn, D Ullmann, H Jakubke. Design and synthesis of fluorogenic trypsin peptide substrates based on resonance energy transfer. Anal Biochem 265(2):225–231, 1998.

75. CD Carroll, M Orlowski. Screening aspartyl proteases with combinatorial libraries. Adv Exp Med Biol 436:375–380, 1998.

76. J Stebbins, C Debouck. A microtiter colorimetric assay for the HIV-1 protease. Anal Biochem 248(2):246–250, 1997.

77. LJ Hyland, BD Dayton, ML Moore, AY Shu, JR Heys, TD Meek. A radiometric assay for HIV-1 protease. Anal Biochem 188(2):408–415, 1990.

78. E Shedletzky, C Unger, DP Delmer. A microtiter-based fluorescence assay for (1,3)-beta-blucan synthases. Anal Biochem 249(1):88–93, 1997.

79. CS Taft, CS Enderlin, CP Selitrennikoff. A high throughput in vitro assay for fungal (1,3)beta-glucan synthase inhibitors. J Antibiot (Tokyo) 47(9):1001–1009, 1994.

80. IC King, M Feng, JJ Catino. High throughput assay for inhibitors of the epidermal growth factor receptor-associated tyrosine kinase. Life Sci 53(19):1465–1472, 1993.

81. R Seethala, R Menzel. A fluorescence polarization competition immunoassay for tyrosine kinases. Anal Biochem 255(2):257–262, 1998.

82. R Seethala, R Menzel. A homogenous, fluorescence polarization assay for src-family tyrosine kinases. Anal Biochem 253(2):210–218, 1997.

83. LM Levin, ML Michener, MV Toth, BC Holwerda. Measurement of specific protease activity utilizing fluorescence polarization. Anal Biochem 247(1):83–88, 1997.

84. DB Lackey. A homogeneous chemiluminescent assay for telomerase. Anal Biochem 264(1):47–52, 1998.

85. V Shyamala, H Khoja, ML Anderson, JX Wang, H Cen, WM Kavanaugh. High-throughput screening for ligand-induced c-fos mRNA expression by branched DNA assay in Chinese hamster ovary cells. Anal Biochem 266(1):140–147, 1999.

86. S Roychoudhry, SM Collins, BA Hynd, CN Parker. High throughput autophosphorylation assay for bacterial protein histidine kinases. J Biomol Screening 2:85–90, 1997.

87. K Kyono, M Miyashiro, I Taguchi. Detection of hepatitis C virus helicase activity using the scintillation proximity assay system. Anal Biochem 257(2):120–126, 1998.

88. DA Armbruster, RH Schwarzhoff, ED Hubster, MK Liserio. Enzyme immunoassay, kinetic microparticle immunoassay, radioimmunoassay, and fluorescence polarization immunoassay compared for drugs-of-abuse screening. Clin Chem 39(10):2137–2146, 1993.

89. L Trinh, R Ziegler, D Watling, RM Snider, E Croze. Development of an immunofluorometric, high-capacity cell-based assay for the measurement of human type I and type II interferons. J Biomol Screening 4:33–37, 1999.

90. N Takeshita, N Kakiuchi, T Kanazawa, Y Komoda, M Nishizawa, T Tani, K Shimotohno. An enzyme-linked immunosorbent assay for detecting proteolytic activity of hepatitis C virus proteinase. Anal Biochem 247(2):242–246, 1997.

91. Y Tian, L-H Wu, F-Z Chung. High throughput 96-well plate assay for receptor-mediated phosphatidylinositol turnover. J Biomol Screening 2:91–99, 1998.

92. A Marziali, TD Willis, RW Davis. An arrayable flow-through microcentrifuge for high-throughput instrumentation. Proc Natl Acad Sci USA 96(1):61–66, 1999.

93. BL Marrs, S Delagrave, DJ Murphy. Novel approaches for discovering industrial enzymes. Current Opinion Microbiol 2:241–245, 1999.
94. DC Demirjian, PC Shah, F Moris-Varas. Screening for novel enzymes. Topics Current Chemistry 200:1–30, 1999.
95. DE Robertson, EJ Mathur, RV Swanson, BL Marrs, JM Short. SIM News 46:3–8, 1996.
96. JM Short. Recombinant approaches for accessing biodiversity. Nature Biotechnol 13:1322–1333, 1997.
97. FH Arnold. Design by directed evolution. Accounts Chem Res 31:125–131, 1998.
98. FH Arnold, AA Volkov. Directed evolution of biocatalysts. Current Opinion Chem Biol 3:54–59, 1999.
99. H Joo, Z Lin, FH Arnold. Laboratory evolution of peroxide-mediated cytochrome P450 hydroxylation. Nature 399:670–673, 1999.

10

Screening Strategies for Ion Channel Targets

Thomas H. Large
Eli Lilly and Company, Indianapolis, Indiana

Martin W. Smith
Eli Lilly and Company, Research Triangle Park, North Carolina

I. INTRODUCTION

Ion channels play a central role in human physiology by helping regulate cellular ion homeostasis, shaping the electrical activity of nerve and muscle cells, and controlling the release of transmitters and hormones. The application of molecular biology and genetic approaches coupled with electrophysiology techniques has greatly facilitated the characterization of ion channel function. One important outcome of these advances has been the ability to understand a growing list of nervous system, cardiovascular, and metabolic disorders as specific defects in ion channel function [1–3]. The corollary of this observation is that modulation of ion channel activity represents an effective therapeutic strategy for a wide variety of disorders. Although small molecules and peptide toxins [4] that interact with ion channels have been known for some time, the technology for identifying new classes of lead compounds by high-throughput screening (HTS) has lagged behind the rapid advances in ion channel characterization. This review will summarize established as well as recently developed technology for configuring, validating, and running HTS campaigns for ion channel targets. However, continuing discoveries in ion channel function and the desire for further miniaturization and expedited sample screening means that utilizing the emerging technologies in fluorescent reporters and cell-based functional assays will be critical to the success of future lead discovery efforts.

II. ION CHANNEL FUNCTION

Transport of ions and small molecules across the cell membrane is accomplished by several classes of integral proteins, including ion channels, passive transporters, and active transport pumps that establish chemical ion gradients. The action of ion pumps, e.g., the Na^+-K^+ ATPase, consumes up to 30% of the energy expenditure of cells and allows the cell to store potential energy in the form of an ion gradient. This energy can then be used to transmit electrical signals, e.g., the action potential, or to drive the transport of small molecules. A fundamental property of ion channels is their ability to form a narrow aqueous pore spanning the membrane that selectively allows the passage of one or a few classes of ions, e.g., Na^+, K^+, Ca^{2+}, or Cl^-. Because ion flux is passive, the rate and direction of the ion flux is determined by the electrochemical gradient, a combination of the ionic chemical gradient and the membrane electrical potential. The membrane potential, typically -50 to -80 mV at rest, is determined by the relative conductances across the membrane for individual ions. An important feature of ion channels is that they are gated pores, allowing controlled, rather than continuous, ion flux. An exception appears to be inward rectifiers and "leak channels," which provide tonic low-K^+ conductances and establish a negatively charged (hyperpolarized) resting membrane potential [5]. Gating mechanisms typically involve ligand binding (ligand-gated) or a voltage-dependent mechanism (voltage-gated) [6] but can also include mechanical activation [7,8]. Ligand-gated mechanisms are quite diverse and range from binding of extracellular ligands to intracellular regulation by G proteins [9], ATP [10] and cyclic nucleotides [11], and Ca^{2+} ions and phosphoinositides [12]. A subset, most notably NMDA receptors, exhibit dual gating mechanisms regulated by both ligand binding and membrane potential.

A critical aspect of the opening and closing of ion channels is the resulting change in membrane potential that occurs as ion conductances across the membrane rapidly change. Rather than bulk charge movement, the flux of relatively few ions through open channels can generate relatively large changes in membrane potential. In simplified terms, as channels open for a particular ion, the membrane potential changes in the direction of the equilibrium potential for that ion. A classic example is that of the action potential in nerve and muscle where the sequential opening and closing of voltage-gated Na^+ and K^+ channels results in a self-propagating depolarization and repolarization of the surface membrane [13]. The Na^+ channel also exemplifies another important feature of ion channels, their ability to adopt an inactive or desensitized state that closes the channel and temporarily prevents reactivation. In the case of the action potential, this permits discrete electrical signals to be propagated and prevents continuous electrical stimulation. However, this rapid desensitization feature can pose a serious hindrance to assay development for HTS.

III. CHANNEL STRUCTURE AND ALLOSTERIC MODULATORS

Ion channels are typically homomeric or heteromeric assemblies of subunits. Transmembrane peptide segments determine ion selectivity, and terminal peptide domains often serve to inactivate the channel by blocking the transmembrane pore. Channel subunits typically exist as structurally related members of a large gene family, e.g., K^+ channels [14], but also neurotransmitter receptors for GABA [15], acetylcholine [16], and glutamate [17]. Highly diverse arrays of potential channel configurations can result from either structural variants generated by alternative mRNA splicing or the assembly of different subunits into heteromeric channels. Furthermore, the various combinations of subunits often differ pharmacologically or functionally, exhibiting differences in channel conductance properties or desensitization [18]. In addition, phosphorylation by cytoplasmic kinases is a major post-translational mechanism for regulating channel function. The multisubunit composition of ion channels and the existence of multiple functional states make these ideal targets for allosteric modulators. In many cases, ligand binding sites appear to exist at subunit interfaces, or between separate lobes contained with a single subunit [19]. An extension of this principle implies that allosteric sites also are likely to exist at subunit interfaces, including the segments forming the channel pore [20]. Allosteric modulators have been characterized for a number of ion channels, including the ionotropic glutamate [21], nicotinic [22], and GABAergic [23] receptors. Indeed, benzodiazepines are well-known allosteric modulators of $GABA_A$ receptors and have proven to be valuable therapeutic agents. Although novel agonists and antagonists of ion channels are common HTS goals, allosteric modulators represent an important additional class of molecules.

IV. ION CHANNEL–RELATED DISEASES

Defects in ion channels are either responsible for or strongly implicated in a growing number of diseases, a reflection of their ubiquitous expression and central role in cellular homeostasis. In some cases, the molecular characterization of an ion channel has led to the discovery of genetic defects in channel structure/function. In other cases, genetic investigations of human disease, or more typically mouse and fly mutants, have led to the identification of new members or classes of ion channels. Because there are a number of excellent reviews of "channelopathies" [3], it is sufficient to mention briefly some prominent examples of human disease linked to channel dysfunction. It also is worth noting that although many such diseases are relatively rare, they represent an entry point for understanding more prevalent diseases [24,25]. For example, long-QT syndromes are defects in cardiac K^+ channels that occur in only 1 in 15,000 individuals and result in cardiac arrhythmia and sudden death. However, understanding long-QT may lead to improved therapeutic strategies for ventricular arrhythmia, a more common cause of sudden death [25]. Other diseases associated with channel defects include deafness and hyperinsulinemia [26] (K^+), cystic fibrosis [27] and myopathies [28] (Cl^-), hereditary hypertension [29] and periodic paralysis [30] (Na^+), and malignant hyperthermia [31] (Ca^{++}) and congenital myasthenic syndrome (nAChR) [32]. It is likely that the number of ion channels as drug targets will continue to grow as channel function and dysfunction are better understood.

V. PRIMARY ASSAY STRATEGIES

Ion channels possess a variety of biophysical properties that include single channel conductance, directionality, ion selectivity, gating dependence on the binding of a ligand or membrane potential, and desensitization. Assay strategies can be devised to detect modulation of any of these properties. Often, however, assay development is simply guided by the activation mechanism and ion selectivity of the target channel. For example, ligand-gated ion channels are amenable to assay formats that detect specific ligand binding events. In some cases, allosteric modulators exist for ligand- and voltage-gated channels, and these molecules may be used as surrogates for developing a ligand binding assay [33]. In general, voltage-gated ion channels or channels gated by cytoplasmic ligands pose greater hurdles in assay development. Assays that report changes in ion flux across the membrane or changes in membrane potential would appear to be the formats of choice for these targets. In particular, cell-based functional assays of ion channel activity are well suited to identify allosteric modulators.

Perhaps the greatest challenge in assay development is overcoming rapid desensitization and inactivation, a common feature of voltage-gated and ligand-gated channels. The use of blockers of desensitization, e.g., cyclothiazide for AMPA glutamate receptors or concanavalin A for kainate glutamate receptors [34], can prolong channel open time sufficiently either to measure bulk ion flux or to allow changes in membrane potential to be recorded. Of course, if the screening goal is the identification of novel blockers of desensitization, "clamping" the channel is unnecessary. It is particularly difficult to develop an assay for a target channel when the resting membrane potential is near the equilibrium potential for the conducting ion. HTS methods, such as fluorescent voltage sensors, may have inadequate sensitivity to detect these responses and may require artificially shifting the resting potential in order to provide an increased signal "window." For some assay formats, valinomycin can be used to increase K^+ conductance and hyperpolarize cells [35], increasing the driving force on inward ion flux and allowing a larger change in membrane potential with the opening of depolarizing channels. Alternatively, expression of additional channels could serve to hyperpolarize cells without significantly decreasing input resistance. Additional challenges arise for ion channels whose activation is use-dependent or which have a voltage-dependent component coupled with ligand-mediated activation. Finally, cases exist where the target of interest is the site of action of an endogenous effector [36], and successful assay validation may require the optimization of each functional element in a manner that preserves the known pharmacological profile of the targeted channel.

An important aspect of the screening strategy involves the source of ion channel target and, more specifically, the use of either primary or recombinant cell lines. The use of primary cell lines in a HTS campaign suffers from problems related to the level of expression of the target channel, although methods exist for engineering expression even in post-mitotic neurons [37,38]. Problems also can be encountered in the scale-up of cell or membrane preparations, especially from neuronal material. In addition, primary cells may express a variety of heterologous ligand-gated or voltage-gated channels, confounding the confirmation of compound activity at the target of interest. Nonetheless, there is intrinsic value in assaying channel function in its native cellular background and in the presence of potentially critical accessory proteins.

More commonly, cDNAs are used to express recombinant ion channel proteins either in insect cells for ligand binding [39] and biochemical isolation [40] or in immortalized mammalian cell lines for functional assays [41]. One potential drawback is that heterologous expression in insect cells, frog eggs, or a mammalian cell line may deprive target channels of interactions with in situ proteins important for normal function. Preferably, the host cells exhibit limited endogenous expression of other ion channels. For example, HEK293 cells are common hosts for screening ligand-gated ion channels because of the lack of extensive expression of endogenous channels [42]. Additional published reports include the human IRK-1 and influenza virus M2 channels expressed in yeast [43] or in Xenopus oocytes [44], AMPA receptors expressed in kidney cells [45], sodium channels expressed in CHO cells [46,47], and $GABA_A$/Benzodiazapine receptors expressed in Xenopus oocytes [48]. However, in some cases it may be necessary to "isolate" pharmacologically the ion channel of interest, and biochemical blockade of competing channels to isolate the channel of interest has been used in experimental pharmacology for many years. Screens using such techniques most commonly involve measurement of radioactive ion flux (typically ^{22}Na, ^{45}Ca, or ^{86}Rb) in cultured cells or vesicle preparations stimulated by agonist treatment, field stimulation, or depolarization with potassium ion. As increasingly selective channel blockers are discovered, these agents can be used to isolate more precisely the poorly characterized channels for study and assay development. In principle, biochemical blockade may allow HTS of channel targets in complex biological preparation, e.g., primary neurons, thereby avoiding the need for generating engineered cell lines and, in some cases, providing a work-around for potential patent issues.

Standard methods of expression include transient plasmid transfection [49] and stable expression using plasmid [50] or viral [38,40] vectors, with the goal of high-level channel expression for ligand binding and ion flux assay formats. However, it should be noted that channel overexpression may be unnecessary for assay formats employing voltage sensors. This is because relatively large changes in membrane potential can result from the opening of relatively few surface channels. Flow-activated cell sorting also can be employed, where a fluorescent assay of channel activity is feasible, to isolate clones or subpopulations of cells with improved channel responses. The validation of the recombinant cell line typically depends on its exhibiting an appropriate pharmacological profile for known compounds that have been previously characterized by whole-cell or patch-clamp electrophysiology.

Since many ion channels are heteromeric assemblies, especially ligand-gated channels in the nervous system, configuring an assay for the target often involves choosing to express either a single subunit or multiple subunits for the screening of homomeric or heteromeric forms of the channel, respectively [49]. Some, but not all, single subunits are able to form a functional channel, and the most straightforward approach is to screen against the homomeric assembly. For heteromeric channels, it is rather easy to engineer cell lines that coexpress two or more cDNAs, although it can be difficult to achieve the optimal expression level for each subunit. Furthermore, it is important to determine the composition of the channel populations expressed on the cell surface, i.e., the proportion of homomeric channels or the stoichiometry of the heteromeric species, an issue most commonly addressed using electrophysiology. Finally, deconvoluting the active compounds in terms of the interactions with the various channel populations may be problematic. Nonetheless, a reasonable strategy is to screen cell lines expressing heteromeric channels and devise appropriate secondary assays, including electrophysiology, to characterize the specific site(s) of compound activity.

The choice of chemical diversity to be tested in ion channel screens will not differ greatly from other types of drug targets. Most sizeable organic chemical libraries may provide sources of peptidomimetics, amino acid analogs, and other compounds having significant three-dimensional structure that may interact specifically with channel proteins. The list of plant-derived products known to interact specifically and potently with ion channels is long enough to provide a compelling case for natural product screening for such targets [51]. Toxins produced by arthropods, insects, snakes, and other animals have long provided useful tools for the study of ion channel function and may be legitimate sources of drug development leads [4,52–56]. Since some assays, such as animal studies and ADME/tox profiles, tend to require much larger amounts of compound than others, the involvement of chemistry resources must then be anticipated in the design of the screening funnel. Ion channel screens may be particularly sensitive to interference by ionophoric compounds, agents that disrupt cell membrane integrity or effectors of endogenous channels, or receptors interacting with ion channels. In the case of screening formats using fluorescent readouts, substances having intrinsic fluorescence or quenching properties, or agents capable of disrupting FRET-pair interactions, can result in false positive results.

VI. SECONDARY ASSAY STRATEGIES

In practice, the primary assay format will give rather imprecise answers to questions about channel behavior. For example, assays based on changes in bulk ion flux measured with a radioactive ion will not distinguish between agents capable of changing agonist affinity or its voltage dependence from those affecting the desensitization mechanism of the channel. Further, changes in cell membrane potential measured with fluorescent indicators will not provide information regarding the specific conductances involved in the response. In nearly every case, the active compounds identified in the primary screening format must be tested in a series of secondary assays before their actions can be considered target specific.

A key factor in the choice of secondary assays and how they fit in the screening funnel is the exact criteria for a compound worthy of extended testing, often referred to as a "lead" compound. Such criteria may describe a profile of potency and selectivity for the target channel, limited toxicity and activity in a correlative animal model. Some of the secondary assays performed may be definitive with respect to these criteria, while others may serve to add information about candidate molecules, but may not provide a go/no-go result. Since large drug discovery operations are likely to be involved in screening entire series of related channels or channel subtypes, cross-over studies may provide important selectivity and perhaps cytotoxicity information without the development of new assays. However, caution must be exercised in using biochemical measures of selectivity to prioritize candidate molecules from the screen hit list. In some cases, preclinical or clinical data will be available to show that compound activity at related ion channels will likely produce unacceptable side effects. Most often, however, such data are simply not available and the use of activity at a related channel as a strict selectivity criterion is little more than guesswork. Of course, in the fortunate instance where a long list of interesting hits is available, the discovery team can perhaps afford to be more demanding in its use of such selectivity measures.

The secondary assay strategy applied to ion channel screens will not differ fundamentally from other types of drug targets. Often an initial confirmatory assay using the same methodology as the primary screen will be run using multiple compound concentrations to determine a crude potency value. The secondary assays for confirmed actives may be designed to (1) demonstrate that the activity of the hits discovered in the primary screen are in fact target dependent and not due to interactions with other endogenous channels, (2) assess the selectivity of the hits for the target channel compared with related channels, (3) discover mechanistic information, (4) identify hits with activity in a tissue-based assay dependent on the targeted ion channel, or (5) rank hits based on cytotoxicity or gross pharmacokinetic parameters. Because these secondary assays need not have the throughput of the primary screen, they may use any of the traditional techniques available to the pharmacologist. The most important of these is confirmation of activity by electrophysiology, often in primary cell preparations where the compound can be tested against the target in a native cellular environment. Nevertheless, the choice of secondary assays and the order in which they are performed is rarely trivial. In particular, the complex nature of cell-based assays of ion channels demands careful planning of the secondary assays in order to focus chemistry efforts on the compounds with authentic activity at the target.

VII. DISCOVERY TECHNIQUES NOT SUITED FOR HTS

Without question, electrophysiological methods have revolutionized our understanding of the properties and behavior of ion channels and their interactions with drugs [57]. Patch clamp electrophysiology refers to a technique for forming a tight seal on the cell surface with a pipette containing a small electrode. This allows the current flow through this small patch (several μm^2) to be recorded for analysis of single-channel properties or whole-cell electrical properties. However, traditional electrode-based methods clearly fall short of a testing rate of thousands of compounds per day, owing to challenges involved in providing suitable biological preparations, establishing stable recordings with the desired characteristics, and delivering test compounds in a controlled manner. Several studies have, however, introduced extensions of the traditional methods with the aim of providing sufficient throughput. One study employed capillary electrophoresis for the delivery of putative antagonists across a patch pipet recording NMDA-induced currents from rat brain olfactory bulb membranes [58]. Another involved the development of multielectrode arrays useful for simultaneous multisite recording [59]. Efforts of this sort will stimulate electrophysiologists to devise medium-to-high-throughput applications for their methods. In general, however, electrophysiological methods, although indispensable to our understanding of detailed channel behavior and drug interactions, have not yet taken a prominent role in HTS efforts, except as confirmatory and secondary assays. A variety of creative alternative approaches for detecting ion channel activity have been reported, including increased fluorescence of GFP fused to channel subunits [60], ion flux across planar bilayers [61], capillary electrophoresis [62], site-directed incorporation of a fluorescent dye to report conformational changes [63], and measurement of cell activation using microphysiometer recordings [43]. These technologies are worth monitoring, but in general are at the proof-of-principle stage and require significant further development to become applicable to HTS operations.

VIII. LIGAND-BINDING FORMATS

Binding assays for compounds that interact with ligand-gated ion channels are well tailored for HTS and have been used extensively in discovery programs directed toward GABA-, glutamate-, purine-, serotonin-, inositol-1,4,5-triphosphate-, and nicotine-responsive channels. In addition, voltage-gated ion channels, although lacking endogenous ligand activation, have been studied using binding paradigms. This has been possible where radio- or fluorescence-labeled drugs or toxins bind to the channel at pharmacologically significant sites, such as ligands commonly used for calcium [64], sodium [65], and potassium channel proteins [66–71]. Binding methods have also been useful in the study of coagonist and allosteric sites on channel proteins. Examples include the binding of TBPS or TBOB to benzodiazepine sites on GABA receptors [72,73] and ligands for the glycine binding site on NMDA receptors [74,75]. Screening paradigms where the affinity of the radioligand may depend on the state of channel activation may be viewed as functional assays and thus provide more information than standard radioligand binding studies.

IX. ION FLUX: RADIOTRACERS PLATFORMS

Assays based on the passage of ions through channels are unbiased with respect to the specific sites of action of prospective drugs that modulate channel behavior. Radiotracer ions have been used widely in drug screening and have the potential to provide a high-throughput format with a functional readout. However, major drawbacks include safety concerns and the relatively high cost of purchasing reagents and disposing of radioactive waste. Applications of direct radiotracers used in drug screening include (^{22}Na) uptake to measure action-potential channels in neuroblastoma cell lines [76], (^{42}K) efflux via Na/K pump activity in astrocytes [77], and GABA receptor activation measured using (^{36}Cl) [78] and (^{45}Ca) transport as a measure of channel activity and to study channel antagonists [79,80]. The specificity of such assays is generally validated and controlled by the use of selective agonists [47] or by selective blockade of nontarget conductances

[77]. As an alternative to the radiolabeled ''native'' ion, surrogate ions also are commonly used and applications include (^{86}Rb) as a potassium substitute [77,81] and (^{14}C-guanidinium) as a sodium surrogate [46,47,82]. Additional studies have involved, for example, lucifer yellow spreading via gap junction channels as a measure of their function [83,84] and induction of mitogenesis by lithium uptake through cation channels [85]. These techniques share the advantage of measuring channel function in a potentially high-throughput format without the development of special technologies. Radiotracer screening platforms are sufficiently well established that contract screening companies employ them to test small and large collections of substances [86–88].

X. TRANSPORTER TARGETS

Transporters are a class of drug targets closely related to ion channels and are amenable to HTS by ligand binding and radiotracer accumulation assays. Competitive binding assays using high-affinity radiolabeled substrates provide a sensitive method for detecting blockade of transport and take advantage of current improvements in radioligand binding and detection methodologies. These assays are available from contract screening companies and have been used to target transporter systems for serotonin [89], dopamine [90,91], adenosine [92], norepinephrine [93], choline [94], and GABA [95]. Although binding assays for uptake proteins have been used to discover useful inhibitors, accumulation assays have the benefit of measuring function, facilitating the discovery of second-site or ''catalytic'' effectors of uptake activity. Accumulation assays have been developed to monitor the cloning of transporter genes, and for biochemical, kinetic, and pharmacological studies of transporter properties. Examples include transporters for glutamate [96], serotonin [97,98], epinephrine [99,100], dopamine, glycine [101,102], proline [103], and GABA [104]. Accumulation assays have been applied to cells expressing cloned transporters and endogenously expressed transporters, and they generally involve incubation of cells with a radiolabeled substrate, washing, solubilization of the cells, and liquid scintillation counting. This format is very well suited to HTS operations where multiwell pipettors, washers, and plate readers can be employed in semihomogeneous assays using the Cytostar-T™ [105] or Flashplate™ [106] technologies.

It is also interesting to note that many transporters, e.g., neurotransmitter transporters, are weakly electrogenic, since they utilize ion concentration gradients to cotransport small molecules. Most of the neurotransmitter transporters can be studied using traditional electrophysiological methods [102,107,108]. Indeed, recent work indicates neurotransmitter transporters are capable of adopting a channellike state exhibiting enhanced ion and transmitter flux [109,110]. This allows the design of HTS formats based on either the ion conductances accompanying transmitter transport or changes in membrane potential (see below) [111].

XI. ION FLUX: FLUORESCENCE FORMATS

Ion channel targets are increasingly being developed for HTS as functional assays with fluorescence readouts, a trend observed more generally throughout HTS operations. Fluorescent dyes are available for monitoring Na^+ [112] and Cl^- [113,114] ion concentration, although sensitivity problems limit their application for HTS. However, fluorescent dye–based assays for intracellular Ca^{2+} mobilization have provided important tools for the study of ion channel physiology [115], particularly by nonelectrophysiologists, and have been widely adapted to HTS assays [35]. Calcium-sensitive dyes have been most commonly used to monitor GPCR-activated signal transduction leading indirectly to calcium flux from intracellular stores [116,117] or directly via coupling to surface Ca^{2+} ion channels. However, voltage-gated calcium channels (VGCCs) and many ligand-gated channels pass sufficient Ca^{2+} to be easily assayed by fluorescent dyes. Alternatively, VGCCs also can be employed as reporters for target channels that do not pass sufficient Ca^{2+}, e.g., depolarizing ligand-gated channels [117,118]. Surprisingly, fluorescent detection of intracellular Ca^{2+} can be more sensitive than electrophysiological methods in cases where channels pass small Ca^{2+} currents, but inactivate very slowly [35]. The dyes of choice for discovery and flow cytometry applications tend to be the radiometric dyes Fura-2 and Indo-1, while Fluo-3, a single-excitation-wavelength dye, is commonly used for developing HTS assays.

Increasingly sensitive and versatile instrumentation has expedited assay development and screening throughput in recent years. It is not our purpose to review comprehensively the entire scope of the instrumentation available for fluorescence-based cell readouts; we shall only describe the instruments most suitable for HTS. Light or laser-scanning fluorescence microscopy coupled either with a photomultiplier or a camera to collect light can be very powerful for the discovery or characterization of dynamic changes in cell calcium, but they suffer from insufficient throughput to support HTS. Likewise, flow cytometry has been effectively applied to study calcium responses, but its throughput is currently limited by the complexity of sample delivery, although attempts are being made to automate flow systems [119]. Newer instruments possessing sophisticated optics, configuration for high-density plate formats, and integrated multichannel pipetting are allowing screens based on calcium- and voltage-sensitive dyes to achieve much higher throughputs. HTS of target channels using Fura-2 measurements of Ca^{2+} have been performed on a custom plate-imaging fluorimeter from SIBIA Neurosciences/Science Applications International Corporation [35]. Introduced in 1996, the fluorometric imaging plate reader (FLIPR™) [120] was designed specifically to allow accurate measurement of changes in cell fluorescence in high-density plate formats and has been widely used for calcium-based [121] and voltage-based [122] assays in HTS campaigns. The FLIPR utilizes an argon laser for dye excitation, integrated liquid handling, and CCD detection optics to provide real-time kinetic data on cell responses to compounds and ligands. Because the entire plate, rather than individual wells, is imaged, throughput is increased significantly in moving from a 96-well to a 384-well format. One powerful application of this technology is the use of multiple solution additions coupled with kinetic profiling of target activity to identify agonist, antagonist, and allosteric modulators in a single test well.

XII. MEMBRANE POTENTIAL: FLUORESCENCE FORMATS

Assay formats that detect changes in membrane potential are valuable alternatives to formats that measure ligand binding or ion flux events. The appeal of this approach is that relatively small changes in ion conductance generate relatively large changes in membrane potential. One prerequisite for successful assay development is that the sensitivity of the indicator is adequate over the predicted dynamic range of the assay. This approach typically employs fluorescent indicators and is desirable from an HTS standpoint because of the relatively low cost and the potential for incorporating automation. However, past work with single-wavelength fluorescent dyes has revealed problems adapting this approach to HTS. Impermeant dyes that rapidly change their distribution within the membrane in response to changes in membrane potential suffer from low sensitivity, typically a 2–10% change in fluorescence per 100 mV change in membrane potential [123]. Conversely, permeant dyes that partition into the surface membrane and cytoplasm in response to cell depolarization are reasonably sensitive, yet have response times in minutes rather than milliseconds. Nonetheless, HTS screening of channel targets using potential-sensing dyes have been performed using instrumentation such as the FLIPR.

An important recent improvement has been the development of fluorescence resonance energy transfer (FRET) dyes as sensors exhibiting increased voltage sensitivity over traditional single-dye oxonols [124], fast temporal response, and radiometric output [125]. The FRET-based voltage sensor is composed of two partners, a fixed energy donor and a mobile energy acceptor. The most sensitive FRET pair are a coumarin-tagged phospholipid (CC2-DMPE) donor, which binds to the outer leaflet of the plasma membrane, and a negatively charged oxonol [DiSBAC$_2$(3)] acceptor, which partitions across the plasmalemma as a function of membrane potential. The coumarin and oxonol pairs are chosen so that the emission spectrum of the coumarin donor overlaps with the excitation spectra of the oxonol acceptor. As with all FRET systems, proximity of the two dyes allows excitation of the acceptor dye by the donor dye. At negative membrane potentials (hyperpolarized), typical of most resting cells, the oxonol is localized on the outer surface near the coumarin donor. Excitation of coumarin results in energy transfer and elevated emission from oxonol (red wavelength). At depolarized or more positive membrane potentials, e.g., as a result of the influx of Na^+ ions, the negatively charged oxonol rapidly (< 0.4 ms) moves to the inner leaflet. The breaking of the FRET pairing results in a decreased oxonol emission and increased emission from coumarin (blue wavelength). Data analyzed as a change in the ratio of blue fluorescence to red fluorescence is particularly valuable in assay conditions where cell number is difficult to control. The sensitivity of voltage sensor dyes has been determined in various cell lines using calibration curves obtained with voltage ramps. Ratio changes per 100 mV change in membrane potential are typically in the range of 40–60% per 100 mV, but they can be as high as 80%, representing the largest voltage-sensitive optical signal reported to date. These levels of sensitivity and response time are sufficient for the development of assays for most depolarizing ion channel targets and, conceivably, electrogenic transporters. More problematic is the application to hyperpolarizing channels, e.g., Cl^- channels, because of the difficulty of engineering cell lines that produce a strong hyperpolarization response. However, THP-1 cells, which have a relatively high resting membrane potential [126], have been assayed on FLIPR using single oxonol dyes to examine both depolarizing and hyperpolarizing channel function. Aurora Biosciences also has developed the voltage ion probe reader (VIPR™), an instrument for measuring the voltage sensor dyes, to facilitate HTS of ion channel targets in 96-well plates [127].

XIII. AUTOMATION CONSIDERATIONS

Moving from a bench-top "analytical" scale to HTS requires a moderate to high level of process automation. The elements requiring automation will depend on the desired throughput, but may involve (1) the delivery of test samples, (2) plate handling, (3) liquid handling, (4) readout, and (5) data processing. These elements may or may not be highly integrated, but in either case they require accurate scheduling and sample tracking. The constraints that automation places on screens directed toward ion channel targets will not necessarily differ from other types of drug targets. Because ion channels function within the context of biological membranes, many ion channel assays will be cell based and will require propagating cell lines, scheduling transfections, and maintaining quality assurance. Nevertheless, decisions regarding the use of modular or integrated robotics and the use of 96-, 384-, or higher order multiwell formats will depend more on the specifications of the assay itself than on the nature of the target.

XIV. CONCLUSIONS

Defects in ion channel function clearly underlie a variety of clinically and economically important human diseases. The corresponding importance of ion channels as drug targets is reflected in the heightened attention these proteins have received in drug screening programs in recent years. Unfortunately, electrophysiology is the technology of choice for the detailed characterization of ion channel behavior but is not well suited for HTS. Instead, traditional methods that have less sensitivity and elegance, including radioligand binding and radioactive/surrogate ion uptake, are most commonly used in large-scale ion channel screening campaigns. More recent functional assays based on fluorescent dyes for measuring ion concentration and fluorescent membrane voltage sensors promise to provide richer data on active compounds. All of these methods are being improved as the demands of drug screening are focused on ion channel targets, and new techniques are emerging that will allow more detailed questions to be asked at the level of the high-throughput screen. These latter techniques include gene reporters of ion channel activity, microfluorimetry, various miniaturization technologies, and a new level of sophistication in measuring cell fluorescence. In principle, the HTS campaign strategy for ion channel targets does not differ from other types of target. The primary screen will commonly be followed by a battery of confirmatory and secondary assays, including electrophysiology, with the aim of identifying compounds that most closely match the hit/lead criteria established for the project. Unique properties of ion channels such as rapid desensitization and voltage dependence, however, may require unusual manipulation of the target in order to generate a robust signal. The combination of high technology, molecular biology, biochemistry, and pharmacology has driven the current advances in the field and will continue to bring unique strategies, approaches, and success to ion channel screening efforts. The ingenuity with which investigators apply these tools likely will determine the number and quality of the lead compounds to emerge from the screening project and, ultimately, the success or failure of the drug discovery efforts.

REFERENCES

1. MH Meisler, LK Sprunger, NW Plummer, A Escayg, JM Jones. Ion channel mutations in mouse models of inherited neurological disease. Ann Med 29:569–574, 1997.
2. B Vafa, PR Schofield. Heritable mutations in the glycine, GABAA, and nicotinic acetylcholine receptors provide new insights into the ligand-gated ion channel receptor superfamily. Int Rev Neurobiol 42:285–332, 1998.
3. EC Cooper, LY Jan. Ion channel genes and human neurological disease: recent progress, prospects and challenges. Proc Natl Acad Sci USA 96:4759–4766, 1999.

4. JM McIntosh, BM Olivera, LJ Cruz. Conus peptides as probes for ion channels. Methods Enzymol 294:605–624, 1999.

5. JM Quayle, MT Nelson, NB Standen. ATP-sensitive and inwardly rectifying potassium channels in smooth muscle. Physiol Rev 77:1165–1232, 1997.

6. WA Catterall. Structure and function of voltage-gated ion channels. Ann Rev Biochem 64:493–531, 1995.

7. VS Markin, AJ Hudspeth. Gating-spring models of mechanoelectrical transduction by hair cells of the internal ear. Ann Rev Biophys Biomol Struct 24:59–83, 1995.

8. OP Hamill, DW McBride Jr. The pharmacology of mechanogated membrane ion channels. Pharmacol Rev 48:231–252, 1996.

9. R Andrade, RC Malenka, RA Nicoll. A G protein couples serotonin and GABAB receptors to the same channels in hippocampus. Science 234:1261–1265, 1986.

10. M Schwanstecher, C Schwanstecher, F Chudziak, U Panten, JPt Clement, G Gonzalez, L Aguilar-Bryan, J Bryan. ATP-sensitive potassium channels. Methods Enzymol 294:445–458, 1999.

11. JT Finn, ME Grunwald, KW Yau. Cyclic nucleotide-gated ion channels: an extended family with diverse functions. Ann Rev Physiol 58:395–426, 1996.

12. LY Jan, YN Jan. Tracing the roots of ion channels. Cell 69:715–718, 1992.

13. A Hodgkin, A Huxley. A quantitative description of membrane current and its application to conduction and excitation in nerve. J Physiol 117:500–544, 1952.

14. C Bargmann. Neurobiology of the *Caenorhabditis elegans* genome. Science 282: 2028–2033, 1998.

15. SJ Russek. Evolution of GABA(A) receptor diversity in the human genome. Gene 227:213–222, 1999.

16. S Heinemann, J Boulter, E Deneris, J Conolly, R Duvoisin, R Papke, J Patrick. The brain nicotinic acetylcholine receptor gene family. Prog Brain Res 86:195–203, 1990.

17. M Hollmann, S Heinemann. Cloned glutamate receptors. Ann Rev Neurosci 17: 31–108, 1994.

18. R Chittajallu, SP Braithwaite, VR Clarke, JM Henley. Kainate receptors: subunits, synaptic localization and function. Trends Pharmacol Sci 20:26–35, 1999.

19. JP Changeux, SJ Edelstein. Allosteric receptors after 30 years. Neuron 21:959–980, 1998.

20. GH Hockerman, BZ Peterson, BD Johnson, WA Catterall. Molecular determinants of drug binding and action on L-type calcium channels. Ann Rev Pharmacol Toxicol 37:361–396, 1997.

21. D Bleakman, D Lodge. Neuropharmacology of AMPA and kainate receptors. Neuropharmacology 37:1187–1204, 1998.

22. C Lena, J Changeux. Allosteric nicotinic receptors, human pathologies. J Physiol Paris 92:63–74, 1998.

23. RL Macdonald, RW Olsen. GABAA receptor channels. Ann Rev Neurosci 17:569–602, 1994.

24. LJ Ptacek. The place of migraine as a channelopathy. Curr Opin Neurol 11:217–226, 1998.

25. MJ Ackerman, DE Clapham. Ion channels—basic science and clinical disease. N Engl J Med 336:1575–1586, 1997.

26. PM Thomas, GJ Cote, N Wohllk, B Haddad, PM Mathew, W Rabl, L Aguilar-Bryan, RF Gagel, J Bryan. Mutations in the sulfonylurea receptor gene in familial persistent hyperinsulinemic hypoglycemia of infancy. Science 268:426–429, 1995.

27. JM Pilewski, RA Frizzell. Role of CFTR in airway disease. Physiol Rev 79:S215–255, 1999.

28. MC Koch, K Steinmeyer, C Lorenz, K Ricker, F Wolf, M Otto, B Zoll, F Lehmann-Horn, KH Grzeschik, TJ Jentsch. The skeletal muscle chloride channel in dominant and recessive human myotonia. Science 257:797–800, 1992.

29. PM Snyder, MP Price, FJ McDonald, CM Adams, KA Volk, BG Zeiher, JB Stokes, MJ Welsh. Mechanism by which Liddle's syndrome mutations increase activity of a human epithelial Na^+ channel. Cell 83:969–978, 1995.

30. SC Cannon. Sodium channel defects in myotonia and periodic paralysis. Ann Rev Neurosci 19:141–614, 1996.

31. MA Wingertzahn, RS Ochs. Control of calcium in skeletal muscle excitation-contraction coupling: implications for malignant hyperthermia. Mol Genet Metab 65:113–120, 1998.

32. K Boonyapisit, HJ Kaminski, RL Ruff. Disorders of neuromuscular junction ion channels. Am J Med 106:97–113, 1999.

33. J Denyer, J Worley, B Cox, G Allenby, M Banks. HTS approaches to voltage-gated ion channel drug discovery. Drug Discovery Today 3:323–332, 1998.

34. K Ogita, Y Yoneda. Signal transduction through ion channels associated with excitatory amino acid receptors. Methods Enzymol 294:385–410, 1999.

35. G Velicelebi, KA Stauderman, MA Varney, M Akong, SD Hess, EC Johnson. Fluorescence techniques for measuring ion channel activity. Methods Enzymol 294:20–47, 1999.

36. KO Holevinsky, Z Fan, M Frame, JC Makielski, V Groppi, DJ Nelson. ATP-sensitive K^+ channel opener acts as a potent chloride channel inhibitor in vascular smooth muscle cells. J Membrane Biol 137:59–70, 1994.

37. D Lo, A McAllister, L Katz. Neuronal transfection in brain slices using particle-mediated gene transfer. Neuron 13:1263–1268, 1994.

38. MU Ehrengruber, M Lanzrein, Y Xu, MC Jasek, DB Kantor, EM Schuman, HA Lester, N Davidson. Recombinant adenovirus-mediated expression in nervous system of genes coding for ion channels and other molecules involved in synaptic function. Methods Enzymol 293:483–503, 1998.

39. DB Carter, DR Thomsen, WB Im, DJ Lennon, DM Ngo, W Gale, HK Im, PH Seeburg, MW Smith. Functional expression of GABAA chloride channels and benzodiazepine binding sites in baculovirus infected insect cells. Biotechnology 10: 679–681, 1992.

40. K Radford, G Buell. Expression of ligand-gated ion channels using Semliki Forest virus and baculovirus. Methods Enzymol 293:459–483, 1998.

41. R Ashley. Ion Channels: A Practical Approach. London: IRL Press, 1996.

42. BJ Hamilton, DJ Lennon, HK Im, WB Im, PH Seeburg, DB Carter. Stable expression of cloned rat GABAA receptor subunits in a human kidney cell line. Neurosci Lett 153:206–209, 1993.

43. K Hahnenberger, M Krystal, K Esposito, W Tang, S Kurtz. Use of microphysiome-

try for analysis of heterologous ion channels in yeast. Nat Biotechnol 14:880–883, 1996.

44. K Giffin, R Rader, M Marino, R Forgey. Novel assay for the influenza virus M2 channel activity. FEBS Letters 357:269–274, 1995.

45. M Hennegriff, A Arai, M Kessler, P Vanderklish, M Mutneja, C Rogers, R Neve, G Lynch. Stable expression of recombinant AMPA receptor subunits: binding affinities and effects of allosteric modulators. J Neurochem 68:2424–2434, 1997.

46. MC Maillard, ME Perlman, O Amitay, D Baxter, D Berlove, S Connaughton, JB Fischer, JQ Guo, LY Hu, RN McBurney, PI Nagy, K Subbarao, EA Yost, L Zhang, GJ Durant. Design, synthesis and pharmacological evaluation of conformationally constrained analogues of N,N′-diaryl- and N-aryl-N-aralkylguanidines as potent inhibitors of neuronal Na^+ channels. J Med Chem 41:3048–3061, 1998.

47. NL Reddy, W Fan, SS Magar, ME Perlman, E Yost, L Zhang, D Berlove, JB Fischer, K Burke-Howie, T Wolcott, GJ Durant. Synthesis and pharmacological evaluation of N,N′-diarylguanidines as potent sodium channel blocker and anticonvulsant agents. J Med Chem 41:3298–3302, 1998.

48. AH Tang, MW Smith, DB Carter, WB Im, PF VonVoigtlander. U-90042, a sedative/hypnotic compound that interacts differentially with the GABAA receptor subtypes. J Pharmacol Exp Ther 275:761–767, 1995.

49. AL Eertmoed, YF Vallejo, WN Green. Transient expression of heteromeric ion channels. Methods Enzymol 293:564–585, 1998.

50. TM Shih, RD Smith, L Toro, AL Goldin. High-level expression and detection of ion channels in Xenopus oocytes. Methods Enzymol 293:529–556, 1998.

51. SH Snyder. Drugs and the Brain. Scientific American Library. New York: W.H. Freeman, 1986.

52. BM Olivera, GP Miljanich, J Ramachandran, ME Adams. Calcium channel diversity and neurotransmitter release: the omega-conotoxins and omega-agatoxins. Ann Rev Biochem 63:823–867, 1994.

53. WF Hopkins, M Allen, BL Tempel. Interactions of snake dendrotoxins with potassium channels. Methods Enzymol 294:649–661, 1999.

54. RD Murrell-Lagnado. Potassium ion channel inactivation peptides. Methods Enzymol 294:640–649, 1999.

55. ML Garcia, M Hanner, HG Knaus, R Slaughter, GJ Kaczorowski. Scorpion toxins as tools for studying potassium channels. Methods Enzymol 294:624–639, 1999.

56. RJ French, SC Dudley Jr. Pore-blocking toxins as probes of voltage-dependent channels. Methods Enzymol 294:575–605, 1999.

57. D Colquhoun. Neher and Sakmann win Nobel Prize for patch-clamp work. Trends Pharmacol Sci 12:449, 1991.

58. K Jardemark, C Farre, I Jacobson, R Zare, O Orwar. Screening of receptor antagonists using agonist-activated patch clamp detection in chemical separations. Anal Chem 70:2468–2474, 1998.

59. U Egert, B Schlosshauer, S Fennrich, W Nisch, M Fejtl, T Knott, T Muller, H Hammerle. A novel organotypic long-term culture of the rat hippocampus on substrate-integrated multielectrode arrays. Brain Research Protocols 2:229–242, 1998.

60. M Siegel, E Isacoff. A genetically encoded optical probe of membrane voltage. Neuron 19:735–741, 1997.

61. TA Mirzabekov, AY Silberstein, BL Kagan. Use of planar lipid bilayer membranes for rapid screening of membrane active compounds. Methods Enzymol 294:661–674, 1999.

62. O Orwar, K Jardemark, C Farre, I Jacobson, A Moscho, JB Shear, HA Fishman, SJ Lillard, RN Zare. Voltage-clamp biosensors for capillary electrophoresis. Methods Enzymol 294:189–208, 1999.

63. BA Griffin, SR Adams, RY Tsien. Specific covalent labeling of recombinant protein molecules inside live cells. Science 281:269–272, 1998.

64. H Rehm, M Lazdunski. Purification and subunit structure of a putative K^+-channel protein identified by its binding properties for dendrotoxin I. Proc Natl Acad Sci USA 85:4919–4923, 1988.

65. K Gaines, S Hamilton, A Boyd. Characterization of the sulfonylurea receptor on beta cell membranes. J Biol Chem 263:2589–2592, 1988.

66. K Dickinson, R Baska, R Cohen, C Bryson, M Smith, K Schroeder, N Lodge. Identification of (3H)P1075 binding sites and P1075-activated K^+ currents in ovine choroid plexus. Eur J Pharmacol 345:97–101, 1998.

67. J Vasquez, P Feigenbaum, V King, G Kaczorowski, M Garcia. Characterization of high affinity binding sites for charybdotoxin in synaptic plasma membranes from rat brain. J Biol Chem 265:15564–15571, 1990.

68. C Mourre, M Hugues, M Lazdunski. Quantitative autoradiographic mapping in rat brain of the receptor of apamin, a polypeptide toxin specific for one class of Ca^{2+}-dependent K^+ channels. Brain Research 382:239–249, 1986.

69. H Im, W Im, J Pregenzer, N Stratman, P VonVoigtlander, E Jacobsen. Two imidazoquinoxaline ligand for the benzodiazepine site sharing a second low affinity site on rat $GABA_A$ receptors but with the opposite functionality. British J Pharmacol 123:1490–1494, 1998.

70. A Lewin, B de Costa, K Rice, P Skolnick. meta- and para-isothiocyanato-t-butylbicycloorthobenzoate: irreversible ligands of the g-aminobutyric acid-regulated chloride ionophore. Mol Pharmacol 35:189–194, 1989.

71. B Siegel, K Sreekrishna, B Baron. Binding of the radiolabeled glycine site antagonist [3H]MDL 105,519 to homomeric NMDA-NR1a receptors. Eur J Pharmacol 312:357–365, 1996.

72. B Felzen, G Berke, P Gardner, O Binah. Involvement of the IP3 cascade in the damage to guinea-pig ventricular myocytes induced by cytotoxic T lymphocytes. Pflugers Arch 433:721–726, 1997.

73. JL Morel, N Macrez, J Mironneau. Specific Gq protein involvement in muscarinic M3 receptor-induced phosphatidylinositol hydrolysis and Ca^{2+} release in mouse duodenal myocytes. Br J Pharmacol 121:451–458, 1997.

74. JA Watson, AC Elliott, PD Brown. Serotonin elevates intracellular Ca^{2+} in rat choroid plexus epithelial cells by acting on 5-HT2C receptors. Cell Calcium 17:120–128, 1995.

75. SE Pedersen, MM Lurtz, RV Papineni. Ligand binding methods for analysis of ion channel structure and function. Methods Enzymol 294:117–135, 1999.

76. Y Jacques, M Fosset, M Lazdunski. Molecular properties of the action potential

Na$^+$ ionophore in neuroblastoma cells. Interactions with neurotoxins. J Biol Chem 253:7383–7392, 1978.

77. O Chassande, C Frelin, D Farahifar, T Jean, M Lazdunski. The Na/K/Cl cotransport in C6 glioma cells. Eur J Biochem 171:425–433, 1988.

78. Y Liu, R Deirich. Role of GABA in the actions of ethanol in rats selectively bred for ethanol sensitivity. Pharmacol Biochem Behavior 60:793–801, 1998.

79. J Church, T Zsoter. Calcium antagonistic drugs. Mechanism of action. Can J Physiol Pharmacol 58:254–264, 1980.

80. D Kiang, R Kollander, H Lin, S LaVilla, M Atkinson. Measurement of gap junctional communication by fluorescence activated cell sorting. In Vitro Cell Dev Biol Anim 30A:796–802, 1994.

81. I Budunova, G Williams. Cell culture assays for chemicals with tumor-promoting or tumor-inhibiting activity based on the modulation of intercellular communication. Cell Biol Toxicol 10(2):71–116, 1994.

82. L Smith, M Price-Jones, K Hughes, J Anson, J Jensen, F Poulsen, F Wiberg. Measurement of ion flux through the GluR6 ion channel using Cytostar-T scintillating microplates. In: Real Case Studies on Developing Effective Assays. Southborough, PA, 1997.

83. S Rogers, L Gahring, R Papke, S Heinemann. Identification of cultured cells expressing ligand-gated cationic channels. Protein expression and purification 2:108–116, 1991.

84. J Brown, V Dissanayake, A Briggs, M Milic, N Gee. Isolation of the [3H]gabapentin-binding protein/a$_2$d Ca^{++} channel subunit from porcine brain: development of a radioligand binding assay for a$_2$d subunits using [^{3}H]leucine. Anal Biochem 255: 236–243, 1998.

85. E McNeal, G Lewandowski, J Daly, C Creveling. [^{3}H]Batrachotoxinin A 20a-benzoate binding to voltage-sensitive sodium channels: a rapid and quantitative assay for local anesthetic activity in a variety of drugs. J Med Chem 28:381–388, 1984.

86. Panlabs. Pharmacology Services, 1999.

87. Novascreen. NovaScreen Contract Research Organization, 1999.

88. Cerep. Cerep Pharmacological Model Catalogue. T. Jean, ed. Poitiers, France, 1999.

89. J Marcusson, M Bergstrom, K Eriksson, S Ross. Characterization of (^{3}H)paroxetine binding in rat brain. J Neurochem 50:1783–1790, 1988.

90. J Vignon, V Pinet, C Cerruti, J Kamenka, R Chicheportiche. (^{3}H)N-(1-(2-benzo (b)thiophenyl)cyclohexyl)piperidine ((^{3}H)BTCP): a new phencyclidine analog selective for the dopamine uptake complex. Eur J Pharmacol 148:427–436, 1988.

91. Z Pristupa, J Wilson, B Hoffman, S Kish, H Niznik. Pharmacological heterogeneity of the cloned and native human dopamine transporter: disassociation of (^{3}H)WIN 35,428 and (^{3}H)GBR 12,935 binding. Mol Pharmacol 45:125–135, 1994.

92. P Marangos, M Finkel, A Verma, M Maturi, J Patel, R Patterson. Adenosine uptake sites in dog heart and brain; interaction with calcium antagonists. Life Science 35: 1109–1116, 1984.

93. S Tejani-Butt, G Ordway. Effect of age on (^{3}H)nisoxetine binding to uptake sites for norepinephrine in the locus coeruleus of humans. Brain Res 583:312–315, 1992.

94. T Vickroy, W Roeske, H Yamamura. Sodium-dependent high-affinity binding of

(^{3}H)hemicholinium-3 in the rat brain: a potentially selective marker for presynaptic cholinergic sites. Life Science 35:2335–2343, 1984.

95. R Shank, W Baldy, L Mattucci, F Villani. Ion and temperature effects on the binding of gamma-aminobutyrate to its receptors and the high-affinity transport system. J Neurochem 54:2007–2015, 1990.

96. JA Meaney, VJ Balcar, JD Rothstein, PL Jeffrey. Glutamate transport in cultures from developing avian cerebellum: presence of GLT-1 immunoreactivity in Purkinje neurons. J Neurosci Res 54:595–603, 1998.

97. RD Blakely, HE Berson, RT Fremeau, MG Caron, MM Peek, HK Prince, CC Bradley. Cloning and expression of a functional serotonin transporter from rat brain. Nature 354:66–70, 1991.

98. DT Wong, FP Bymaster, DA Mayle, LR Reid, JH Krushinski, DW Robertson. LY248686, a new inhibitor of serotonin and norepinephrine uptake. Neuropsychopharmacol 8:23–33, 1993.

99. T Pacholczyk, RD Blakely, SG Amara. Expression cloning of a cocaine- and antidepressant-sensitive human noradrenaline transporter. Nature 350:350–354, 1991.

100. RD Blakely, LJ De Felice, HC Hartzell. Molecular physiology of norepinephrine and serotonin transporters. J Exp Biol 196:263–281, 1994.

101. K Sakata, K Sato, P Schloss, H Betz, S Shimada, M Tohyama. Characterization of glycine release mediated by glycine transporter 1 stably expressed in HEK-293 cells. Molecular Brain Res 49:89–94, 1997.

102. B Lopez-Corcuera, R Martinez-Maza, E Nunez, M Roux, S Supplisson, C Aragon. Differential properties of two stably expressed brain-specific glycine transporters. J Neurochem 71:2211–2219, 1998.

103. RT Fremeau, MG Caron, RD Blakely. Molecular cloning and expression of a high affinity L-proline transporter expressed in putative glutamatergic pathways or rat brain. Neuron 8:915–926, 1992.

104. M Liu, RL Russell, L Beigelman, RE Handschumacher, G Pizzorno. Beta-alanine and alpha-fluoro-beta-alanine concentrative transport in rat hepatocytes by GABA transporter GAT-2. Am J Physiol 276:G206–210, 1999.

105. R Graves, R Davies, P Owen, M Clynes, I Cleary, G O'Beirne. An homogeneous assay for measuring the uptake and efflux of radiolabelled drugs in adherent cells. J Biochem Biophys Methods 34:177–187, 1997.

106. NENLifesciences, FlashNews Index. NEN Lifesciences, 1999.

107. SG Amara, MJ Kuhar. Neurotransmitter transporters: recent progress. Ann Rev Neurosci 16:73–93, 1993.

108. JI Wadiche, JL Arriza, SG Amara, MP Kavanaugh. Kinetics of a human glutamate transporter. Neuron 14:1019–1027, 1995.

109. WA Fairman, RJ Vandenberg, JL Arriza, MP Kavanaugh, SG Amara. An excitatory amino-acid transporter with properties of a ligand-gated chloride channel. Nature 375:599–603, 1995.

110. MB Robinson. The family of sodium-dependent glutamate transporters: a focus on the GLT-1/EAAT2 subtype. Neurochem Int 33:479–491, 1998.

111. B Billups, M Szatkowski, D Rossi, D Attwell. Patch-clamp, ion-sensing, and glutamate-sensing techniques to study glutamate transport in isolated retinal glial cells. Methods Enzymol 296:617–632, 1998.

112. R Haugland. Handbook of Fluorescent Probes and Research Chemicals. M Spence, ed. Eugene, OR: Molecular Probes, 1996.
113. GR Ehring, YV Osipchuk, MD Cahalan. Swelling-activated chloride channels in multidrug-sensitive and -resistant cells. J Gen Physiol 104:1129–1161, 1994.
114. MK Mansoura, J Biwersi, MA Ashlock, AS Verkman. Fluorescent chloride indicators to assess the efficacy of CFTR cDNA delivery. Hum Gene Ther 10:861–75, 1999.
115. RY Tsien. Fluorescent probes of cell signaling. Ann Rev Neurosci 12:227–253, 1989.
116. PC Sternweis, AV Smrcka. Regulation of phospholipase C by G proteins. Trends Biochem Sci 17:502–506, 1992.
117. J Hescheler, G Schultz. Heterotrimeric G proteins involved in the modulation of voltage-dependent calcium channels of neuroendocrine cells. Ann NY Acad Sci 733:306–312, 1994.
118. E Stetzer, U Ebbinghaus, A Storch, L Poteur, A Schrattenholz, G Kramer, C Methfessel, A Maelicke. Stable expression in HEK-293 cells of the rat alpha3/beta4 subtype of neuronal nicotinic acetylcholine receptor. FEBS Lett 397:39–44, 1996.
119. Y Cambet, E Sebille, C Bradshaw, J-P Aubry, D Church, A Bernard. A fluorescence activated cell sorter (FACS) as an automated readout for high throughput cell-based screening assays. San Diego, CA: Society for Biomolecular Screening, 1997.
120. K Schroeder, B Neagle. FLIPR: a new instrument for accurate, high throughput optical screening. J Biomol Screening 1:75–80, 1996.
121. E Sullivan, EM Tucker, IL Dale. Measurement of $[Ca^{2+}]$ using the fluorometric imaging plate reader (FLIPR). Methods Mol Biol 114:125–133, 1999.
122. D Church, Y Cambet, A Bernard, D Estoppey, A Surprenant, R North, E Kawashima. Fluorometric identification of purinergic P2X7, P2X2/3 and P2X4 receptor antagonists. San Diego, CA: Society for Biomolecular Screening, 1997.
123. TJ Ebner, G Chen. Use of voltage-sensitive dyes and optical recordings in the central nervous system. Prog Neurobiol 46:463–506, 1995.
124. JE Gonzalez, RY Tsien. Voltage sensing by fluorescence resonance energy transfer in single cells. Biophys J 69:1272–1280, 1995.
125. JE Gonzalez, RY Tsien. Improved indicators of cell membrane potential that use fluorescence resonance energy transfer. Chem Biol 4:269–277, 1997.
126. SY Kim, MR Silver, TE DeCoursey. Ion channels in human THP-1 monocytes. J Membr Biol 152:117–130, 1996.
127. JE Gonzalez, K Oades, Y Leychkis, A Harootunian, PA Negulescu. Cell-based assay and instrumentation for screening ion-channel targets. Drug Discovery Today 4:431–439, 1999.

11
High-Throughput Screening Assays for Detection of Transcription

Mohanram Sivaraja
CURIS, Cambridge, Massachusetts

I. INTRODUCTION

On average, about one-third of the human genome is actively transcribed. However, the identities of the transcribed genes differ in various cell types and at distinct stages of differentiation and development [1]. The quantitation of gene expression at the cellular level is of central importance for understanding cellular biology and its role in human disease. Recent advances in DNA sequencing, and a variety of other techniques to examine the differential expression of mRNA levels in different cell and tissue types, together with advances in microarray technologies, have contributed to the identification of novel genes and their roles in cellular physiology and disease processes [2–4].

A number of pathological processes have been shown to result from inappropriate gene expression. For example, the over-expression of the selectin family of genes and their regulation by a host of cytokines (for example, interleukin 1 and tumor necrosis factor) are implicated in acute and chronic inflammation [5]. The over-expression of immunoglobulin IgE leads to hypersensitivity responses to specific allergans, which result in allergic diseases [6]. Amplification of a number of oncogenes such as bcl-2 and erb-2, or the inhibition of tumor suppressor genes, leads to the development of a variety of cancers [7–9]. Finally, expression of the genomes of pathogenic viruses such as human immunodeficiency virus (HIV) lead to chronic and fatal infection [10]. The ability to monitor the product of in vitro transcription reactions and mRNA expression in intact cells in a simple automated format would make possible the development of high-throughput screening (HTS) assays that permit the identification of small molecules that function as transcriptional regulators.

II. RNA DETECTION METHODS

A. Techniques Used for Detection of the RNA Product of In Vitro Transcription Reactions or RNA Isolated from Cells

Traditional RNA detection methods involve ethidium bromide staining following gel electrophoresis, RNA (Northern) blotting techniques, RNase protection, and direct radiolabeling of RNA [11]. None of these traditional methods for RNA detection are suitable for automation and HTS. More recently, a number of novel methods have been developed to detect small amounts of specific RNA. The most commonly used methods involve the polymerase chain reaction (PCR) [12,13] and include amplification of the specific RNA product and subsequent quantitation of that product using fluorescence, chemiluminescence, or radioactivity [14–18]. All of these techniques involve RNA isolation, reverse transcription-PCR (RT-PCR) amplification, and subsequent detection by gel-based or ELISA-type assays. While certain steps of these assays can be automated, the entire procedures are cumbersome and therefore not ideally suited for HTS.

RNA amplification can also be achieved with the use of nucleic acid sequence based amplification (NASBA) [19]. The NASBA technique, unlike assays involving PCR, is isothermal and uses the concurrent enzymatic activities of reverse transcriptase, RNase H, and T7 RNA polymerase, along with two complementary o-nt primers. NASBA-amplified RNA can then be detected by heterogenous methods, such as an enzyme-linked gel assay (ELGA) [20], or electrochemiluminescence [21], or by homogenous methods such as fluorescence correlation spectroscopy [22] or molecular beacon probes [23].

A number of available methods for RNA detection do not involve product amplification prior to detection. For example, Wu et al. [24] described an in vitro transcription HTS filtration assay that incorporates multiple radioactive labels in the product transcript. This assay was used to perform a HTS of nonspecific transcriptional elongation by E. coli RNA polymerase (RNAP). In addition, several sandwich hybridization methods have been described that use chemiluminescence or bioluminescence detection [25,26]. One variation of the sandwich hybridization technique is the sensitive branched DNA (bDNA) signal amplification assay for RNA detection [27]. The bDNA assay utilizes a DNA hybridization probe that is linked to a branched o-nt complex whose ends are labeled with alkaline phosphatase (AP). Thus the readout signal is amplified by the presence of multiple AP labels per hybridization probe. This technique, while automatable, is time-consuming and expensive.

A number of homogenous methods have been described for RNA detection. One such method is the combination of RNase protection and scintillation proximity assay technology [28]. Other homogenous techniques include nucleic acid binders used in the development of assays that allow quantitation of RNA using either fluorescence (e.g., thiazole orange and oxazole yellow) [29,30] or chemiluminescence detection using acridinium esters [31]. RNA synthesis can also be measured by incorporating fluorescently labeled nucleotides into the nascent transcript [32].

B. Techniques Used for RNA Quantitation in Cell-Based Assays

The most commonly used methods for cell-based HTS of mRNA expression involve reporter gene assays. Typically, the promotor region and other DNA sequence elements thought to be important for appropriate regulation of the gene of interest are linked to a reporter gene. This reporter gene construct is either stably or transiently introduced into a suitable cell line. This type of assay has the drawback that the construct may not contain all of the essential endogenous elements that regulate the gene of interest in its physiological milieu. Stable knock-in cell lines remedy this shortcoming by inserting a reporter gene into the intact endogenous gene. However, the generation of these knock-in cell lines is extremely tedious and often not possible. Furthermore, establishing stable cell lines is a time-consuming process, and appropriate gene regulation is not guaranteed in transformed cell lines.

Commonly used reporter genes used to monitor gene expression are those encoding proteins for which substrates yielding luminescent products are available, and include luciferase, β-galactosidase, chloramphenicol acetyltransferase, and AP [33]. In most cases, the choice of reporter gene is dictated by the nature of the host cell and the level of sensitivity required in the assay. One of the most exciting recent developments in the development of reporter genes has been the use of green fluorescent protein (GFP) from the jellyfish *Aequorea victoria* [34]. GFP emits green light upon exposure to UV or blue light. Unlike other bioluminescent molecules it does not require other cofactors or substrates, hence simplifying the assay and making it more amenable to HTS. Besides the wild-type GFP, a number of mutants are now commercially available that emit light at different wavelengths, making it possible to perform multiple gene expression assays from the same cells or mixture of cells [35]. Furthermore, because detection of GFP is noninvasive it is well suited to monitor kinetics of gene expression.

Radioactive and nonradioactive in situ hybridization (ISH) is widely used to measure cellular mRNA abundance and localization [36,37]. However, current ISH methods are technically not suitable for high-volume applications, due to the extensive sample manipulation involved. Recently, Harris et al. [38] described a microtiter plate ISH assay utilizing radiolabeled riboprobes that is more sensitive than conventional Northern blots. However, the assay requires a number of steps and is time-consuming (overnight incubation is required for hybridization of the riboprobe to the target mRNA). This protocol is thus not ideally suited for HTS.

Despite these tremendous advances in the field of nucleic acid detection, very few assay formats allow the sensitive detection of RNA in an automated, high-throughput manner. In general the available methods for RNA detection generally lack sensitivity, require extensive manipulation, and are quite expensive. There is clearly a need for methods that allow direct monitoring of mRNA expression in a HTS format without RNA amplification or the introduction of reporter genes.

A novel, sensitive, and simple assay for the detection of specific in vitro transcription reaction products and a facile ISH assay using anti-RNA:DNA hybrid antibodies for the direct detection of mRNA in intact cells is described below. Anti-RNA:DNA antibodies have been widely used for the detection of nucleic acids, particularly in clinical applications [39–41]. Recently, Tropix Inc. has introduced a commercial mRNA HTS assay that utilizes anti-RNA:DNA antibodies to detect specific mRNA expression in cells, which does not need amplification, purification of mRNA, or the development of stable cell lines, which is required with reporter assays. This method involves cell lysis and denaturation, and subsequent hybridization of the target mRNA to complementary DNA-probe (biotinyated o-nt). The DNA/RNA complex is then captured in a sterptavidin-coated microplate. An AP conjugated anti-RNA:DNA antibody is added, which binds to the RNA/DNA complex. AP activity is detected using chemiluminescent AP substrate.

Various strategies have been employed to raise monoclonal and polyclonal antibodies that specifically recognize RNA:DNA heteroduplexes independent of the nucleic acid sequence, while not binding to single-stranded or double-stranded RNA or DNA [42–45]. Information from these studies was used to develop the HTS assay for RNA detection described herein.

III. DETECTION OF PURIFIED RNA BY ANTI-RNA:DNA ANTIBODIES

RNA detection assay was optimized and validated by quantifying purified RNA generated from the in vitro transcription of interlukin-8 (IL-8 plasmid containing the human IL-8 cDNA IL-8) and G-less cassette encoding plasmids (G-less plasmid and LTR#5 plasmids contain the HIV LTR promotor with TAR sequences followed by a 450 nucleotide (nt) G-less cassette). The procedure used for measuring the purified in vitro transcribed RNA is optimized for a number of parameters, including hybridization buffer contents, hybridization time, o-nt concentration, antibody concentration, and wash buffer. Luminescence signal is measured by reading the plates in a luminometer (Victor from Wallac Inc., Gaithersburg, MD).

Data that validate the assay procedure and show that it can detect specific RNA sequences is shown in Figure 1. Panel A shows that the maximum luminescence signal was obtained only when both the target RNA and its corresponding complementary DNA (cDNA) oligonucleotides (o-nts) were added to the reaction mixture. When either the target RNA or o-nts were omitted, the signal was eliminated. Treatment with RNase A, after hybridization of the o-nts to the target RNA (panel A, column 4) resulted in a very small loss in signal relative to no RNase A treatment (column 1). However, treatment with RNase A prior to addition of the o-nts (column 5) completely eliminated the signal as RNase A is known to degrade selectively single-stranded RNA and not RNA:DNA heteroduplexes. These results indicate that RNA:DNA heteroduplex formation is necessary for signal detection. Figure 1, panel B shows that the assay detects specific RNA sequences. The maximum signal was obtained when the target RNA was hybridized to its DNA o-nts (columns 1 and 3); no signal is detected when noncomplementary DNA o-nts were used (columns 2 and 4).

The dose–response curve obtained with IL-8 RNA (Fig. 2) shows that the assay is sensitive enough to detect as little as 100 attomoles and is linear up to at least 100 fmoles of RNA (data for the upper end of the range is not shown). The sensitivity and linearity range of the assay make it highly suitable for quantitating the RNA product of in vitro transcription reaction. The yield of an in vitro transcription reaction is typically in the low attomole to fmole range. Assay sensitivity is presumably dependent on the amount of RNA:DNA heteroduplex formed, as this creates binding sites for the anti-RNA:DNA antibody. The extent of RNA:DNA heteroduplex formation is dependent on the length and sequence of the target RNA and DNA o-nts, and on the hybridization conditions. RNA sequences typically have extensive secondary structure, which could prevent effective hybridization at temperatures below the melting temperature (Tm). The total number of distinct DNA o-nts used and the regions of the target RNA to which they hybridize were found to be critical for maximizing signal intensity. In order to achieve the maximum signal and yet not have to optimize the sequences of the DNA o-nts used for each target RNA, a number of short DNA o-nts (35-mers) that were contiguous and complementary to the target RNA sequence, and spanned at least three-quarters of the target RNA sequence, are used. Another parameter found to be important for maximizing signal was the hybridization temperature. The hybridization reactions for all of the in vitro transcription experiments are carried out at room temperature, to avoid extra steps of placing

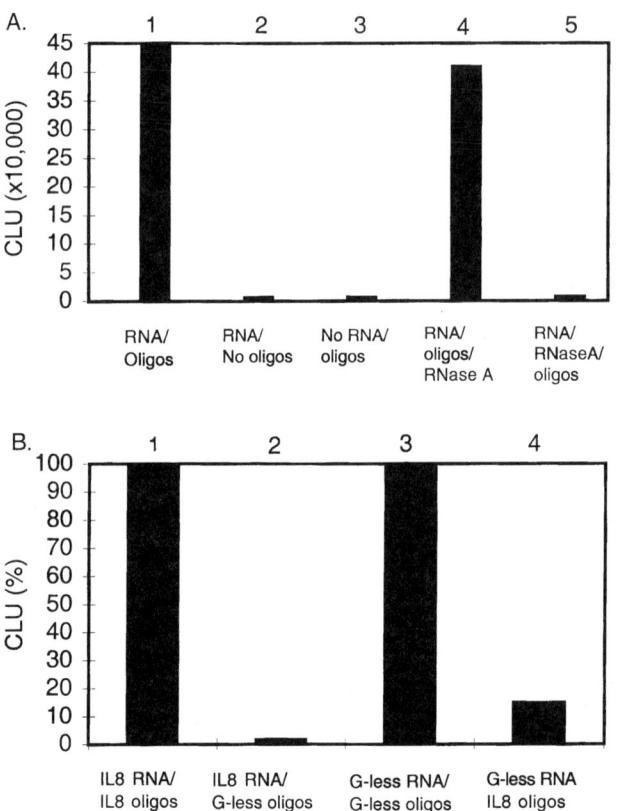

Figure 1 Dependence of specific target RNA detection on cDNA o-nts. These data represent the detection of purified IL-8 or G-less RNA under various assay conditions. (A) Column 1, both the IL-8 RNA (5 fmoles) and IL-8 cDNA o-nts (o-nts) present; column 2, IL-8 RNA (5 fmoles) alone with no cDNA o-nts present; column 3, IL-8 o-nts alone with no IL-8 RNA present; column 4, hybridization of the IL-8 cDNA o-nts to IL-8 RNA (5 fmoles) and subsequent treatment with RNase A; column 5, treatment of the IL-8 RNA (5 fmoles) with RNase A prior to hybridization with the IL-8 o-nts. (B) Column 1, both the IL-8 RNA (5 fmoles) and IL-8 cDNA o-nts present; column 2, IL-8 RNA (5 fmoles) and cDNA G-less o-nts present; column 3, G-less RNA (5 fmoles) and cDNA G-less o-nts present; column 4, G-less RNA and IL-8 cDNA o-nts present.

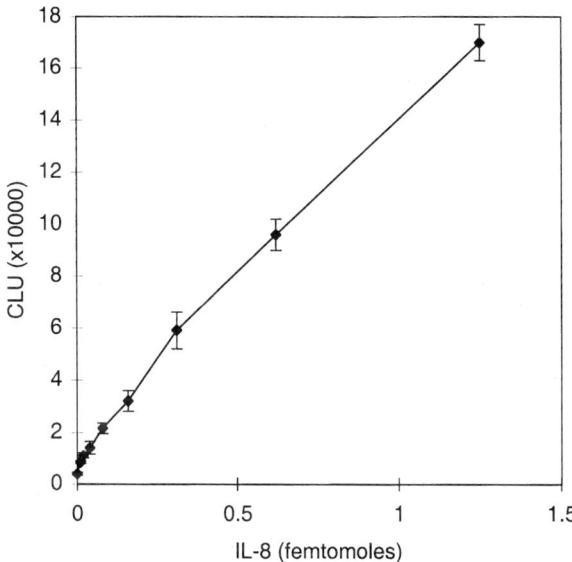

Figure 2 Dose–response of purified IL-8 RNA. Error bars represent the standard deviation of three experiments.

the samples in an incubator to automate the assay for increased throughput of the screen.

IV. IN SITU mRNA DETECTION BY ANTI-RNA:DNA ANTIBODIES

This ISH technique was applied for direct detection of mRNA expression in permeabilized cells. The assay was validated by measuring the expression of the IL-8 gene in uninduced and IL-1-induced endothelial cells. A schematic of the assay procedure is given in Figure 3. The assay takes about 4 hr to complete and is readily automatable for HTS. Figure 4, columns 1 and 2 show the signal obtained for uninduced and IL-1-induced cells, respectively. The luminescence signal in IL-1-treated cells was increased nearly threefold over that of uninduced cells. When IL-8 o-nts were omitted from the hybridization step of the assay (column 3), the IL-1-treated cells yielded a signal identical to that obtained for uninduced cells. Clearly, the IL-8 o-nts are required for detection of the IL-1-induced signal, indicating that formation of RNA:DNA hybrids is necessary for

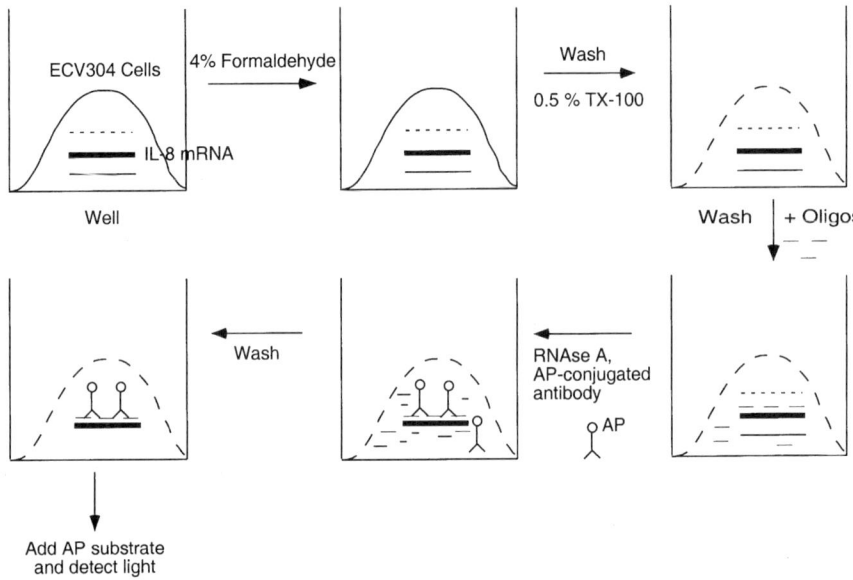

Figure 3 Schematic diagram of the HTS cell-based mRNA detection assay. Dotted line and thin solid line, mRNA besides IL-8 mRNA; thick line, IL-8 mRNA; TX-100, triton X-100; Oligos, IL-8 cDNA o-nts.

signal production. Luciferase reporter gene assays, and Northern (RNA) blot analyses [46] typically show a tenfold induction of IL-8 mRNA after treatment with IL-1. The lower level of induction observed here with the RNA:DNA duplex assay method (3-fold induction by IL-1) is most likely a result of lower sensitivity of the assay. The observation that there is no difference between the signal obtained for uninduced cells and the background signal level (no IL-8 cDNA o-nts in the hybridization step) indicates that the assay is unable to detect any IL-8 mRNA produced by uninduced cells. Hence the fold stimulation seen in the presence of IL-1 may not be a true quantitative reflection of IL-8 RNA levels following IL-1 stimulation.

The assay was validated further with respect to the need for RNA:DNA hybrid formation (Fig. 4, column 4). When the cells were treated with RNase A prior to the addition of IL-8 o-nts, the IL-1-induced signal was at background levels. RNase A presumably degraded the single-stranded IL-8 mRNA, thus preventing the formation of RNA:DNA hybrids. The assay specifically detects the desired target mRNA, as adding a control o-nt unrelated in sequence to IL-8 mRNA (Fig. 4, column 5) to the IL-1-treated cells yielded no signal.

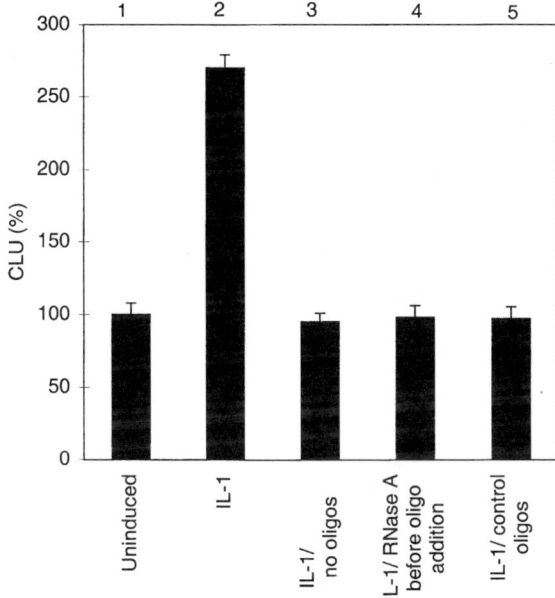

Figure 4 Detection of IL-8 mRNA in endothelial cells. Column 1, uninduced cells; column 2, cells treated with the cytokine IL-1; column 3, cells treated with IL-1, and no IL-8 cDNA o-nts; column 4, cells treated with IL-1 and subjected to RNAseA treatment prior to incubation with IL-8 cDNA o-nts; column 5, cells treated with IL-1 and incubated with control cDNA o-nts. Unless otherwise mentioned, all the experiments were performed with 30,000 cells. Error bars represent the standard deviation obtained from four experiments.

The assay was optimized for a number of parameters, such as cell permeablization conditions, hybridization conditions, assay steps, and incubation times. Figure 5A shows the result of permeablizing uninduced and IL-1-induced cells by incubation in (1) 2X SSC for 10 min, (2) 100% methanol for two min, (3) 0.5% Triton X-100 in 2X SSC for 10 min. The largest IL-1-induction was observed in cells treated with Triton X-100 (0.5%). Figure 5B shows the effect of temperature (25 to 55°C) on hybridization of IL-8 o-nts to IL-8 mRNA. Because RNA is known to contain extensive secondary structure, an increase in the hybridization temperature might be expected to enhance RNA:DNA hybrid formation and hence increase the sensitivity of the assay. However, increasing the hybridization temperature > 37°C increased hybridization of induced mRNA

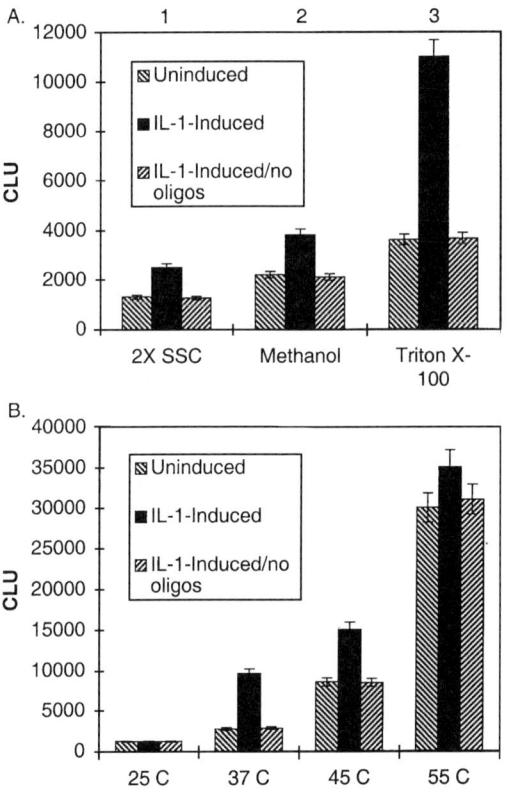

Figure 5 Optimization of cell permeablization conditions (panel A) and cDNA o-nt-mRNA hybridization temperature (panel B). IL-8 mRNA expression was monitored for uninduced cells (right inclined stripes), IL-1-induced cells (solid) and IL-1-induced cells that were not treated with IL-8 cDNA o-nts (left inclined stripes). (A) column 1, cells were incubated with 2X SSC for 10 min during the permeablization step; column 2, cells were incubated with 100% methanol for 2 min; column 3, cells were incubated with 0.5% (v/v) Triton X-100 in 2X SSC for 10 min. (B) The effect of variation of hybridization temperature on signal production.

and also increased the background signal level. The maximal fold induction by IL-1 was obtained at hybridization incubation temperature of 37°C.

An increase in the cell number yielded a nearly linear increase in the IL-1-induced signal, while the uninduced signal remained unchanged (curve A, Fig. 6) at cell numbers above 5000. Below 5000 cells in the assay the increase in IL-8 mRNA upon IL-1 induction was not detectable, and optimal signal of threefold induction of the signal was observed with IL-1 stimulation with 30,000

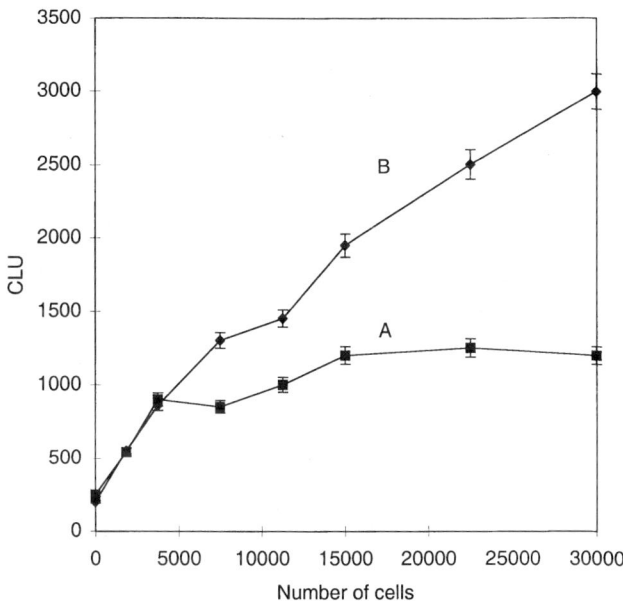

Figure 6 Dose–response obtained with increasing numbers of cells. Curve A, uninduced; curve B, IL-1 induced.

cells. Increasing the number of cells beyond 30,000 did not increase significantly the IL-1-induced signal.

Using a HTS luciferase reporter gene assay, a selective inhibitor (compound T339142) of IL-8 mRNA expression induced by IL-1 was identified. This inhibitor was further characterized by Northern blot analysis. The inhibition by T339142 of IL-1-induced IL-8 mRNA expression was compared with the luciferase reporter gene assay (Fig. 7, curve B) and the direct IL-8 mRNA detection method described in this review (Fig. 7, curve A). Both methods revealed similar inhibitory properties of the compound. This data further validates the technique described herein and illustrates that it is suitable for the identification of inhibitors of gene expression.

This assay measures directly endogenous gene transcription and therefore can be used for identifying drugs that regulate the expression of selected genes. In addition, the method eliminates some of the key disadvantages associated with the more commonly used HTS cell-based assay formats (reporter gene assays and knock-in assays). The assay is sensitive enough to measure moderate- to high-abundance mRNA.

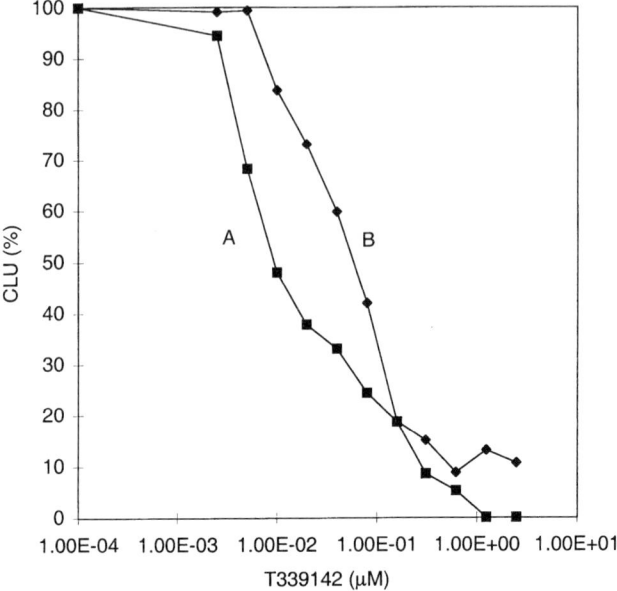

Figure 7 The effect of increasing concentrations of T339142 on IL-8 mRNA expression. IL-8 mRNA expression was monitored by the luciferase gene expression assay (curve A) and the direct anti-RNA:DNA antibody method described in this chapter (curve B). T339142 was added to the reaction mixture about 1 hr prior to IL-1 treatment. The compound was added in 10 μL aliquots containing 2% DMSO in phosphate-buffered saline to about 30,000 plated cells in 90 μL of media to yield the final desired compound concentration in 0.2% DMSO.

V. HTS ASSAY FOR DETECTION OF IN VITRO TRANSCRIPTION REACTION PRODUCT BY ANTI-RNA:DNA ANTIBODIES

This RNA:DNA duplex formation assay was further developed for a HTS assay to measure the product of a eukaryotic in vitro transcription reaction. HIV-1 basal and activated transcription was used as a test system. HIV-1 and other related lentiviruses encode an essential regulatory protein called Tat [47]. Tat is a transcriptional elongation factor that activates the expression of HIV-1 genes by binding to the transactivation-responsive region (TAR), a stem-loop structure at the 5′ end of all HIV-1 mRNA transcripts. Tat consists of a basic RNA binding region and a transcriptional activation domain that interacts with RNA polymerase II

(RNAPII) and possibly other cellular proteins. Small molecules that inhibit the activity of Tat would likely be effective drug candidates for the treatment of HIV infection. The complete assay procedure is shown schematically in Figure 8.

Figure 9 shows the signal obtained for RNA product generated by in vitro transcription from the HIV LTR promotor using crude HeLa cell nuclear extracts [44]. The sequence and number of steps in the assay were optimized to achieve maximum throughput. Figure 9A shows that the product transcript G-less RNA obtained from (1) basal transcription, (2) Tat-activated transcription, and (3) treatment with the RNAPII inhibitor α-amanitin. Generation of the signal was completely dependent on the presence of complementary DNA o-nts, indicating that RNA:DNA heteroduplex formation is necessary for detection of the RNA product. The signal obtained from the basal transcription reaction was ~ 1.5-fold greater than that obtained from a sample treated with the RNAPII inhibitor α-amanitin (defined as background). Addition of Tat to the in vitro transcription reaction resulted in a six-fold increase in signal over the basal level. When the

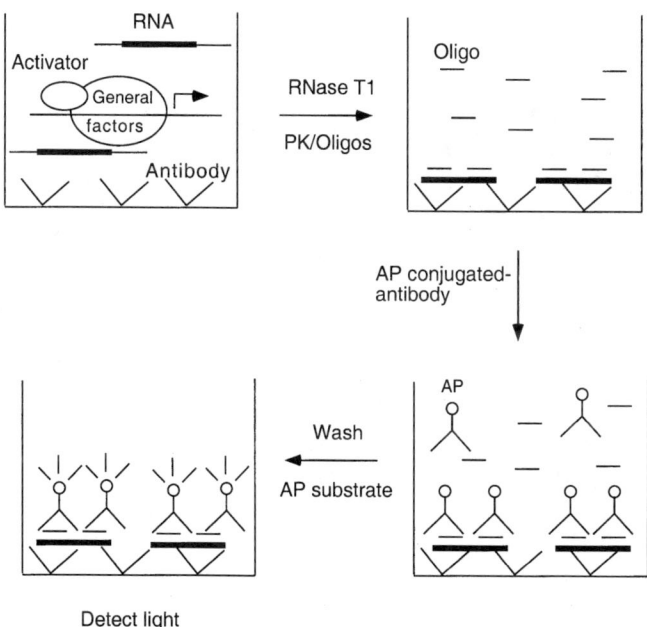

Figure 8 Schematic diagram of the HTS in vitro transcription assay. Activator, Tat; General factors, general transcription accessory proteins; Antibody, anti-RNA:DNA; PK, proteinase K; Oligos, DNA o-nts complementary to target RNA.

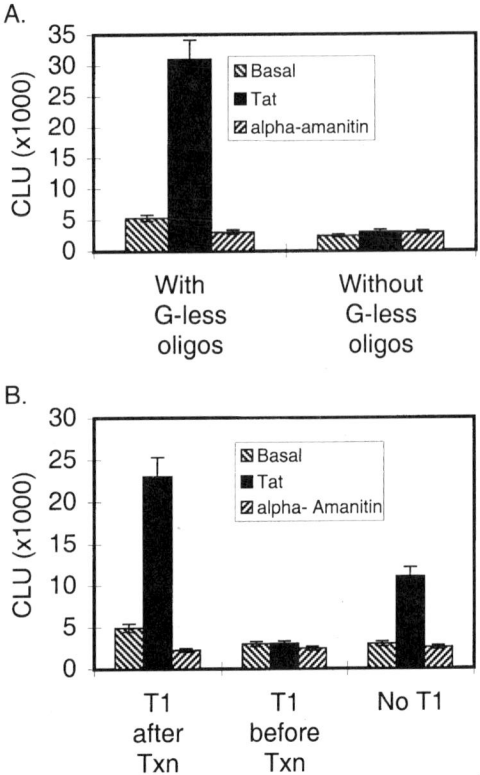

Figure 9 The figure shows data for three different transcription reactions: (1) basal (right inclined stripes), (2) Tat-activated (solid), and (3) α-amanitin treated (left-inclined stripes). (A) Dependence of the in vitro transcription signal on the addition of cDNA o-nts. (B) Dependence of the Tat-activated signal on RNase T1. First set of columns, RNase T1 was added to the reaction mixture after transcription took place; second set of columns, RNase T1 was added to the reaction mixture prior to the addition of nuclear extract and Tat; third set of columns, RNase T1 was omitted from the reaction. Error bars represent the standard deviation of three experiments.

G-less cDNA o-nts were not added to the reaction mixture, the signals obtained for both basal and Tat-activated transcription were identical to background levels (that is, the α-amanitin treated sample).

Tat activates the transcription of HIV-1 genes by binding to the TAR se-quence present at the 5′ end of nascent HIV-1 transcripts. Therefore degradation of TAR-RNA by RNAse T1 treatment should abolish transcriptional activation by Tat. RNase T1 degrades the RNA at guanosine residues. Figure 9B shows that the Tat signal is dependent on the presence of the TAR element. Addition of RNase T1 to the in vitro transcription reaction after transcription took place yielded the maximal Tat-induced signal, and addition of RNase T1 before the transcription reaction reduced the signal to basal transcription level. In the absence of RNase T1, the overall basal and Tat-activated signals were reduced. This signal reduction presumably results from the formation of secondary structure in the RNA transcript, which interferes with hybridization to the cDNA o-nts. Thus, in order to achieve the maximal signal, all assays were performed with RNase T1 treatment following the transcription reaction. Further confirmation of the dependence of the Tat signal on the TAR element was obtained by measuring the signal generated with a mutant version of the LTR promotor from which the TAR element was removed. Transcription from this mutated promotor yielded a signal equivalent to basal levels.

Tat induced transcription to maximal levels at about 100 nM, and this signal was completely abolished by α-amanitin in the reaction (Fig. 10A). The Tat concentration chosen (40 nM) for the HTS assays was in the linear region of the curve. The nucleoside analog DRB (panel B), which is a well-characterized inhibitor of Tat activation [45], inhibits the Tat signal with an IC_{50} of 1 μM (Fig. 10B). A similar IC_{50} has been obtained with a gel-based transcription assay. Maximal Tat-activated and basal signals were obtained with 150 μM of each of the four NTPs (Fig. 10C). The signal obtained in the absence of NTPs was identical to the background levels observed with α-amanitin-treated samples. All HTS assays were performed with 250 μM of each NT. The signal reached saturation when the DNA template (LTR#5) amounts greater than 0.25 μg were added to the in vitro transcription reaction (Fig. 10D). For the HTS assays, 0.75 to 1 μg of the DNA template was used.

A. Robotic Assay of RNAPII In Vitro Transcription

After optimization of the assay conditions, a HTS robotic assay designed to identify specific inhibitors of Tat function was performed. The robotic assay was performed on a Zymark robot. The assay consists of seven reagent addition steps and one wash step. The additions were performed by the Zymark pipettor arm, the Zymark RAS-RAM unit, or a Titertek multidrop from ICN (Costa Mesa, CA). The wash was performed with a 96-well microtiter plate washer from Bio-Tek Instruments (Winooski, VT). During all incubation steps, the microtiter reaction plate was placed on a shaker. About fifty 96-well plates could be assayed in less than 10 hr.

Figure 11 shows representative data obtained from a single robotic run. The signals observed in the first eight wells of the microtiter plate corresponded

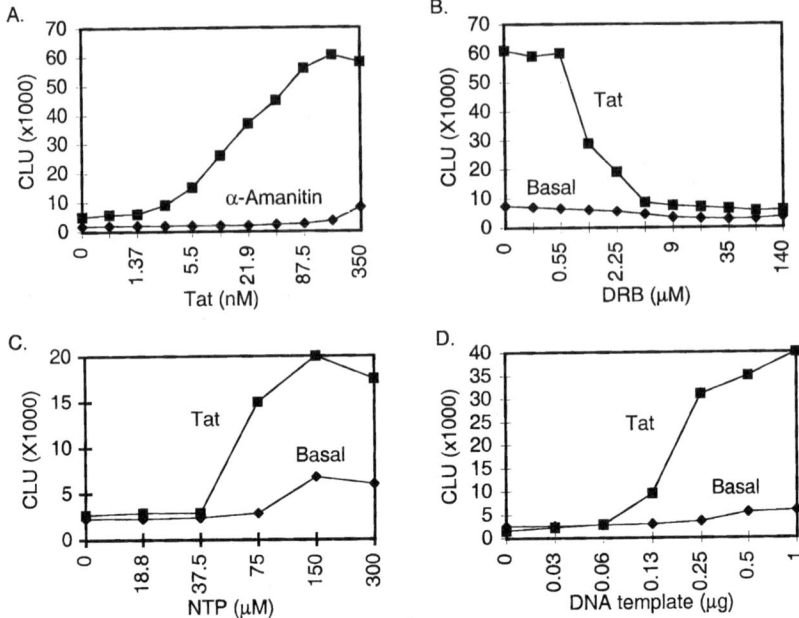

Figure 10 Effect of various assay parameters on Tat activation of HIV-1 in vitro transcription. (A) Dose–response of Tat with and without α-amanitin. (B) Inhibition by DRB of basal and Tat-activated (40 nM) transcription signal. (C) Dose–response of NTP mix (mix of ATP, CTP, GTP, and UTP) on basal and Tat-activated (20 nM) transcription. (D) Dose–response of the DNA template (LTR#5) on basal and Tat-activated (20 nM) transcription.

to basal transcription reaction. The signal observed in the last eight wells (wells 89 to 96) corresponded to α-amanitin-treated samples. The other 80 wells received all of the reagents necessary for Tat-activated transcription, along with a different test drug compound in each well at a concentration of 10 μM (dissolved in DMSO). The DMSO concentration in all the wells was 5%. Typically, a five- to tenfold window was observed between the wells with (activated) and without (basal) Tat. The standard deviation from the mean was generally less than 20% for assays containing pure chemicals. These features make the assay suitable for automated HTS. Over 200,000 chemicals were screened using this technology. The assay format described herein is a general one and is completely independent of the transcriptional apparatus. Therefore it can in principle be used to develop a HTS assay for any transcriptional system.

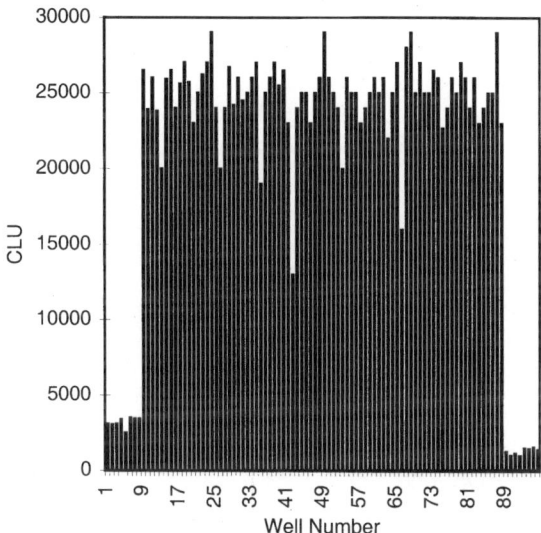

Figure 11 Robotic data obtained from in vitro transcription reactions performed in a 96-well microtiter plate containing 80 random pure chemicals. Columns 1 to 8 correspond to basal transcription reactions; columns 89 to 96 correspond to α-amanitin-treated samples. The other 80 columns correspond to wells that received all of the reagents necessary for Tat-activated transcription along with a pure chemical at a concentration of 10 μM. The DMSO concentration in all the wells was 5%.

VI. CONCLUSIONS

Recent advances in functional genomics are providing a wide variety of novel targets for drug discovery. The regulation of these novel genes at the transcriptional level is an important strategy for identifying drug lead candidates. HTS formats for transcription processes that are simple, quick, and inexpensive will likely play an increasing role in the identification of novel lead compounds. Existing RNA detection formats suffer from a number of drawbacks that make them less than ideal for HTS. These include lack of sensitivity and the use of cumbersome, slow, and expensive formats. In this review a novel RNA detection assay that is suitable for automation and HTS is described. The format has been validated for the detection of in vitro transcription reaction products in the subfemtomole region, as well as for direct measurement of mRNA in cells.

ACKNOWLEDGMENTS

I thank Kelly LaMarco for her outstanding work in editing the manuscript. I also thank Vijay Baichwal, Lalo Flores, Gary Lee, Patrick Baeuerle, Greg Peterson, and Uli Schindler for their helpful discussions and for providing a number of the reagents.

REFERENCES

1. F Antequera, A Bird. Predicting the total number of human genes. Nature Genet 7: 345–346, 1994.
2. AJ Schafer, RJ Hawkins. DNA variation and the future of human genetics. Nature Biotech 16:33–39, 1998.
3. I Vietor, LA Huber. In search of differentially expressed genes and proteins. Biochem Biophys Acta 1359:187–199, 1997.
4. KG Weinstock, EF Kirkness, NH Lee, JA Earle-Huges, CJ Venters. cDNA sequencing: a means of understanding cellular physiology. Curr Opin Biotechnol 5:599–603, 1994.
5. MP Bevilacqua, RM Nelson. Endothelial-leukocyte adhesion molecules in human disease. Ann Rev Med 45:361–378, 1994.
6. ST Holgate, J Corne, P Jardieu, RB Flick, CH Heusser. Treatment of allergic airways disease with anti-IgE. Allergy 53:83–88, 1998.
7. GJ Zhang, I Kimijima, A Tsuchiya, R Abe. The role of bcl-2 expression in breast carcinoma. Oncol Rep 5:1211–1216, 1998.
8. H Meden, W Kuhn. Overexpression of the oncogene c-errbB-2 (HER2/neu) in ovarian cancer: a new prognostic factor. Eur J Obster Reprod Biol 71:173–179, 1997.
9. L Wolff. Contribution of oncogenes and tumor suppressor genes to myeloid leukemia. Biochem Biophys Acta 1332:67–104, 1997.
10. AD Frankel, JA Young. HIV-1: fifteen proteins and an RNA. Ann Rev Biochem 16:1–25, 1998.
11. J Sambrook, EF Fritsch, T Maniatis. Molecular Cloning: A Laboratory Manual. 2d ed. Cold Spring Harbor: Cold Spring Harbor Laboratory Press, 1989.
12. RK Saiki, S Scharf, F Faloona, KB Mullis, GT Horn, HA Erlich, N Arnheim. Enzymatic amplification of beta-globin genomic sequences and restriction site analysis for diagonis of sickle cell anemia. Science 230:1350–1354, 1985.
13. AM Wang, MV Doyle, DF Mark. Quantitation of mRNA by the polymerase chain reaction. Proc Natl Acad Sci USA 86:9717–9721, 1989.
14. YS Lie, CJ Petropoulos. Advances in quantitative PCR technology: 5' nuclease assays. Curr Opin Biotechnol 9:43–48, 1998.
15. C Orlando, P Pinzani, M Pazzagli. Developments in quantitative PCR. Clin Chem Lab Med 36:255–269, 1998.
16. BA Arnold, RW Hepler, PM Keller. One-step fluorescent probe product-enhancement reverse transcriptase assay. Bio Techniques 25:98–106, 1998.
17. S Su, RG Vivier, MC Dickson, N Thomas, MK Kendrick, NM Williamson, JG

Anson, JG Houston, FF Craig. High-throughput RT-PCR analysis of multiple transcripts using a microplate RNA isolation procedure. Bio Techniques 22:1107–1113, 1997.

18. S Bortolin, TK Christopoulos. Detection of BCR-ABL transcripts from the Philadelphia translocation by hybridization in microtiter wells and time-resolved immunofluorometry. Clin Chem 41:693–699, 1995.

19. R Sooknanan, B VanGemen, LT Malek. In: DL Wiedbrauk, DH Farkas, eds. Molecular Methods for Virus Detection. San Diego: Academic Press, 1995, pp 261–285.

20. B Van Gemen, T Kievits, P Nara, HG Huisman, S Jurrians, J Goudsmit, P Lens. Qualitative and quantitative detection of HIV-1 RNA by nucleic acid sequence-based amplification. AIDS 7:107–110, 1993.

21. B Van Gemen, R Van Beuningen, A Nabbe, D Van Strijp, S Jurrians, P Lens, T Kievits. A one-tube quantitative HIV-1 RNA NASBA nucleic acid amplification assay using electrochemiluminescent (ECL) labeled probes. J Virol Methods 49:157–168, 1994.

22. F Oehlenschlager, P Schwille, M Eigen. Detection of HIV-1 RNA by nucleic acid sequence-based amplification combined with fluorescence correlation spectroscopy. Proc Natl Acad Sci USA 93:12811–12816, 1996.

23. G Leone, H Van Schijndel, B Van Gemen, FR Kramer, CD Schoen. Molecular beacon probes combined with amplification by NASBA enable homogeneous, real-time detection of RNA. Nucleic Acids Res 26:2150–2155, 1998.

24. P Wu, S Daniel-Issakani, K LaMarco, B Strulovici. An automated high-throughput filtration assay: application to polymerase inhibitor identification. Anal Biochem 245:226–230, 1997.

25. JK Ishii, SS Ghosh. Bead-based sandwich hybridization characteristics of o-nt-alkaline phosphatase conjugated and their potential for quantitating target RNA sequences. Bioconjugate Chem 4:34–41, 1992.

26. B Galvan, TK Christopoulos. Bioluminesence hybridization assays using recombinant aequorin. Application to the detection of prostrate-specific antigen mRNA. Anal Chem 68:3545–3550, 1996.

27. FS Nolte, J Boysza, C Thurmond, WS Clark, JL Lennox. Clinical comparison of an enhanced-sensitivity branched-DNA assay and reverse transcription-PCR for quantitation of human immunodeficiency virus type 1 RNA in plasma. J Clin Microbiol 36:716–720, 1998.

28. MK Kenrick, L Jiang, CL Potts, PJ Owens, DJ Shuey, JG Econome, JG Anson, EL Quinet. A homogeneous method to quantify mRNA levels: a hybridization of RNase protection and scintillation proximity assay technologies. Nucleic Acids Res 25:2947–2948, 1997.

29. T Ishiguro, J Saitoh, H Yawata, M Otsuka, T Inoue, Y Sugiura. Fluoresence detection of specific sequence of nucleic acids by oxazole yellow-linked oligonucleotides. Homogeneous quantitative monitoring of in vitro transcription. Nucleic Acids Res 24:4992–4997, 1996.

30. HS Rye, JM Dabora, MA Quesada, RA Mathies, AN Glazer. Fluorometric assay using dimeric dyes for double- and single-stranded DNA and RNA with picogram sensitivity. Anal Biochem 208:144–150, 1993.

31. NC Nelson. Rapid detection of genetic mutations using the chemiluminescent hy-

bridization protection assay (HPA): overview and comparison with other methods. Crit Rev Clin Lab Sci 35:369–414, 1998.

32. KS Dunkak, MR Otto, JM Beechem. Real-time fluorescence assay system for gene transcription: simultaneous observation of protein/DNA binding, localized DNA melting, and mRNA production. Anal Biochem 243:234–244, 1996.

33. J Alam, JL Cook. Reporter genes: application to the study of mammalian gene transcription. Anal Biochem 188:245–254, 1990.

34. M Chalfie, Y Tu, G Euskirchen, WW Ward, DC Prasher. Green fluorescent protein as a marker for gene expression. Science 263:802–805.

35. BP Cormack, RH Valdivia, S Falkow. FACS-optimized mutants of the green fluorescent protein (GFP). Gene 173:33–38, 1996.

36. RH Singer, JB Lawrence, CA Villnave. Optimization of in situ hybridization using isotopic and non-isotopic detection methods. Bio Techniques 4:230–250, 1986.

37. RW Dirks. RNA molecules lighting up under the microscope. Histochem Cell Biol 106:151–166, 1996.

38. DW Harris, MK Kenrick, RJ Pither, JG Anson, DA Jones. Development of a high-volume in situ mRNA hybridization assay for the quantification of gene expression utilizing scintillation microplates. Anal Biochem 243:249–256, 1996.

39. G Schmitz, A Dotzauer. Proof of hepatitis A virus negative-sense RNA by RNA/DNA-hybrid detection: a method for specific detection of both viral negative- and positive-strand RNA species. Nucleic Acids Res 26:5230–5232, 1998.

40. I Fliss, M St-Laurent, E Emond, RE Simard, R Lemieux, A Ettriki, S Pandian. Anti-DNA:RNA antibodies: an efficient tool for non-isotopic detection of *Listeria* species through a liquid-phase hybridization assay. Appl Microbiol Biotechnol 43:717–724, 1995.

41. DB Casebolt, CB Stephensen. Monoclonal antibody solution hybridization assay for detection of mouse hepatitis virus infection. J Clin Microbiol 30:608–661, 1992.

42. F Coutlee, RH Yolken, RP Viscidi. Nonisotopic detection of RNA in an enzyme immunoassay using a monoclonal antibody against DNA-RNA hybrids. Anal Biochem 181:153–162, 1989.

43. I Fliss, M St-Laurent, E Emond, R Lemieux, RE Simard, A Ettriki, S Pandian. Production and characterization of Anti-DNA-RNA monoclonal antibodies and their application in *listeria* detection. Applied Environ Microbiol 59:2698–2705, 1993.

44. Y Kitagawa, BD Stollar. Comparison of poly(A).poly(dT) and poly(I).poly(dC) as immunogens for the detection of antibodies to RNA-DNA hybrids. Mol Immunol 19:413–420, 1982.

45. CO Yehle, WL Patterson, SJ Boguslawski, JP Albarella, KF Yip, RJ Carrico. A solution hybridization assay for ribosomal RNA from bacteria using biotinylated DNA probes and enzyme-labeled antibody to DNA:RNA. Mol Cell Probes 1:177–193, 1987.

46. RM Strieter, SL Kunkel, HJ Showell, DG Remick, SH Phan, PA Ward, RM Marks. Endothelial cell gene expression of a neutrophil chemotactic factor by TNF-α, LPS, and IL-1β. Science 243:1467–1469, 1989.

47. KA Jones, BM Peterlin. Control of RNA initiation and elongation at the HIV-1 promotor. Ann Rev Biochem 63:717–743, 1994.

48. G Lee, J Wu, P Luu, P Ghazal, O Flores. Inhibition of the association of RNA

polymerase II with the preinitiation complex by a viral transcriptional repressor. Proc Natl Acad Sci USA 93:2570–2575, 1996.

49. JD Dignam, RM Lebovitz, RG Roeder. Accurate transcription initiation by RNA polymerase II in a soluble extract from isolated mammalian nuclei. Nucleic Acids Res 11:1475–1489, 1983.

50. HSY Mancebo, G Lee, J Flygare, J Tomassini, P Luu, Y Zhu, J Peng, C Blau, D Hazuda, D Price, O Flores. P-TEFb kinase is required for HIV Tat transcriptional activation in vivo and in vitro. Genes Dev 11:2633–2644, 1997.

12

Screening of Combinatorial Biology Libraries for Natural Products Discovery

Christopher J. Silva and Paul Brian
Cubist Pharmaceuticals, Vancouver, British Columbia, Canada

Todd Peterson
Genicon Sciences Corporation, San Diego, California

I. INTRODUCTION

The filamentous soil bacteria actinomycetes are a rich source of natural product antibiotics. The structural variety and broad range of associated biological activities of the molecules isolated from these organisms are astonishing. The diverse geographical locations from which these organisms have been isolated, exotic and not so exotic, attest to the keen interest the pharmaceutical industry has for these soil bacteria [1]. Over the past 50 years natural product antibiotics have revolutionized the practice of modern medicine so that it is difficult to imagine a world without them. Surprisingly, the actinomycetes (family: Actinomycetales) are a relatively small and closely related group of bacteria given the molecular diversity of the compounds they produce.

In the past, researchers combed the world to discover greater microbial and molecular diversity. Screening actinomycetes isolated from soil samples from Easter Island and Puerto Rico resulted in the discovery of immunosupressant rapamycin and erythromycin, respectively. A strain from Texas yielded tetracycline. Other strains that produced penicillin N, clavulonic acid, and the enediyne antitumor agent esperamycin A1 were found in South America (Fig. 1).

Over time, this traditional approach to natural products discovery became perceived as one of diminishing return due to inherent problems associated with

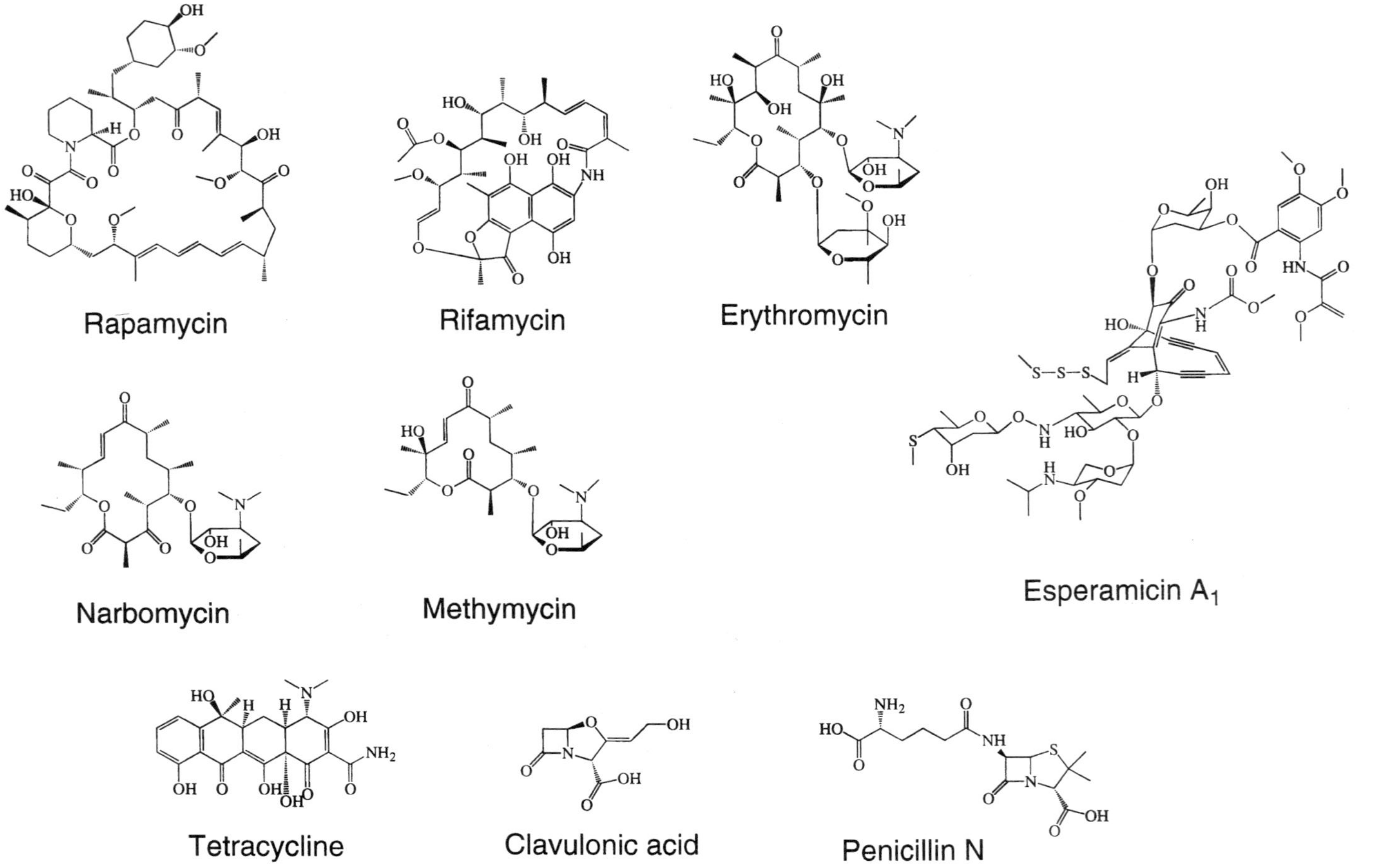

Figure 1 Structure of some actinomycete natural products.

primary environmental isolates, namely, the isolation of active compounds from complex fermentation mixtures, the arduous dereplication necessary to prevent rediscovery of known compounds, and production instability in primary environmental isolates. The emergence of high throughput screening (HTS), genomics technologies, and combinatorial chemistry allows the rapid screening of large collections of structurally diverse synthetic compounds against a variety of novel disease targets [2]. This combination seems to provide a diversity of natural products but without their inherent problems. In light of these significant changes in the pharmaceutical industry, newer approaches for natural products discovery were clearly necessary.

A hint of a potential new approach occurred as a result of the search for novel structures [3]. Actinomycetes could be used to biotransform structurally complex pharmacophores. Indeed, actinomycetes are used in a variety of important biotransformations. They have been used to functionalize carbon skeletons, transfer amino sugars, and modify other structures that they do not naturally produce [4–6]. This implies that many of the enzymes involved in natural product biosynthesis do not necessarily have absolute substrate specificity and can modify a range of structures beyond those they naturally produce. This observation suggests that it may be possible to produce structurally novel secondary metabolites by combining enzymes from different biosynthetic pathways. Thus by applying genetic methods for combining biosynthetic enzyme activities from different actinomycetes, it may be possible to do exactly this [7,8].

The intervening 50 years have shown that the vast majority of bacteria present in the environment are ''uncultivable'' [9]. By uncultivable we mean that under standard laboratory conditions these organisms will not grow and are therefore unavailable for screening. This is particularly important among the soil bacteria, where less than 1% from a given soil sample appear to be cultivable. This implies that there is the potential for a 100-fold increase in microbial and molecular diversity, if the biosynthetic potential present in uncultivable organisms could be accessed (Table 1).

The tools necessary to exploit these potential sources of diversity have been developed in the fields of molecular biology and microbial genetics. It is a matter of routine to clone and heterologously express genes. In fact, enzymes of entire pathways have been cloned and heterologously expressed. Thus, the genes from a variety of pathways can be combined, cloned, and expressed in a host organism. This is equivalent to combining biosynthetic enzymes from a variety of biosynthetic pathways and has the potential of allowing these biosynthetic enzymes to modify substrates that are not their natural ones. Additionally, genes from uncultivable organisms can be cloned and heterologously expressed, resulting, paradoxically, in the production of natural products from hosts that can not be cultivated. The application of modern genetic techniques to the field of natural products drug discovery has the potential to revolutionize the field [7,8].

Table 1 Estimated Percentage of
Cultivable Microorganisms from Selected
Environmental Samples

Habitat	Cultivability
Seawater	0.001–0.1
Freshwater	0.2
Mesotrophic lake	0.01–1
Unpolluted estuarine waters	0.1–3
Activated sludge	1–15
Sediments	0.2
Soil	0.3

Source: Ref. 9.

II. MOLECULAR BIOLOGY AND HIGH-THROUGHPUT SCREENING (HTS)

Modern molecular biology has significantly changed the practice of natural products drug discovery through the introduction of high-throughput screening (HTS). Genetic manipulation is routinely employed to provide defined molecular and cell-based assay targets. These targets can be produced cheaply and in such abundance that it is possible to engage in HTS. If an assay is dependent upon a difficult to isolate enzyme, then one need only clone and overexpress the DNA that encodes the enzyme of interest to render the once precious enzyme quite abundant. It is now a matter of routine to generate essentially unlimited amounts of desired enzymes, receptors, and other screening targets in isolated molecular form or in an engineered cellular environment for HTS.

Over the past 15 years, the field of molecular biology has dramatically advanced our genetic, molecular, and biochemical understanding of natural product biosynthesis. The approaches described in this chapter embrace and apply these advances to the field of natural products drug discovery. This new approach to natural products discovery is called combinatorial biology [7,8]. In this process, new sets of biosynthetic genes are generated by deliberately ''shuffling'' genes from distinct biosynthetic pathways. In addition, it is now possible to obtain biosynthetic genes from previously inaccessible microbial sources, such as uncultivable organisms or symbionts. These gene sets are cloned and heterologously expressed to yield libraries of recombinant organisms bearing novel sets of biosynthetic enzymes. This genetic process is called combinatorial biology. Complemented by an integrated bioassay system to identify and isolate recombinant clones efficiently, we can produce natural products with a wide range of biological activities and a high probability of structural novelty.

III. COMBINATORIAL BIOLOGY

As previously mentioned, Streptomyces and other actinomycetes produce an amazing variety of natural product structures with remarkable structural diversity and a broad array of important biological activities. Numerous actinomycete-derived natural product drugs have been successfully developed over the past 50 years and are currently in use as human and veterinary pharmaceuticals. Since the actinomycetes are a closely related group of filamentous bacteria, it is very likely that genes from actinomycetes will be expressed in heterologous actinomycete hosts. Given this and the importance of actinomycete-derived natural products in pharmaceutical discovery, our initial combinatorial biology efforts are focused on actinomycetes to capitalize upon the growing understanding of actinomycetes and to harness their tremendous biosynthetic diversity and potential.

A. Molecular Genetics and Biochemistry of Natural Product Biosynthesis

Cubist's combinatorial biology technology involves genetic manipulation of genomic DNA from microorganisms that synthesize natural products. The starting collection of microorganisms can include fully characterized microbial isolates, partially characterized or uncharacterized collections, or DNA isolated directly from marine or terrestrial environments [7]. In order to understand the essential concept of combinatorial biology, some basic molecular genetic features of natural product biosynthesis must be appreciated.

An essential general characteristic of the natural product biosynthetic pathways in bacteria that facilitates combinatorial biology is the clustering of genes that provide all genetic instructions for production of the specific metabolite (Fig. 2) [10,11]. A typical natural product biosynthetic pathway is comprised of

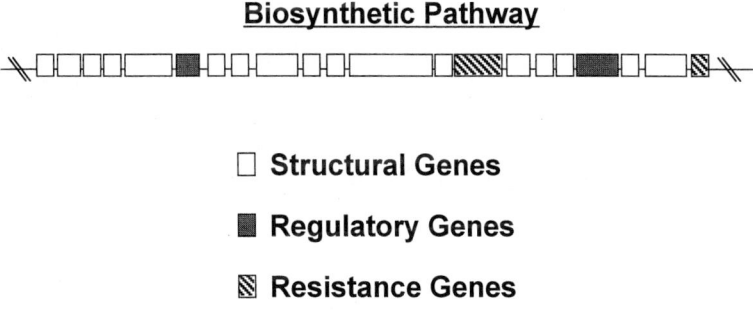

Figure 2 Natural product biosynthetic gene clusters.

5 to 100 linked genes (or more, depending on the complexity of the molecule) that together regulate expression and encode the biosynthetic enzymes responsible for natural product assembly and structural modification. In addition, natural product gene clusters include one or more genes that encode enzymes for cellular self-protection and extracellular drug transport. Typically, a complete pathway includes a cluster of genes encompassing 20–150 kilobases (kb). In actinomycetes, such as Streptomyces, the typical genome is about 8 megabases [12,13]. For combinatorial biology to successfully generate novel natural products, it must be possible to transfer efficiently rather large segments of DNA (>40 kb) from natural product-generating donor microorganisms into an appropriate engineered expression host. That is an essential aspect of Cubist's technological proficiency [7].

B. Combinatorial Biology: General Features of a New Drug Discovery Technology

Combinatorial biology is a proprietary leading edge drug discovery technology that was developed recently from advances made in molecular genetic manipulation and enzymology of natural product biosynthesis in actinomycetes, fungi, and other microbial systems [14]. This technology involves transfer of metabolic pathways from natural secondary metabolite-producing microorganisms (e.g., Streptomyces, Micromonspora, Actinomodura) to an engineered host (e.g., *S. lividans*) that allows control over timing and level of expression of natural product biosynthetic genes [7].

There are several ways that new natural products are created by combinatorial biology approaches. (1) It is now possible to transfer entire metabolic pathways from a donor strain to an engineered host using diverse molecular genetic tools (e.g., mobilizable vectors) [14]. This process allows a single pathway to be isolated and manipulated genetically to create libraries of recombinants capable of producing modified forms of a particular natural product. (2) Transfer of metabolic pathways from a natural donor strain to an engineered host can lead to expression of "silent" pathways that are not normally expressed under the culture conditions used to grow the donor strain [14]. Thus combinatorial biology can harness previously undetected metabolic pathways that result in discovery of new natural products. (3) Cubist has developed proprietary technology that involves efficient reassembly or "shuffling" of natural product biosynthetic pathway genes in a process of "combinatorial pathway" construction [7].

C. Combinatorial Biology Technology

The starting point for library construction is high molecular weight genomic DNA isolated from cultured donor strains or directly from the environment. Once genomic DNA is isolated and suitably prepared, it is used to create two basic types of libraries. The first type of library, the natural pathway library (Fig. 3), provides metabolic pathways wherein the genes that comprise the pathway have their native linear configuration. This allows production of the same or similar molecules prescribed in the original donor strain [7]. Natural pathway library clones provide rapid access to the desired biosynthetic gene cluster and an immediate strategy for construction of ''biased'' combinatorial libraries and for screening such libraries using an engineered Streptomyces expression host in conjunction with the macrodroplet screening system (Fig. 3). The second type of library involves prior enrichment for DNA that contains natural product biosynthetic genes, followed by ''shuffling'' of the genes associated with donor biosynthetic pathways in a process that mimics natural processes of genetic recombination [7]. In both library construction approaches, DNA specifying production of natural products is introduced into an expression vector and transferred through conjugation or transformation into an appropriate host that provides effective production of natural products. The recombinant microorganisms created in this way can be screened directly by encapsulation in a gel macrodroplet that contains a target organism or reporter-based assay system. The integration of combinatorial biology expression libraries and HTS relieves a significant bottleneck in the natural product drug discovery process (Figs. 3, 4).

In addition to the examples of combinatorial biology library formats described above, other approaches for the discovery of novel natural products are available. An important example involves generation of ''biased'' combinatorial libraries (Fig. 5) [8]. In this library format, a selected group of microorganisms is chosen for their specific ability to produce an important class of natural products. Through genetic manipulation, a library of recombinant microorganisms is generated that produce variously modified forms of the particular natural product chemotype. This approach is the recombinant version of enzymatic biotransformation and akin to medicinal chemistry on a known pharmacophore. However, since libraries are efficiently generated and screened in an engineered, characterized expression host, new natural products may be discovered more quickly and subjected to immediate fermentation scale-up once a promising lead has been discovered (Fig. 5).

D. Vector Systems for Expression of Natural Products

Combinatorial biology technology development programs involve construction of advanced molecular and genetic tools that provide increasingly efficient access to natural products for different donor and expression host systems. Toward this, two basic types of vector expression systems that accommodate a broad size range of DNA fragments to be cloned, mobilized, and expressed have been developed for several systems. One type of vector expression system includes shuttle

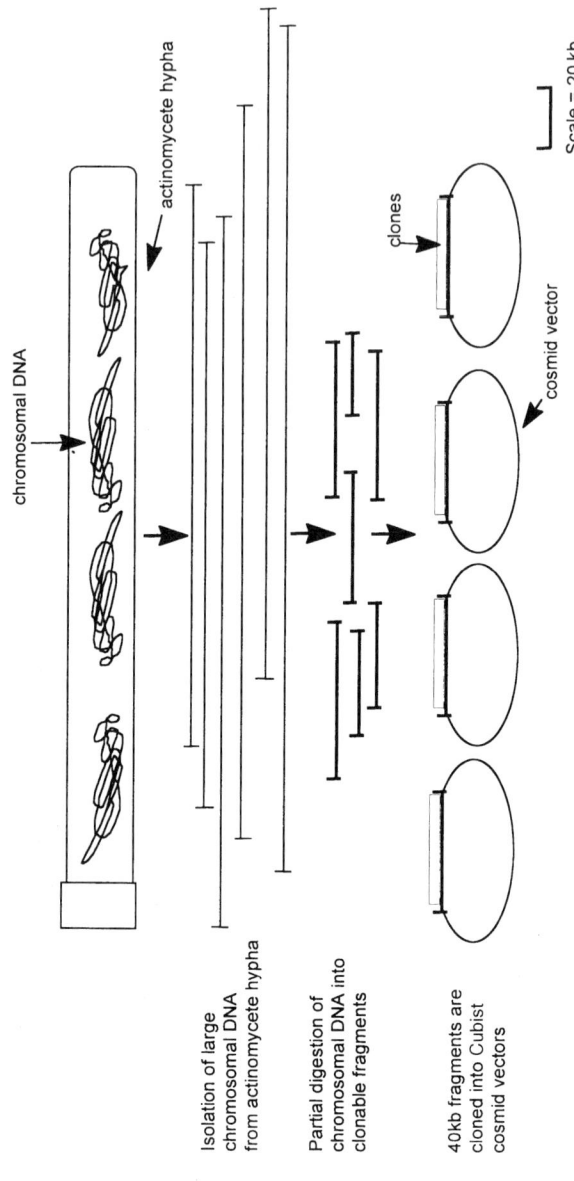

Figure 3 Natural pathway library construction.

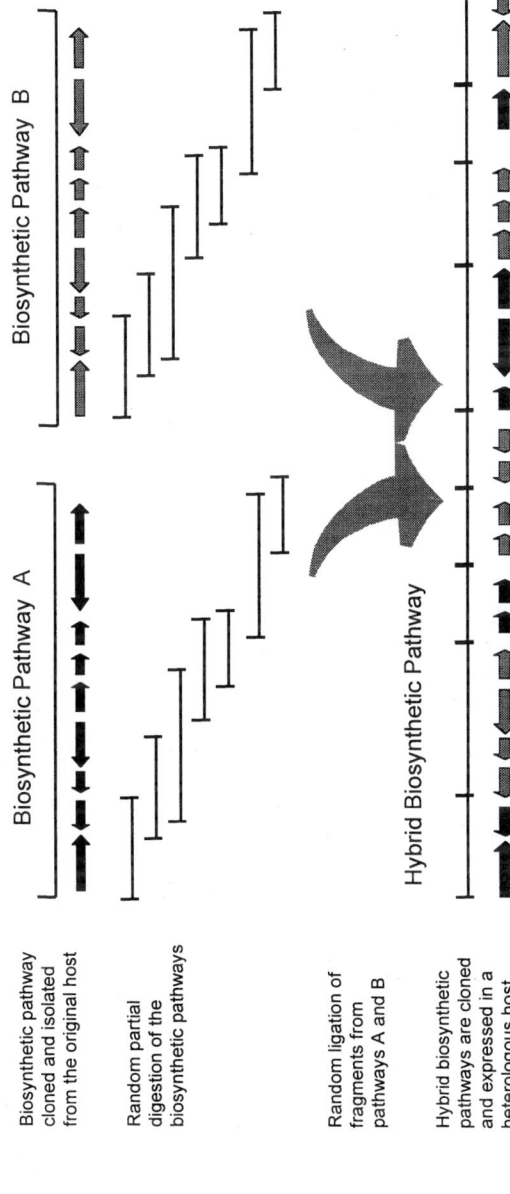

Figure 4 Combinatorial pathway library construction.

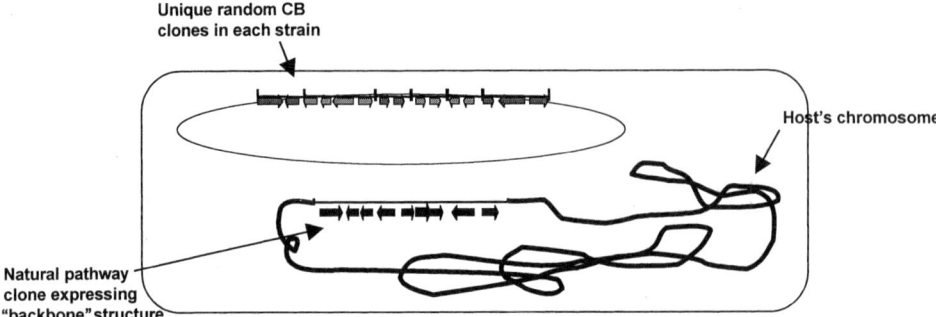

Figure 5 Combinatorial pathway biased library for lead optimization and "biological medicinal chemistry."

cosmids. These allow efficient cloning of up to 50-kb fragments of DNA from a diverse range of microbes. These vectors have the capability to be mobilized (via interspecies conjugation), from *E. coli* directly into a broad range of potential expression hosts, and many are designed for efficient integration within the host chromosome. This strategy is being pursued in order to enhance long-term genetic stability of the cloned biosynthetic pathways. This capability is particularly important when high-level production of a combinatorial biology-derived natural product is required.

A second type of vector expression system derived from bacterial artificial chromosome (BAC)-based plasmids is a key combinatorial biology genetic tool [15]. These vectors are designed for cloning of DNA fragments greter than 120 kb. This feature is important because many complex natural product biosynthetic pathways are large clusters of genes that cannot be accommodated on cosmids due to the size limitation imposed by lambda packaging. Here again, flexible features are being built in to these vectors to allow facile mobilization, gene expression, and chromosomal integration.

E. Bacterial Expression Hosts

A critical aspect in development of the genetic toolbox for combinatorial biology involves the use of engineered bacterial host systems for stable expression of donor DNA containing genes for natural product biosynthesis. Engineered host systems have been developed (or are under development) for *Streptomyces lividans*, *Streptomyces fradiae*, and *Myxococcus xanthus*. For Streptomyces, a series of advanced-engineered strains are being developed in which indigenous metabolic pathways (e.g., actinorhodin, undecylprodigiosin for *S. lividans*) have been disabled or deleted from the genome. This simplifies the potential baseline product profile and removes competing endogenous pathways that may drain metabolic flux potential away from the desired heterologous system. Further modifications to these expression hosts will include genetic manipulation of regulatory genes involved in antibiotic production to increase yields from the heterologous pathways [16–18].

IV. SCREENING NUMBERS

The combinatorial biology approaches described above differ significantly from other more directed practices for pathway manipulation. The random elements of pathway isolation and gene shuffling in combinatorial biology that allow the merger of rare and unexpected biosynthetic capabilities also, by necessity, generate a large number of clones. To make combinatorial biology a practical method of natural products drug discovery, a screening system that could screen large numbers of recombinant organisms has to be developed. First, consider the number of clones that need to be screened in order to isolate a single biosynthetic pathway in an actinomycete. This calculation is based upon the work of Clarke and Carbon and is summarized in the equation $N = ln[(1 - p)/ln(1 - (o/g)]$ where N is the number of clones necessary to guarantee, with a probability of p, that at least one of the N clones contains an intact pathway of size W cloned into a vector whose insert size is $W+$ an overhang of o base pairs [19]. The size of the genome of the source organism is g. A number of cloning vectors that can handle inserts of varying sizes have been developed. For statistical purposes $p = 0.95$ is considered to be acceptable, a typical overhang is 10–15 kb, and the sizes of some sequenced biosynthetic pathways (W) are shown in Table 2. Table 3

Table 2 Size of Sequenced Biosynthetic Pathway for Selected Actinomycete-Derived Natural Products

Pathway	Size (kB)
Rapamycin	120
Rifamycin	95
Erythromycin	60
Methymycin	55
Narbomycin	55
Tetracyline	30

Source: Refs. 20–25.

Table 3 Number of Clones Needed to be Screened to
Insure at Least One Will Possess a Complete Biosynthetic
Pathway Based Upon the Size of the Overhang and the
Confidence Level

Overhang (kb)	$p = 0.5$	$p = 0.75$	$p = 0.95$
1	5500	11,000	24,000
5	1100	2,000	4,800
10	550	1,100	2,400
15	370	740	1,600
20	280	550	1,100
25	220	440	960
50	110	220	480

summarizes these values for several probabilities and overhang sizes. Typically,
between 2,000 and 3,000 clones must be screened to isolate a single pathway
from a single actionmycete, provided the cloning vector can accommodate an
insert of size $W+$ 10–15 kb.

Early in the development and conception of combinatorial biology, it was
evident that screening would be an essential component of the technology
[7,8,26]. The preceding example quantifies this observation for a single organism.
The numbers soon become daunting, considering that a typical soil sample contains hundreds or thousands of individual bacterial species. Thus, adequately to
assay a natural pathway expression library derived from a single soil sample
involves screening at least hundreds of thousands of clones. This effort will yield
the raw genetic material in the form of expressed pathways producing biologically active natural products, which in turn may be combined and manipulated
to produce combinatorial biology libraries of greater complexity. Experience has
shown the difficulty in predicting exactly how these combined pathway genes
and their resultant biosynthetic enzymes will interact. This further emphasizes
the need for efficient screening systems.

V. SCREENING OF NATURAL PRODUCTS

The potential of combinatorial biology was evaluated in a pharmaceutical discovery environment [27,28] with 34 actinomycete strains. The natural products produced by each of these donor strains under defined culture conditions were well
characterized with respect to their chemical properties and biological activities.

DNA from each donor species was isolated, pooled, cloned, and transferred into the expression host *S. lividans*. Both natural (*vide supra*) and combinatorial (*vide supra*) pathway expression clones were constructed for the project. A total of 10,400 isolated combinatorial biology expression clones representing both natural pathway (8,200 clones) and combinatorial pathway (2,200 clones) formats were generated and screened using a traditional approach.

Each clone was screened independently and individually cultivated on a small scale for approximately two weeks, the culture medium was extracted with butanol, butanol solution was removed, and the resulting residue was redissolved in a small volume of DMSO. The resulting DMSO solution was screened for activity in antimicrobial plate assays against a panel of eight target microorganisms. Of the 8,200 natural pathway clones screened 205 (2.5%) clones produced metabolite(s) that had significant biological activity when compared to control cultures of *S. lividans* harboring cloning vector without donor inserts. Initial dereplication analysis of an early subset of natural pathway clones indicated the presence of at least 5 known natural products produced by several donor strains. This early result supported the original project premise for facile heterologous expression of biosynthetic pathways in these closely related microorganisms.

For the combinatorial pathway format set of clones, 71 (3.2%) of 2,200 clones produced biologically active compound(s) relative to controls. Many of the active clones from both the natural and the combinatorial pathway sets of clones produced potentially novel natural products based upon their biological activity profiles and chemical properties. Twenty-one natural pathway clones and 19 combinatorial pathway clones were selected for dereplication and chemical analysis based upon specific criteria. The results from dereplication and structural elucidation included two novel compounds, BMS-246784 and BMS-240777, a known but unusual metabolite (BMS-246781), and two known but unusual prodigiosins.

The effort to screen the combinatorial biology clones for this early project required considerable resources. The 10,400 250-mL flasks needed for the initial culture of this number of clones fill a space of 10 cubic meters. Furthermore, more than 1000 liters of medium and 500 liters of butanol were required for the primary fermentations and extractions. Butanol solvent had to be removed from each of the 10,400 from each extraction. Considerable additional resources and effort were required for scale-up and analysis of selected active clones.

Over the years, pharmaceutical companies have developed an encompassing traditional process to screen fermentation broths [28]. Of necessity, this process reflects an expected diversity of growth and production requirements learned from screening soil isolates composed of organisms that have distinct inherently unpredictable biological properties. The natural products screening approach used reflected this historical experience. Combinatorial biology library clones do not share these characteristics, as each clone is genetically identical except for the relatively small amount of donor DNA that has been introduced. This important property of these approaches can be taken advantage of to develop high throughput for screening combinatorial biology library clones. At Cubist, an encapsulation approach to primary biological activity screening was developed to exploit the uniform growth properties and temporal production characteristics of clones and facilitate the natural products drug discovery process. In this HTS system recombinant expression library clones are encapsulated to generate individual assay units that serve as culture flask, medium, and extract on a millimeter scale.

VI. HTS USING ALGINATE MACRODROPLETS

Sodium alginate is the water-soluble salt of alginic acid, a copolymer of 1–4 linked α-D-mannuronic and β-L-guluronic acids, which are extracted from various seaweeds. Sodium alginate solutions have the desirable property of gelling upon contact with divalent cations, such as Ca^{2+} [26,29]. Ca-alginate has been used as a convenient means of encapsulating cells and enzymes. The physical properties, such as gel strength and porosity, can be controlled by appropriate formulation of the starting alginate solution and the choice and concentration of Ca^{2+}.

Ca-alginate gel spheres were used in the screening program. The spheres, called macrodroplets, are formulated by adding sodium alginate to a solution containing growth medium and then adding this formulation dropwise to a solution containing Ca ions. The slightly viscous Na-alginate/medium solution becomes a solid gel upon contact with Ca^{2+} (Fig. 6). The macrodroplets are quite firm, yet they are permeable to atmospheric oxygen, small molecules, and proteins according to the pore size of the gel. Permeability is primarily determined by the alginate concentration. In practice, a 1% Na-alginate solution is permeable to all essential medium components.

In order to encapsulate actinomycete cells, actinomycete cells are added in the form of spores, protoplasts (cells enzymatically treated to remove cell walls), or mycelial fragments to an alginate solution containing growth medium [30]. The alginate solution and cells are thoroughly mixed to ensure a uniform cell suspension. This cell suspension is placed in a reservoir and extruded through an appropriately sized orifice. The cellular suspension leaves the orifice in the shape of a teardrop and becomes spherical as it falls. Upon contact with the solution of calcium ions, the suspension instantly gels to form a rigid sphere that gently encapsulates the cells and retains their viability.

Sodium alginate
solution with cells

Droplet size determined
by orifice size

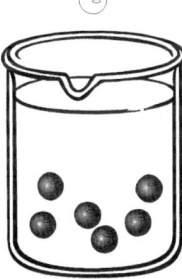

Calcium chloride solution with
gelled spherical macrodroplets
containing cells

Figure 6   Formation of macrodroplets.

A.   Development of Macrodroplet Bioassays

A wide variety of cells, among them mammalian, fungal, bacterial, and actinomycete, were encapsulated. All remain viable after encapsulation, so that this approach is useful for a wide range of bioassay applications. The precise formulation, diameter, and configuration of macrodroplets vary and are dependent upon
the desired bioassay. The number of cells encapsulated within an individual macrodroplet is determined by the initial density of the cell suspension and the
volume of the generated macrodroplet. Though various size macrodroplets can
be generated, those in the range of 2 to 4 mm in diameter are found to be the
most useful. After sieving to remove them from the $CaCl_2$ solution, the macrodroplets are placed on a grid in a plastic tray to separate them from one another.
The tray is placed in a humid incubator at an appropriate temperature for an
empirically determined length of time (Fig. 7). For actinomycete combinatorial
biology library clones, a 10 to 14 day incubation period at 30°C is typically
used. During this time, the actinomycetes grow into fully developed colonies and
produce natural products. Importantly, natural products produced by the encapsulated colonies accumulate and are retained within the matrix of the macrodroplet.

Figure 7 Time course of growth for *Streptomyces lividans* in macrodroplets (diameter 1.5 mm) from initial encapsulation to 1 week (right to left, bottom to top).

The macrodroplet screening system is a remarkably flexible assay format, and three specific formats were tested. The plate assay format involves placing a number of macrodroplets, which encapsulate 1 or 2 actinomycete clones, on a plate. The clones are allowed to ferment and grow for a period of 7 to 14 days. After the fermentation period, the entire plate is overlaid with soft nutrient agar that contains the desired assay organism (Fig. 8). In the double encapsulation format, macrodroplets containing actinomycete clones are fermented in the same way as the plate assay (Fig. 8). Once the fermentation period has passed, the macrodroplets are encapsulated in a second layer of Ca-alginate that contains the desired assay organism. In this way the assay lawn is wrapped around the macrodroplet. The third format is coencapsulation, where we take the actinomycete clone and encapsulate it in a macrodroplet with the assay organism (Fig. 8). In this way the macrodroplet contains the assay lawn. Examples of each of these formats will be discussed below.

B. Uses of the Plate Assay Format

Recently we isolated the heterologously expressed genes for the tetracycline pathway [31,32]. Genomic DNA was prepared from *Streptomyces avellaneus*, an acti-

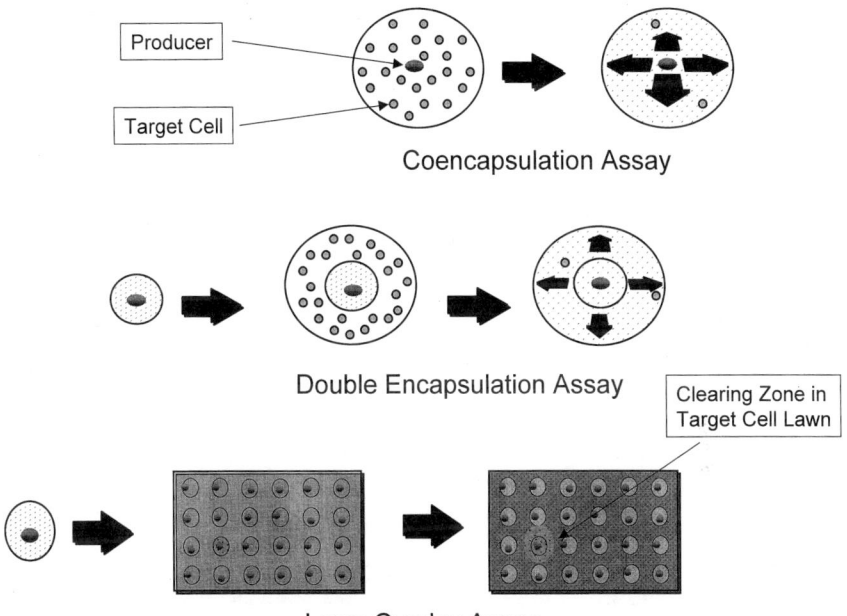

Figure 8 Three macrodroplet assay formats.

nomycete known to produce tetracycline, to generate a cosmid library. This natural pathway library was transferred to *S. lividans*, and recombinant spores were encapsulated for activity screening. The size of the tetracycline pathway is about 30 kb, and the size of the insert for vector is 40–45 kb, resulting in one producing clone per 2,000 clones screened. Approximately 12,000 clones were screened using the macrodroplets in a plate assay format. Each macrodroplet contained 1 or 2 clones, and about 10,000 macrodroplets were screened. About 150–200 macrodroplets were placed on each plate (9 × 13 × 2 cm), and about 60 of these plates were used. Each plate was incubated at 30°C for 10 days in a humid atmosphere (to prevent the macrodroplets from drying out). After this period of fermentation, the plates of macrodroplets were overlaid with a soft nutrient agar containing a strain of *E. coli*. After an overnight incubation at 37°C, the plates were examined for the presence of clearing zones, which indicate the presence of a clone, which produces tetracycline (Fig. 9). The plates contained 12 recombinant clones that produced clearing zones. Six were selected for further analysis, which confirmed the heterologous production of tetracycline. The isolation of these six clones demonstrated the power of the macrodroplet approach to screening. The entire screening of 10,000 macrodroplets occupied a portion of a shelf

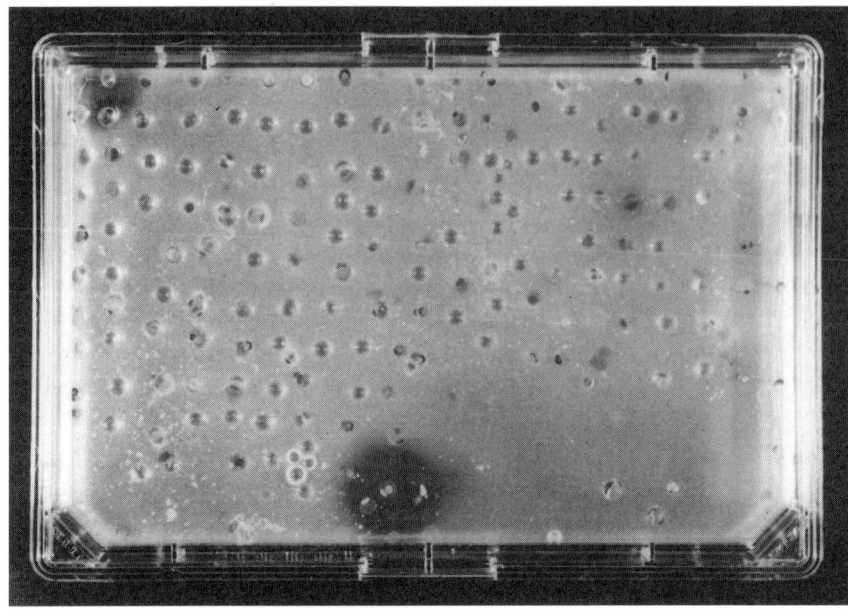

Figure 9 Assay plate showing clearing zone (center bottom) due to an encapsulated clone producing tetracycline on a lawn of *E. coli*.

of a standard laboratory incubator, in contrast to the effort to screen the 10,400 clones from our previous combinatorial biology experiment [27,28].

Alginate encapsulation could also be used to encapsulate photocleavable resins [33]. A resin bound penicillin V was prepared, which upon illumination of light at 365 nm will be cleaved. This wavelength of blue light does not interfere with normal growth of *E. coli*. The resin bound penicillin V in an alginate matrix was encapsulated, and the macrodroplets were placed on lawns of *E. coli* prepared in soft nutrient agar. Without illumination the lawns grew luxuriantly around the macrodroplets. With illumination the lawns grew luxuriantly on the plates but showed zones of clearing around the macrodroplets. The zones increased in size as the time of illumination increased. Since the light source was directed from above, approximately half of the resin remained unexposed to the light and retained the bound antibiotics. These macrodroplets were dissolved with sodium citrate and the resin was recovered. The resin was washed and exposed to blue light at 365 nm, which resulted in the liberation of penicillin V as measured by UV absorption and verified by mass spectrometry.

C. Double Encapsulation

Already formed macrodroplets can be encapsulated in a second, outer layer of alginate. In this way a macrodroplet containing a developed actinomycete colony can be coated with an outer layer that contains an assay organism such as *E. coli* (Fig. 10). If the actinomycete produces an antibiotic, that kills *E. coli*, then it will diffuse into this added outer layer and kill the *E. coli*. In this case, the outer layer will remain clear. If the actinomycete does not produce an antibiotic, then the *E. coli* in the outer layer will grow and the outer layer will be cloudy. This approach is a further miniaturization of the lawn assay (Fig. 10).

An advantage of this approach is that it allows organisms with very different growth rates to be assayed in the same macrodroplet. Double encapsulation was used to make macrodroplets containing actinomycetes in the core and *E. coli*, fungi, or mammalian cells in the second layer of alginate. Those actinomycetes in the core of a doubly encapsulated macrodroplet that produced antifungal compounds killed the fungi in the outer layer and it remained clear. In doubly encapsulated macrodroplets with an actinomycete in the core that produced antibiotic compounds, the *E. coli* in the outer layer was killed and it remained clear. Those actinomycetes in the core of a doubly encapsulated macrodroplet that produced no antibiotics failed to kill fungi or *E. coli* and their outer layer was filled with fungal or *E. coli* colonies. In principle, enzymes could also be included in the outer layer as well. The diffusion of the antibiotic out of the macrodroplet during the application of the second layer may limit the usefulness of this approach. Preliminary work measuring the diffusion of tetracycline from macrodroplets en-

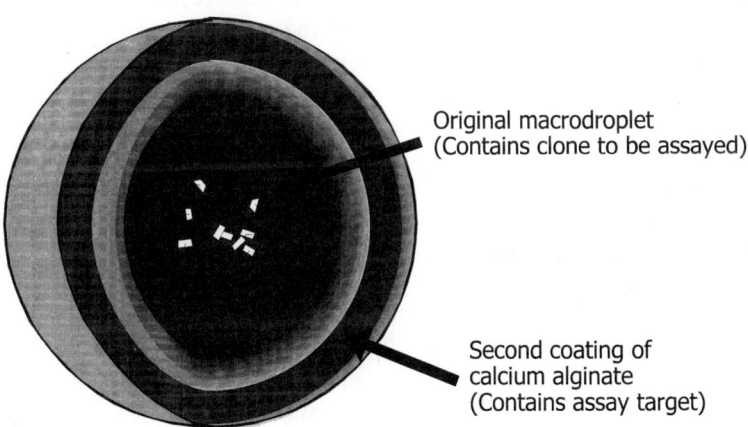

Figure 10 Doubly encapsulated macrodroplet.

capsulating tetracycline producing clones suggests that the amount of antibiotic lost in the time it takes to doubly encapsulate 10,000 macrodroplets is not significant enough to affect the assay.

D. Coencapsulation: The Screening of Chemical Libraries

Based on these results, another, less potent, antibiotic naladixic acid was bound to the same resin and tested on *E. coli* lawns as described earlier. Again, the lawns grew luxuriantly around the macrodroplets. Upon illumination, the lawns grew luxuriantly on the plates and around the macrodroplets. Analysis of a sample of the unilluminated naladixic acid bound resin after exposure to blue light at 365 nm showed that naladixic acid was cleaved from the resin in relation to the length of exposure to the light. Clearly, the concentration of naladixic acid was insufficient to prevent the growth on the plates.

Instead of encapsulating the antibiotic bound resin in alginate and then placing in on an assay lawn compared with coencapsulation them in the same alginate matrix and illuminating the resulting macrodroplet. A suspension of resin bound naladixic acid was prepared in 1% Na-alginate in LB medium, *E. coli*, and an indicator that turns blue (X-gal) in the presence of live *E. coli* and galactose. This suspension was added dropwise into a solution of 135 mm $CaCl_2$ in LB broth to yield a number of calcium alginate macrodroplets which coencapsulated *E. coli* and resin bound naladixic acid. The macrodroplets were divided into six groups. Each group was placed on a plastic petri plate and illuminated by light at 365 nm for varying lengths of time. Those macrodroplets illuminated for less than 10 min turned blue, which indicated the presence of live bacteria. Those macrodroplets illuminated for more than 15 minutes remained clear, indicating that there were no live bacteria in the macrodroplet. Encapsulated *E. coli* can be illuminated by this light source for 45 minutes without affecting the growth of *E. coli*. This work illustrates the power of the macrodroplet as a "closed" assay vessel.

The possibility of coencapsulating an actinomycete and a target organism in the same macrodroplet was investigated. In this way the macrodroplet replaces flask, medium, extraction solvent, and assay lawn. Actinomycete colonies and *E. coli* were successfully coencapsulated in the same macrodroplet. These macrodroplets were incubated over night at 37°C and examined. Those macrodroplets containing developed colonies that came from actinomycetes that produced antibiotics resulted in macrodroplets that had no *E. coli* colonies. Those macrodroplets containing developed colonies of actinomycetes that produced no antibiotics were filled with *E. coli* colonies. We did analogous experiments with fungi and obtained the same results. Unfortunately, actinomycetes are slow growing and typically take 3 or 4 days to begin producing antibiotics, and *E. coli* colonies are fully developed in the macrodroplets long before the actinomycete begins to produce the antibiotics. This difference in growth rates makes coencapsulation less attractive for a simple killing assay.

VII. ENCAPSULATING MACROORGANISMS

In addition to a number of microorganisms, we have encapsulated several macro-organisms. The eggs of the commercially important insect pests *Plutella xylostella* and *Helicoverpa zea* were successfully encapsulated. The sterilized eggs were encapsulated in an alginate macrodroplet and were placed in an incubator (Fig. 11). A similar number of unsterilized and unencapsulated eggs were placed in the same incubator. Both groups of eggs hatched in similar numbers. This is a clear demonstration of the potential of the macrodroplet to encapsulate macro as well as microorganisms.

The insect eggs can be used in the coencapsulation assay format. It is not difficult to imagine screening a combinatorial chemistry library, synthesized on

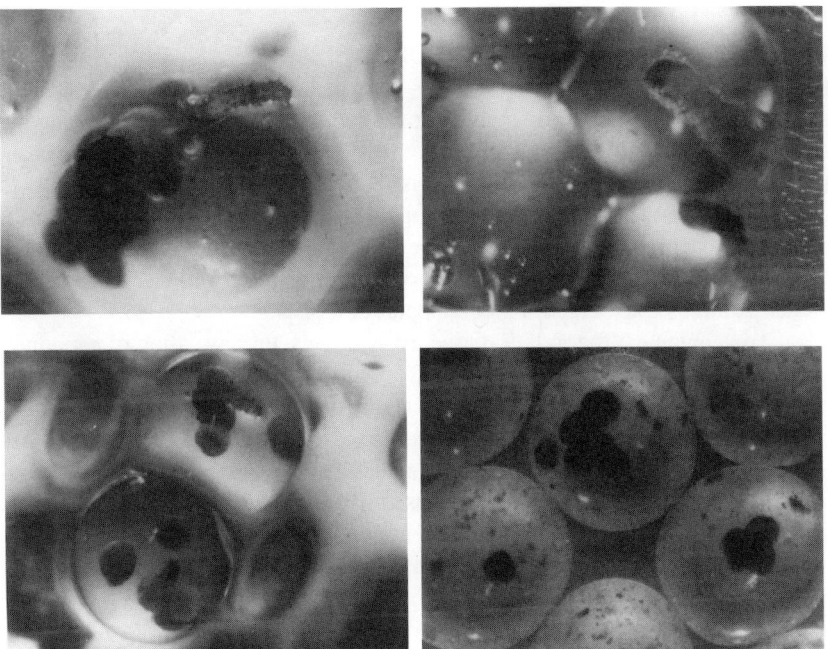

Figure 11 Growth of encapsulated *Helicoverpa zea* from eggs to larvae (right to left, bottom to top).

the photocleavable resin, in the macrodroplets for compounds that are larvicidal, inhibit hatching, or inhibit feeding. When bacteria and larvae were coencapsulated in the same macrodroplet, both grew. The small molecule source in this case is a clone, rather than a member of a chemical library. Alternately, the clone can be part of an engineered library designed to produce proteins such as variants of BT toxin. In this case the screen would be to look for larvae affected by consumed bacteria, for those bacteria would be producing biologically active protein. About 100,000 of these macrodroplets would fill only two shelves of an incubator. This same approach can be applied to other macroorganisms in the form of seeds or spores in an analogous manner.

Insect eggs in a plate assay format were explored by suspending the sterilized eggs in soft agar and overlaying this egg-containing lawn on a petri plate. Larvae hatched at the same frequency on the lawn as those not grown on the lawn. Now macrodroplets with actinomycete clones can be placed and one can assay these clones for their ability to interfere with the hatching or life cycle of an insect larva. The same could be done with a library of synthetic compounds bound to a photocleavable resin. In principle this approach can be extended to lawns of plant seeds.

VIII.　ENCAPSULATING OTHER ORGANISMS

The three macrodroplet formats are shown to work effectively with actinomycetes. Application of this approach to other ''biosynthetically talented'' organisms such as fungi and myxobacteria is explored. Both are useful sources of chemical diversity and much is known about their genetics. We have done some preliminary combinatorial biology on both types of organisms. Since both types are used to produce natural products through fermentation, the macrodroplet approach is adopted to screen the combinatorial clones.

To supplement the *E. coli*–based assay targets, encapsulation of actinomycetes has been used in the plate assay format to produce zones of clearing on lawns of fungal spores. We have used lawns of other prokaryotes in the plate format including *Staphylococcus aureus*, various engineered *E. coli* strains, *Enterococcus faecalis*, *Sarcina aurantica*, and *Bacillus subtilis*. In addition, engineered mammalian cell lines were also used in the plate assay format in this way. The macrodroplets can be used to screen clones or chemical libraries against a range of lawns from enzymes to insects and almost anything in between.

IX.　FUTURE DEVELOPMENTS IN SCREENING

Given the capacity for modern molecular genetic tools and methods to generate recombinant DNA libraries, it was appreciated at an early stage that the efficient practice of combinatorial biology would require the screening of large numbers of clones. For example, a single genomic library from a single donor strain may contain greater than 10^5 individual clones. In view of such sample numbers, cost, miniaturization, and throughput become critical factors for screening technology development. Having demonstrated the macrodroplet approach for a variety of microbial target and producer systems, consideration of additional uses of encapsulated organisms in screening is of interest. From a conceptual and practical standpoint, automation quickly becomes an important element for generating macrodroplets and scoring activities. Additional applications include screening of chemical libraries in alginate matrices and encapsulating macroorganisms such as insect pest larvae.

Large numbers of macrodroplets of near uniform size and shape are routinely made. In manipulating these macrodroplets, a number of labor-saving devices are constantly being devised, and we explore the role of automation in this area. Macrodroplets are essentially storage spheres for any secondary metabolites that the encapsulated organism should produce. Currently, each clone is screened against a single target and extended to screen each against a number of targets by the use of robotics to move plates of macrodroplets on and off different lawns simultaneously. In this way a set of macrodroplets (between 96 and 384) is lifted from the incubation plate, placed on an assay lawn, and allowed to stand for between 5 and 15 minutes to allow a portion of any secondary metabolites to diffuse out. This process can be repeated a number of times but is limited by the amount of secondary metabolite present in the macrodroplet. In practice this can be done for three lawns at most.

Three parameters are selected to optimize in order to obtain maximal secondary metabolite production. The simplest is to make larger macrodroplets and determine if they hold more secondary metabolites. Then the fermentation medium is optimized. Lastly, optimization of the host's ability to produce secondary metabolites is explored. Ideally, we would like to use each macrodroplet in eight assays. In this way, each clone could be assayed with a number of targets, and we could use the results to aid in dereplication. This work is in progress.

X. CONCLUSION

Cubist recognizes the essential role of screening in its practice of combinatorial biology. Using encapsulated clones, it was shown that the process of preliminary screening could be miniaturized dramatically. The bulk of the screening efforts were focused in the more challenging and rewarding arena of dereplication and structure elucidation. The near genetic uniformity of clones and the consequent near uniform chemical background were exploited to simplify dereplication. Simplification of the process of exploring natural products diversity was achieved by streamlining and miniaturizing the initial screening and simplifying the dereplication process through the use of clones. These methods complement the natural products drug discovery process.

ACKNOWLEDGMENTS

We wish to acknowledge Tina Legler for her kind assistance in preparing some of the figures in this manuscript. We also acknowledge the members of ChromaXome, Nicole Nasby, Alex Cantafio, Mary Sorensen, Angela Hansen, Pattie Evans, Howard Xu, John Zhu, Nina Aronson, Ke Li, and Heather Elbert.

REFERENCES

1. American type culture collection (ATCC). Catalogs. CD ROM. Rockville, Maryland: ATCC and Folio Infobase, 1997.
2. CR Hutchinson. Drugs synthesized by genetically engineered microorganisms. Bio/ Technology 12:375–380, 1994.
3. A Nakagawa, S Omura. Biosynthesis of bioactive microbial metabolites and its application to the structural studies and production of hybrid compounds. J Antibiotics 49:717–741, 1996.
4. I Maezawa, A Kinumaki, M Suzuki. Biological glycosylation of macrolide aglycones. I. Isolation and characterization of 5-O-mycaminosyl narbonolide and 9-hihydro-5-mycaminosyl narbonolide. J Antibiotics 29:1203–1208, 1976.
5. I Maezawa, A Kinumaki, M Suzuki. Biological glycosylation of macrolide aglycones. II. Isolation and characterization of desosaminyl-platenolide I. J Antibiotics 31:309–318, 1978.
6. S Omura, N Sadakane, Y Tanaka, H Matsubara. Chimeramycins: new macrolide antibiotics produced by hybrid biosynthesis. J Antibiotics 36:927–930, 1983.
7. KA Thompson, LM Foster, TC Peterson, NM Nasby, P Brian. Methods for generating and screening novel metabolic pathways. U.S. patent 5,824,485.
8. TC Peterson, LM Foster, P Brian. Methods for generating and screening novel metabolic pathways. U.S. patent 5,783,431.
9. RI Amann, W Ludwig, K-H Schleifer. Phylogenetic identification and in situ detection of individual microbial cells without cultivation. Microb Rev 59:143–169, 1995.
10. BA Rudd, DA Hopwood. Genetics of actinorhodin biosynthesis by Streptomyces coelicolor A3(2). J Gen Microbiol 114:35–43, 1979.
11. BA Rudd, DA Hopwood. A pigmented mycelial antibiotic in Streptomyces coelicolor: control by a chromosomal gene cluster. J Gen Microbiol 119:333–340, 1980.
12. HM Kieser, T Kieser, DA Hopwood. A combined genetic and physical map of the Streptomyces coelicolor A3(2) chromosome. J Bacteriol 174:5496–5507, 1992.
13. M Redenbach, HM Kieser, D Denapaite, A Eichner, J Cullum, H Kinashi, DA Hopwood. A set of ordered cosmids and a detailed genetic and physical map for the 8 Mb Streptomyces coelicolor A3(2) chromosome. Mol Microbiol 21:77–96, 1996.

14. RH Baltz. Genetic manipulation of antibiotic-producing Streptomyces. Trends in Microbiol 6:76–83, 1998.

15. H Shizuya, B Birren, UJ Kim, V Mancino, T Slepak, Y Tachiiri, M Simon. Cloning and stable maintenance of 300-kilobase-pair fragments of human DNA in *Escherichia coli* using an F-factor-based vector. Proc Natl Acad Sci USA 89:8794–8797, 1992.

16. RH Baltz, TJ Hosted. Molecular genetic methods for improving secondary-metabolite production in actinomycetes. Trends Biotechnol 14:245–250, 1996.

17. WC Champness, KF Chater. Regulation and integration of antibiotic production and morphological differentiation in Streptomyces spp. In: P Piggot, CP Moran, P Youngman, eds. Regulation of Bacterial Differentiation. Washington DC: American Society for Microbiology, 1994, pp 61–93.

18. KF Chater, MJ Bibb. Regulation of bacterial antibiotic production, In: H Kleinkauf, H von Doren, eds. Products of Secondary Metabolism. VCH Weinheim, Germany: Bio/Technology, Vol 6, 1997, pp 57–105.

19. L Clarke, J Carbon. A colony bank containing synthetic Col E1 hybrid plasmids representative of the entire *E. coli* genome. Cell 9:91–99, 1976.

20. I Molnar, JF Aparicio, SF Haydock, LE Khaw, T Schwecke, A Konig, J Staunton, PF Leadlay. Organisation of the biosynthetic gene cluster for rapamycin in *Streptomyces hygroscopicus*: analysis of genes flanking the polyketide synthase. Gene 169:1–7, 1996.

21. JF Aparicio, I Molnar, T Schwecke, A Konig, SF Haydock, LE Khaw, J Staunton, PF Leadlay. Organization of the biosynthetic gene cluster for rapamycin in *Streptomyces hygroscopicus*: analysis of the enzymatic domains in the modular polyketide synthase. Gene 169:9–16, 1996.

22. PR August, L Tang, YJ Yoon, S Ning, R Muller, TW Yu, M Taylor, D Hoffmann, CG Kim, X Zhang, CR Hutchinson, HG Floss. Biosynthesis of the ansamycin antibiotic rifamycin: deductions from the molecular analysis of the rif biosynthetic gene cluster of *Amycolatopsis mediterranei* S699. Chem Biol 5:69–79, 1998.

23. S Donadio, D Stassi, JB McAlpine, MJ Staver, PJ Sheldon, M Jackson, SJ Swanson, E Wendt-Pienkowski, Y-G Wang, B Jarvis, CR Hutchinson, L Katz. Recent developments in the genetics of erythromycin formation. In: RH Baltz, GD Hegeman, PL Skatrud, eds. Industrial Microorganisms: Basic and Applied Molecular Genetics. Washington DC: American Society for Microbiology, 1993, pp 257–265.

24. Y Xue, L Zhao, HW Liu, DH Sherman. A gene cluster for macrolide antibiotic biosynthesis in *Streptomyces venezuelae*: architecture of metabolic diversity. Proc Natl Acad Sci USA 95:12111–12116, 1998.

25. MJ Ryan, JA Lotvin, N Strathy, SE Fantini. Cloning of the biosynthetic pathway for chlorotetracycline and tetracycline formation and cosmids useful therein. U.S. patent 5,589,385.

26. NM Nasby, TC Peterson. Methods for screening compounds using encapsulated cells. PCT Patent Publication Number WO 98/41869.

27. TC Peterson, P Brian, LM Foster, K Li, RJ Fielding, KA Thompson, G McClure, LC Rupar, SW Mamber, KW Brooksire, R Belval, E Pack, K Gugliotti, S Forenza. Diverse bioactivities from actinomycete combinatorial biology libraries. In: R Baltz, G Hegemen, P Skatrud, eds. Proceedings from the 1996 Genetics and Molecular

Biology of Industrial Microbiology Conference. Fairfax, Virginia: Society for Industrial Microbiology Press, 1997, pp 71–76.

28. DJ Hook, CM More, JJ Yacobucci, G Duby, S O'Connor. An integrated biological physiochemical system for the identification of antitumor compounds in fermentation broths. J Chromatog 385:99–108, 1987.

29. NM Nasby, TC Peterson. Immobilization for high-throughput screening. In: Immobilized Cells. Heidelberg: Springer-Verlag, in press.

30. A Cantafio, N Aronson, N Nasby, T Legler, K Thompson. Unpublished observations.

31. NM Nasby, T Legler, P Brian, P Evans, C Silva, A Hansen, P Murphy, H Xu, T Peterson. Streptomyces protoplast transformation, regeneration and natural product production in alginate macrodroplets: high throughput screening of combinatorial biology libraries. Manuscript in preparation.

32. NM Nasby, T Legler, P Brian, P Evans, C Silva, A Hansen, P Murphy, H Xu, T Peterson. Streptomyces protoplast transformation, regeneration and natural product production in alginate macrodroplets: high throughput screening of combinatorial biology libraries. Poster Society for Industrial Microbiology meeting, Denver, Colorado, August 9–12, 1998.

33. CP Holmes. Model studies for new o-nitrobenzyl photolabile linkers: substituent effects on the rates of photochemical cleavage. J Org Chem 62:2370–2380, 1997.

13

Higher-Throughput Screening Assays With Human Hepatocytes for Hepatotoxicity, Metabolic Stability, and Drug–Drug Interaction Potential

Albert P. Li
In Vitro Technologies, Inc., Baltimore, Maryland

I. INTRODUCTION

Besides pharmacological activity, pharmacokinetic and toxicological properties are critical to the success of a drug candidate in the clinic. As the time and cost involved in clinical studies is substantial, it would be ideal if the human pharmacokinetic and toxicological properties could be determined during the pre-clinical phase of drug development. In the past, preclinical studies were predominantly performed using (nonhuman) laboratory animals as surrogates for humans, yielding mixed results. The frequent inability of studies with laboratory animals to predict human clinical findings is consistent with the known species differences in pharmacological and toxicological effects of xenobiotics and is often attributed to species differences in biotransformation [1–4].

A major factor for species differences in xenobiotic biotransformation is the difference in drug metabolizing enzymes, especially the cytochrome P450 (CYP) isoforms [5–7]. As the liver represents the major organ for drug metabolism, human liver–derived experimental systems have been used extensively for the evaluation of human-specific drug properties. Experimental systems derived from the liver include systems with intact viable cells (intact cell systems) such as hepatocytes and liver slices, as well as cell-free systems such as liver homogenates, postmitochondrial supernatants (S9), and microsomes. The strengths and limitations of these different in vitro hepatic models have been reviewed recently [8,9]. The intact cell systems, with full complements of enzymes and cofactors at physiological levels and natural spatial orientations, should be more representative of the liver in vivo than cell-free systems with disrupted membranes and incomplete cofactors and enzymes. The presence of the intact plasma membrane allows for the modeling of differences between intracellular and extracellular concentrations, potentially resulting from active uptake and excretion. Furthermore, cytotoxicity studies can be performed with the intact cells, allowing investigations on toxic mechanisms, including the relationship between metabolism and toxicity [8].

Human hepatocytes therefore represent a physiologically relevant experimental model for the evaluation of liver-related human drug properties such as hepatotoxicity, metabolism, and drug–drug interaction potential [8–11]. As relatively few laboratories have access to fresh human livers for hepatocyte isolation, human hepatocytes are not yet a universally available experimental system. Cryopreservation, if successful, would greatly enhance the availability of human hepatocytes. For instance, hepatocytes can be routinely prepared from one laboratory that has access to fresh human livers, stored as cryopreserved cells, and shipped as needed to other laboratories for experimentation.

II. XENOBIOTICS AND THE HUMAN LIVER

The liver is the key organ for biotransformation of xenobiotics that enter the human body, either intentionally (e.g., pharmaceuticals) or unintentionally (e.g., environmental pollutants). All blood-borne xenobiotics are firstly metabolized by the liver, a process known as first-pass metabolism. The major consequence of the metabolic transformation is the formation of water-soluble metabolites that are removed from the body. In general, the xenobiotic is firstly oxidized (phase I metabolism) and then conjugated to highly polar molecules such as glucose, sulfate, cysteine, and glutathione (phase II metabolism). The highly polar metabolites are then transported directly from the hepatocytes into the biliary canaliculi (phase III metabolism) to be excreted as bile, or are released back into the bloodstream to be excreted into the urine via the kidneys. In the intestine, the metabolites may be deconjugated by gut bacterial flora and reabsorbed, leading to a repeat of the metabolic processes (enterohepatic recirculation). The sequential oxidation–conjugation of xenobiotics is commonly known as metabolic detoxification, a process that should rid the body of xenobiotic toxicants. However, it is now also known that many innocuous chemicals can be metabolically "activated" into highly reactive metabolites with toxicological consequences. The liver, being the organ where this metabolic "activation" occurs, is therefore often the target organ of chemical toxicants.

The major enzymes for biotransformation pathways, phase I oxidation and phase II conjugation, are present in the liver in significant amounts. It is now known that human drug metabolizing enzymes, especially those related to CYPs, can be substantially different from those for nonhuman animals. This species difference in xenobiotic metabolism is one of the major reasons that results from laboratory animals may not be directly applicable to man. Human hepatocytes, which contain most of the drug-metabolizing enzymes present in the human liver, represent an experimental system with which human-specific drug metabolism can be studied. The key CYP isoforms in human liver include CYP1A2, CYP2A6, CYP2C9, CYP2C19, CYP2D6, CYP2E1, and CYP3A4. Of these isoforms, CYP3A4 is the isoform present in the highest quantity and is believed to catalyze a large number of xenobiotics [12]. It is generally believed that over 50% of current human pharmaceuticals are substrates of CYP3A4.

III. ISOLATION AND CULTURING OF HUMAN HEPATOCYTES

The liver is estimated to be approximately 2.6% of the human body weight, consisting of both parenchymal and nonparenchymal cells. The parenchymal cells, commonly known as hepatocytes, comprise the majority of cells of the liver (approx. 60% by cell number and 80% by weight). These cells are the major sites of xenobiotic metabolism and often are the cells injured by chemical toxicants. The nonparenchymal cells include the endothelial cells, which line the sinusoidal space; kupffer cells, which are the stationary macrophages believed to be responsible for the scavenging of endotoxins, and lipocytes or Ito cells, which normally serve as vitamin A storing cells but upon liver injury differentiate into collagen-producing and rapidly proliferating fibroblasts, leading to fibrosis and, upon chronic injury, cirrhosis. Endothelial cells and kupffer cells are also cytokine-producing cells believed to play major roles in both the regeneration of the liver and the progression of liver diseases.

Procedures for the isolation of human hepatocytes are well established and are not significantly different from those used to isolate hepatocytes from laboratory animals. The major difficulty is the availability of fresh human livers to allow hepatocyte isolation. Liver is treated by a two-step collagenase digestion procedure that has been previously reported in detail [13]. The entire lobe or a three-side encapsulated liver segment is perfused with an EGTA-containing isotonic solution at physiological pH (pH 7.2–7.5) to clear the blood. The EGTA serves to bind divalent ions and thereby prevent clotting. Also, EGTA loosens cell junctions to facilitate the subsequent enzyme digestion step. After the first perfusion, the relatively blood-free liver segment is then perfused with a collagenase solution for cell dissociation. After dissociation, the digested liver is dissected and agitated or combed to release the dissociated cells. The fibrous connective tissue is separated from the dissociated cells by filtering through either a nylon screen or gauze. The hepatocytes can be harvested by low speed centrifugation (50 × g). Most of the nonparenchymal cells stay in the supernatant. If necessary, the viable hepatocytes can be further purified from nonviable hepatocytes and the nonparenchymal cells by density gradient sedimentation.

The critical factors for successful isolation of human hepatocytes are the condition of the human liver, the collagenase employed, and the length of digestion. In general, the longer the length of cold storage and the higher the fat content of a liver, the lower the hepatocyte yield and viability. A liver that has a storage duration of less than 24 hr with a fat content of less than 50% is preferred. It is important to select a lot of collagenase that is not cytotoxic to human hepatocytes. However, this has become less of a problem since commercially available collagenase now is in general noncytotoxic to human hepatocytes. The timing of the digestion is critical. Both under- and overdigestion would compromise the yield and viability of hepatocytes. In general, the viability of the isolated hepatocytes is over 80%. The yield of hepatocytes is approximately 10 million cells per gram of human liver. For instance, liver biopsies weighing 10 g would yield approximately 100 million viable hepatocytes. A large segment of human liver weighing 500 g would yield approximately 5 billion viable hepatocytes. In Vitro Technologies Inc. is routinely able to procure relatively large liver segments that yield 5–20 billion hepatocytes per isolation. This high yield of human hepatocytes is one of the major motivating forces for the optimization of cryopreservation procedures (described below).

After isolation, human hepatocytes can be used as a suspension culture for short-duration (hours) studies or as attached cultures for longer term (days or weeks) studies. In suspension, hepatocytes would rapidly loose viability: usually 50% after approximately 12 hr and over 90% after 24 hr. As adherent cultures, hepatocytes remain viable for weeks. Most investigators culture hepatocytes on collagen-coated surfaces. While basement membrane extract (e.g., Matrigel) can be used to prolong the maintenance of differentiated properties for rat hepatocytes, it appears to have fewer effects on human hepatocytes than on rodent hepatocytes. When cultured on collagen-coated substratum, hepatocytes exhibit an epithelial morphology with prominent nuclei. The hepatocytes are often binucleated. When cultured as confluent cultures or couplets, bile canaliculi can be observed between hepatocytes. On Matrigel the cells remain rounded. Overlaying hepatocytes attached on collagen with either collagen (collagen–gel sandwich) or Matrigel is believed to maintain the liver functions of hepatocytes. Though in earlier studies hepatocytes cultured on collagen were used, the collagen–gel overlaying procedure is now routinely used [18].

IV. CRYOPRESERVATION OF HUMAN HEPATOCYTES

For obvious reasons, cryopreservation can significantly enhance the convenience and ease of experimentation with human hepatocytes. During the past decades, it was generally concluded that cryopreservation of hepatocytes was possible but not reproducible. In earlier days, it was generally believed that rodent hepatocytes were quite easily cryopreserved while human hepatocytes were more challenging. Now we believe that it is just a matter of the quality of the hepatocytes. Human hepatocytes isolated from fresh (less than 24 hr post-clamp time) livers can be cryopreserved as well as hepatocytes from nonhuman animals. DMSO is the most frequently used cryopreservant [14,15]. Frequently, after cryopreservation, viability based on trypan blue exclusion remains high and the cells are metabolically active as suspension cultures [16–18]. However, in general only a small percentage (< 50%) of the cells attach, and in some cases the cells initially attach and detach after overnight culturing. When the majority of the cryopreserved hepatocytes remain viable and able to attach after thawing, normal hepatocyte functions apparently are retained. The functions retained include metabolic activation of promutagens [14] and response to CYP inducers [17]. Efforts are underway in our laboratory to improve the attachment efficiency of cryopreserved human hepatocytes.

For cryopreserved human hepatocytes to be used routinely, the cells should have acceptable viability and metabolic activities. The cryopreserved human hepatocytes after thawing consistently exhibit acceptable viability, though it is slightly lower than the values of freshly isolated hepatocytes [18]. Also the viability and yield of viable cells are not decreased with cryopreservation durations for 12 months and longer [18]. Storage of cryopreserved hepatocytes is only possible in the liquid or vapor phase of liquid nitrogen, at a temperature of ≤150° C. Storage of cryopreserved hepatocytes at higher temperatures (e.g., at −70°C) leads to a rapid decrease in viability.

The major concerns about the use of cryopreserved hepatocytes in drug metabolism studies are that the specific drug-metabolizing enzymes may be compromised by cryopreservation, and that cryopreservation may lead to the selection of a specific subpopulation of hepatocytes that may have properties significantly different from those of the freshly isolated hepatocytes. Results show that there is no apparent basis for these concerns [18]. Hepatocytes isolated and cryopreserved from multiple donors were shown to be competent in the major pathways of drug metabolism. The major CYP isoforms (CYP1A2, CYP2A6, CYP2C9, CYP2C19, CYP2D6, CYP2E1, and CYP3A4) as well as the phase II enzymes

Table 1 Viability and Drug-Metabolizing Enzyme Activities of Freshly Isolated (Fresh) and Cryopreserved (Cryo.) Human Hepatocytes Isolated from 10 Donors

Lot No.	Status	Viability (%)	CYP1A2	CYP2A6	CYP2C9	CYP2C19	CYP2D6	CYP2E1	CYP3A4	7HC-G	7HC-S	ECOD
72	Fresh	91	0.07	202	13	30.6	68	18	40	522	17	66
	Cryo.	65	0.06	187	13	39.5	72	39	112	495	10	53
73	Fresh	97	0.31	119	30	30.5	54	17	59	474	14	43
	Cryo.	73	0.43	95	19	21.5	42	35	93	277	16	32
74	Fresh	93	0.03	127	16	0.8	53	25	90	307	9	30
	Cryo.	66	0.05	140	16	0.5	57	29	70	175	8	21
75	Fresh	96	0.01	78	21	6.3	29	114	47	316	56	56
	Cryo.	85	0.02	55	25	6.0	33	150	44	235	28	40
76	Fresh	86	1.31	25	23	16.0	42	152	32	513	33	70
	Cryo.	71	1.49	24	17	20.0	46	300	45	380	27	67
77	Fresh	96	0.20	62	23	2.4	42	8	101	517	53	33
	Cryo.	56	0.22	74	20	2.2	36	11	90	388	25	24
78	Fresh	97	0.31	119	30	31.0	54	17	59	474	14	43
	Cryo.	71	0.50	99	19	24.0	48	31	100	314	8	31
79	Fresh	96	0.03	185	47	9.1	49	122	—	315	109	85
	Cryo.	76	0.04	175	69	9.4	51	156	—	265	140	79
80	Fresh	87	0.03	28	15	—	12	233	25	171	44	64
	Cryo.	71	0.01	56	8	—	16	502	24	137	41	70
Mean	Fresh	93	0.26	105	24	15.8	45	78	57	401	39	54
	Cryo.	70	0.31	101	23	15.4	45	139	72	296	34	46
(Cryo./Fr)		75.57	122.61	96	94	97.1	100	178	128	74	87	85.10

Results of 10 sequential isolations are shown. Viability was determined based on trypan blue exclusion. Enzyme activities were CYP1A2, ethoxyresorufin O-deethylation; 2A6, coumarin 7-hydroxylation; 2C9, tolbutamide 4-hydroxylation; 2C19, mephenytoin 4-hydroxylation; 2D6, dextromethorphan O-demethylation; 3A4, testosterone 6β-hydroxylation; UDPGT, umbelliferone glucuronidation; ST, umbelliferone sulfation; ECOD, 7-ethoxycoumarin-O-deethylation. Activity units are pmol/min/10⁶ viable hepatocytes.

UDPGT and ST are present in cryopreserved human hepatocytes. No consistent differences in activities were found between freshly isolated and cryopreserved hepatocytes (Table 1). The major drug-metabolizing pathways were not significantly altered due to cryopreservation, which supports the validity of cryopreserved human hepatocytes as an experimental system for drug metabolism studies.

V. APPLICATIONS OF HUMAN HEPATOCYTES IN DRUG METABOLISM

A. Advantages of Primary Hepatocytes in Drug Metabolism

Hepatocytes represent a physiologically-relevant experimental model of the liver because of the following properties:

1. Physiological Enzyme, Cofactor, and Substrate Concentrations

Freshly isolated hepatocytes contain complete, undisrupted enzymes and cofactors at the same level as the liver in vivo, therefore allowing the generation of data similar to hepatic metabolism in vivo.

2. Relevant Drug Concentrations

Hepatocytes in culture retain active uptake as well as biliary excretion similar to hepatocytes in the liver in vivo. Drugs that are bioaccumulated will have a higher intrahepatocyte concentration than plasma concentration, or be actively excreted, and will therefore have a lower intrahepatocyte concentration than plasma concentration would behave similarly in cultured hepatocytes. For these drugs that have differential plasma and intracellular concentrations, values critical to in vitro–in vivo extrapolation of drug–drug interaction, e.g., Ki values, derived from hepatocytes, would be more relevant to those obtained with microsomes.

3. Inducible CYP

CYP inhibition and induction are the two major mechanisms of drug–drug interactions. While CYP inhibition can be studied with microsomes and hepatocytes, CYP induction can only be studied in a living cell, i.e., hepatocytes. Induction protocols for CYP induction, especially for CYP1A and 3A, are well established [19–21]. Known human in vivo inducers such as rifampin, phenobarbital, and phenytoin are all potent CYP inducers in primary human hepatocytes. Hepatocytes in culture appear to retain species-specific response to CYP inducers. For instance, rifampin, a potent CYP3A inducer for humans in vivo, induces CYP3A4 in human hepatocytes but not in rat hepatocytes. On the other hand, dexamethasone is significantly more potent in the induction of CYP3A in rat hepatocytes than in human hepatocytes [22].

B. Limitations of Primary Hepatocytes

The limitations of the use of human hepatocytes are in general related to the use of freshly isolated hepatocytes and can be overcome by the use of cryopreserved cells. The use of isolated hepatocytes suffers disadvantages of all in vitro systems: lack of host factors.

1. Limited Availability

The major limitation of primary human hepatocytes is that viable hepatocytes can only be obtained from relatively fresh human livers, usually within 24 hr after liver isolation. Human hepatocytes can be prepared in a laboratory that has access to human livers, cryopreserved and then transported to other laboratories for experimentation. However, freshly isolated hepatocytes are still required for enzyme induction studies.

2. Limited Storage Options

Freshly isolated hepatocytes contain metabolism activities similar to the liver. After culturing, due to the lack of endogenous CYP inducers, the activity of the inducible isozymes such as 1A and 3A would decrease. While cultured hepatocytes are useful for induction studies, only freshly isolated hepatocytes or cryopreserved hepatocytes are appropriate for the evaluation of metabolite profile and metabolic rate.

   Because of intact cell properties, at least theoretically, the use of hepatocytes should provide data more reflective of the in vivo data than cell-free systems such as microsomes. After reviewing literature values of in vivo metabolic clearance data and intrinsic clearance obtained using different in vitro hepatic systems, Houston concluded that the best correlation with in vivo data was made using hepatocytes, followed by liver slices, and then by liver microsomes [23]. Hepatocytes performed better than liver slices, probably because of the artefactual presence of cell barrier due to the multiple cell layers in the liver slices. The importance of the use of intact cells is illustrated by the fact that both hepatocytes and liver slices perform better than liver microsomes in the prediction of intrinsic clearance.

C. Sex and Species Differences in Drug Metabolism
     in Hepatocytes

The validity of hepatocyte as a metabolic model is further substantiated by studies on specific chemicals with known in vivo sex and species differences in metabolism. In general, results with hepatocytes are consistent with in vivo results. Amphetamine (AMP) is metabolized by the liver into two major metabolites: para-hydroxyamphetamine (pHA) via aromatic hydroxylation, and benzoic acid (BA) via oxidative deamination. Species differences in AMP metabolism have been established in vivo. The rate of metabolism is rabbit > rat > monkey > human. In the rat, aromatic hydroxylation leading to pHA is the major pathway, while in all other species, side chain oxidation leading to BA is the predominant pathway. The results on AMP in primary hepatocytes from rat, rabbit, rhesus monkey, and human also showed species differences in both rate of metabolism and pathway preference, as with the in vivo observations [8].

Both sex and species differences in phase II metabolism found in vivo are also reproduced by primary hepatocytes. In rodents, sex differences in hepatocyte metabolism of acetaminophen (AAP) are known. The male rat has higher AAP sulfotransferase activity than the female rat. The difference is believed to be a result of the stimulatory influence of testosterone and the suppressive effect of estrogen on one of the two AAP sulfotransferases. Upon AAP administration, the male rat excretes more AAP sulfate conjugate and less AAP glucuronide conjugate than the female rat. Both male and female humans excrete more glucuronide than sulfate. Primary hepatocyte cultures from male and female human and rat were found to reproduce accurately the known sex and species differences [24,25]. Female rat hepatocytes produced similar quantities of glucuronide and sulfate from AAP, while male rat hepatocytes produced predominantly more of the sulfate than the glucuronide. Increasing AAP concentration led to increases in glucuronide formation, while sulfate formation plateaued, suggesting the early saturation of the sulfation pathway, which is a well-established phenomenon in vivo [24]. In humans, this sex difference is not observed in vivo nor in cultured hepatocytes. When male and female patients were administered AAP before abdominal surgery, urine AAP metabolites were compared to metabolite generated from hepatocytes from the same patients. Similar results were observed both in vivo and in vitro; AAP glucuronidation predominated over AAP sulfation with no sex differences [25.]

VI. APPLICATION OF HUMAN HEPATOCYTES IN THE EVALUATION OF DRUG–DRUG INTERACTIONS

A. Mechanism and Clinical Significance of Pharmacokinetic Drug–Drug Interactions

Multiple drug therapy is widely practiced either to treat a medical disorder or to treat several concurrently existing ailments in the same patient. It is now known that drugs may interact, resulting in serious pharmacological and/or toxicological consequences. Drugs interact mainly by a phenomenon known as pharmacokinetic drug–drug interactions—alteration of the metabolic clearance of a drug by a coadministered drug. While pharmacokinetic drug–drug interactions can occur during absorption, metabolism, disposition, and elimination phases after initial drug administration, interference with drug metabolism appears to be the predominant mechanism. The interference can occur via inhibition or induction of the metabolism of one drug by a coadministered drug. Both mechanisms of pharmacokinetic drug–drug interactions can have serious clinical consequences.

Inhibition of drug metabolism results in an increase in plasma/tissue drug concentration, which, especially for drugs with a narrow therapeutic index, can lead to toxicity. Pharmacokinetic drug–drug interactions via the inhibitory mechanism therefore are a safety concern. A well-established example of drug–drug interaction via an inhibitory mechanism is the occurrence of torsades de pointes ventricular arrhythmia in patients receiving concomitant therapy with the nonsedating antihistamine terfenadine and azole antifungals such as ketoconazole, itraconazole, clotrimazole, and macrolide antibiotics such as erythromycin [26,27]. Terfenadine is believed to be metabolized by CYP3A4 [28]. The adverse drug interactions observed with terfenadine are related to the inhibition of its metabolism by the coadministered drugs, leading to the accumulation of parent terfenadine to a cardiotoxic level [29].

Conversely, induction of drug metabolism can result in a decrease in drug concentration to a level that is no longer efficacious. A significant drug–drug interaction via an induction mechanism is the interaction between rifampin and oral contraceptives [30–32]. Rifampin is an antimicrobial agent that is a known inducer of CYP3A4 in human [33] and in human hepatocytes [12,19–21]. The key active ingredients of oral contraceptives, estrogen and progesterone analogs, are substrates of CYP3A4 [34]. Rifampin administration was found to lead to uterine bleeding and pregnancies in women taking oral contraceptives, which is believed to be due to the enhanced clearance of the active ingredients [35]. Besides oral contraceptives, rifampin also was reported to decrease the therapeutic effects and/or plasma concentration of other drugs, including verapamil [36], cyclosporin [37], doxycycline [38,39], itraconazole [40], prednisolone [41], and zidovudine [42]. While drug–drug interaction due to enzyme inhibition is a safety concern, therapeutic failure is the major concern for drug–drug interaction due to enzyme induction.

B. Mechanism-Based Approaches to the Evaluation of Drug–Drug Interactions

In the past, drug–drug interaction potential for specific drugs was discovered only because of the occurrence of adverse reactions in patients. It would be more prudent to evaluate the drug–drug interaction potential of new drugs before the drugs were used in humans. As it would be impossible to evaluate the potential interaction of a new drug with all existing drugs, a more practical approach is to define the drug–drug interaction potential of the new drug based on mechanistic evaluations [9], followed by clinical trials specifically designed to further define the mechanistic observations.

The key mechanism of pharmacokinetic drug–drug interactions is the ability of one drug to affect the metabolism of another drug. This can be due to inhibition or induction of the drug-metabolizing enzymes. At present, emphasis is placed on the induction and inhibition of CYP isoforms.

1. CYP Induction

Primary human hepatocytes represent the most relevant in vitro human system in which pharmacokinetic drug–drug interactions can be studied via the induction mechanism. Microsomes are nonliving, so induction studies are not possible. The viability of liver slices cannot be maintained for the duration (usually over 48 hr) that is required for induction.

Induction of CYP1A by polybrominated biphenyl in primary human and rat hepatocytes showed that the rat is the more sensitive species [43]. A 10–1000-fold higher concentration of the inducer found to be inducing in rat hepatocytes was required to induce human hepatocytes, which suggests that human hepatocytes are similarly less susceptible to the induction effects of polybrominated biphenyl.

Using primary human hepatocytes, the effects of rifampin, a known CYP3A inducer, on lidocaine, a known CYP3A substrate metabolism, have been reported [44]. Treatment of primary human hepatocytes with rifampin led to a significant dose-dependent induction of lidocaine metabolism. There was a significant increase in V_{max} with little effect in K_m, an observation consistent with increased expression of the CYP3A isozyme protein rather than a change in affinity for the substrate. These results suggest that the dosage of lidocaine administration to patients on rifampin should be altered from that designed for normal patients. In a CYP induction assay using primary human hepatocyte cultures to evaluate the rifamycin analogs rifampin, rifapentine, and rifabutin, it was found that rifampin and rifapentine were more potent inducers of CYP3A than rifabutin, a finding that reflects what was observed in humans in vivo [21]. This assay has been extended to evaluate all CYP isoforms.

2. CYP Inhibition

Human hepatic microsomes are most frequently used as an experimental model to evaluate the inhibitory mechanism of drug–drug interactions. The most effective application is to use isozyme-specific inhibitors to determine which CYP isozymes are responsible for drug metabolism. The principle is that drugs metabolized by the same isozymes would have inhibitory effects on the metabolism of each other. Primary hepatocytes represent an experimental system that can substantiate findings with microsomes. For a drug to inhibit the metabolism of another drug, it needs to be present in the hepatocytes in vivo. The intracellular concentration is dependent on both membrane permeability and the activity of the multiple drug resistance protein that actively pumps xenobiotics out of the hepatocytes. It is therefore necessary to confirm findings in microsomes with an intact cell system such as hepatocytes and liver slices.

Terfenadine has been associated with several adverse drug interactions. The metabolism of terfenadine was studied using primary human and rat hepatocytes and compared with results from an immortalized cell line, HepG2 [11,19]. While all the three-cell systems extensively metabolize terfenadine, human hepatocytes were found to produce both C-oxidation and N-dealkylation products. Rat hepatocytes and HepG2 cells produce only C-oxidation products. Further, the known inhibitors of CYP3A, the isozyme believed to be mainly responsible for terfenadine metabolism: ketoconazole, erythromycin, and troleandomycin, inhibited terfenadine metabolism in human but not in rat hepatocytes. These results suggest that primary human hepatocytes represent a more appropriate experimental model than rat hepatocytes or HepG2 cells for the evaluation of drug–drug interactions [11,18,19]. Intact human hepatocytes are now routinely used for the evaluation of the potential of xenobiotics to inhibit CYP isoforms. Results with known CYP inhibitors provide the expected responses. The potent and specific CYP inhibitors for intact human hepatocytes are furafylline for CYP1A2, quinidine for CYP2D6, and ketoconazole for CYP3A4 [18].

VII. APPLICATIONS OF HUMAN HEPATOCYTES IN TOXICOLOGY

A. Hepatocytes as a Critical Cell Population for Hepatotoxicity Studies

A large number of naturally occurring and man-made chemicals are hepatotoxins. In many cases, the toxicity is a function of the metabolic conversion (bioactivation) of the parent compound into highly reactive metabolites. AAP, carbon tetrachloride, dimethylnitrosamine, and halothane are examples of xenobiotics that are "bioactivated" by CYP mixed-function monooxygenases in the liver. Species differences in xenobiotic metabolism are therefore important factors contributing to the known species differences in chemical toxicity. Most of the xenobiotic metabolic enzymes of the liver reside in the parenchymal cells or hepatocytes. The hepatocytes therefore are usually the first cell types that are damaged upon hepatotoxic insult. Primary hepatocytes therefore represent a useful experimental system to evaluate acute hepatic injury. The nonparenchymal cells are important in the progression of pathological events. For instance, inflammation of the liver is primarily a function of the cytokine production in the endothelial cells and kupffer cells. Hepatic fibrosis and cirrhosis are believed to be due to the "activation" of lipocytes to collagen-synthesizing fibroblasts.

B. Advantages and Disadvantages

In vitro toxicology, the evaluation of toxicological properties of physical and chemical agents using isolated cells and tissues, is an area that has evolved rapidly in the past decade. The initial purpose is to achieve the so-called 3Rs: replacement (of whole animals); reduction (of animal use); and refinement (of toxicity assays). Pragmatically, in vitro toxicology is mostly useful in the rapid screening of chemicals and in the mechanistic evaluation of toxicological phenomena. Hepatocytes are an ideal in vitro system to evaluate hepatotoxicity. The advantages are the retainment of species-specific metabolism, the defined experimental conditions common to most in vitro systems, and the requirement of relatively low amount of test materials. Molecular approaches now are often coupled with hepatocyte culture systems for mechanistic evaluation.

The disadvantages are the lack of host factors and the lack of nonparenchymal cells. Classical toxicologists are often skeptical of in vitro data, probably due to the years of practicing toxicology as an empirical science. The common belief among "classical" toxicologists is that no matter how much we know about the in vitro properties of an agent, we still will not know what toxicological effects, especially upon chronic exposure, it would have in an animal in vivo. This is a true and often valid concern. The in vitro–in vivo gap can only be closed if the in vivo toxicological mechanisms behind the toxicological consequence can be studied using primary hepatocytes. While in vivo toxicological studies are often performed without any prior mechanistic knowledge, the application of in vitro systems requires a thorough understanding of the limitations of the experimental system used.

While in vitro systems, as of now, cannot yet replace most critical in vivo toxicology assays such as the chronic bioassay, they are extremely useful for mechanistic evaluations. This is probably the most important aspect of in vitro toxicology. Via the elucidation of mechanisms one can extrapolate from high to low doses, from one species to another, and from acute to chronic exposure. The importance of in vitro systems such as the primary hepatocyte culture may not be in the prediction of human toxicity per se, but in bridging the gap between laboratory animals and humans to allow a better prediction of human toxicity based on whole animal data (see reviews on the application of hepatocytes in toxicology, Refs. 8, 20, 45, and 46).

VIII. HIGHER-THROUGHPUT SCREENING (HrTS) WITH HUMAN HEPATOCYTES

Higher-throughput screening (HrTs) and human hepatocytes are not compatible, due to the difficulty most laboratories have in procuring human livers for research. The success in cryopreservation, however, has dramatically enhanced the availability of human hepatocytes for research. Currently, cryopreserved human hepatocytes are commercially available. Furthermore, as cryopreserved hepatocytes can be used at the convenience of the investigator, HrTS with human hepatocytes is now a reality. The typical protocols for HrTS assays are given below [18].

A. Thawing of the Cryopreserved Human Hepatocytes

As the HrTS assays all require the use of cryopreserved human hepatocytes, the procedures for thawing of the cells are described here. This thawing procedure needs to be followed carefully, as deviations may result in lowered viability. On the day of assay, the vials containing the cryopreserved hepatocytes are removed from liquid nitrogen storage. The cells are thawed by gently shaking the vials in a 37°C water bath. As soon as the contents are thawed (approx. 75–90 sec), the vials are transferred to an ice bucket. The hepatocyte suspensions (approximately 1 mL) are then transferred into a 50 mL centrifuge tube on ice. Cold medium (15 mL) is then added to the cell suspension slowly. After dilution, the cells are centrifuged at 50 × g for 3 min. The resulting pellet is resuspended in 2 mL of medium. Viability in general can be determined based on trypan blue exclusion [13,15].

B. HrTS for Metabolic Stability

The cryopreserved human hepatocytes are resuspended in an isotonic buffer at physiological (e.g., Krebs–Hensleit buffer) pH of 7.2–7.5 in a cell density of 2×10^6 cells/mL and seeded onto 96-well filter plates (Millipore, MultiScreen) in a volume of 50 μL (0.1×10^6 cells). Aliquots of 50 μL of buffer containing 2X concentrations of the chemical to be tested are added to the wells preseeded with the hepatocytes to achieve a final 1X concentration. A pragmatic endpoint is the quantification of the disappearance of the parent chemical (a final concentration between 1 and 10 μM). The plates are incubated at 37°C for 4–6 hr. At the end of the incubation period, 100 μL of ice-cold organic solvent (methanol or acetonitrile) is added to each well to stop the reaction. The resulting samples are filtered by centrifugation into a recipient 96-well plate for later analysis. At present, LC/MS analysis is the most efficient. Incubation of test chemical with killed hepatocytes (freeze–thaw cycle at −20°C or heat-inactivation) under identical conditions is used as control. Metabolic stability is expressed as a percentage of parent disappearance using the equation

$$\% \text{ Disappearance} = \left(\frac{\text{peak area (live hepatocytes)}}{\text{peak area (dead hepatocytes)}} \right) \times 100$$

C. HrTS for Cytotoxicity

Cryopreserved human hepatocytes are thawed and resuspended in incubation medium at a density of 1.25×10^6 cells/mL. A volume of 40 μL/well of the cell suspension is loaded into 96-well plates, followed by a 40 μL/well addition of solutions of test chemicals containing twice the desired final concentrations. The plates are incubated at 37°C for 4 hr, after which an assay for viability is initiated by adding 20 μL/well of a solution of (3-[4,5-dimethylthiazol-2-yl]-2,5-diphenyl-tetrazolium bromide (MTT, 5 mg/mL in incubation medium). The plates are returned to the incubator for another 3 hr. After the incubation, 150 μL acidified isopropanol (isopropanol containing 0.04 M HCl) is added directly into the wells and incubated in an incubator with shaking (50 rpm) for the extraction of the blue MTT product. The blue color developed is quantified at 570 nm with reference to 650 nm. Dead cells (killed by one freeze–thaw cycle at -20°C) treated with the same concentrations of the test chemicals are used as negative controls. Results are expressed as percent relative survival:

$$\text{Relative survival} = \left(\frac{\text{Absorbance (treatment)}}{\text{Absorbance (negative control)}} \right) \times 100$$

D. HrTS for Inhibitory Drug–Drug Interactions

After thawing, the cryopreserved human hepatocytes are resuspended in KHB at a cell density of 2×10^6 cells/mL. A volume of 25 μL of the cell suspension/well is added into 96-well filter plates containing 25 μL/well of the CYP inhibitors. The plates are incubated at 37°C for 15 min (preincubation). After the preincubation, aliquots of 50 μL of each CYP isoform-specific substrate (see below) are added to the samples and incubated for 2 hr. 100 μL of cold methanol is added to the samples to stop the reaction. The resulting aliquots are filtered by centrifugation into another 96-well plate for later HPLC analysis. Six known CYP inhibitors [target isozyme(s)/final concentration] are used as positive controls: diethyldithiocarbamate (2A6, 2E1/50 μM), furafylline (1A2/50 μM), ketoconazole (3A4/1 μM), quinidine (2D6/10 μM), sulfaphenazole (2C9/10 μM), and tranylcypromine (2A6, 2C19/10 μM). The CYP isoform-specific substrates (target isozyme/final concentrations) used are phenacetin (1A2/100 μM), coumarin (2A6/100 μM), tolbutamide (2C9/100 μM), (s)-mephenytoin (2C19/100 μM), dextromethorphan (2D6/100 μM), chlorzoxazone (2E1/100 μM), and testosterone (3A4/100 μM). KHB is used as a negative control. Results are expressed as % inhibition.

$$\% \text{ Inhibition} = \left(\frac{\text{Activity (vehicle control)-Activity (treatment)}}{\text{Activity (vehicle control)}} \right) \times 100$$

IX. CONCLUDING REMARKS

Screening of new chemical entities for human pharmacokinetic and toxicological properties using human hepatocytes should greatly enhance the drug development process. Cryopreserved human hepatocytes retain adequate viability and drug-metabolizing enzyme activities [18], and human hepatocyte-based HrTS assays have been developed. A HrTS assay by definition should have a high capacity and a rapid turnaround time and therefore is not possible with freshly isolated human hepatocytes, as they have limited and unpredictable availability. On the other hand, HrTS assays are compatible with precharacterized cryopreserved human hepatocytes. HrTS assays in 96-well plates are being developed for major drug properties that are critical to the clinical success of a drug candidate: metabolic stability, toxicological potential, and inhibitory drug–drug interaction potential. The use of the 96-well plate format allows automation and minimizes the amount of experimental materials (both chemicals and hepatocytes) required.

As HrTS for cytotoxicity requires live cells, of the several in vitro liver models, hepatocytes, liver slices, and liver microsomes, only hepatocytes and slices can be used. As liver slices require a fresh human liver, this experimental system cannot be used routinely for screening. Cryopreserved human hepatocytes represent the only human liver in vitro system that can be used in this application. Experience so far shows that acute drug toxicity can be screened with this system [18]. In our laboratory we are in the process of defining toxicological end points (e.g., apoptosis, gene expression) that may reflect chronic hepatic toxicity and hepatocarcinogenicity.

HrTS assays for metabolic stability studies are performed in some laboratories using human liver microsomes [47]. A major concern with the use of liver microsomes is that only phase I oxidation pathways are evaluated. Results with microsomes thus may not be relevant if enzymes that are not present in the microsomes, for instance, cytosolic phase II conjugating enzymes, are major pathways of metabolism. Human hepatocytes, with their complete enzyme pathways and cofactors, should represent a more appropriate system for the evaluation of metabolic stability than microsomes.

Currently, evaluation of inhibitory pharmacokinetic drug–drug interaction potential is routinely performed using human liver microsomes [48]. Results with intact hepatocytes may be more relevant than those generated using microsomes for evaluation of CYP inhibitory potential. The apparent Ki value for the inhibition of terfenadine metabolism by ketoconazole was significantly lower for intact human hepatocytes than that for human liver microsomes [11,19]. This could be due to the bioaccumulation of ketoconazole in the hepatocytes, thereby leading to a lower apparent Ki value, and/or the nonspecific binding of chemicals to microsomes, leading to a higher apparent Ki value. It was convincingly shown that the effective free-drug concentrations would decrease with increasing microsomal concentration due to artefactual ubiquitous binding of drugs to microsomal membranes [49].

CYP induction studies with human hepatocytes represent the most relevant preclinical experimental system for the evaluation of human CYP induction. As of this writing, all drugs known to induce CYP isoforms in humans in vivo are also inductive in human hepatocytes. We have recently observed that rifampin has induction potential towards the sulfation of ethynyl estradiol, the active ingredient of oral contraceptives [50]. Our findings suggest that rifampin/oral contraceptive interaction may involve the induction of estrogen sulfotransferase activity. This finding is consistent with the known extensive sulfation of ethynyl estradiol in vivo. Our results also suggest that human hepatocytes can be used to evaluate drug–drug interactions involving phase II metabolism pathways.

The HrTS assays with human hepatocytes described here are made feasible because of the success in cryopreservation. The advantages of cryopreserved human hepatocytes over freshly isolated hepatocytes include the following:

1. Experiments with cryopreserved human hepatocytes can be planned, unlike experiments with freshly isolated human hepatocytes. The need for a fresh liver was a major drawback in the past for experiments with human hepatocytes. With cryopreserved human hepatocytes this is no longer a problem.

2. Studies can be performed using hepatocytes that have been previously characterized. That hepatocytes from specific individuals with desired drug-metabolizing activities could be selected for experimentation from the cryopreserved human hepatocyte bank is a definite advantage. A bank of hepatocytes from poor metabolizers for polymorphic CYP isoforms such as CYP2D6 and CYP2C19 is being created so that experiments can be designed for the comparison of metabolic fate and/or toxicity between poor metabolizers and extensive metabolizers.

3. Assays can be performed at different times with cryopreserved hepatocytes from the same donor, thereby allowing repeat experimentation. This is not possible with freshly isolated hepatocytes.

4. Hepatocytes can be pooled from multiple donors for experimentation. Pooling of microsomes from different donors to provide a representation of a ''generic'' human is a common practice, and the same approach can be applied to cryopreserved human hepatocytes.

The success in cryopreservation of human hepatocytes represents a major technological advance. It allows research on human drug metabolism and toxicity to be routinely performed in various laboratories, including those that currently have no access to human livers. The application of cryopreserved human hepatocytes to the screening of new chemical entities using the HrTS assays described here can be implemented with relative ease in laboratories dealing with drug metabolism and toxicity and should improve the current practices in the selection of drug candidates for development.

In our laboratory, research is currently underway to define limitations and develop new applications of the applications of human hepatocytes in drug development. Our latest findings include the definition of the effects of organic solvents in xenobiotic metabolism [51], and the induction of Phase II conjugative metabolism [50]. We believe that human hepatocytes represent an important experimental system for the evaluation of human drug properties.

REFERENCES

1. DJ Jenden. Difficulties in using animal data to predict pharmacological response in man. Neurosci Biobehav Rev 15:105–107, 1991.
2. T Green. Species differences in carcinogenicity: the role of metabolism in human risk evaluation. Teratog Carcinog Mutagen 10:03–113, 1990.
3. JH Lin. Species similarities and differences in pharmacokinetics. Drug Metab Dispos 23:1008–1021, 1995.
4. HW Ruelius. Extrapolation from animals to man: predictions, pitfalls and perspectives. Xenobiotica 17:255–265, 1987.
5. DF Lewis, C Cloannides, DV Parke. Cytochrome P450 and species differences in xenobiotic metabolism and activation of carcinogens. Environ Health Perspect 106: 633–641, 1998.
6. S Wrighton, AM VandenBranden, BJ Ring. The human drug metabolizing cytochrome P450. J Pharmacokinet Biopharm 24:461–473, 1996.
7. FP Guengerich. Comparisons of catalytic selectivity of cytochrome P450 subfamily enzymes from different species. Chem Biol Interact 106:61–182, 1997.
8. AP Li. Primary hepatocyte cultures as an in vitro experimental model for xenobiotic metabolism and toxicology. Comments Toxicol 6:99–219, 1998.
9. AP Li. The scientific basis of drug–drug interactions: mechanism and preclinical evaluation. Drug Information Journal 32:657–664, 1998.
10. K Lee, Y Vandenberghe, M Herin, R Herin, D Cavalier, D Beck, AP Li, N Verbekes, M Lesne, J Roba. Comparative metabolism of SC-42867 and SC-51089, two PGE$_2$ antagonists, in rat and human hepatocyte cultures. Xenobiotica 24:25–36, 1994.
11. M Jurima-Romet, HS Huang, DJ Beck, AP Li. Evaluation of drug interactions in intact hepatocytes: inhibition of terfenadine metabolism. In Vitro Toxicol 10:655–663, 1996.
12. AP Li, A Rasmussen, DL Kaminski. Substrates of human hepatic cytochrome P450 3A4. Toxicology 104:1–8, 1995.
13. AP Li, MA Roque, DJ Beck, DL Kaminski. Isolation and culturing of hepatocytes from human liver. J Tiss Culture Methods 14:139–146, 1992.
14. LJ Loretz, AGE Wilson, AP Li. Promutagen activation by freshly isolated and cryopreserved rat hepatocytes. Environmental Molecular Mutagenesis 12:335–341, 1988.
15. LJ Loretz, AP Li, MW Flye, AGE Wilson. Optimization of cryopreservation procedures for rat and human hepatocytes. Xenobiotica 19:489–498, 1989.
16. SM Moerlein, RA Weisman, DL Beck, AP Li, MJ Welch. Metabolism in vitro of radioiodinated N-isopropyl-para-iodoamphetamine by cultured hepatocytes. Nucl Med Biol 20:49–56, 1993.
17. CE Ruegg, PM Silber, RA Mughal, J Ismail, C Lu, AP Li. P450 induction in primary human hepatocytes after cryopreservation. In Vitro Toxicol 10:217–222, 1997.
18. AP Li, C Lu, JA Brent, C Pham, A Fackett CE Ruegg, P Silber. Cryopreserved human hepatocytes: characterization of drug metabolizing activities and applications in higher throughput screening assays for hepatoxicity, metabolic stability, and drug–drug interaction potential. Chem Biol Interact 121:17–35, 1999.

19. AP Li, M Jurima-Romet. Applications of primary human hepatocytes in the evaluation of pharmacokinetic drug–drug interactions: evaluation of model drugs terfenadine and rifampin. Cell Biol Toxicol 13:365–374, 1997.

20. AP Li, P Maurel, MJ Gomez-Lechon, LC Cheng, M Jurima-Romet. Preclinical evaluation of drug–drug interactions: present status of the application of primary human hepatocytes in the evaluation of cytochrome P450 induction. Chemico-Biological Interactions 107:5–16, 1997.

21. AP Li, MK Reith, A Rasmussen, JC Gorski, SD Hall, L Xu, DL Kaminski, LK Cheng. In vitro evaluation of drug–drug interaction potential: a comparison of rifampin, rifapentine, and rifabutin in cytochrome P450 3A induction potential in primary human hepatocytes. Chemico-Biological Interactions 107:17–30, 1997.

22. AP Li. Evaluation of drug–drug interactions in primary human hepatocytes. Advances Pharmacol 43:103–130, 1997.

23. JB Houston. Utility of in vitro drug metabolism data in predicting in vivo metabolic clearance. Biochem Pharmacol 47:1469–1479, 1994.

24. RE Kane, J Tector, JJ Brems, AP Li, D Kaminski. Sulfation and glucuronidation of AAP by cultured hepatocytes reproducing in vivo sex differences in conjugation on Matrigel and Type 1 collagen. In Vitro Cell Dev Biol 27A:953–960, 1991.

25. RE Kane, AP Li, DR Kaminski. Sulfation and glucuronidation of AAP by human hepatocytes cultured on matrigel and type 1 collagen reproduces conjugation in vivo. Drug Metab Disp 23:303–307, 1995.

26. BP Monahan, CL Ferguson, ES Killeavy, BK Lloyd, J Troy, LR Cantilena Jr. Torsades de pointes occurring in association with terfenadine use. JAMA 264:2788–2790, 1990.

27. S Pohjala-Sintonen, M Viitsalo, L Toivonen, P Neuvonen. Torsades de pointes after terfenadine-itraconazole interaction. Brit Med J 306:186, 1993.

28. CH Yun, A Okerholm, FP Guengerich. Oxidation of the antihistamine drug terfenadine in human liver microsomes. Role of cytochrome P4503A(4) in N-dealkylation and C-hydroxylation. Drug Metab Disp 21:403–409, 1993.

29. RL Woosley, Y Chen, JP Freiman, RA Gillis. Mechanism of the cardiotoxic actions of terfenadine. JAMA 269:1532–1536, 1993.

30. PF D'Arcy. Drug interactions with oral contraceptives. Drug Intell Clin Pharm 20: 353–362, 1986.

31. JM Grange, PA Winstanley, PD Davies. Clinically significant drug interactions with antituberculosis agents. Drug Saf 11:242–251, 1994.

32. MS Benedetti, P Dostert. Induction and autoinduction properties of rifamycin derivatives: a review of animal and human studies. Environ Health Perspect 102:101–105, 1994.

33. C Ged, JM Roullion, L Pichard, J Combalbert, N Bressot, P Bories, H Michel, P Beaune, P Maurel. The increase in urinary excretion of 6β-hydroxycortisol as a marker of human hepatic cytochrome P450 IIIA induction. Br J Clin Pharmacol 38: 373–387, 1989.

34. FP Guengerich. Metabolism of 17α-ethynylestradiol in humans. Life Sciences 47: 1981–1988, 1990.

35. D Reimers, L Nocke-Finck, H Breuer. Rifampicin and the "pill" do not go well together. JAMA 227:608, 1974.

36. MF Fromm, D Busse, HK Kroemer, M Eichelbaum. Differential induction of prehepatic and hepatic metabolism of verapamil by rifampin. Hepatology 24:796–801, 1996.

37. MF Herbert, JP Roberts, T Prueksaritanont, LZ Benet. Bioavailability of cyclosporine with concomitant rifampin administration is markedly less than predicted by hepatic enzyme induction. Clin Pharmacol Ther 52:453–457, 1992.

38. R Lang, B Shasha, E Rubinstein. Therapy of experimental murine brucellosis with streptomycin alone and in combination with ciprofloxacin, doxycycline, and rifampin. Antimicrob Agents Chemother 37:2333–2336, 1993.

39. JD Colmenero, LC Fernandez-Gallardo, JA Agundez, J Sedeno, J Benitez, E Valverde. Possible implications of doxycycline-rifampin interaction in the treatment of brucellosis. Antimicrob Agents Chemother 38:2798–2802, 1994.

40. J Drayton, G Dickinson, MG Rinaldi. Coadministration of rifampin and itraconazole leads to undetectable levels of serum intraconazole (letter). Clin Infect Dis 18:266, 1994.

41. KH Lee, JG Shin, WS Chong, S Kim, JS Lee, IJ Jang, SG Shin. Time course of the changes in prednisolone pharmacokinetics after co-administration or discontinuation of rifampin. Eur J Clin Pharmacol 45:287–289, 1993.

42. KD Gallicano, J Sahai, VK Shulakla, I Seguin, A Pakuts, D Kwok, BC Foster, DW Cameron. Induction of zidovudine glucuronidation and amination pathways by rifampicin in HIV-infected patients. Br J Clin Pharmacol 48:168–179, 1999.

43. JC Merrill, DJ Beck, DA Kaminski, AP Li. Polybrominated biphenyl induction of cytochrome P450 mixed function oxidase activity in primary rat and human hepatocytes. Toxicol 23:147–152, 1995.

44. AP Li, L Xu, A Rasmussen, DL Kaminski. Rifampicin induction of lidocaine metabolism in cultured human hepatocytes. J Pharmacol Exp Therap 274:673–677, 1995.

45. AP Li. Primary hepatocyte culture as an in vitro toxicological system. In: Shayne Gad, ed. In Vitro Toxicology. Boca Raton, FL: CRC Press, 1994, pp 195–220.

46. MJ Gomez-Lechon, A Monloya, P Lopez, T Donato, A Larrauri, JV Castell. The potential use of cultured hepatocytes in predicting the hepatotoxicity of xenobiotics. Xenobiotica 18:725–735, 1988.

47. MH Tarbit, J Berman. High-throughput approaches for evaluating absorption, distribution, metabolism and excretion properties of lead compounds. Curr Opin Chem Biol 2:411–416, 1998.

48. AD Rodrigues, SL Wong. Application of human liver microsomes in metabolism-based drug–drug interactions: in vitro–in vivo correlations and the Abbott Laboratories experience. In: AP Li, ed. Drug–drug Interactions: Scientific and Regulatory Perspectives. San Diego, CA: Academic Press, 1997, pp 65–101.

49. RS Obach. Nonspecific binding to microsomes: impact on scale-up of intrinsic clearance to hepatic clearance as assessed through examination of warfarin, imipramine and propranolol. Drug Metab Dispos 12:1359–1369, 1997.

50. AP Li, NR Hartman, C Lu, JM Collins, JM Strong. Effects of cytochrome P450 inducers on 17α-ethinyloestradiol (EE$_2$) conjugation by primary human hepatocytes. Br J Clin Pharmacol 48:733–742, 1999.

51. J Easterbrook, C Lu, Y Sakai, AP Li. Effects of organic solvents on the activities of the cytochrome P450 isoforms, UDP-dependent glucuronyl transferase, and phenol sulfotransferase in human hepatocytes. Drug Metab Dispos 29:141–144, 2001.

14

High-Throughput Screening for Metabolism-Based Drug–Drug Interactions

Vaughn P. Miller and Charles L. Crespi
GENTEST Corporation, Woburn, Massachusetts

I. INTRODUCTION

The commercial success of a new drug entity (NDE) depends on its pharmacological activity and several ADME (absorption, distribution, metabolism, and excretion) properties. ADME properties influence the amount and frequency of NDE administration and population variability in pharmacokinetics. One important property is the ability of the NDE to cause metabolism-based pharmacokinetic drug–drug interactions. In such an interaction the NDE inhibits the metabolism of a comedication. As a result, the circulating concentrations of the comedication are increased and, if the comedication has a narrow therapeutic index, an adverse reaction can occur. A NDE which causes drug–drug interactions can be more time-consuming and costly to develop, suffer decreased market acceptance, and in some cases can lead to product withdrawal. As a recent example, shortly after approval Posicor [1] (mibefradil) was withdrawn from the marketplace due to drug–drug interactions at the level of cytochrome P450 (CYP) metabolism. Testing for drug–drug interactions has been the subject of a recent FDA guidance document [2]. These competitive and regulatory pressures have created a need to move this testing into the lead optimization phase of drug development.

The kidney is a key organ in the clearance of xenobiotic molecules. Xenobiotic compounds that are too lipophilic to be filtered by the kidney are directly metabolized by the body to more hydrophilic compounds by phase I and phase II reactions. CYP enzyme system is the most important enzymes involved in

403

phase I reactions, though there are a number of other oxidative enzymes capable of metabolizing xenobiotic compounds. The majority of drug–drug interactions are metabolism based, i.e., two or more drugs compete for metabolism by the same enzyme, and the majority of these interactions involve CYP [3,4]. The CYP isoenzymes belong to a superfamily of membrane-bound, heme-containing mixed-function oxygenases that are a principal enzyme system for the metabolism of drugs. These enzymes are expressed in many tissues but in mammals are found at the highest levels in the liver. The liver and many other tissues contain several different CYP forms with different substrate specificities. CYP enzyme system is pivotal in drug clearence. CYP principally function to introduce oxygen into a molecule to increase the hydrophilicity of the product and hence the ease with which the product can be eliminated from the body. If this metabolism is rate limiting for the elimination of the drug, then inhibition of CYP metabolism can inhibit elimination and increase circulating concentrations of the drug.

There are 11 xenobiotic-metabolizing CYPs that are expressed in a typical human liver (CYP1A2, CYP2A6, CYP2B6, CYP2C8/9/18/19, CYP2D6, CYP2E1, and CYP3A4/5). Comprehensive reviews of each of the CYP subfamilies have been recently published [5]. A subset of these enzymes, CYP1A2, CYP2C9, CYP2C19, CYP2D6, and CYP3A4, appear to be responsible for the metabolism of most drugs [6,7]. The importance of these enzymes in drug metabolism is due to their mass abundance (e.g., CYP3A4 is the most abundant CYP in human liver) and their preference for chemical structures commonly found in drugs (e.g., CYP2D6 preferentially binds and metabolizes drugs with basic amine functionalities). In addition, several of these enzymes (e.g., CYP2C9, CYP2C19, and CYP2D6) are polymorphic in humans with a significant percentage of the population either lacking the enzyme or carrying a variant form [8–10].

The rationale for in vitro screens is the single enzyme paradigm for drug–drug interactions. In this paradigm it is assumed that if a NDE inhibits the metabolism of one probe substrate for an enzyme then it inhibits all substrates of that enzyme. Therefore potential drug–drug interactions can be tested on an enzyme-by-enzyme basis with a limited number of probe substrates. The classical approach for in vitro CYP inhibition analysis is to use a drug as a probe substrate and measure inhibition over a range of substrate and inhibitor concentrations. Quantitative measures of inhibition potential, apparent K_i or IC_{50} at a given substrate concentration, are calculated. In lead optimization, a higher throughput mode of operation, it is not always necessary to determine an apparent K_i or IC_{50}. For example, as an initial screen, a single inhibitor concentration can be tested as a means to identify potent (<1 μM) and/or weak inhibitors (>50 μM). More detailed analyses would be performed in follow-up testing.

Traditionally, analysis of the metabolism of a drug probe substrate usually involves HPLC separations that are by definition not high throughput. A number of improvements and refinements to the assay methodology have been reported.

For example, Rodrigues et al. [11–13] have used a radiolabeled drug molecule and measured the formation of the radiolabeled metabolites, either formaldehyde or acetaldehyde. The simple sample workup (for example, charcoal extraction) facilitates parallel processing in a multiwell plate. Similarly, Wynalda and Wienkers [14] have reported the use of radiolabeled substrates with a rapid HPLC analysis.

Much higher data acquisition rates (true high throughput) can be achieved by using direct fluorometric CYP substrates in a multiwell plate format. However, suitable substrates for most of the human drug-metabolizing cytochromes P450 were originally unavailable. Now, the great increase in understanding of the substrate requirements of human CYPs made it possible to develop higher throughput CYP assays. Recently, microtiter plate-based, fluorometric assays for the activities of the five principal drug metabolizing enzymes, CYP1A2, CYP2C9, CYP2C19, CYP2D6, and CYP3A4, have been reported [15]. These assays are based on CYP catalyzed O-dealkylation reactions, which generate an easily detectable fluorescent product. The CYP enzymes can be introduced into the assay as a single cDNA-expressed enzyme or as enzyme mixtures (i.e., tissue fractions such as human liver microsomes). If tissue fractions are to be used, the probe substrate must be selectively metabolized by the enzyme of interest and not by other enzymes.

Since our initial report [15] a number of new substrates and refinements in the assay methodology have been achieved. A summary of the current fluorometric assay methods is a focus of this article. (Note: The most recent information regarding this assay can be found at the internet web site: *www.gentest.com.*)

II. HIGH-THROUGHPUT SCREENING (HTS)-CYP ASSAY

A. Substrates

The following CYP substrates are available with GENTEST Corporation (Woburn, MA): 3-[2-(*N,N*-diethyl-*N*-methylammonium)ethyl]-7-methoxy-4-methylcoumarin (AMMC), 3-[2-(*N,N*-diethyl-*N*-methylammonium)ethyl]-7-hydroxy-4-methylcoumarin (AMHC), 7-benzyloxyquinoline (BQ), 7-hydroxyquinoline (HQ), 7-benzyloxy-4-trifluoromethylcoumarin (BFC), 7-hydroxy-4-trifluoromethylcoumarin (HFC), and 7-methoxy-4-trifluoromethylcoumarin (MFC). The microsome preparations of CYPs also are available with GENTEST: CYP1A2 (catalog # P203), CYP2C9 (catalog # P258), CYP2C19 (catlog # P259), CYP2D6 (catlog # P217), and CYP3A4 (catlog # P202). The catalog numbers are provided because these enzymes are available in multiple formats. The other chemicals, 7-ethoxy-3-cyanocoumarin (CEC), 7-hydroxy-3-cyanocoumarin (CHC), 7-benzyloxyresorufin (BzRes), furafylline, sulfaphenazole, and ketoconazole, are from Ultrafine Chemicals (Manchester, UK).

B. Assay Design

Several different assay designs are possible and appropriate for different applications. The determination of an IC_{50} with eight inhibitor concentrations is described below. However, in a screening mode with large numbers of chemicals it may be desirable to test a limited number of inhibitor concentrations in order to classify a potential NDE as either a potent or a weak inhibitor (for example inhibition at 1 μM or no inhibition at 50 μM). Actual study design is influenced by not only the number of potential NDEs to be tested and the available resources but also the therapeutic area, likely comedications, and the inhibition potential of competing medications. It may not be necessary to screen all five CYP enzymes. For example, there are currently few clinically significant drug–drug interactions involving CYP2C19. Therefore a subset of these enzymes may be sufficient to meet the goals of the program. In addition, for some therapeutic indications (multidrug therapy with drugs of poor oral bioavailablity), drug–drug interactions may be desirable as it can increase the bioavailability of comedications.

The following general parameters affect assay and program design. In order to detect competitive inhibitors, a substrate concentration should be chosen that is close to or below the apparent K_m. Assays should be conducted under initial rate conditions, the formation of the metabolite should be linear with respect to enzyme concentration and incubation time, and the total consumption of the substrate should be less than 20%. All of the substrates can be used in a discrete mode (stopped by the addition of acetonitrile and Tris base). All substrates except AMMC can also be used in a continuous mode at pH 7.4 where the assay is conducted at 37°C in the plate scanner with continuous data acquisition. This approach assures assay linearity but occupies the instrument during the entire incubation period (i.e., lowers throughput). The fluorescence of the AMMC metabolite, AMHC, requires an elevated pH, which prohibits its use in a continuous mode.

The user should be aware that some potential NDEs may bind extensively to microsomal protein [16] and thus the free inhibitor concentration may be less than the nominal concentration. Accordingly, the user may wish to keep the protein concentration as low as possible and standardize the protein concentration by the addition of "control" microsomes (microsomes devoid of CYP). Some potential NDEs may interfere with the assay by either fluorescing at the wavelengths being used to monitor metabolite formation or quenching the fluorescence of the metabolite. In general, the substrates with higher excitation and emission wavelengths are less sensitive to these interferences. However, concurrent (or follow-up) testing for possible interference may be desirable.

Most organic solvents, especially DMSO, are inhibitory to cytochromes P450 [17,18]. Therefore organic solvent concentrations should be kept as low as possible. Use of up to 2% acetonitrile, 1% methanol, and 0.2% DMSO are compatible with the assays described herein. Some enzyme/substrate pairs (CYP1A2/CEC, CYP2D6/AMMC, and CYP3A4/BQ) permit the use of higher DMSO concentrations (up to 0.5%) without substantial enzyme inhibition.

Finally, users should be aware that for CYP3A4 (unlike CYP1A2, CYP2C9, CYP2C19, and CYP2D6) the extent of inhibition is substrate dependent. For some inhibitors, up to 300-fold differences in IC_{50} values can be obtained depending on the substrate being used in the assay. This property is not fully understood, but it appears to be related to the unique ability of CYP3A enzymes to bind multiple molecules often in their active site [19]. As a practical matter, it is prudent to use multiple probe substrates in any screen for CYP3A4 inhibition, and some level of uncertainty will remain regarding the potential to inhibit CYP3A4.

C. Plates and Equipment

96-well microtiter plates are currently the most common plate format. Denser plate formats are also suitable with this assay. However, some adjustment in assay parameters may be needed in order to achieve adequate signal-to-noise ratios. With any in vitro methodology one should remain cognizant that properties of the plates can influence the results. For example, some NDEs may bind to some plastics, and this binding can reduce the free NDE concentration and hence the apparent inhibition potency. In addition, the plate material can affect background fluorescence and hence the signal-to-noise ratio in the assay. In general, black plates tend to have the best signal-to-noise ratio for coumarin derivatives, while white plates have a slightly better signal-to-noise ratio for resorufins.

Liquid handling steps can be performed using most liquid handing systems. We have established the IC_{50} methodology described below on a Packard Multiprobe Ilex liquid handing station for plate setup. Initiation and stopping the incubations can be performed with manual pipettors/liquid handlers (8 or 96 channel) which are more rapid than using the Multiprobe Ilex liquid handling station. Small numbers of chemicals can be entirely tested as a manual operation using multichannel pipettes. However, the repetitive nature of manual testing quickly becomes tedious.

The model of fluorescence plate readers can dramatically affect assay performance. Commercially available fluorescence plate readers vary substantially in sensitivity (>10-fold) and price is a poor indicator of performance. In addition, excitation and emission filters can vary in performance. These properties of the instrumentation have a profound effect on assay sensitivity and hence reproducibility. Some current models of instruments from some manufacturers are unsuitable for this assay. The assay methods described in this article were initially developed on a Cytofluor 2350 and have recently been adapted to a more sensitive

BMG Fluostar model 403. More sensitive fluorescence plate readers, Spectramax-Gemini from Molecular Devices, HTS 7000 from Perkin Elmer, Analyst and Acquyest from LJL biosystems, and many other fluorescence readers from different vendors are available. Users of these methods are well advised to evaluate the sensitivity of an instrument prior to purchase. As a practical matter, a signal-to-noise ratio greater than 3 is required for adequate assay reproducibility.

III. IC_{50} DETERMINATION

The method for the determination of IC_{50} values using an eight-point curve is given in Figure 1. The highest inhibitor concentration is an experimental variable, but it is typically ≥ 100 μM if aqueous solubility permits. Other experimental designs are possible and may be more desirable. Table 1 provides the assay parameters routinely used in our laboratory. The positive control inhibitors (highest inhibitor concentration) for the different enzymes are as follows: CYP1A2, furafylline (100 μM); CYP2C9, sulfaphenazole (10 μM); CYP2C19, tranylcypromine (500 μM); CYP2D6, quinidine (0.5 μM); and CYP3A4, ketoconazole (5 μM).

The substrates are initially prepared in acetonitrile. The final concentrations of the substrates were chosen to be approximately the apparent K_m with the exception of BFC, where the apparent K_m is above the limit of aqueous solubility and hence the concentration chosen is below the apparent K_m.

After addition of buffer, cofactors, and inhibitor (100 μL per well), the plates are prewarmed to 37°C. Incubations are initiated by the addition of 100 μL prewarmed enzyme and substrate. For all enzymes except CYP2D6, the final cofactor concentrations are 1.3 mM NADP+, 3.3 mM glucose-6-phosphate, and 0.4 U/mL glucose-6-phosphate dehydrogenase. For CYP2D6, the final cofactor concentrations are 8.1 μM NADP+, 0.41 mM glucose-6-phosphate, and 0.4 U/mL glucose-6-phosphate dehydrogenase. The lower NADP+ concentration is required in order to avoid interference from the fluorescence of NADPH. Incubations are carried out for 30 min [CYP1A2, CYP2C19, and CYP3A4 (BFC and BQ)] or 45 min [CYP2C9, CYP2D6, and CYP3A4 (BzRes)] and stopped by the addition of 65 μM of 80% acetonitrile and 20% 0.5M Tris base. The CEC metabolite CHC is measured using an excitation wavelength of 409 nm and an emission wavelength of 460 nm. The AMMC metabolite AMHC is measured using an excitation wavelength of 390 nm and an emission wavelength of 460 nm. The BzRes metabolite resorufin is measured using an excitation wavelength of 530 nm and an emission wavelength of 590 nm. The BQ and BFC metabolites HQ and HFC are measured using an excitation wavelength of 409 nm and an emission wavelength of 530 nm. Data can be exported and analyzed using an Excel spreadsheet. The IC_{50} values are calculated by linear interpolation. Figure 1 contains a representative inhibition curve for tranylcypromine and CYP2C19.

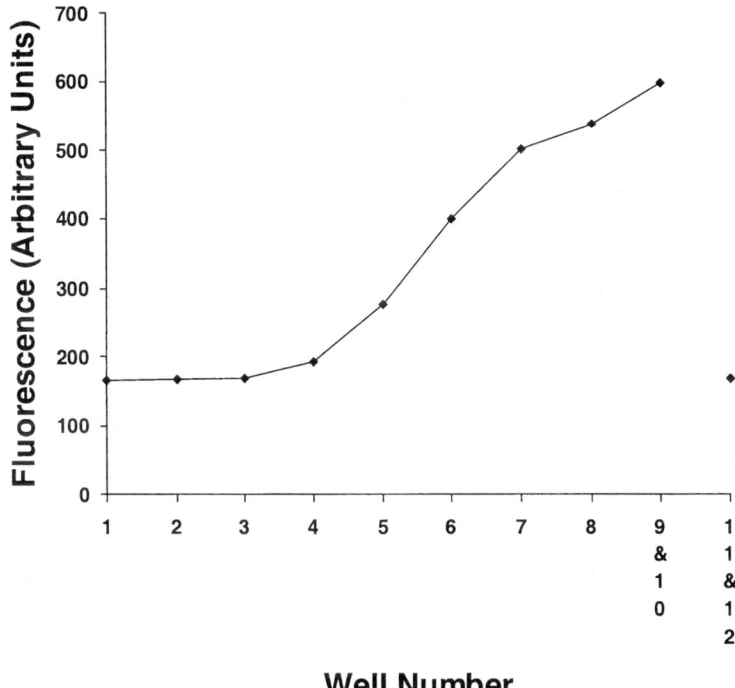

Well Number

Figure 1   Inhibition curve for tranylcypromine inhibition of CYP2C19. The assay was conducted as described in Table 1. The experimental design can be varied as desired. The 12 wells in the row of a 96-well plate can be used for one inhibition curve. Wells 1 to 8 contain serial $1:3$ dilutions of the inhibitors (50 µL transferred between wells containing 100 µL). Wells 9 and 10 contain no inhibitor, and rows 11 and 12 are blanks for background fluorescence (a stop solution is added before the enzyme). Data was acquired using a Cytofluor model 2350 fluorescent plate scanner.

IV.   REFINEMENTS IN THE METHODOLOGY

There are several important refinements to the assay procedure described in this article relative to the originally published method [15]. The purpose of these refinements was to increase the signal-to-noise ratio in the assay, which allowed a reduction in the amount of enzyme necessary to obtain a robust signal. The CYP2C9 and CYP2C19 enzyme preparations have been changed to contain the CYP redox partner cytochrome b_5. Cytochrome b_5 stimulates the rates of metabolite formation for some cytochromes P450 and was originally present with CYP3A4. Incorporation of cytochrome b_5 into the CYP2C9 and CYP2C19 micro-

Table 1 Summary of Assay Conditions

Enzyme	CYP1A2	CYP2C9	CYP2C19	CYP2D6	CYP3A4	CYP3A4	CYP3A4
Substrate	CEC	MFC	CEC	AMMC	BzRes	BQ	BFC
Substrate concentration	5 µM	75 µM	25 µM	1.5 µM	50 µM	40 µM	50 µM
mole enzyme per well	0.5	2	1	1.5	5	5	1
Pot phos buffer concentration	100 mM	25 mM	50 mM	100 mM	200 mM	200 mM	200 mM
Metabolites	CHC	HFC	CHC	AMHC	Resorufin	HQ	HFC
Ex/Em λ	409/460	409/530	409/460	390/460	530/590	409/530	409/530

some preparations increased the rates of CHC formation 4- to 6-fold. This allows a reduction in the amount of enzyme needed in the assay. The presence or absence of cytochrome b_5 does not affect observed IC_{50} values.

The original assay method used 0.1 M Tris (pH 7.5) for CYP2C9 and 0.1 M potassium phosphate (pH 7.4) for the other enzymes. The optimal concentrations are listed in Table 1. Optimization of the buffer concentrations increased metabolite formation by 20 to 50% for CYP2C9, CYP2C19, and CYP3A4.

New substrates have been developed. These include (1) AMMC as a substrate for CYP2D6. Relative to the original substrate, CEC, AMMC has threefold better signal-to-noise ratio per unit enzyme. AMMC O-demethylation is CYP2D6 selective in human liver microsomes (CEC is not enzyme selective). Therefore this substrate can be used with either cDNA-expressed CYP2D6 or human liver microsomes (HLM). However, the low CYP2D6 content in HLM requires the use of high protein concentrations. (2) BQ and BFC as substrates for CYP3A4. Relative to BzRes, both BQ and BFC have higher aqueous solubility, and BFC has a 15-fold better signal-to-noise ratio per unit enzyme. Both BQ and BFC are highly selective for CYP3A in HLM. These substrates can be readily used with either cDNA-expressed CYP3A4 or HLM as an enzyme source.

V. VALIDATION OF IC_{50} VALUES

For CYP1A2, CYP2C9, CYP2C19, and CYP2D6 enzymes, a good correlation was observed between IC_{50} values obtained with the traditional method and with the fluorometric substrates [20]. This has been observed both retrospectively with a panel of 20 established inhibitors and in follow-up traditional testing of potential NDEs, which showed high potency in the fluorometric assays. Similarly, Palamanda et al [21] have found that for 62 compounds, IC_{50} values determined with cDNA-expressed CYP2D6 and the original fluorescent substrate CEC for CYP2D6 were in good agreement with IC_{50} values determined with dextromethorphan and HLMs. However, for 9 of the 62 compounds the difference between the two systems was more than fivefold, and the authors recommend follow-on testing using traditional methodology in HLMs.

The good correlation in IC_{50} values among fluorometric substrates and traditional substrates for CYP1 and CYP2 enzymes does not extend to CYP3A4. This enzyme demonstrates marked substrate specificity in IC_{50} values among different probe substrates. This effect is observed from both traditional substrates and fluorometric substrates. This poor correlation between substrates cannot be easily attributed to differences in experimental conditions. For example, all substrates were used at a concentration around the apparent K_m, and the enzyme was often present at identical concentrations. When BQ and BFC were used to determine IC_{50} values with HLMs and cDNA-expressed CYP3A4, the IC_{50} values obtained were in good agreement. Therefore these differences in IC_{50} values appear not to be related to the source of the enzyme.

Human CYP3A4 has been demonstrated to bind and metabolize simultaneously multiple compounds in its active site [19]. Cooperativity, activation, and complex inhibition kinetics [22,23] are much more common with CYP3A4 than with enzymes of the CYP1 and CYP2 families. For example, a common probe substrate for CYP3A4, testosterone, does not inhibit but activates or stimulates BzRes and BQ dealkylation. At this time, the full extent of the substrate dependence in CYP3A4 inhibition is unknown, and additional research is needed. A prudent approach would be to use multiple CYP3A4 probe substrates in a screening mode and follow-up studies with likely comedications.

VI. TEMPORAL CHANGES IN IC_{50} VALUES

Some CYP inhibitors covalently modify and inactivate the enzymes. Such mechanism-based inhibitors may not be potent competitive inhibitors and thus may not be detected in brief assays for competitive inhibition. Different experimental approaches (typically time- and NADPH-dependent inhibition) are needed to identify this class of inhibitors. Two aspects of the experimental design facilitate the detection of mechanism-based inhibitors. First, the incubation times are relatively long, therefore there is an opportunity for time- and NADPH-dependent inactivation to occur. Second, most of the assays are direct, therefore the plates can be scanned at multiple time points and temporal changes in IC_{50} analyzed. With this approach, a 1000-fold decrease in IC_{50} values was observed for tienilic acid [24] inhibition of CYP2C9 (Fig. 2). Such a substantial time-dependent decrease in IC_{50} values implies that mechanism-based inhibition is occurring. Similar time-dependent decreases were observed for CYP1A2 and furafylline, another

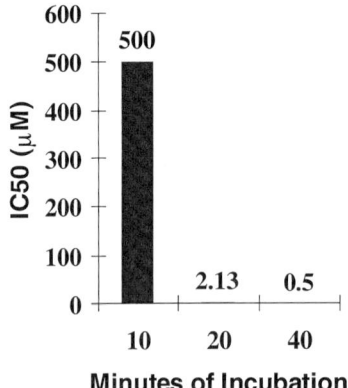

Figure 2 Time-dependent inhibition of CYP2C9 by tienilic acid. The assay was conducted as described in the materials and methods section. Tienilic acid was dissolved in water, and the highest concentration was 1 mM. At the indicated times the plate was scanned and the IC_{50} value calculated. The values above the individual bars are the calculated IC_{50} values.

mechanism-based inhibitor [24]. This type of observation indicates that more detailed studies may be appropriate.

VII.   CONCLUSIONS

Inhibition of drug metabolism by cytochromes P450 (CYP) is a principal mechanism for pharmacokinetic drug–drug interactions. In vitro methods for quantitatively measuring the extent of CYP inhibition have been utilized for several years. Classical methods use drug molecules as substrates and time-consuming, HPLC-based analysis. There has been a need to develop methodologies that do not require HPLC separations for data acquisition.

Considerable progress has been made in increasing the throughput of measured CYP-related properties of a NCE. This progress has come through novel experimental designs like the development of sensitive fluorometric assays for the drug-metabolizing CYPs. Microtiter plate-based, direct, fluorometric assays for the activities of the five principal drug-metabolizing enzymes are now available, and parameters for the use of these substrates to measure CYP inhibition have been established. This methodology is quantitative, rapid, reproducible, and compatible with existing HTS instrumentation. Further refinements in the assays are likely. In particular, a higher fluorescence output of the metabolite and more rapid substrate turnover would further reduce the time and materials needed for conducting these assays. The complexity of CYP3A4 inhibition analysis indicates that the use of a single substrate is inappropriate. Therefore it is desirable to identify additional fluorometric CYP3A4 substrates and an appropriate approach for decision making based on the results of experiments with multiple substrates.

REFERENCES

1. ME Mullins, BZ Horowitz, DHJ Linden, GW Smith, RL Norton, J Stump. Life-threatening interactions of mibefradil and beta-blockers with dihydropyridine calcium channel blockers. J Amer Med Assoc 280:157–158, 1998.

2. Anon. Guidance for Industry. Drug metabolism/drug interaction studies in the drug development process: studies in vitro. USFDA 1997:1–10, 1997.

3. M Murray. P450 enzymes. Inhibition mechanism, genetic regulation and effects of liver disease. Clin Pharmacokinet 23:132–146, 1992.

4. FP Guengerich. Role of cytochrome P450 enzymes in drug–drug interactions. In: AP Li, ed. Drug–Drug Interactions: Scientific and Regulatory Perspectives. San Diego: Academic Press, 1997, pp 7–35.

5. C Ioannides. Part II. Cytochrome P450 Families/Subfamilies. In: C Ioannides, ed. Cytochromes P450: Metabolic and Toxicological Aspects. New York: CRC Press, 1996, pp 77–300.

6. M Spatzenegger, W Jaeger. Clinical importance of hepatic cytochrome P450 in drug metabolism. Drug Metab Rev 27:397–417, 1995.

7. S Rendic, FJ Di Carlo. Human cytochrome P450 enzymes: a status report summarizing their reactions, substrates, inducers and inhibitors. Drug Metab Rev 29:413–580, 1997.

8. FJ Gonzalez, RC Skoda, S Kimura, M Umeno, UM Zanger, DW Nebert, HV Gelboin, JP Hardwick, UA Meyer. Characterization of the common genetic defect in human deficient for debrisoquine metabolism. Nature 331:442–446, 1988.

9. SA Wrighton, JC Stevens, GW Becker, M VandenBranden. Isolation and characterization of human liver cytochrome P450 2C19: correlation between 2C19 and S-mephenytoin 4′-hydroxylase. Arch Biochem Biophys 306:240–245, 1993.

10. H Furuya, P Fernandez-Salguero, W Gregory, H Taber, A Steward, FJ Gonzalez, JR Idle. Genetic polymorphism of CYP2C9 and its effect on warfarin maintenance dose requirement in patients undergoing anticoagulation therapy. Pharmacogenetics 5:389–392, 1995.

11. AD Rodrigues, MJ Kukulka, BW Surber, SB Thomas, JY Uchic, GA Rotert, G Michael, B Thome-Kromer, JM Machinist. Measurement of liver microsomal cytochrome (CYP2D6) using [O-*methyl*-¹⁴C]-dextromethorphan. Anal Biochem 219: 309–320, 1994.

12. AD Rodrigues, MJ Kukulka, EM Roberts, D Ouellet, TR Rodgers. (O-methyl C-14)naproxen O-demethylase activity in human liver microsomes—evidence for the involvement of CYP1A2 and P4502C9/10. Drug Metab Disp 24:126–136, 1996.

13. AD Rodrigues, BW Surber, Y Yao, SL Wong, EM Roberts. [O-Ethyl 14C]Phena-

cetin O-deethylase activity in human liver microsomes. Drug Metab Disp 25:1097–1100, 1997.

14. MA Wynalda, LC Wienkers. Assessment of potential interactions between dopamine receptor agonists and various human cytochrome P450 enzymes using a simple in vitro inhibition screen. Drug Metab Disp 25:1211–1214, 1997.

15. CL Crespi, VP Miller, BW Penman. Microtiter plate assays for inhibition of human, drug-metabolizing cytochromes P450. Anal Biochem 248:188–190, 1997.

16. RS Obach. Nonspecific binding to microsomes: impact on scale-up of in vitro intrinsic clearance to hepatic clearance as assessed through examination of warfarin, imipramine and propranalol. Drug Metab Disp 25:1359–1369, 1997.

17. N Chauret, N Gauthier, DA Nicoll-Griffith. Effect of common solvents on in vitro cytochrome P450-mediated metabolic activities in human liver microsomes. Drug Metab Disp 26:1–4, 1998.

18. D Hickman, J-P Wang, Y Wang, JD Unadkat. Evaluation of the selectivity of *in vitro* probes and suitability of organic solvents for the measurement of human cytochrome P450 monooxygenase activities. Drug Metab Dis 26:207–215, 1998.

19. KR Korzekwa, N Krishnamachary, M Shou, A Ogai, RA Parise, AE Rettie, FJ Gonzalez, TS Tracy. Evaluation of atypical CYTOCHROME P450 kinetics with two-substrate models: evidence that multiple substrates can simultaneously bind to the cytochrome P450 active sites. Biochem 37:4137–4147, 1998.

20. CL Crespi, VP Miller, BW Penman. High throughput screening for inhibition of cytochrome P450 metabolism. Med Chem Res 8:457–471, 1998.

21. JR Palamanda, L Favreau, C-C Lin, AA Nomeir. Validation of a rapid microtiter plate assay to conduct cytochrome P450 2D6 enzyme inhibition studies. Drug Discovery Today 3:466–470, 1998.

22. RW Wang, DJ Newton, TD Scheri, AYH Lu. Human cytochrome P450 3A4-catalyzed testosterone 6(beta)-hydroxylation and erythromycin *N*-demethylation: competition during catalysis. Drug Metab Disp 25:502–507, 1997.

23. YF Ueng, T Kuwabara, YJ Chun, FP Guengerich. Cooperativity in oxidation catalyzed by cytochrome P450 3A4. Biochem 36:370–381, 1997.

24. A Mancy, P Broto, S Dijols, PM Dansette, D Mansuy. The substrate binding site of human liver cytochrome P450 2C9: an approach using designed tienilic acid derivatives and molecular modeling. Biochem 34:10365–10375, 1995.

15

The ATCG of Drug Discovery

Prabhavathi B. Fernandes
Ricerca, LLC, Concord, Ohio

I. INTRODUCTION

A. ... And Now We Know Our ATCGs ...

The publication of human and microbial genomic sequences has resulted in the availability of large amounts of data that are directly relevant to the discovery of novel and useful drug targets. The elucidation of the function of new genes is not a simple matter and human genetic studies have been among the most useful in identifying genes implicated in disease, although homologs, or orthologs, of human genes in mouse, worm, fly, and yeast genomes have also been useful in deciphering gene function. The advent of genomics has led to the study of protein structure and function under the rubric of proteomics. Proteins function in a cell mostly by interacting with other proteins, and protein interaction maps of cellular circuits are now available. Screening strategies to address protein-protein interactions are being developed, and many drug targets in the future are expected to be directed towards these interactions. In addition to using genetic and analytical approaches for finding new drug targets, chemical libraries can be used to inhibit the activity of new proteins, and thus reveal function. The combination of high-throughput screening with testing compounds for absorption, distribution, metabolism, and excretion (ADME) and toxicity will help in the early clarification of clinical utility of both new drug targets, as well as drug candidates.

With the completion of the sequencing of the human genome, strategic thinking relative to drug discovery has shifted towards expected clinical utility of new targets that can be validated [1–3]. Many believe that knowledge of the

Reprinted from *Current Opinion in Molecular Therapeutics* by permission of the publisher, Pharma-Press Ltd. Copyright PharmaPress Ltd, 2000.

sequences of the full complement of human genes has set a limit to the potential number of new drug targets. However, any figure for the number of drug targets derived from gene numbers is bound to be an underestimate, since a simple extrapolation of the number of proteins from the calculated number of gene sequences does not take into account alternative splice variants, post-translational modifications proteins undergo, or that some proteins simply function differently depending upon the identity of their interacting partners. Announcements that the new genomic sequences will trigger a tsunami of new targets that will overwhelm the research and development capabilities of drug discovery companies are premature. It is also apparent that old fashioned biology will still be required to identify and validate those genes that could represent novel drug targets.

The risk involved in drug research and development is multiplied severalfold when the target is novel, since each new target, in addition to the chemical compound itself, must be validated in man. Even in the face of genetic evidence, blocking a protein function with a drug could have different effects from blocking gene function. Knockout mutants and regulated gene expression may not reveal whether a gene represents a good target or not since *drugs block proteins and not genes*. In addition, drugs do not generally block the function of the target protein completely or permanently. A historically-informed perspective relevant to this issue can be obtained from the study of the successful development of angiotensin converting enzyme (ACE) inhibitors for the treatment of hypertension. Two isoforms of ACE are expressed from the same gene but from distinct promoters, and knockout mutants of ACE are embryonic lethals, have severe structural abnormalities in the vascular system and kidneys, demonstrate electrolyte imbalance, and cause male infertility; drugs that block ACE have, however, been proven to be safe and effective in man [4]. These discrepancies between gene knockout and drug-action knockout effects can be explained by the fact that drugs do not block gene function but reversibly block protein function, and during therapy not all of the protein is inhibited all the time. Therefore, scientists have turned to studying protein structure and function as the means of determining the utility of each new protein as a drug target [5,6]. Sequence homology, orthology, and divergence between the genome and protein sequence of other organisms and humans can usefully be employed for determining function [7].

This chapter will cover methods for identifying drug targets from gene sequences, aspects of validation of a new gene as a drug target, and explain how new genes are being used in drug discovery with a special emphasis on determining protein-protein interactions and designing drugs to block protein-protein interactions. In brief, the genomic/proteomics initiative in drug discovery may be defined by three terms: opportunity, complexity, and expected clinical utility.

II. METHODS FOR IDENTIFYING DRUG TARGETS FROM GENE SEQUENCES

New genes are identified as potential drug targets based upon sequence similarity and recognizable motifs compared with targets of known protein classes for which drugs have been successfully developed. For example, many thousands of sequences similar to G protein-coupled receptors (GPCRs), single transmembrane tyrosine kinase receptors, and proteases have been identified in the human genome sequence. Although all of these paralogs and similar proteins may not be useful as drug targets, it is assumed that a fair number will be as useful as their predecessors.

Several issues need to be addressed before considering new gene targets in drug discovery:

1. Which biochemical pathway is the target involved in?
2. Is there evidence of target malfunction in the particular disease?
3. Is there evidence of the target being organ- or tissue-specific?
4. Should the target protein be inhibited or activated?
5. Are there subtypes of the target protein that will need to be isolated to obtain selective drugs?
6. Can the target protein be expressed?

Among the thousands of GPCRs identified many are involved with smell and taste. They could have utility in improving the quality of life in aged patients, but are not useful for discovering drugs against other acute or chronic diseases. Many genes that have essential functions have redundant genes that can compensate for the malfunctioning gene. Many target proteins or their subtypes are distributed widely in a variety of organs and tissues. Thus, blocking the target at all organ sites may be detrimental to the host, and subtype-selective drugs may be required. It is easier to find antagonists than agonists, inhibitors than activators. For example, after much effort, the gene for Werner's syndrome was isolated and the protein identified to be a DNA helicase, a DNA metabolism enzyme that is mutated in disease [8,9]. It is unlikely that we will be able to activate this mutant enzyme. Similarly, mutant p53 must be reverted to the active form in order for a cancer cell to undergo apoptosis. After many years of effort, a small molecule compound that can select for the wild-type p53 conformation was reported and this compound has yet to be developed into a drug [10]. Tissue localization helps in characterization of the target. However, even this can be misleading. For example, an orphan GPCR, later called the orexin receptor, was localized in the hypothalamus. Using established methods of biochemistry and affinity purification a natural ligand for this receptor was identified. Since the receptor was localized in the region of the hypothalamus where a number of receptors for controlling feeding and satiety are located, the receptor was called orexin and was believed to control feeding [11]. More recently, through genetic studies in dogs with narcolepsy, it was found that the orexin receptor was identical to hypocretins that were mutated in narcoleptic dogs [12–15].

III. CHARACTERIZING FUNCTIONS OF NEW GENES

Differential expression of genes from cells derived from diseased tissue versus normal tissue has been used extensively to identify genes that could be involved in disease [16,17]. In one example, differential gene expression was used to identify genes expressed during the aging process. Interestingly, aging induces the expression of stress response genes that can be inhibited by caloric restriction [18]. Although useful to some degree, differential display of expressed genes does not reveal the specific gene or genes implicated in disease, since the expression patterns of genes associated with the genesis of the disease state, as well as of those genes affected by the disease process, are changed. For example, over 300 genes could be differentially expressed in prostate cancer cells relative to normal prostate cells, and still painstaking biological studies on each of these genes could take years to identify the relevant gene(s). In using differential mRNA expression profiles of cells to identify useful genes, it is important to differentiate those mRNA that are actually translated into proteins. A method to separate mRNA that is associated with single ribosomes versus polysomes has been described [19]. This method can be used to identify those genes that are translated to proteins with high probability.

Several mutant human genes have been identified as being responsible for disease or at least associated with disease [20,21]. In the absence of known mutations in human genes, mutant genes can be identified in animals and yeast to establish the function of new genes. Tagging and mutagenesis of the yeast genome has been useful in analyzing the functional effects of each gene [22]. Gene fusions are also used to identify protein function in *Esherichia coli*, and the interactions of proteins of unknown function with proteins of known function can lead to the elucidation of the role of the protein in question [23]. The functional link can be determined by co-expression of mRNA as well as by domain fusions. Comparative databases are being developed to analyze protein families, domain fusions, protein interactions and protein expression patterns in order to better predict the function of new proteins [24,25]. Thus, a variety of methods are needed to characterize gene function.

IV. VALIDATION OF NEW TARGETS

Genetic methods have been applied to the validation of genes as drug targets. Koch's postulates, devised as a set of criteria for determining whether a given microbe is the cause of a particular infectious disease, can also be applied to the validation of new genes as drug targets. In applying these postulates to the field of genomics, if a new gene putatively implicated in disease is isolated, a mutant form of it should result in the disease phenotype, and reintroduction of the normal gene should revert the phenotype to wild-type. Mice carrying a mutant *p21* gene and lacking p21 protein are no longer susceptible to renal hypertrophy or hyperplasia after renal ablation. The lack of a functional *p21*$^{wafl/cip1}$ gene ameliorates progression to chronic renal failure [26]. Thus, p21 could be a drug target for renal hyperplasia. In the case of polygenic diseases, the complexity of proving the association of two or more genes with a single disease is complicated and family studies have been useful.

Geneticists are able to take advantage of dominant-negative mutations in order to block the function of wild-type genes, in which the introduction dominant-negative mutation results in the phenotype of a functional knockout. In one example, transdominant peptides that block the yeast pheromone pathways were used [27]. This method for genetic selection can be used for validation of targets in cellular systems. Examining each protein in isolation can be misleading in context of the cell as well as the tissue that it is expressed in [28]. For example, proteins such as p53, NFκB and RAS can have growth-promoting or apoptotic functions depending upon the situation they are in. Databases that combine information from protein structure and function analysis, genetics and cellular pathways will be needed to understand the physiological functioning of each protein in the cell.

V. USING GENOME SEQUENCES OF SIMPLE ORGANISMS

The obstacles previously faced in obtaining and analyzing mammalian DNA and protein sequences has for a long time driven researchers to study organisms more amenable to experimental manipulation such as *Saccharomyces* [29], *Caenorhabditis elegans*, *Drosophila*, and the Zebra fish. *C. elegans* has many features that make it attractive as an experimental model organism: it has a short life-cycle that is completed in days as opposed to months or years in mammals, it is able to grow on culture plates, and is amenable to both classical and molecular genetic studies. These advantages make *C. elegans* a useful tool for studying the complexity of cellular interactions, which are broadly similar in *C. elegans* and other multicellular animals. Mutations in orthologs of human genes can provide a good starting point to identifying gene function, with the caveat that there are significant differences between worms and man. A consortium has been created to knock out each of the 19,099 genes of *C. elegans* in order to identify the complete function of the genome [31]. Interestingly, many gene knockouts show no deleterious phenotype. Also, in many cases orthologs may not be identified from comparisons with the worm or fly due to evolutionary divergence.

VI. MICROBIAL TARGETS

Gene sequencing has opened new possibilities for treating infectious diseases. Microbial drug resistance is on the increase and the complete genome sequences of over two dozen microbes are now available. Surprisingly, the function of about half of *E. coli* genes are not known and new pathways are being revealed [32,33]. Since bacterial genes occur in operons, polycistronic message analysis as well as identification of neighboring genes may provide hints of the function of the new genes. Clustered genes can be studied using bacterial artificial chromosomes (BACs) [34]. Finally, the cloning and expression of bacterial genes and proteins is not a difficult task and structural analysis can reveal potentially useful information about function. Many new metabolic pathways have been elucidated, and the sequences of the genes encoding the components of these pathways have been compared across species, including bacterial organisms; genes that do not share significant similarity with mammalian genes, which are essential for viability, and which are found in a wide range of bacterial species have been used in drug discovery as new anti-infective targets. Interestingly, targets such as dihydrofolate reductase and DNA gyrase that have orthologs in yeast and human cells have yielded selective and safe drugs. This enigma, in being able to identify selective inhibitors in spite of gene sequence similarity, extends to proteins that exist in yeast and human cells, such as the shared 37% sequence similarity in the gene for 14α-lanosterol demethylase. Therefore, the structure of the protein target is possibly of even greater importance than the gene sequence.

One approach to validating targets is to develop peptide libraries that can bind and block function. Screens can be developed to identify inhibitors, simply by over- or underexpressing the new gene. Differential growth inhibitory activity is expected with a change in gene dosage. Thus, new genes are used in drug discovery without knowing function. In addition to cellular assays and biochemical assays, some laboratories prefer to isolate the protein and use affinity selection methods to identify compounds that bind to the target protein using analytical techniques, such as high-pressure liquid chromatography and mass spectrometry. It is assumed that among the compounds that bind will be some compounds that can affect function. As with cell-free enzyme-based screens, this cell-free approach will not address cell permeability. A cell-based method for developing screens is to use the concept that the expression levels of the gene are affected by accumulation of intermediates in the pathway. This method uses easy-to-read colorimeteric reporters such as β-galactosidase or promoter fusions with *mudlac*. Yet, another time-tested method is to under- or overexpress the target gene to develop a screen based upon differential sensitivity to inhibitors. These methods have the advantage of identifying molecules that not only penetrate the cell but also inhibit the target.

The development of bacterial genomics has faced a number of challenges, such as unstable mRNA and the expression of different genes in vivo compared to in vitro [33]. Affymetrix Inc (CA) has developed a system for fluorescence labeling of nonpolyadenylated mRNA, thus enabling more accurate analysis of gene expression using oligonucleotide microarray. Several methods have been used for identifying genes that are expressed only in vivo and are essential for function. Simply described, these methods involve labeling all genes and differentiating those genes that are expressed only in vivo. Genes that are essential for viability in animals can be determined by random insertions of the mariner-based transposon TnAraOut into essential genes [36]. Genome sequences will help in finding new drugs to combat difficult-to-treat bacterial infections, such as tuberculosis. The genome sequence of *Mycobacterium tuberculosis* has revealed new enzymes involved with lipid biosynthesis that may provide selective antimycobacterial targets [37]. In addition to this traditional type of target, the genome sequence of *M. tuberculosis* has revealed novel glycine-rich proteins that may be responsible for its genetic variation and the ability of the organism to remain dormant in the host cell until the host is immunocompromised. Four types of polypeptide synthases have been found and the metabolites of these enzymes may be involved in immunosuppression, just as cyclosporin has immunosuppressive activity. These proteins could prove to be the Achilles heel of this virulent microorganism.

VII. DETERMINING PROTEIN–PROTEIN INTERACTION

Understanding the interactions of a given protein with other proteins and macromolecules within a cell goes a long way towards understanding the biology of the cell and also towards understanding the role of individual proteins [38–43]. Protein interaction maps are being developed by using the yeast two-hybrid technology and a complete protein interaction map of *Saccharomyces cerevisiae* has been published [44,45]. In order to obtain a complete analysis, 6000 ORFs were fused to the Gal4 activation domain and also to the DNA-binding domains to find all possible interactions. Interestingly, some interactions were identified only when fused to the DNA-binding domain or to the activation domain in the yeast two-hybrid system. This map has helped in placing some genes of unknown function into functional pathways. Mapping the interactions from the protein of interest can lead to proteins of known function. The function of the new gene is then believed to be in the pathway or function of the gene with known function.

In order to have a cellular environment in which proteins undergo posttranslational modifications similar to those in a human cell, a novel mammalian cell-based two-hybrid system was developed [46]. Instead of enzyme detection, cells with interacting proteins can be directly measured using a fluorescence activated cell sorter. Also, two different bacterial two-hybrid systems have been described that could circumvent some of the disadvantages of the yeast system, i.e., slow growth, low permeability, and also the nuclear location of the interacting proteins [47,48]. The newly developed bacterial two-hybrid systems allow for genetic selection. One system is useful for homodimeric proteins and utilizes the λ repressor protein that binds its operator as a homodimer at its N-terminal DNA-binding domain. The C-terminal of the λ repressor is replaced with the homodimerizing protein of interest, so that when the molecules are blocked from dimerizing by a peptide or small molecule, the repression is lifted and expression of the fused lacZ-tet reporters are expressed. The second system also uses a fused transcriptional system with the DNA-binding domain of one protein made of three zinc fingers fused to the first homo- or heterodimerizing protein of interest. The partner of the homo- or heterodimerizing protein is fused to the *E. coli* RNA polymerase α subunit. The selection is provided by a novel reporter in *E. coli* that has been previously used for selections in *Saccharomyces*. This reporter is constructed by deleting the chromosomal *hisB* gene and using the yeast *HIS3* gene in the plasmid construct. *HIS3* can complement the *hisB* defect in *E. coli*. Dimerization of proteins is detected by growth on medium lacking histidine. In addition to two-hybrid methods, a variety of technologies such as 2-D gel electrophoresis, mass spectrometry, NMR, peptide fingerprinting of proteins and high-throughput screening with functional assays have been described as proteomics [39,42,43]. Linking these protein databases to genomic sequence databases will add value, but eventually relevant clinical information must be linked with novel proteins to make them useful in drug discovery.

VIII. DESIGNING INHIBITORS AGAINST NEW TARGETS

Designing drugs against new targets is a time consuming and expensive process and serious consideration must be given to determining the validity of the protein as a useful drug target [49]. Based upon homology and distinctive motifs belonging to classes of proteins that have traditionally been used as drug targets, thousands of GPCRs, proteases, and kinases have been identified [50]. Many of these newly identified genes code for proteins whose function is unknown and which are known as orphan targets. These orphan targets are useful for drug discovery only if their functions can be defined, and genetic means of elucidating function are time consuming and expensive. Therefore, a more rapid means of identifying function has been described whereby diverse chemical libraries are used to find selective inhibitors on the basis that an inhibitor is able to obtain a phenotype in a cell [51]. The inhibitor is chemically optimized based upon its desired activity. This method, called chemical genomics [52,53] identifies compounds that are specific inhibitors of new proteins that can be used in the validation of new targets in animal models.

Approximately 1 in 12 lead molecules that reach phase I clinical trials are expected to survive through clinical development. Failures are mostly unrelated to the compound's lack of selectivity for the target protein, but are more often the consequence of toxicity and bioavailability issues [54,55]. It can be assumed that compounds that bind the same target site have similar structure; thus, diversification of chemical structures to find different types of pharmacophores to build analogs will reduce the risk that is associated with the unpredictable nature of drug toxicity and bioavailability of a single type of molecule. Advanced techniques to identify the structure of proteins will help in identifying alternate sites that could affect the structure and therefore the function of the protein.

It is believed that many drugs fail during clinical development because of the heterogeneity of drug metabolism in patient populations [56]. One approach to taking differences within populations into consideration while optimizing lead molecules is to test compound libraries designed around lead molecules in cytochrome P450 enzyme assays. These predictive studies may help in decreasing the attrition rate of drugs in development. Chemical libraries have been used to find potent and selective inhibitors of kinases [57]. Searching the human genome reveals about 2000 kinases. The purine scaffold was used to build chemical libraries to find inhibitors of cell cycle kinases. It is proposed that specific and selective inhibitors can be used to determine the potential utility of these kinases as targets for drugs. The inhibitors bind to the ATP site of the kinases and with the vast variety of kinases the scaffold may have to be varied to find even more selective compounds. Although simplistic in approach, in actual fact the involvement of each of the kinases in a particular human disease would have to be defined as well as the possibility that some of the kinases may act in cooperation with another protein to cause disease. Keeping in mind these limitations, chemical genomics can be used to identify specific inhibitors of new targets.

Proteins interact with other proteins, such as ligands with receptors, through large hydrophobic pockets. The hydrophobic pocket can be designed to identify small molecule ligands by mutagenizing the receptor and ligand interacting site to reduce their affinity [58]. A small-molecule indole-containing library was screened using a human growth hormone mutant receptor and mutant human growth factor ligand to find molecules that increased the binding affinity of the mutant receptor to mutant ligand. This strategy could be generalized to finding small molecules that modulate protein-protein interactions in a variety of cellular systems. In another strategy, disulfide-containing small-molecule ligands were tethered to the target protein of interest if the small molecule shows inherent affinity to the protein [59]. This method was validated by identifying a potent and selective inhibitor of thymidylate synthase.

As with kinases, protease motifs are common in the genome sequence. It is estimated that about 2% of coding sequences code for proteases [60]. One method to identify the substrate specificity of new proteases is to use a mixture of complex substrates in solution with a combinatorial library of fluorogenic substrates [60]. In this method, cleavage of fluorescent substrates can be monitored readily and the substrate characterized rapidly.

Many methods are under development for the high-throughput determination of protein structure [43,61–63]. Databases are being developed that include 3-D protein structure and protein fold information derived from NMR, computational structure and molecular surface analysis, and X-ray diffraction data. High-flux synchroton radiation sources have allowed the rapid collection of data [64]. Many hundreds of new structures are being added each year to the protein databases (e.g., Brookhaven Protein Database). The Argonne National Laboratories (Chicago, IL) are amongst the front runners in determining protein structure. It is predicted that there are 5000 to 10,000 different folds by which all human proteins can be classified [65]. The expression of large amounts of protein, nucleation and crystallization of proteins, especially membrane proteins, has, however, been limited. Bioinformatics is being developed to automate data retrieval and to filter, cluster, analyze, and visualize data. Representation of each fold, and different sequences that can give the same fold, will be revealed by these data. Such analysis will help in understanding how two proteins with dissimilar sequences can perform similar functions as well as make better predictions of structure from DNA sequence.

Eventually, it is expected that inhibitors will be designed using protein fold information. Analytical software for reconstruction of functional organization of a microorganism based upon whole genome sequences has been designed (WIT-Pro). Other software is used for similarity, phylogenetic, functional and spatial clustering of genes and proteins [66]. The Argonne National Laboratories is developing an integrated system using DNA sequence and protein data from 38 genomes to understand higher-level functional systems and the physiology of the whole organism.

IX. ASSAY SYSTEMS FOR ADDRESSING NEW TARGETS

Assays can be developed for orphan GPCRs, proteases, kinases, and phosphatases without knowing the cellular function of the new protein. For other classes of proteins, especially proteins with unknown activity, alternative technologies are being developed. Figure 1 summarizes how new gene targets and protein technologies are being coupled with combinatorial chemistry, microarrays for gene expression and in vivo studies to develop drugs.

The yeast two-hybrid system is being used extensively to develop screens to find inhibitors of protein-protein interactions [67]. Microbes have served as surrogate systems to study new mammalian genes in a cellular milieu. An advantage of microbial systems is that microbes can be arrayed on microchips. A method to array yeast cells by coating them on microscope slides mixed with silicon dioxide has been described. As the spread dries the yeast cells are displayed in hexagonal arrays in the porous silica support that allows molecules to

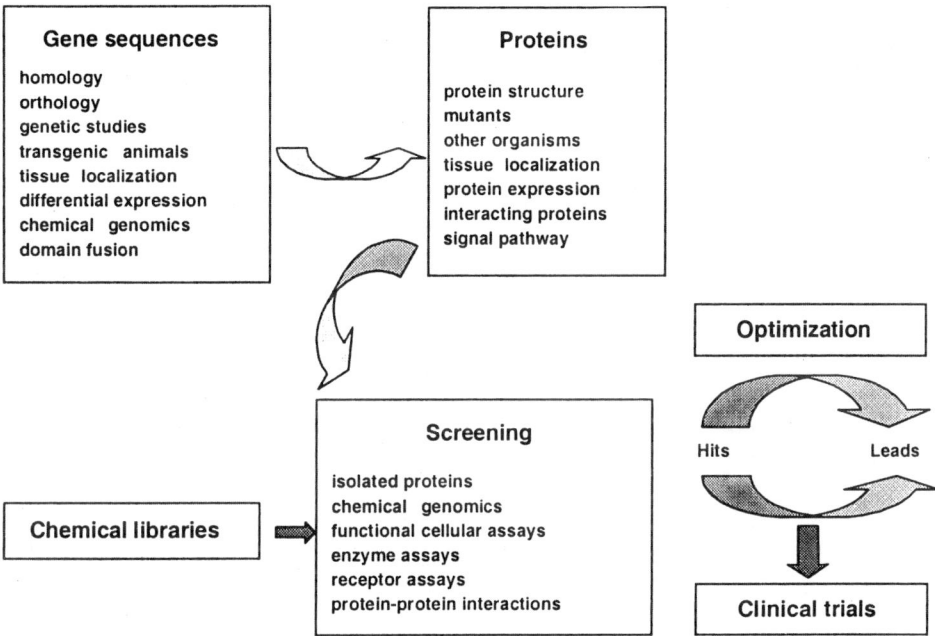

Figure 1 The path from genomics to drugs.

penetrate readily and can be used as biosensors. A fluorescent DNA probe that penetrated the silicon layer was used to show the practicality of this array approach for studying heterologously expressed genes in yeast to examine the function of relevant human targets [68].

Methods to identify the activation of whole pathways are being developed. In one method, the signaling pathways resulting from activation of T-lymphocytes was determined by using promoterless β-lactamase reporter fusion to a number of genes [69]. A fluorogenic substrate is used to detect the activation or repression of individual reporter tagged genes in the T-cells. This is a powerful method for detecting functional pathways that respond to external stimuli. Proteins in pathways can also be identified by activation of the receptor involved with the pathway of interest, separation of proteins using 2-D gel electrophoresis and then detection of phosphorylated proteins using phosphotyrosine and phosphoserine antibodies [70]. This is a reliable, but tedious, means of identifying the pathways in which new proteins are involved. Alternatively, multiple fluorescent-labeled cellular proteins can be used to screen for inhibitors that act at specific pathways and targets [71]. Compounds that act in these systems are expected to have high value as they show selectivity to specific pathways and work in cellular systems.

In addition to cell-based technologies, many technologies are available to identify small molecules that bind with high affinity to new proteins [72]. These methods utilize classical affinity techniques coupled to mass spectrometry, NMR, scanning calorimery and other physical methods to identify the bound molecules; those that are identified must later be tested in cellular systems to show that they do have functional activity. Structural approaches can be used to design drugs to block protein-protein interactions. The limiting factor to the use of NMR is the inability to get concentrations of 0.1 mM of protein in solution. Recently, an NMR technique called transverse relaxation optimized spectroscopy (TROSY) has been described. TROSY can be a powerful tool in studying protein-protein interactions and for studying the structure of proteins larger than 100 kDa. Transverse relaxation optimized spectroscopy does not use the traditional 2-D NMR spectra, but focuses on one of the four component lines and uses it as the correlation peak [73,74]. Microarrays have been used to identify DNA-binding proteins that interact with oligonucleotides. In this method, 40 mer-long double stranded oligonucleotides are immobilized by their 3' end in arrays. The DNA is labeled during the hybridization and priming steps and was used to bind methylases and restriction enzymes [75].

X. CONCLUSION

Genetic studies have helped in identifying the genes involved with many monogenic diseases, such as cystic fibrosis, Duchenne muscular dystrophy, hemophilia A, and *BRCA1* and breast cancer [76]. Identifying the disease genes in many of these cases has not simplified the discovery of drugs to treat these diseases. Genotyping of DNA variations of individuals, called single nucleotide polymorphisms (SNPs), is expected to help in developing drugs that are effectively tailored to individual needs. Genomics has led to proteomics and to studies for validating new proteins as useful drug targets. However, there are so many new and interesting proteins that the challenge is to focus on those with the most promise to yield drugs [77]. More than 90% of new drug candidates that enter clinical development fail because of unpredicted toxicities or lack of efficacy [20]. Genotyping the patient population to identify known risk and metabolism factors will help in designing clinical trials and identifying the patients with the gene mutation. This process replicates the genetic identification of the drug target that is part of the early drug discovery process. Finally, the cloudy issue of patents that do not allow the use of new genes for commercial purposes, including the expression of the protein for developing screens, could restrict the use of many new genes that have been revealed [78]. It could indeed become as restrictive as to say that the oak tree was patented after finding the acorn [79]. Finding a gene is as far from a non-protein small drug as an acorn is from an oak tree, if not further.

ACKNOWLEDGMENTS

To my numerous colleagues, too many to name individually, who I have worked with and with whom I have learned many lessons in drug discovery.

REFERENCES

1. SAJR Aparicio. How to count . . . human genes. Nature Genet 25:129–130, 2000.
2. RH Reeves. Recounting a genetic story. Nature 405:283–284, 2000.
3. A Pandey, M Mann. Proteomics to study genes and genomes. Nature 405:837–846, 2000.
4. SP Kessler, TM Rowe, JB Gomos, PM Kessler, GC Sen. Physiological nonequivalence of the two isoforms of angiotensin-converting enzyme. J Biol Chem 275: 26259–26264, 2000.
5. A Dove. Proteomics: translating proteomics into products? Nature Biotechnol 17: 233–236, 1999.
6. AR Mendelsohn, R Brent. Postgenomic protein analysis: the next bend in the river. Nature Biotechnol 16:520–521, 1998.
7. SA Chervitz, L Aravind, G Sherlock, CA Ball, EV Koonin, SS Dwight, MA Harris, K Dolinski, S Mohr, T Smith, S Weng, JM Cherry, D Botstein. Comparison of the complete protein sets of worm and yeast: orthology and divergence. Science 282: 2022–2028, 1998.
8. E Pennisi. Premature aging gene discovered. Science 272:193–194, 1996.
9. CE Yu, J Oshima, YH Fu, EM Wijsman, F Hisama, R Alisch, S Matthews, J Nakura, T Miki, S Ouais, GM Martin, J Mulligan, GD Schellenberg. Positional cloning of the Werner's syndrome gene. Science 272:258–262, 1996.
10. BA Foster, HA Coffey, MJ Morin, F Rastinejad. Pharmacological rescue of mutant p53 conformation and function. Science 286:2507–2510, 1999.
11. T Sakurai, A Amemiya, M Ishii, I Matsuzaki, RM Chemelli, H Tanaka, SC Williams, JA Richardson, GP Kozlowski, S Wilson, JR Arch, RE Buckingham, AC Haynes, SA Carr, RS Annan, DE McNulty, WS Liu, JA Terrett, NA Elshourbagy, DJ Bergsma, M Yanagisawa. Orexins and orexin receptors: a family of hypothalamic neuropeptides and G protein-coupled receptors that regulate feeding behavior. Cell 92: 573–585, 1998.
12. JM Siegel. Narcolepsy: a key role for hypocretins. Cell 98:409–412, 1999.
13. RM Chemelli, JT Willie, CM Sinton, JK Elmquist, T Scammell, C Lee, JA Richardson, SC Williams, Y Xiong, Y Kisanuki, TE Fitch, M Nakazato, RE Hammer, CB Saper, M Yanagisawa. Narcolepsy in orexin knockout mice: molecular genetics of sleep regulation. Cell 98:437–451, 1999.

14. MS Mondal, M Nakazato, S Matsukura. Orexin hypocretins: novel hypothalamic peptides with divergent functions. Biochem Cell Biol 78:299–305, 2000.

15. S WK amson, ZT Resch. The hypocretin/orexin story. Trends Endocrinol Metab 11:257–262, 2000.

16. P Spence. Obtaining value from the human genome: a challenge for the pharmaceutical industry. Drug Disc Today 3:179–188, 1988.

17. MJ Dunn. Studying heart disease using the proteomic approach. Drug Disc Today 5:76–84, 2000.

18. CK Lee, RG Klopp, R Weindruch, TA Prolla. Gene expression profile of aging and its retardation by caloric restriction. Science 285:1390–1393, 1999.

19. Q Zong, M Schummer, L Hood, DR Morris. Messenger RNA translation state: the second dimension of high-throughput expression screening. Proc Natl Acad Sci USA 96:10632–10636, 1999.

20. PW Kleyn, ES Vesell. Genetic variation as a guide to drug development. Science 281:1820–1821, 1998.

21. DA Haber, ER Fearon. The promise of cancer genetics. Lancet 351Suppl 2:1–8, 1998.

22. P Ross-MacDonald, PS Coelho, T Roemer, S Agarwal, A Kumar, R Jansen, KH Cheung, A Sheehan, D Symoniatis, L Umansky M Heidtman, FK Nelson, H Iwasaki, K Hager, M Gerstein, P Miller, GS Roeder, M Snyder. Large-scale analysis of the yeast genome by transposon tagging and gene disruption. Nature 402:413–418, 1999.

23. AJ Enright, I Illiopoulos, NC Kyrpides, CA Ouzounis. Protein interaction maps for complete genomes based on gene fusion events. Nature 402:86–90, 1999.

24. EM Marcotte, M Pellegrini, MJ Thompson, TO Yeates, D Eisenberg. A combined algorithm for genome-wide prediction of protein function. Nature 402:83–86, 1999.

25. MY Galperin, EV Koonin. Who's your neighbor? New computational approaches for functional genomics. Nature Biotechnol 18:609–613, 2000.

26. J Megyesi, PM Price, E Tamayo, RL Safirstein. The lack of a functional p21WAF1/CIP1 gene ameliorates progression to chronic renal failure. Proc Natl Acad Sci USA 96:10830–10835, 2000.

27. G Caponigro, MR Abedi, AP Hurlburt, A Maxfield, WI Judd, A Kamb. Transdominant genetic analysis of a growth control pathway. Proc Natl Acad Sci USA 95:7508–7513, 1998.

28. S Huang. The practical problems of post-genomic biology. Nature Biotechnol 18:471–472, 2000.

29. GA Offeau, BG Barrell, H Bussey, RW Davis, B Dujon, H Feldmann, F Galibert, JD Hoheisel, C Jacq, M Johnston, EJ Louis, HW Mewes, Y Murakami, P Philippsen, H Tettelin, SG Oliver. Life with 6000 genes. Science 274:546–567, 1996.

30. The *C. elegans* Sequencing Consortium. Genome sequence of the nematode *C. elegans*: a platform for investigating biology. Science 282:2012–2018, 1998.

31. Plasterk RHA. Hershey heaven and *Caenorhabditis elegans*. Nature Genet 21:63–64, 1999.

32. FR Blattner, G Plunkett 3rd, CA Bloch, NT Perna, V Burland, M Riley, J Collado-Vides, JD Glasner, CK Rode, GF Mayhew, J Gregor, NW Davis, HA Kirkpatrick,

MA Goeden, DJ Rose, B Mau, Y Shao. The complete genome sequence of *Escherichia coli* K-12. Science 277:1453–1474, 1997.

33. DT Moir, KJ Shaw, RS Hare, GF Vovis. Genomics and antimicrobial drug discovery. Antimicrob Agents Chemother 43:439–466, 1999.
34. MR Rondon, SJ Raffel, RM Goodman, J Handelsman. Towards functional genomics in bacteria: Analysis of gene expression in *Escherichia coli* from a bacterial artificial chromosome library of *Bacillus cereus*. Proc Natl Acad Sci USA 96:6451–6455, 1999.
35. TR Gingeras, C Rosenow. Studying microbial genomes with high-density oligonucleotide arrays. Am Soc Microbiol News 66:463–469, 2000.
36. N Judson, JJ Mekalanos. TnAraOut, a transposon-based approach to identify and characterize essential bacterial genes. Nature Biotechnol 18:740–745, 2000.
37. ST Cole, R Brosch, J Parkhill, T Garnier, C Churcher, D Harris, SV Gordon, K Eiglmeier, S Gas, CE Barry 3rd, F Tekaia, K Badcock, D Basham, D Brown, T Chillingworth, R Connor, R Davies, K Devlin, T Feltwell, S Gentles, N Hamlin, S Holroyd, T Hornsby, K Jagels, A Krogh, J McLean, S Moule, L Murphy, K Oliver, J Osborne, MA Quail, MA Rajandream, J Rogers, S Rutter, K Seeger, J Skelton, S Suares, JE Sulston, K Taylor, S Whitehead, BG Barrell. Deciphering the biology of mycobacterium tuberculosis from the complete genome sequence. Nature 393: 537–544, 1998.
38. MJ Grossel, H Wang, B Gadea, W Yeung, PW Hinds. A yeast two-hybrid system for discerning differential interactions using multiple baits. Nature Biotechnol 17: 1232–1233, 1999.
39. SP Gygi, GL Corthals, Y Zhang, Y Rochon, R Aebersold. Evaluation of two-dimensional gel electrophoresis-based proteome analysis technology. Proc Natl Acad Sci USA 97:9390–9395, 2000.
40. P Uetz, RE Hughes. Systematic and large-scale two-hybrid screens. Curr Opin Microbiol 3:303–308, 2000.
41. A Goffeau. Genomic-scale analysis goes upstream? Nature Biotechnol 16:907–908, 1998.
42. A Persidis. Proteomics. Nature Biotechnol 16:393–394, 1998.
43. S Borman. Proteomics: Taking over where genomics leaves off. Chem Eng News 31:31–37, 2000.
44. P Uetz, L Giot, G Cagney, TA Mansfield, RS Judson, JR Knight, D Lockshon, V Narayan, M Srinivasan, P Pochart, A Qureshi-Emilli, Y Li, B Godwin, D Conover, T Kalbfleisch, G Vijayadamodar, M Yang, M Johnston, S Fields, JM Rothberg. A comprehensive analysis of protein-protein interactions in saccharomyces cerevisiae. Nature 403:623–627, 2000.
45. S Oliver. Guilt-by-association goes global. Nature 403:601–603, 2000.
46. M Fotin-Mleczek, M Rottman, G Rehg, S Rupp, FJ Johannes. Detection of protein-protein interactions using a green fluorescent protein-based mammalian two-hybrid system. Biotechniques 29:22–26, 2000.
47. JK Joung, El Ramm, CO Pabo. A bacterial two-hybrid selection system for studying protein-DNA and protein-protein interactions. Proc Natl Acad Sci USA 97:7382–7387, 2000.

48. S-H Park, RT Raines. Genetic selection for dissociative inhibitors of designated protein-protein interactions. Nature Biotechnol 18:847–851, 2000.
49. DF Veber, FH Drake, M Gowen. The new partnership of genomics and chemistry for accelerated drug development. Curr Opin Chem Biol 1:151–156, 1997.
50. DA Jones, FA Fitzpatrick. Genomics and the discovery of new drug targets. Curr Opin Chem Biol 3:71–76, 1999.
51. PG Schultz. Bringing biologic solutions to chemical problems. Proc Natl Acad Sci USA 95:14590–14591, 1998.
52. SL Schreiber. Target-oriented and diversity-oriented organic synthesis in drug discovery. Science 287:1964–1969, 2000.
53. AB Martin, PG Schultz. Opportunities at the interface of chemistry and biology. Trends Cell Biol 9:M24–M28, 1999.
54. A Krantz. Protein-site targeting diversification of the drug discovery process. Nature Biotechnol 16:1294, 1998.
55. JF Waring, RG Ulrich. The impact of genomics-based technologies on drug safety evaluation. Annu Rev Pharmacol Toxicol 40:335–352, 2000.
56. J Hodgson, A Marshall. Pharmacogenomics will the regulators approve? Nature Biotechnol 16:243–246, 1998.
57. NS Gray, L Wodicka, AM Thunnissen, TC Norman, S Kwon, FH Espinoza, DO Morgan, G Barnes, S LeClerc, L Meijer, SH Kim, DJ Lockhart, PG Schultz. Exploiting chemical libraries, structure, and genomics in the search of kinase inhibitors. Science 281:533–538, 1998.
58. Z Guo, D Zhou, PG Schultz. Designing small-molecule switches for protein-protein interactions. Science 288:2042–2045, 2000.
59. DA Erlanson, AC Braisted, DR Raphael, M Randal, RM Stroud, EM Gordon, JA Well. Site-directed ligand discovery. Proc Natl Acad Sci USA 97:9367–9372, 2000.
60. JL Harris, BJ Backes, F Leonetti, S Mahrus, JA Ellman, CS Craik. Rapid and general profiling of protease specificity by using combinatorial fluorogenic substrate libraries. Proc Natl Acad Sci USA 97:7754–7759, 2000.
61. K Garber. The next wave of the genomics business. Technol Rev 47–56, 2000.
62. SK Burley, SC Almo, JB Bonanno, M Capel, MR Chance, T Gaasterland, D Lin, A Sali, FW Studier, S Swaminathan. Structural genomics: beyond the human genome project. Nature Genet 23:151–157, 1999.
63. J Skolnick, JS Fetrow. From genes to protein structure and function: Novel applications of computational approaches in the genomic era. Trends Biotechnol 18:34–39, 2000.
64. T Gaasterland. Structural genomics: Bioinformatics in the driver's seat. Nature Biotechnol 16:625–627, 1998.
65. J Skolnick, JS Fetrow, A Kolinski. Structural genomics and its importance for gene function analysis. Nature Biotechnol 18:283–287, 2000.
66. CA Ouzounis, PD Karp. Global properties of the metabolic map of *Escherichia coli.* Genome Res 10:568–576, 2000.
67. M Vidal, H Endoh. Prospects for drug screening using the reverse two-hybrid system. Trends Biotechnol 17:374–381, 1999.
68. S Chia, J Urano, F Tamanoi, B Dunn, JI Zink. Patterned hexagonal arrays of living cells in sol-gel silica films. J Am Chem Soc 122:6488–6489, 2000.

69. M Whitney, E Rockenstein, G Cantin, T Knapp, G Zlokarnik, P Sanders, K Durick, FF Craig, PA Negulescu. A genome-wide functional assay of signal transduction in living mammalian cells. Nature Biotechnol 16:1329–1333, 1998.

70. V Soskic, M.Gorlach, S Poznanovic, FD Boehmer, J Godovac-Zimmermann. Functional proteomics analysis of signal transduction pathways of the platelet-derived growth factor β receptor. Biochemistry 38:1757–1764, 1999.

71. RN Ghosh, Y-T Chen, R DeBiasio, RL DeBiasio, BR Conway, LK Minor, KT Demarest. Cell-based, high content screen for receptor internalization, recycling and intracellular trafficking. Biotechniques 29:170–175, 2000.

72. GR Lenz, HM Nash, S Jindal. Chemical ligands, genomics and drug discovery. Drug Disc Today 5:145–156, 2000.

73. S Borman. Advances in NMR of macromolecules. Chem Eng News 10:855–856, 1998.

74. K Wuthrich. Protein recognition by NMR. Nature Struct Biol 7:188–189, 2000.

75. ML Bulyk, E Gentalen, DJ Lockhart, GM Church. Quantifying DNA-protein interactions by double-stranded DNA arrays. Nature Biotechnol 17:573–577, 1999.

76. GJB Van Ommen, E Bakker, JT den Dunnen. The human genome project and the future of diagnostics, treatment, and prevention. Lancet:SL5–SL10, 1999.

77. R Langreth. Awash in genes. Forbes 3:136–137, 2000.

78. RS Crespi. Patents on genes. Clarifying the issues. Nature Biotechnol 18:683–684, 2000.

79. M Wadman. US court tests the breadth of patent protection on proteins. Nature 404:532, 2000.

16
Genomics/Functional Proteomics for Identification of New Targets

A. Donny Strosberg
Hybrigenics, Paris, France

I. INTRODUCTION

In recent years, genomics have conquered an ever more important place in the tool kit for modern drug discovery [1,2]. This has led most of the largest pharmaceutical, agrofood, and biotechnology companies to become subscribers to access sequence databases. By now, all major life sciences companies have linked up with private or public providers of new DNA sequences, and with additional genomic information databases. Moreover, consortia have emerged to take better advantage of the wealth of rapidly accumulating new raw data. This data initially consisted of mostly expressed cDNAs, the so-called expressed sequence tags (ESTs) to account for the rather limited sequence information made available through the first sequencing approaches. A more recent campaign, waged both by private groups and public consortia, now aims at actually sequencing the whole human genome, thus including large stretches of so-called noncoding regions. In addition, the genomic sequences are now being scrutinized for the presence of single-nucleotide polymorphisms (SNPs), to attempt to link these changes with the presence of genes associated with diseases. The race to map all the SNPs present in the human genome was started by a few small biotechnology companies, but more recently a consortium was formed composed of eight of the largest pharmaceutical and public research laboratories in the world.

Furthermore, the genomes of a number of unicellular or multicellular organisms, such as the yeast *Saccharomyces cerevisiae* and the nematode *Caeranobdibtis elegans* have now also been sequenced, and sequencing has been started on animals, such as *Drosophila melanogaster* or the laboratory mouse, commonly used as models in biological experimentation. Vast amounts of sequence information is thus rapidly becoming available, often in public databases, but frequently it is devoid of essential annotation that has simply not been done for lack of time, lack of skills, or lack of available means. Information overload is thus threatening those scientists not wired to appropriate annotation groups.

While sequences are thus accumulating at truly amazing speed, far greater than anticipated only a few years ago, the corresponding basic annotation is not, and neither is information about expression of genes, synthesis, and most importantly function of the corresponding proteins [3]. The main justification of sequencing, i.e., discovery of new pharmacological targets, thus remains essentially an unfulfilled promise that will have to await considerable additional work by bona fide protein chemists, now given the more fashionable name of "proteomics specialists". Salvation is however on its way, and the combination of a number of truly genome-scale technologies combined with development of user-friendly bioinformatics is expected to provide the pharmaceutical and agrofood industries with the Holy Grail of vast new numbers of validated targets for drug discovery and active product development. The various advantages of the new technologies available to the present-day protein chemists for unraveling protein function at the cell-wide scale will be discussed.

II. DEDUCING FUNCTION FROM COMPARISON OF SEQUENCED GENOMES

A. If You Cannot Study Function, Study Structure . . . and Compare

It has always been useful to accumulate structural data well before function could be approached. Hundreds of cytochrome C or variants of hemoglobins were thus sequenced to try to relate even single amino acid substitutions with modifications in functional properties, and many myeloma proteins' primary structures were determined before these immunoglobulins could actually be shown, in rare cases, to function as antibodies. In fact, Per Edman himself used the first protein sequencer to determine the sequence of Bence-Jones proteins, later identified as immunoglobulin light chains. With the advent of DNA sequencers, the number of sequences has dramatically increased, whereas the capacity to study function has not. One supposedly easy way around this dramatically increasing problem is to compare the sequences, preferably with those of known genes, and corresponding proteins.

B. Immunoglobulins, Antibodies, and T-Cell Receptors

This approach has of course been very fruitful with myeloma proteins and the corresponding antibodies. Using comparative analyses, immunologists were thus able to distinguish first variable from constant regions, then hypervariable regions from framework stretches. Later these hypervariable regions were identified as the "complementarity determining regions" responsible for the recognition of antigens. Starting from immunoglobulins, immunologists were able to apply similar analyses to study structure–function relationships in T-lymphocyte, membrane bound receptors.

C. Seven Transmembrane/G-Protein Coupled Receptors (GPCRs)

Similar approaches have now abundantly been used to make predictions about the function of sequenced unknown genes. These studies are most fruitful when applied to large families of highly homologous proteins, as was the case with immunoglobulins. A good example is provided by members of the seven transmembrane domain receptor family, also often described as the G-protein coupled receptors (GPCRs). In this vast group of membrane proteins, mostly characterized by the gene sequences, many subfamilies exist. Usually these subfamilies are characterized by greater homology. One may thus identify biogenic amine receptors such as those for adrenaline, dopamine, histamine, or serotonin from receptors binding peptides such as angiotensin, opioids, or neurotensin, solely on the basis of the presence of a number of conserved amino acid residues at a very limited set of positions throughout the proteins. New pharmacological subtypes were thus assigned a presumed function, often on the basis of strong sequence homology with members of a given subgroup of receptors, as in the case of β3-adrenoreceptor, which was identified as a noradrenaline binding protein on the basis of the sequence homology with the known β1 and β2 subtypes [4].

Olfactory receptor genes have thus been grouped together not only on the basis of their expression in olfactory neurons but also because of structural homology, although most of these presumed receptors have still not been assigned any binding properties. The limitation of this approach is clearly seen in the fact that there are still numerous so-called "orphan" receptors, for which no homology was found to be revealing enough actually to suggest function. With decreasing homology however, the accuracy in predicting function also wanes.

Nevertheless, comparative structural analyses continue to be useful and are used to a great extent in annotating sequence databases. With more sequences accumulating steadily, comparisons between genomes of various organisms should indeed reveal sufficient similarities to allow increasing numbers of identifications. The discovery of nearly 50% of totally "new" sequences in the sequenced genome of abysmal bacteria should certainly prepare us for many more orphan proteins.

III. GENE EXPRESSION

Several complementary procedures have been developed to survey expression patterns of genes in cells grown in various conditions. DNA/RNA display techniques compare mRNAs extracted from cells grown in different circumstances. These RNAs are then separated by electrophoresis and revealed by cross-hybridization with known DNAs. DNA chips also utilize known oligonucleotide sequences, and because of the truly impressive capacity of these silicon wafers, membranes, or glass plates, important portions of genomes can be applied to the solid phase in one single operation and hybridization revealed concurrently. These hybridization-based methods may provide gene expression data for individual known genes. They are, however, not as useful in comparing the abundance of particular transcripts.

Serial analysis gene expression (SAGE) is yet another approach to the evaluation of individual gene expression [5]. This method is based on the integration of two principles: the use of short oligonucleotide sequence tags ($\sim$15 to 20 nucleotides) containing sufficient information to identify uniquely any individual transcript and the concatenation of the tags in serial fashion to allow for increased efficiency in a sequenced-based analysis [6].

IV. PROTEIN EXPRESSION STUDIES

Several systems have now been developed to explore protein expression patterns obtained from the analysis of cells exposed to various growth conditions.

A. Polyacrylamide Gel Separation-Based Methods

1. Protein Separation and Visualization

Usually, protein mixtures are extracted in standardized conditions and separated by polyacrylamide gel electrophoresis in two dimensions: the first dimension is a size separation in SDS buffer, the second an isoelectric separation using ampholytes, i.e., a mixture of synthetic polymers that form a pH gradient, thus allowing proteins to migrate until they reach the position corresponding to their isoelectric point, where they become neutral. Individual protein ''spots'' are visualized by one of several staining procedures (e.g., silver staining, Coomassie Blue or Ponceau Red staining) or by autoradiography after radioactive labeling by growing cells in radioisotope-containing medium or by extraneous labeling using such as I^{125}.

2. Protein Identification

Identification of protein spots requires sequencing, which until recently was mostly done by extraction followed by blotting on a glass-fiber membrane that was directly introduced in the amino acid sequencer. Recently, however, easy-to-use mass spectrometers allow sequencing by recognition of patterns of masses of peptide fragments. This considerably reduces the amounts necessary for sequencing, and facilitates sequence identification, now that fragmentation databases on a large variety of possible peptide sequences become available.

3. Reproducibility and Databases

The next difficulty is to make this whole procedure reproducible. At this time, scientists concur that at least three independent two-dimensional gels should be run with the same type of sample to obtain a reliable result. For comparative purposes, a basic pattern is obtained by combining results obtained with ten separate presumably identical samples. Using such a combination of extraction/separation/sequencing procedures, combined with NASA-derived analysis of 2-D gel patterns have resulted in a flurry of publications that presented patterns of protein expression maps related to a variety of cells and conditions. The latest example was presented by scientists working at Oxford Glycosciences, which described protein expression maps corresponding to breast cancer cells. Over 10,000 individual proteins were identified, of which a few dozens seemed to be breast cancer–specific.

B. Alternative Method for Profiling of Protein Expression

A method to describe the profile of expressed cellular proteins, by taking advantage of the rare presence of cysteine residues, was recently described [7]. All cysteine residues of cells grown in two different conditions are labeled, by utilizing one of two available isotopes. The two populations of labeled proteins are then mixed, these proteins are digested with proteolytic enzymes, and the relative abundance of the isotopes is compared for each labeled peptide sequenced by mass spectrometry. This method thus allows us to find out which protein is present in the cellular extract, provided it contains at least one cysteine residue, and the procedure provides a level of quantitation related to the condition in which the cell was grown.

V. MOTIF RECOGNITION STUDIES

For many years, it has been the goal of protein chemists to try to identify motifs of protein folding based on amino acid sequences. Various approaches have been used for this purpose.

A. Sequence Motifs

Motif-discovery and pattern-matching algorithms rapidly become central to a number of applications in computational biology (for a brief review of recent work, see Ref. 8). The aim of any of the reviewed algorithms is theoretically to discover all of the patterns occurring a predefined number of times in any set of sequence data. Motifs are, e.g., for known substrate-binding or active sites in enzymes.

B. Remote Homologies, Functional Motifs

Several computational approaches have recently been described for detecting structural and even functional evolutionary relationships between proteins with few identical residues. Since protein structure appears to vary less in evolution than actual sequence, using information related to function may help identify residues conserved for the preservation of essential folding motifs. Several databases have thus been assembled that collect all the known crystal structures of members of functional families.

C. Protein Family Motifs

The compilation of new genome sequences provides even more opportunities to classify additional members of families with recognized functional properties, thus allowing faster and better quality annotation of sequence databases. Orthologous groups of proteins, i.e., those conserved across genomes, will soon be grouped in a new type of database.

D. Protein Structure Motifs

Finally, attention is being paid to the attempt to use known structural motifs for predicting protein folds based on homology with experimentally determined structures: SWISS-MODEL is a new database using comparative modeling structure prediction strategies to build 3-D structure models for all protein entries in the well-known SWISS-PROT and trEMBL databases by using structural data about homologous proteins extracted from the PDB (http://www.rcsb.org/pdb/).

Motif and pattern recognition are extensively used to group proteins and predict function. While this approach has been useful to annotate genomes, it is also the main vehicle for dissemination of inaccurate functional inference. Thus a wide effort is expected from experts to classify motifs and patterns and clarify those linked to structural conservation, functional active site for binding or catalytic activity, or phylogenetic proximity. Such effort for curation is efficiently pursued at the SBI with the development of the Swiss-Prot database.

VI. WHOLE CELL PROTEIN INTERACTION MAPPING

In the cell, proteins do not act in isolation, but usually form transitory or stable complexes in order to participate in pathways and act in networks. Protein–protein interactions thus constitute an essential aspect of the normal workings of the living cell. These interactions intervene in every cellular process including cell cycle control, energy metabolism, cell division, and even cell death. Unraveling the various interactions in which individual proteins are involved thus constitutes an invaluable way to understand protein function.

A. Biochemical Methods

Recently, a high-throughput genome-wide version of the two-hybrid yeast method was developed to create what have since become known as protein interaction maps (PIMs) for whole cells [9]. While an early systematic matrix approach had led Bartels et al. [10] to discover 22 interactions between the 53 expressed proteins of bacteriophage T7 expressed in *Escherichia coli*, the automated generic version of the procedure of Fromont-Racine et al. is rapidly becoming the method of choice for mapping whole proteomes [9a]. In less than 15 months, a scientific team at Hybrigenics (*www.hybrigenics.fr*) has thus linked half of the 1600 proteins of *Helicobacter pylori* into still partial protein interaction maps for this ulcer-provoking bacterium [9a] and has partially completed PIMs for several other microorganisms as well as mammalian cells. In a recent article, Uetz et al. reported data obtained by the matrix approach both in an academic low-throughput although quite efficient setting [11] and in a high-throughput version [Curagen (www.curagen.com)]. This chapter clearly demonstrates the limitation of the matrix-derived high-throughput procedure, characterized by high levels of false negatives due to the use of predefined full-length proteins and a predefined unique experimental selective system for all potential interactions [only 15% of the open reading frames (ORFs) do interact with at least another protein]. This strategy might also lead to many false positives, since most (over 60%) positives are not reproducibly found when the experiment is repeated twice. In contrast, the Hybrigenics approach strongly reduces the occurrence of false negatives and considerably decreases false positives by providing a heuristic value for each interaction. Hybrigenics developed the automated PIM Rider, essentially another type of web browser, thus provides a quantitative measurement of the predictive biological value of the detected protein–protein interactions. This was achieved by developing an algorithm called the predicted biological score (PBS). This PBS tells how strong the interaction is and thus allows a classification that results in the definition of the preferred pathway. The PIM Rider also provides selected interacting domains (SIDs) involved in the various protein–protein interactions listed in the database [9a]. These SIDs are the direct products of the automated genome-wide approach developed by Fromont-Racine et al. [9]. Indeed, in this procedure, every ORF may be represented by many different partial fragments, by selecting the parts that are common; one can thus define the actual domain responsible for the interaction with a given partner.

Since one protein may interact with various proteins, several different domains can be defined for a single ORF. Even the PIM Rider is however not complete; residues key for the interaction cannot be identified, and pathway identification still requires biological experimentation.

Eilbeck et al. have developed an object oriented protein interaction database called INTERACT (at http://bioinf.man.ac.uk/interactso.html) that aims at providing the appropriate architecture to store, query, and analyze protein interaction data [12]. A first version contains over a thousand interactions gathered form three different types of source; data from the MIPS [13], from scientists submitting results over the Web, and results from published literature. Eilbeck et al., have been able to create a web-based browser that allows biologists to answer a number of queries about interacting proteins [12].

These questions have been listed by the authors in a table that is reproduced here in a modified form (Table 1), since it very nicely summarizes what protein chemists would like to know about their favorite molecules. Not all these questions can be answered by INTERACT. Two other sites provide a view on protein–protein interactions, one from Myriad (ProNet), the other from Curagen. A clear definition of the initial set of data that populate the database and information on its accuracy are now available at this time. Currently, several protein sequence databases provide annotations for describing protein–protein interactions. Unfortunately, the compilation of many results, most of which were obtained through different, not quality controlled or standardized procedures, leads to the listing of quite a number of false positive interactions. It will be a challenge for the coming years to offer to the scientific community access to accurate protein–protein interaction data, the same being true for most large-scale genomic approaches (expression profile, polymorphisms, mutant phenotypes), for which neither standards for experimental procedures nor standards for dataset format have been defined. This is particularly important since these sets of data are then used by bioinformatics algorithms for various predictions on function and inference of interactions in other organisms. We see here a typical example of a possible propagation of many specific data based ultimately on wrong experimental results.

Table 1   Information Needed to Assign a Function to a Protein

---

The partners a protein interacts with (including itself, in dimer formation)
The strength, the reproducibility of the interaction in different conditions,
   in different cell types
The domains (motifs?) involved in the protein–protein interactions
The pathway(s) the protein is involved in
The location in the cell where the interaction occurs

B. Predictive Methods Based on Bioinformatics

Several groups have proposed to predict protein function by applying computational methods.

1.   Annotation by sequence similarity to known proteins is based on the assumption that homology in structure results in conservation of activity [14]. At least three factors limit such an approach however; sequence similarity has to be continuous over a sufficient stretch of DNA, and has to be readily recognized, homologues with known function should already be available for comparison, and finally, activity of the protein has to be correlated with the parts that are structurally conserved.

2.   Prediction of protein interaction based on gene fusion events was proposed by Enright et al. [15] as a more sophisticated version of sequence comparison, by focusing on homologies between DNA sequences that encode single proteins in one genome, and apparently "fused" domains in another genome. The assumption is made that fused domains necessarily interact with each other, and therefore that independent proteins homologous to these domains should interact with a high probability. Enright et al. have actually documented their approach by "linking" 215 proteins in three different prokaryotic genomes, with very few estimated false positives [14]. This approach addresses one of the main criticism of the "random" sequence comparison but does not replace actual analysis of function.

3.   Correlative functional linking through a combination of methods was described by Marcotte et al. [16]. These authors purport to replace biochemical analysis of protein function by integrating results from correlated evolution, correlated messenger RNA expression patterns, and patterns of domain fusion. Marcotte et al. showed that the number of links proposed by each of the methods applied to the yeast genome. Most types of links used by Marcotte et al. rely of course on a number of assumptions that will need to be verified experimentally [16]. Correlated evolution between several genomes assumes that proteins sharing a similar phylogenetic profile are expected to be functionally linked. Correlated mRNA expression patterns were obtained for the yeast genome by comparing 97 publicly available DNA chip data sets corresponding to changes of expression levels of mRNA (not proteins!) during a variety of physiological conditions. Here the assumptions are first, that all mRNA levels accurately and similarly reflect protein levels, and second, that apparently correlated expression levels reflect functional linkage. The resulting proposed functions are still very broadly defined: protein functions are thus referred to as involved in "metabolism" or "transcription," which would probably describe a large proportion of randomly selected proteins anyway. Nevertheless, this kind of approach, as imprecise as it may be, does present the considerable advantage of providing information and hypotheses for at least the yeast genome. It is easy to validate or

invalidate, by actually performing experiments such as genome-wide protein interaction mapping, as proposed by Fromont-Racine et al. [9], or by systematic gene knockout analysis, as presently performed by the group of Ronald Davis at Stanford and independently by the European Yeast Genome Sequencing Consortium (EUROFAN).

VII. WHOLE CELL OR WHOLE ANIMAL STUDIES FOR PROTEIN FUNCTION

A number of systems are now available to identify protein function in whole cells or whole animals by analyzing the biological effects of mutations in the corresponding genes either expressed naturally, for instance, in disease, or artificially, by deliberate gene alteration. Specialized reviews discuss these systems in great detail. These are listed here with discussion of relative advantages and disadvantages.

A. Yeast Knockouts

The first sequenced whole genome was that of yeast, and its 6,217 ORFs are now scrutinized by a variety of approaches. At this time, at least 2,557 of these ORFs still correspond to uncharacterized proteins [13,17–19]. Two groups have independently decided to knock out every one of the 6,217 ORFs: the laboratory of Ronald Davis and EUROFAN. For the latter, each of the viable mutant yeasts is subjected to ten different growth conditions to evaluate changes induced by the invalidation of a single gene. Results are not yet available, but it is already clear that a number of knockouts do not yield any information, because of the lack of effect in any of the ten selected conditions.

A more recent approach was described by Ross-MacDonald et al. [18], in which gene disruption was performed by targetted transposon tagging together with insertion of the lacZ gene which encodes β-galactosidase. Over 11,000 strains were thus generated, each carrying a transposon inserted within a region of the genome expressed during vegetative growth and/or sporulation. These insertions affect nearly 2,000 annotated gene disruption phenotypes, which were determined for nearly 8,000 strains using 20 different growth conditions. Over 300 previously nonannotated ORFs were identified. Over 1,300 transposon-tagged proteins were analyzed by indirect immunofluorescence.

B. Fruit Fly Mutants

Well before the yeast sequence became available, a number of groups involved in the study of development had collected a large variety of fruit fly (Drosophila) mutants, for which accurate genetic studies had pinpointed presumed mutated genes. The Drosophila genome is now sequenced and should soon become available on the Web. Preliminary indications suggest that annotations of the ORFs seemed to be quite valuable and may thus help considerably in assigning function to a number of presumed proteins.

C. Nematode Mutants

Although nematode (e.g., *C. elegans*) mutants have been much less studied than fruit flies, the recent availability of the entirely sequenced, but unfortunately sparsely annotated, genome has spurred considerable interest in collections of *C. elegans* containing one-to-one knocked out genes [20]. Specialists in nematode phenotypes have become in high demand to try to link the mutations with physiologic response. Some groups thus claim to be able to differentiate over 500 different behaviors (www.DevGen.com). This is certainly a challenging statement, which by itself summarizes the hope placed in such microscopic worms to go from gene to integrated biological function.

D. Mice Mutants

Naturally occurring mice mutants, or animals knocked out for specific genes, are widely available, but the genome is not yet sequenced. It may however become accessible in the not too distant future and would certainly provide for a very attractive system to link gene sequence with protein function.

E. Inherited Diseases in Humans

By far the most exciting results linking protein defects with disease have of course been obtained in humans. Families in which a disease appears to be inherited in a strictly Mendelian manner have thus yielded correlations between mutations occurring in a single gene and presence of the disease. Because of the increasing availability of data describing the sequence of the human genome, and the presence of genetic markers, correlations now become much easier; and the assignment of a role in the disease, and thus deduction of normal function, have a much better scientific basis.

VIII. CONCLUSION

Considerable progress has thus been accomplished towards the complete structural analysis of whole genomes. At the time of writing, 53 genomes have been nearly completely sequenced, of which about 6 are "closed," meaning that no gaps are left. Most of the sequences have been completed once, with the exception of *H. pylori*, for which two different strains have been sequenced. In that single case, differences in sequence have been ascribed to differences in the isolates. It is likely that in the near future complete sequences will become available for many other genomes including the human and murine genomes, but with possible unexplained sequence variations. Beyond genomics, however, a much more challenging field of research and development opens up. Indeed, on average less than 40% of the sequenced genes have known functions, and the obvious avenue to unravel functions will be the study of the encoded proteins. Methods developed so far to study proteins are both too specific and too tedious to allow rapid progress. Generic technologies, such as cell-wide, genome-wide protein interaction mapping will be applied systematically.

The assignment of proteins in functional pathways will necessarily have to be completed by cellular studies, including the analysis of the effects of antisense treatment, or of gene inactivation. Methods will have to be devised to perform such studies generically, on a large scale. Just as necessary will be high-throughput procedures for determining the three-dimensional structure of domains of proteins found to be involved in protein interactions. Current methods for crystallization or structural analysis by NMR seem to become more amenable to automation. One may thus reasonably expect that soon biologists will benefit from the knowledge of all structures of proteins, and from the roles of these proteins in physiologic pathways. It will then be a challenge to integrate all this knowledge to understand the functioning of complex structures such as the brain. This will probably be the assignment for the next decade.

REFERENCES

1. PD Karp, M Krummenacker, S Paley, J Wagg. Integrated pathway-genome databases and their role in drug discovery. TIBTECH 17:275–281, 1999.
2. MY Galperin, EV Koonin. Searching for drug targets in microbial genomes. Curr Op Biotech 10:571–578, 1999.
3. S Zozulya, M Lioubin, RJ Hill, C Abram, ML Gishizky. Mapping signal transduction pathways by phage display. Nature Biotech 17:1193–1198, 1999.
4. AD Strosberg. Structure, function and regulation of adrenergic receptors. Protein Sci 12:1198–1209, 1993.
5. AH Bertelsen, VE Velculescu. High-throughput gene expression analysis using SAGE. Drug Discovery Today 3:152–115, 1998.
6. RJ Cho, M Fromont-Racine, L Wodicka, B Feierbach, T Stearns, P Legrain, DJ Lockhart, RW Davis. Use of whole genome oligonucleotide arrays for highly parallel analysis of genetic selections. Proc Natl Acad Sci USA 95:3752–3757, 1998.
7. SP Gygi, B Rist, SA Gerber, F Turecek, MH Gelb, R Aebersold. Quantitative analy-

sis of complex protein mixtures using isotope-coded affinity tags. Nature Biotech 17:994–999, 1999.

8. H Khalak, V Di Francesco. Analysing biological molecules: onward to function. Trends Biotechnol 17:262–264, 1999.

9. M Fromont-Racine, JC Rain, P Legrain. Toward a functional analysis of the yeast genome through exhaustive two-hybrid screens. Nature Genetics 16:277–282, 1997.

9a. JC Rain, L Selig, H De Reusse, V Battaglia, C Reverdy, S Simon, G Lenzen, F Petel, J Wojcik, V Schachter, Y Chemana, A Labigne, P Legrain. The protein–protein interaction map of *Helicobacter pylori.* Nature 409:211–215, 2001.

10. PL Bartel, JA Roecklein, D SenGupta, SA Fields. A protein linkage map of *Escherichia coli* bacteriophage T7. Nat Genet 1:72–77, 1996.

11. P Uetz, L Giot, G Cagney, TA Mansfield, RS Judson, JR Knight, D Lockshon, V Narayan, M Srinivasan, P Pochart, A Qureshi-Emili, Y Li, B Godwin, D Conover, T Kalbfleisch, G Vijayadamodar, M Yang, M Johnston, S Fields, JM Rothberg. A comprehensive analysis of protein–protein interactions in *Saccharomyces cerevisiae.* Nature 403:623–627, 2000.

12. K Eilbeck, A Brass, N Paton, C Hodgman. *INTERACT*: an object-oriented protein–protein interaction database. Proceedings of the Seventh International Conference on Intelligent Systems for Molecular Biology (ISMB'99).

13. HW Mewes, K Heumann, A Kaps, K Mayer, F Pfeiffer, S Stocker, D Frishman. MIPS: a database for genomes and protein sequences. Nucleic Acids Res 27:44–48, 1998.

14. EV Koonin. Genomic microbiology: right on target? Nature Biotech 16:821–822, 1998.

15. AJ Enright, I Iliopoulos, NC Kyrpides, CA Ouzounis. Protein interaction maps for complete genomes based on gene fusion events. Nature 402:86–90, 1999.

16. EM Marcotte, M Pellegrini, MJ Thompson, TO Yeates, D Eisenberg. A combined algorithm for genome-wide prediction of protein function. Nature 402:83–86, 1999.

17. T Ito, K Tashiro, S Muta, R Ozawa, T Chiba, M Nishizawa, K Yamamoto, S Kuhara, Y Sakaki. Toward a protein–protein interaction map of the budding yeast: a comprehensive system to examine two-hybrid interactions in all possible combinations between the yeast proteins. Proc Natl Acad Sci USA 97:1143–1147, 2000.

18. P Ross-MacDonald, PS Coelho, T Roemer, S Agarwal, A Kumar, R Jansen, K-H Cheung, A Sheehan, D Symoniatis, L Umansky, M Heidtman, FK Nelson, H Iwasaki, K Hager, M Gerstein, P Miller, P Roeder, M Snyder. Large-scale analysis of the yeast genome by transposon tagging and gene disruption. Nature 402:413–418, 1999.

19. M Vidal, P Legrain. Yeast forward and reverse 'n'-hybrid systems. Nucleic Acids Res 27:919–929, 1999.

20. JM Walhout, R Sordella, X Lu, JL Hartley, GF Temple, MA Brasch, N Thireey-Meg, M Vidal. Protein interaction mapping in *C. elegans* using proteins involved in vulval development. Science 287:116–122, 2000.

17

Bioinformatics

Identification of Novel Targets and Their Characterization

Chandra S. Ramanathan
Bristol-Myers Squibb Company, Wallingford, Connecticut

Bioinformatics is a scientific discipline that encompasses all aspects of biological information; acquisition, processing, storage, distribution, analysis, and interpretation. It combines the tools and techniques of biology, computer science, and mathematics with the aim of understanding the biological significance of a variety of data [1]. This discipline consists of two main interdisciplinary subfields: research and development work required to build the software and database infrastructure, and computation-based research devoted to understanding and solving biological questions [2]. The emphasis of this chapter will be on the application of bioinformatics tools and databases to identify and characterize novel gene targets, which is of interest to pharmaceutical companies. The classical drug discovery paradigm (Fig. 1) depends on the knowledge of the functional activity of a protein. This functional knowledge is used either to isolate a target or to develop a screen against the target.

The recent advances in genomics and genomic technologies have changed the way of performing drug discovery. A molecular sequence is an information-rich source of data that is currently at the core of a revolution in genomics. Sequence data are universal, are now inexpensive to obtain, and are being generated at a prodigious rate. The current paradigm is genecentric, and identification of a novel or orphan gene is the starting step in a drug discovery process (Fig. 2).

The last decade saw an explosion in the generation and availability of genomic data. The human genome project (http://www.nhgri.nih.gov/HGP/) is generating sequence data at an exponential rate. Many microbial genomes, and com-

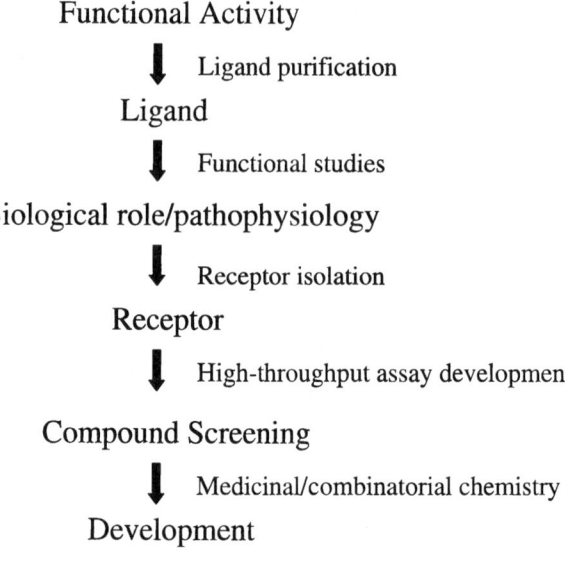

Figure 1 "Classical" drug discovery.

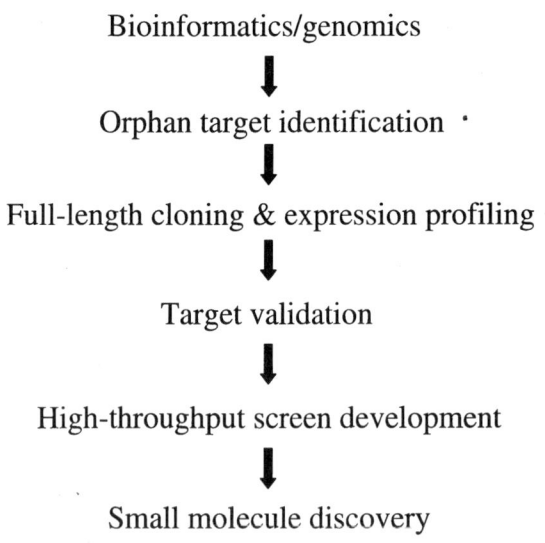

Figure 2 Bioinformatics/genomics-based drug discovery.

plete genomes of yeast, the fruit fly, the worm, and the human (working draft), have been completed during this period. In addition to simple sequences, we are beginning to see a deluge of data on (1) gene expression, (2) polymorphisms/ mutations, (3) protein interactions, and (4) regulatory networks. In addition, molecular biological data have to be integrated with biochemical, chemical, and clinical data. This has resulted in an information overflow, and the winner will be one who can extract the knowledge from this complex mesh of data. The knowledge of interest to drug companies is the identification of drug targets, and the functional characterization and elucidation of gene–disease relationships.

This chapter deals with the discovery of eukaryotic drug targets by bioinformatics methods. In this context, a ''protein'' or ''target'' of interest to a company involved in drug discovery can be

1. A novel protein related to a protein or family of proteins of interest
2. An unknown protein that has a desirable expression profile
3. A known or unknown protein that has been shown to be differentially expressed in a biological condition of therapeutic importance
4. A mutation in the gene associated with a disease
5. A human homologue of a model organism (yeast, fruit fly, or worm) protein that exhibits an interesting phenotype

Currently, there are around 500 such molecular targets [3]. It is expected that this number will increase tenfold, and advances in genomics can be used to exploit these targets for future drug therapy. This chapter deals with the application of bioinformatics tools in the context of genecentric drug discovery.

I. DATABASES FOR MINING NOVEL GENES

Biological databases are key resources for identifying novel drug targets [4–6]. When sequence databases were first created, the amount of data was small and the entries were manually created and entered as text files. As new types of data were captured or created, new data repositories were created using variety of file formats [7]. Over the past ten years, the management of biological information has truly come of age, becoming increasingly integrated into the scientific process. It is now almost impossible to think of an experimental strategy in biomedicine that does not involve some online foray into scientific databases. At the core of this shift is a huge data explosion, most notably in the amount of gene sequence and mapping information.

An example for this growth is illustrated in Figure 3. From its inception in November 1988, the National Center for Biotechnology Information (NCBI; home page URL: http://www.ncbi.nlm.nih.gov/) was charged with providing data access and molecular biology analysis tools to the public domain. Figure 3 (http://

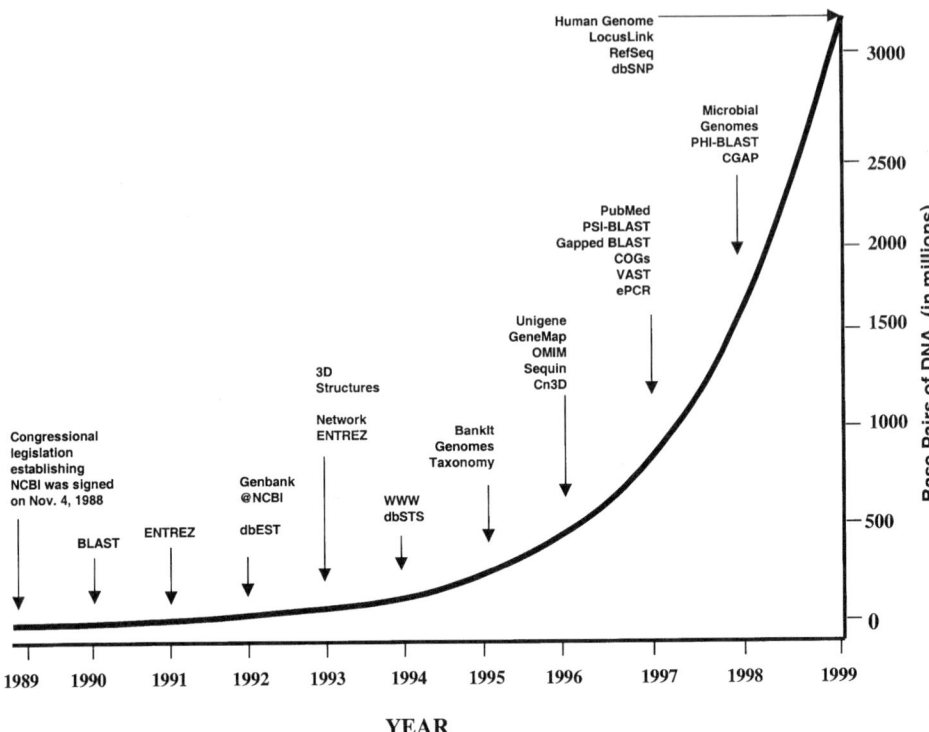

Figure 3   A decade of growth for GenBank and NCBI.

www.ncbi.nlm.nih.gov/) exemplifies the flood of data and analysis tools coming in many new forms in the first ten years of NCBI. This year-by-year tour marks highlights from NCBI's first decade. A detailed description of all the products and services mentioned in Figure 3 is available at their home page.

The flood of sequence data, mainly from completed genomes, has led to problems with quality control and the speed with which the data can be searched and analyzed [8–10]. The majority of biological databases now need to be reengineered and updated to cope with this deluge of information. As stated by Baker and Brass [7], databases can be split into four broad categories based on the source of their data:

1.   *Primary databases*: These contain one principal kind of information (e.g., sequence data), which may be derived from many sources, such as large-scale sequencing projects, individual submissions, the literature, and other databases. Examples for primary databases include protein databases like Protein In-

formation Resource (PIR) (http://www-nbrf.georgetown.edu/pir/) and Swissprot Protein Database (http://expasy.hcuge.ch/sprot).

2. *Secondary databases*: These contain one principal kind of information (e.g., alignment data), which is derived solely from other databases. The data may be a straightforward subset of another database or may be derived by analysis of another database. Examples for secondary databases include motif and pattern databases like BLOCKS (http://www.blocks.fhcrc.org/), derived from PROSITE (http://expasy.hcuge.ch/sprot/prosite.html), and PRINTS (http://www.biochem.-ucl.ac.uk/bsm/dbbrowser/PRINTS/), derived from OWL (http://www.biochem.-ucl.ac.uk/bsm/dbbrowser/OWL/OWL.html). These databases are described in detail in Sec. IV.

3. *Knowledge databases*: These are specialist databases containing related information from many sources, such as the literature, expert input, and other databases. Examples of knowledge databases include structural databases like SCOP (http://scop.mrc-lmb.cam.ac.uk/) from PDB (http://www.pdb.bnl.org/) and the *E. coli* biochemical pathway tool, EcoCyc (http://ecocyc.panbio.com/ecocyc/).

4. *Integrated database systems*: Those are combinations of primary or secondary databases. Examples of integrated databases include corporate genomic databases developed by pharmaceutical and biotechnology companies as single-stop sources for genomic information for their biologists.

The different types of databases described above serve as repositories for storing and accessing biological data. As mentioned earlier, the idea here is to use this data for high-throughput identification of novel targets. Two main databases that are widely used for identifying novel targets are the Expressed Sequence Tag (EST) databases and the High Throughput Genomic sequences from the International Human Genome Project. These two are the high-impact databases, and they are described in the next subsection.

A. Expressed Sequence Tag (EST) Database

ESTs provide the direct window onto the expressed genome [11]. They are single-pass partial sequences from CDNA libraries. The main public source of these EST sequences is dbEST (http://www.ncbi.nlm.nih.gov/dbest/), and the EST clones are available from the I.M.A.G.E. consortium (integrated molecular analysis of genomes and their expression consortium; URL: http://www.bio.llnl.gov/bbrp/image/image.html). As of March 14, 2001, there were 7,550,778 ESTs in the public domain. The number of available ESTs in the public domain for humans, mouse, rat, nematode, fruit fly and zebra fish is given in Table 1.

EST databases and cDNA sequencing are now used widely as part of both academic and commercial gene discovery projects. With the availability of high

Table 1   Number of Available ESTs in the Public
Domain (as of 03/14/2001)

Organism	Number of ESTs
Homo sapiens (human)	3,288,343
Mus musculus and *domesticus* (mouse)	1,955,500
Rattus species (rat)	265,763
Caenorhabditis elegans (nematode)	109,215
Drosophila melanogaster (fruit fly)	116,471
Danio rerio (zebrafish)	85,586

performance and a sharp decline in the cost of computing, large-scale analysis and review of EST sequence data, particularly with regard to the data quality, are now possible. The EST data, in addition to being a source for novel targets, are also linked with other genomic information in databases such as EGAD [12] and XREFDb [13].

The analysis of EST databases has resulted in the identification of several successful novel targets. One of the success stories of using EST databases is the identification and subsequent cloning of a candidate gene for chromosome 1 familial Alzheimer's disease (STM2) [14]. Another example is the identification of human homologue between the yeast equivalent (hMLH1) of the bacterial DNA mismatch repair gene mutL. Missense mutations in hMLH1 has been associated with chromosome-3-linked hereditary nonpolyposis colorecetal cancer [15,16]. This success and the potential for novel gene discovery from EST databases have also been capitalized by biotechnology companies like Human Genome Sciences (http://www.humangenomesciences.com) and Incyte (http://www.incyte.com).

EST sequences are generated by shotgun sequencing methods. The sequencing is random, and a sequence can be generated several times. This results in a huge amount of redundancy in the database. Large-scale bioinformatics and experimental comparative genomics are complex and time-consuming. One challenge is to eliminate the redundancy in the EST databases. Sequence-cluster databases such as UniGene [17], EGAD, and STACK (sequence tag alignment and consensus knowledgebase) [18] address the redundancy problem by coalescing sequences that are sufficiently similar that one may reasonably infer that they are derived from the same gene. Many companies, for example, Celera (http://www.celera.com) and Incyte, have their own clustering software. A commercial software available for clustering is based on the D2 algorithm and is currently marketed by DoubleTwist (http://doubletwist.com/). In addition to EST sequences, there is another major source of sequence data in the public domain. This is the human genomic data and it is the topic of focus in the next section.

B. Human Genomic Database

The main public domain source for the human genome data is from the Human Genome Project (HGP). This is an international research program designed to construct detailed genetic and physical maps of the human genome, to determine the complete nucleotide sequence of human DNA, to localize the estimated 30,000–40,000 genes within the human genome, and to perform similar analyses on the genomes of several other organisms used extensively in research laboratories as model systems. The scientific products of the HGP will comprise a resource of detailed information about the structure, organization, and function of human DNA and the information that constitutes the basic set of inherited ''instructions'' for the development and functioning of a human being. The working draft of the human genome was published on June 26, 2000. The International Human Genome Sequencing Consortium published the article on ''Initial sequencing and analysis of the human genome'' in the journal Nature on February 15, 2001. The sequencing status of HGP as of March 14, 2001 was:

Finished sequence: 1,040,372 kb (32.5% of the genome)
Working draft sequence: 1,951,344 kb (61.0% of the genome)

It is estimated that 3% of the total human genome encodes proteins. The major challenge in the analysis of genomic DNA sequence is to find these protein-coding regions and other functional sites. The current laboratory methods are only adequate for characterizing sequences of a few hundred bases at loci of special interest (e.g., disease genes). They are quite laborious and not suitable for annotating multimegabase long anonymous sequences. Computational methods serve as a potential alternative that can be used to characterize and annotate these megabase long sequences, either in an automated or semiautomated way [19].

Several computational tools have been developed in recent years to tackle the gene prediction problem. A listing of gene identification resources, freely available for academic use, is given in Table 2. It is important to distinguish two different goals in gene finding research. The first is to provide computational methods to aid in the annotation of the large volume of genomic data that is produced by genome sequencing efforts. The second is to provide a computational model to help elucidate the mechanisms involved in transcription, splicing, polyadenylation, and other critical pathways from genome to proteome. No single computational gene finding approach will be optimal for both these goals [38]. Also, no single gene finding tool can claim to be successful in completely identi-

Table 2 A Selected List of Public Gene Identification Resources on the Internet

Gene identification method	URL
GENSCAN [20]	http://genomic.stanford.edu/GENSCANW.html
AAT [21]	http://genome.cs.mtu.edu/aat.html
MZEF [22]	http://sciclio.cshl.org/genefinder
VEIL [23]	http://www.cs.jhu.edu/labs/combio.veil.html
MORGAN [24]	http://www.cs.jhu.edu/labs/compbio/morgan.html
Genie [25]	http://www-hgc.lbl.gov/inf/genie.html
FGENES [26]	http://dot.imgen.bcm.tmc.edu:9331/gene-finder/gf.html
FGENES-M	http://dot.imgen.bcm.tmc.edu:9331/gene-finder/gf.html
FGENESH/HMM	http://dot.imgen.bcm.tmc.edu:9331/gene-finder/gf.html
GeneID [27]	http://www.imim.es/Geneldentification/Geneid/geneid_input.html
GeneParser2 [28]	http://beagle.colorado.edu/~eesnyder/GeneParser.html
GeneLang [29]	http://cbil.humgen.upenn.edu/~sdong/genlang.html
GRAIL-II [30]	http://combio.ornl.edu/Grail-bin/EmptyGrailForm
SORFIND [31]	http://www.rabbithutch.com/
PROCRUSTES [32]	http://www-hto.usc.edu/software/procrustes/index.html
GenView [33]	http://www.itba.mi.cnr.it/Webgene
HMMgene [34]	http://www.cbs.dtu.dk/services/HMMgene/
GeneMark-HMM [35]	http://genemark.biology.gatech.edu/GeneMark/hmmchoice.html
Glimmer [36]	http://www.cs.jhu.edu/labs/compbio/glimmer.html
Xpound [37]	ftp://igs-server.cnrs-mrs.fr/pub/Banbury/xpound

fying all the exons in a gene. The current strategy is to use a combination of different gene finding programs. Comparison of different gene finding programs is available in the literature [39]. One such example is the comparison of some gene finding programs for the 1.4 MB genomic region BRCA2 on human chromosome 13q: http://genomic.sanger.ac.uk/gf/braca2.html. A combination of fgeneshm and genescan and combination of fgeneshm and fgenes have been shown to be superior in gene prediction when compared to other methods (http://genomic.sanger.ac.uk/gf/brac2.html).

The programs listed in Table 2 share a number of limitations. The current methods only detect protein-coding regions in genes. They are not trained to predict noncoding regions, which can be biologically important in some cases. As a consequence, the noncoding part of RNA genes, such as XIST [40], H19 [41], IPW [42], and the newly discovered NTT [43], would have been totally transparent to the current gene prediction programs. The genes listed above are all known to play a key role in transcription inactivation and/or imprinting.

Some of the inaccuracies in these programs stem from the inadequacy to predict accurately the intron/exon boundaries, which in turn results in limitations in identifying the eukaryotic gene structure and a poor performance on a short exon. Furthermore, a large number of false splice site predictions eventually lower the reliability of predicted exons. Because of the different types of processing performed by the various gene prediction programs, it seems plausible to use a combination of gene finding programs and to look for consensus regions produced by the various programs [44]. The ESTs and the predicted exons from the gene finding programs will serve as our input query for identifying novel proteins.

II. COMPUTATIONAL STRATEGY FOR NOVEL GENE DISCOVERY

A simple strategy for high-throughput identification of novel targets is given in Figure 4. This section gives an overview of the steps involved in the identification of novel proteins. The first step is selecting the probe sequence or motif. The probe sequence or motif is selected based on the biological problem of interest. The search sequences can be obtained from Entrez (http://

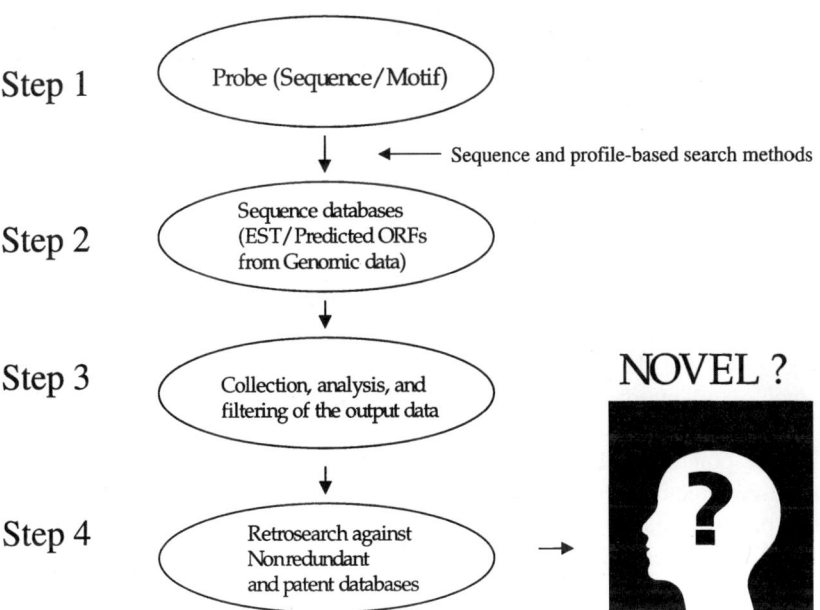

Figure 4 A simple protocol for identifying novel genes.

www.ncbi.nlm.nih.gov/) and the Hidden Markov Model (HMM) profiles from the Pfam database (http://pfam.wustl.edu/). The second step is the mining of this sequence or profile against a sequence database. Usually, the sequence databases are high-impact databases described earlier (ESTs and genomic databases). The quality of this database determines the quality of the predicted novel proteins. Sequencing errors in the data can produce false positive results. The search program to be used depends on what information is to be gained from the query of interest. The third step is the interpretation of these results and identifying the positive hits for subsequent analysis. The promising hits are searched back against the protein databases, and this process is described here as retrosearch. The EST/ genomic hits are retrosearched against the non redundant protein and patent databases to make sure they are novel. The novel ones are evaluated based on bioinformatics analysis and biological knowledge to determine whether they will be drug targets or candidates for further biological evaluation. The search programs and methods for characterizing the database hits are described in detail in the next section.

III. BIOINFORMATICS METHODS FOR IDENTIFYING NOVEL TARGETS

A molecular sequence is not very informative by itself. It is only in the context of other biological sequences that sequence data becomes truly illuminating. Sequence similarity analysis, primarily based on evolutionary homology, is the single most powerful tool for functional and structural inference available to computational biology. The sequence searching methods can be broadly divided into two categories, sequence based and profile based. These two approaches can be used in identifying novel targets for EST/genomic databases and will be the subject of discussion in this section.

A. Sequence-Based Search Methods

The design of a database search requires consideration of what information is to be gained about the query sequence of interest. The rationale here is that the homology (evolutionary relationship) can be inferred from sequence similarity, and from this one may be able to infer function.

As pointed out by Steven Brenner [45], planning a good database search experiment requires an understanding of the method being applied. Fundamentally, database searches are simple operations: a query sequence is aligned with each of the sequences, targets, in a database. Programs such as BLAST [46] and FASTA [47] use heuristics to decide whether to do a complete comparison and speed up the alignment procedure, while the Smith–Waterman algorithm [48] rigorously compares the query sequence with each target in the database. A score is computed from each alignment, and the query–target pairs with the best scores are then reported to the user. Typically, statistics are used to help improve the interpretation of these scores. A detailed explanation of these algorithms can be found in Ref. 49. BLAST is the most widely used and popular tool for sequence comparisons. This program runs 10–100 times faster than Smith–Waterman at the cost of overlooking an occasional similarity. BLAST is widely used to locate quickly high regions of sequence similarity. There is a trade-off between speed and sensitivity, and the user should make the choice of method depending upon the problem to be solved. For drug discovery, FASTA is an underutilized tool. FASTA is more sensitive than BLAST in detecting distantly related sequences. PSI-BLAST is an extension of the BLAST program that is powerful in identifying weak similarities. PSI-BLAST takes as an input a single protein squence and compares it to a protein database, using the gapped BLAST program. Then the program constructs a multiple alignment, and then a profile, from any significant local alignments found. The original query sequence serves as a template for the multiple alignment and profile, whose lengths are identical to that of the query. The profile is compared to the protein database, again seeking local alignments. After a few minor modifications, the BLAST algorithm is directly used for this comparison. The profile construction followed by a BLAST search is repeated an arbitrary number of times or until convergence.

1. Programs and Comparisons

Whether to use a protein or a DNA sequence query depends upon the biological information to be inferred from the analysis. Different types of BLAST programs are listed in Table 3. If a sequence under consideration is a protein or codes for a protein, then the search should be at the protein level, because proteins allow one to detect far more distant homologies than does DNA [50]. Degeneracy in codon usage and comparison of noncoding frames have to be taken into account while using DNA comparisons for identifying similar sequences. The chemical properties of amino acids also help in the degree of similarity to be assessed rather than simple recognition of identity or nonidentity. For these reasons, DNA versus DNA comparison is typically used in cases to find whether a gene has been previously sequenced or to find the splicing junction. On the other hand, if the user is interested in information contained in the untranslated region of a gene, a protein-based search is not useful and the DNA-based method has to be used.

2. Database Search Methods

Which database has to be searched once again depends upon the original search plan. The sequence databases are available. A list of commonly used peptide and

Table 3 BLAST Programs[a]

Program	Query	Database	Comments
blastn	DNA	DNA	Tuned for very-high-scoring matches, not distant relationship, UTR analysis, splicing pattern
blastp	Protein	Protein	To find homologous proteins
blastx	DNA (translated)	Protein	Analysis of new DNA sequences and ESTs for finding matching and novel proteins
tblastn	Protein	DNA (translated)	Querying protein probes of interest against DNA database (e.g., ESTs for finding novel proteins)
tblastx	DNA (translated)	DNA (translated)	Useful for gene structure and EST analysis, computationally intensive

[a] Similar programs are available for FASTA. FASTA programs are available at ftp://ftp.virginia.edu/pub/fasta. In Blastx, tblastn, and tblastx methods, DNA sequences or databases are translated to all six reading frames and the searches are done at the protein level.

nucleic acid databases and a brief description of their contents are given in Tables 4 and 5.

3. Search Parameters

Filtering: Statistics of database searches assume that unrelated sequences will look essentially random with respect to each other. However, there are certain patterns in sequences that violate this rule. The most common exceptions are long runs of the same or similar amino acids (polyalanine, polyproline, hydrophobic

Table 4 List of Peptide Sequence Databases Available at NCBI

Nr	All nonredundant GenBank CDS translations + PDB + SwissProt + PIR + PRF
Month	Subset of nr that is new or modified in the last 30 days
Swissprot	The last major release of the SWISS-PROT protein sequence database (no updates)
Yeast	Yeast (*Saccharomyces cerevisiae*) protein sequences
E. coli	*E. coli* genomic CDS translations
Pdb	Sequences derived from the 3-dimensional structure Brookhaven Protein Data Bank
kabat [kabatpro]	Kabat's database of sequences of immunological interest
Alu	Translations of select Alu repeats from REPBASE

Table 5 List of Nucleotide Sequence Databases Available at NCBI

Nr	All nonredundant GenBank + EMBL + DDBJ + PDB sequences (but no EST, STS, GSS, or phase 0, 1, or 2 HTGS sequences)
Month	Subset of nr that is new or modified in the last 30 days
dbest	Nonredundant database of GenBank + EMBL + DDBJ EST divisions
dbsts	Nonredundant database of GenBank + EMBL + DDBJ STS divisions
htgs	htgs unfinished High Throughput Genomic Sequences: phases 0,1, and 2 (finished, phase 3 HTG sequences are in nr)
Yeast	Yeast (*Saccharomyces cerevisiae*) genomic nucleotide sequences
E. coli	*E. coli* genomic nucleotide sequences
pdb	Sequences derived from the 3-dimensional structure
kabat [kabatnuc]	Kabat's database of sequences of immunological interest repeats from query sequences
epd	Eukaryotic Promoter Database
gss	Genome Survey Sequence, includes single-pass genomic data, exon-trapped sequences, and Alu PCR sequences

stretch). Such regions of sequence could spuriously obtain extremely high scores. For this reason, the default option is to include filtering in the NCBI BLAST server. The SEG [51] program is used for masking proteins, and the DUST [52] program is used for masking DNA sequences. These programs are not guaranteed to filter all the low-complexity sequences. The user has to be careful that sometimes valid hits might be missed if part of the sequence is masked.

4. Other Parameters

It is advisable for users to go with the default matrix and gap parameters. These parameters determine how similarity between two sequences is determined. When two residues are aligned, programs use the matrix to determine whether the amino acids are similar or very different. The default matrix is BLOSUM62 [53]. The user has to understand the evolutionary implications of various matrices before using them in a sequence search. The gap parameters determine how much an alignment is penalized for having gaps. There are other parameters that determine the heuristics that BLAST uses. By altering these numbers, the user can alter the sensitivity and speed of the search. These parameters are complex and beyond the scope of this chapter. It is very rare for users to alter these parameters from the defaults. The FASTA program has one such parameter, which can be beneficial for users. It is called ktup. Searches with ktup = 1 are slower but more sensitive than BLAST; ktup = 2 is fast but less effective. The third set of parameters determines how many matches have to be reported. These numbers can be changed at the user's discretion.

5. Interpretation of Results

This is the most challenging part of the search process. Scores calculated by the program using statistical measures serve as guidelines, and they are useful most of the time. In cases of weak similarity or alignments with low statistical significance, biological knowledge and intuition can help in the interpretation. A biologically significant alignment need not be statistically significant and vice versa. The common way for interpreting the results is to use the E value [54]. The E value of a match measures the expected number of sequences in the database that would achieve a given score by chance. Since these values depend on the number of entries in the database, they will change with the number of sequences in the database. Nonetheless, lower E values (1e-5 and less) can suggest meaningful similarity. It is highly encouraged to look at the actual alignment and check the region of the query matching with the target. If this region happens to be of functional importance and residues shown to have a functional role are matching, then the alignment may be biologically significant even if it has a high E value.

   The presence of sequence similarity allows the inference of homology, and the homology can help to infer whether they share functions. This inferring of function from the homologous matched sequences can be tricky. If the score is good and the alignment matches the entire protein, then there is a very good chance that they share the same or a related function. If only a partial target sequence matches with the query, they might share a domain and only contribute one aspect of overall function. This situation is true with a multidomain protein. One should be cautious before jumping to any functional conclusion. An EST matching with a zinc finger domain of a nuclear hormone receptor need not be a gene encoding NHR. It can be any DNA binding domain shared by many families of proteins. Many sequences have greatly diverged during evolution, and they cannot be detected by simple sequence similarity search methods. Thus failure to find a significant match does not indicate that no homologues exist in the database. This suggests that a more powerful computational tool, that goes beyond the simple pairwise sequence similarity, should be used. This leads us to the other database search method based on profiles.

B. Profile-Based Methods

Analysis of a multiple sequence alignment can reveal gene functions that are not clear from simple pairwise sequence alignment [55]. Software packages are available that can take a multiple sequence alignment and build a profile of it. As stated by Sean Eddy [55], a profile incorporates position-specific information that is derived from the frequency with which a given residue, amino acid, or nucleic acid base is seen in an aligned column. Component residues of active sites or ligand binding pockets or functional motifs tend to be well conserved in sequence families. Using this information, which includes both conserved and less-conserved residues, a sensitive database search is possible.

Much of the new software for profile searches is based on statistical models called hidden Markov models (HMMs) [56]. This section is an introduction to profile-based HMM methods, and more comprehensive reviews are available elsewhere [57]. Profile-based searches can be done in two ways, using the publicly available HMM profiles, and creating a new HMM profile from aligned sequence data.

Pfam is a database of protein domain families. It is available in the public domain at http://www.sanger.ac.uk/Software/Pfam/ and http://www.cgr.ki.se/Pfam/ (Europe) and at http://pfam.wustl.edu/ (USA). Using the publicly available HMM profiles is convenient if the domain of interest is already present in the Pfam database [58].

Pfam contains curated multiple sequence alignments for each family. These multiple sequence alignments are used to create HMM profiles. These profiles are then used to identify protein domains in uncharacterized sequences. Pfam contains functional annotation, literature references, and database links for each family. There are two multiple sequence alignments for each Pfam family, the seed alignment, which contains a relatively small number of representative members of the family, and the full alignment, which contains all members in the database that can be detected. All alignments are taken from pfamseq, which is a nonredundant set composed of the SWISS-PROT and SP-TrEMBL collections of protein family alignments that were constructed semiautomatically using HMMs. Sequences that were not covered by Pfam were clustered and aligned automatically and are released as Pfam-B.

The Pfam distribution contains a number of files: Pfam-A.seed, Pfam-A.full, Pfam, PfamFrag, SwissPfam, Pfam-B, diff, and Pfamseq [59]. Pfam-A.seed and Pfam-A.full contain the seed and full annotation in a marked-up alignment format called the Stockholm format. The Pfam file contains the library of Pfam profile HMMs. Any given sequence can be searched against this file to find any Pfam domain present in the query sequence. The Pfam models are iteratively defined. They start from clear homologues and incorporate increasingly distant family members in the process.

PfamFrag is a library of profile HMMs designed specifically to find matches to protein fragments; SwissPfam is a file containing the domain organization for each protein in the database; Pfam-B contains the data for Pfam-B families in the Stockholm format. Sequences that are not available when Pfam-A is generated are clustered and aligned automatically and are released as Pfam-B; diff is a file containing the changes between releases to allow incremental updates of Pfam-derived data; pfamseq contains the underlying sequence database, in fasta format, that all sequences in Pfam are taken from. The Pfam package contains the above-mentioned files, and executables are available for different operating systems.

Collect nonredundant set of sequences belonging to a novel family
↓
Generate a multiple sequence alignment
↓
HMMbuild to create a profile
↓
HMMcalibrate
(Estimates parameters needed for calculating E value in database searches)
↓
HMMsearch – to search a database using the newly created profile

Figure 5   Steps in creating a new HMM profile.

HMMER is a freely distributed implementation of profile HMM software for protein sequence analysis. It is available at http://hmmer.wustl.edu. There are currently nine programs in the HMMER package. These programs can be used to search a protein database or create a new HMM profile. If the domain of interest is not present in the Pfam database, then the user has to create a new HMM profile for the desired domain. A demonstration of how to create a new profile is presented in Figure 5. The description of the nine programs and an online manual for HMMER is available at the HMMER URL given above. For nucleic acid analysis, a new package called Wise2 is available at the Sanger Center, UK (http://www.sanger.ac.uk/Software/Wise2/index.shtml). It can compare a single protein or a profile HMM to a genomic DNA sequence and predict a gene structure. The genomic sequence analysis algorithm is called Genewise, and the corresponding one for ESTs is called ESTwise.

The earlier sections gave an overview of the sequence-based and profile-based methods. The next step is to use this information in the identification of novel proteins of interest. The next section talks about the application of these methods in novel gene discovery.

C. Identification of a Novel Protein: An Example

This example illustrates the use of database search methods and the application of the strategy for identifying novel genes described above. An example described here is G-protein coupled receptors (GPCRs). They are excellent drug targets, and approximately 50–60% of marketed drugs are GPCRs [60]. With recent advances in genomics, more information on the functional role of GPCRs is available. An increasing number of mutations in GPCRs were found to be associated with disease conditions.

Figure 6 gives an overview of the protocol that can be used to identify novel GPCRs. This method can be applied to any protein family. It is advisable

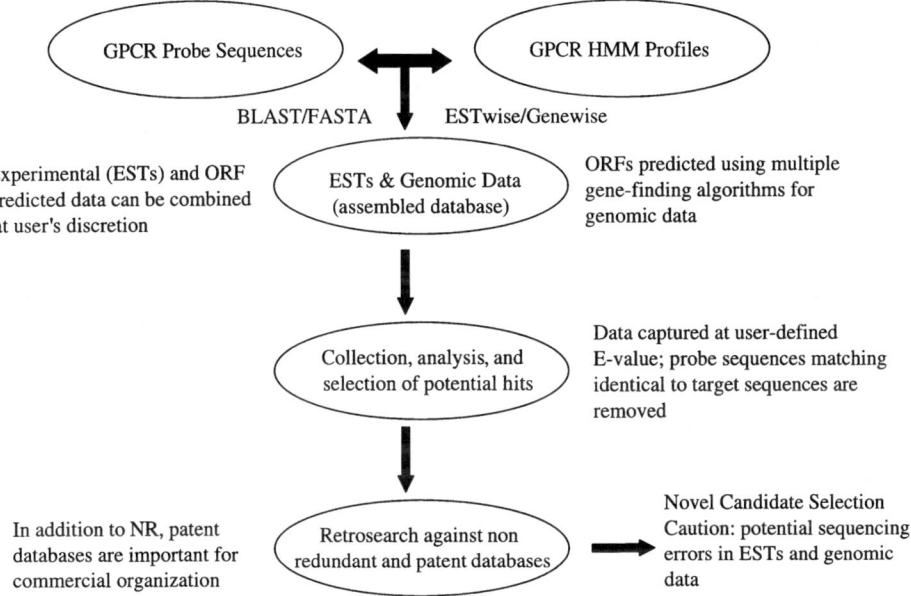

Figure 6 Application of bioinformatics techniques for identifying novel GPCRs.

to use both sequence-based and profile-based search based methods to ensure that the search is as complete as possible, not missing any weak homology hits. This type of process can be easily automated by using PERL scripts [61]. Designing a user-friendly web interface for accessing the search output data would help biologists to browse through the results.

IV. PROTEIN CLASSIFICATION AND FUNCTIONAL ASSIGNMENT

So far, we have discussed the various ways for finding novel proteins related to a query of interest. Another aspect of the problem is when you have a set of sequences and you would like to identify a putative functional assignment. The set of sequences can be from any source, such as a cDNA library, a subtractive hybridization study, transcriptional profiling, a sequence from a chromosomal region implicated with some disease, or genomic data from the Human Genome Project. In the case of genomic data, gene finding tools have to be applied first for finding the exons. The resulting exons can be analyzed for functional assignment. As mentioned earlier, the Genewise suite of programs (http://www.sanger.-

ac.uk/Software/Wise2/index.shtml) can take a genomic sequence as input and
search them against the HMM profile libraries. In this section, users have a set
of sequences and they have no a priori functional knowledge about the sequence.

The analysis pipeline is presented in Figure 7. The input sequences considered here are EST sequences and genomic data. The EST sequences have to
be masked for repetitive elements, and low-quality sequence regions should be
removed. Then they have to be compared against the existing sequence contigs.
The EST will either become a part of an existing contig or be a novel singleton
cluster. Similarly, genomic sequences have to be masked, and the ORFs predicted
from gene finding programs are compared to existing EST databases. As mentioned earlier, the new sequence can merge two clusters in the database, extend
a cluster, or form a new singleton cluster. Combining EST data with predicted
genomic data is at the user's discretion. Many companies may want to keep
experimental data and predicted data separately. The first step is to analyze those
sequences that are new and not present in internal databases, using a sequence-
based approach. Partial matches with database sequences have to be carefully
analyzed, since the possibility of matching to a domain in a multidomain protein
cannot be ruled out. Since ESTs and exons predicted from genomic data are
partial sequences, only partial matches to database sequences can be expected.

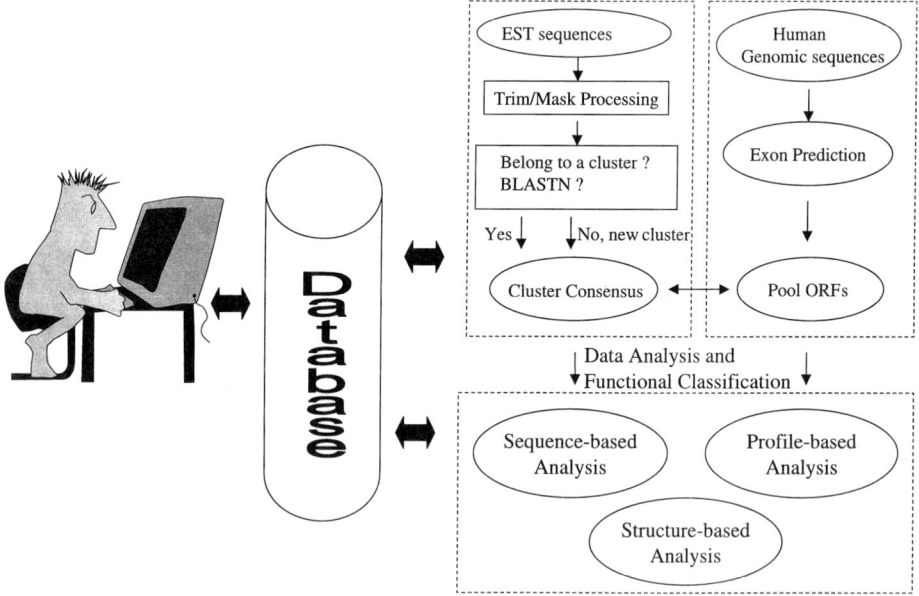

Figure 7   Bioinformatics analysis protocol for an unknown EST/genomic sequence.

These sequences should also be analyzed carefully, since the ESTs and working draft genomic data may have sequencing errors.

In addition to the sequence-based approach, a profile-based approach should also be used. The simple answer to questions like ''Which method I should use?'' or ''Which of the databases I should use?'' is ''Try as many methods and databases as possible.'' When using sequence databases, make sure the databases are current. Since sequence data is being added at an explosive rate, a nightly update of sequence databases is highly recommended. A combination of nonredundant databases (protein and nucleic acids) and patent databases is very good in screening for novel hits.

A. Motif and Profile-Based Methods

Although domain databases have considerable overlap, they all have their particular strengths [62]. Domains important in signal transduction are likely to be found in PROSITE profiles [63,64] or SMART [65]. Pfam [58] excels in extracellular domains, and PROSITE patterns are good at identifying enzyme classes by their active site motif. Even where the domain databases overlap, the apparent redundancy adds to the reliability of the classification. A list of motif databases and their FTP addresses is given in Table 6.

The PROSITE pattern library [63,64] was one of the pioneering efforts in collecting descriptors for functionally important motifs. It is based on regular expression, which emphasizes only the most highly conserved residues in a protein family. A PROSITE pattern does not attempt to describe a complete domain or even protein, but just tries to identify the most diagnostic residue combinations, such as the catalytic site of an enzyme. A benefit of PROSITE is that all motifs have an extensive documentation with references to literature. The main disad-

Table 6 Motif Databases Available in the Public Domain

Databases	FTP site for downloading[a]
PROSITE patterns	ftp://ftp.expasy.ch/databases/prosite
PROSITE profiles	ftp://ftp.isrec.isb-sib.ch/sib-isrec/profiles
pfam	ftp://ftp.sanger.ac.uk/pub/databases/Pfam
pfam (PROSITE format)	ftp://ftp.isrec.isb-sib.ch/sib-isrec/profiles
BLOCKS	ftp://ncbi.nlm.nih.gov/repository/blocks
PRINTS	ftp://ftp.biochem.ucl.ac.uk/pub/prints
PRINTS (in BLOCK format)	ftp://ncbi.nlm.nih.gov/repository/blocks
ProDom	ftp://ftp.toulouse.inra.fr/pub/prodom
SBASE	ftp://ftp.icgeb.trieste.it/pub/SBASE

[a] Use may be restricted for commercial users.

vantage of PROSITE is that its syntax is too rigid to represent highly divergent protein families. PROSITE's short patterns do not contain enough information to yield statistically significant matches in the large databases available now. Since 1995, the PROSITE pattern library has been supplemented by the PRO-SITE profile library to overcome these problems. They are in between a sequence-to-sequence comparison and the matching of a regular expression to a sequence. The unconserved regions in the profile get lower weight scores instead of being totally neglected. PROSITE profiles try to cover complete domains wherever possible. The PRODOM [66,67] was the first comprehensive collection of complete domains. It is constructed from SWISS-PROT in a fully automated manner. Pfam, which evolved from ProDom, is a collection of HMMs that are conceptually related to prosite profiles [68,69]. The Pfam models are iteratively defined. They start from clear homologues and incorporate increasingly distant family members in the process. Because of their information-rich descriptors, both PRO-SITE profiles and Pfam collections can detect even very distant instances of a motif rarely found by any other method. The documentation available with a Pfam profile is minimal but frequently contains pointers to the corresponding PROSITE entry. Pfam models and PROSITE searches are interconvertible, and combination searches are available.

Profile collections focusing on a specific functional family are also available. The SMART database is an independent collection of 181 HMMs focusing on protein domains involved in signal transduction. BLOCKS [70] and PRINTS [71] are two motif databases that represent protein or domain families by several short, ungapped multiple alignment fragments. The PRINTS motifs are also refined iteratively, resulting in an increased sensitivity compared with BLOCKS. Both BLOCKS and PRINTS have good documentation.

In all the discussed databases, the number of available domain descriptors is not an indicator of complete coverage and not to be overestimated as a quality criterion. The motif and profile databases and associated search methods differ in generality and sensitivity. For example, the relatively small prerelease library of PROSITE profiles detected 70 significant domain matches in a collection of 880 unknown yeast genes. Pfam 3.0 reported 36 matches, whereas the PROSITE pattern collection found 114.

Once a novel protein is identified, a challenging analysis is assigning that protein to a particular subfamily. A domain search can reveal whether the query protein belongs to a superfamily such as GPCR or Nuclear Hormone Receptor. The next question is whether there is any subfamily information available or identification of other proteins closely related to the query protein. Characterization of orphan proteins to the subfamily level is of paramount importance to drug discovery scientists. One approach to this problem is to study the evolutionary relationship among the family members by phylogenetics [70]. An introduction to molecular phylogenetics is provided by Hillis et al. [73]. The merits and pitfalls

Table 7 A Selected List of Phylogenetics-Related Programs and Information Available on the Internet

Program	URL
PAUP	http://onyx.si.edu/PAUP/
PHYLIP	http://evolution.genetics.washington.edu/phylip.html
Hennig86	http://www.vims.edu/~mes/hennig/software.html
MEGA/ METREE	http://www.bio.psu.edu/faculty/nei/imeg
GAMBIT	http://www.lifesci.ucla.edu/mcdbio/Faculty/Lake/Research/Programs/
MacClade	http://phylogeny.arizona.edu/macclade/macclade.html
TreeView	http://taxonomy.zoology.gla.ac.uk/rod/treeview.html

of various phylogenetics methods are summarized in useful reviews of phylogenetics [74–77]. There are three common methods used in phylogenetic analysis, the distance matrix, maximum likelihood, and parsimony methods. The distance matrix method is faster and most commonly used in phylogenetic analysis. Parsimony uses the position-specific information in a multiple sequence alignment. It is sensitive but takes a longer time to run compared to the distance matrix method. Maximum likelihood uses a different model. It takes into account every sequence change, and it is slowest of the three methods. Table 7 gives information about the sources of information in the web for some of the phylogenetics programs.

B. Structure-Based Methods

Structure is evolutionarily conserved to a greater extent than sequence. Even if two sequences do not share obvious sequence similarity, they might share the same 3-D structure (fold). Thus, fold recognition methods have tremendous potential in characterizing unknown proteins. The following flowchart gives the consequences of evolution and the physicochemical properties of protein on the number of genes, sequences, and structures (adapted from Ref. 78).

$$> 10^5 \text{ genes} \rightarrow 10^4 \text{ sequence families} \rightarrow \sim 10^3 \text{ folds}$$
$$\rightarrow \sim 10^2 \text{ structures} \rightarrow \sim 10 \text{ supersecondary motifs}$$

Although there are many genes, the number of domain sequence families is much smaller. The number of folds is almost certainly an order of magnitude less, as is the number of architectures. The architecture refers to the packing of sheets and helices in a structure regardless of sequential connectivity. At the end of the equation, the number of structural supersecondary motifs that constitutes the fold is very small. Biological complexity is achieved by using local variations together

Table 8 A List of Selected Databases Pertinent to Genetic Epidemiology

Database	URL	Description
Human Disease Susceptibility Genes		
Online Mendelian Inheritance in Man (OMIM)	http://www.ncbi.nlm.nih.gov/	Detailed information of disease susceptibility genes and inherited phenotypes
GeneCards	http://bioinformatics.weizman n.ac.il/cards/	Database of human genes, their products and their involvement in diseases
Cardiff Human Gene Mutation Database	http://www.uwcm.ac.uk/ uwcm/mg/hgmd0.html	Collection of genetic mutations responsible for human inherited diseases
Statistical Genetics		
Human Genetic Analysis Resource (HGAR)	http://darwin.mhmc.cwru.edu/	Distributes Statistical Analysis for Genetic Epidemiology (SAGE) Software
Lab. of Statistical Genetics—Rockefeller University	http://linkage.rockefeller.edu	Linkage programs
Genome Mapping and Sequencing		
Database of Expressed Sequence Tags (dbESTs)	http://www.ncbi.nlm.nih.gov/	Information of expressed sequences derived from numerous tissues and cell types
Database of Sequence Tagged Sites (dbSTS)	http://www.ncbi.nlm.nih.gov/	Sequencing and mapping data on short genomic landmark sequences

European Molecular Biology Laboratory (EMBL)	http://www.ebi.ac.uk/	Annotated DNA sequence information
GenBank	http://www.ncbi.nlm.nih.gov/	NIH genetic sequence database containing annotated DNA sequence information
Gene Map of the Human Genome	http://www.ncbi.nlm.nih.gov/	Maps expressed genes (ESTs) to specific regions of the genome
IMAGE consortium	http://www.bio.llnl.gov/bbrp/image/image.html	Makes a variety of DNA sequence, EST clones, mapping, and gene expression data available to public
Unique Gene Sequence Collection (UniGene)	http://www.ncbi.nlm.nih.gov/	Assembled DNA sequences to identify and map new human genes
Genetic Markers		
Cooperative Human Linkage Center	http://www.chlc.org/	Contains statistically rigorous genetic maps enriched for highly variable microsatellite markers
Marshfield Medical Research Institute	http://www.marshmed.org/genetics	Mammalian genotyping services to qualified applicants
Whitehead Institute for Biomedical Research—MIT	http://www.genome.wi.mit.edu/	Human Single Nucleotide Polymorphisms (SNPs) database contains genetic maps showing genomic locations of SNPs
Genetic Education		
Centers for Disease Control	http://www.cdc.gov/genetics/	Links to medical genetic literature, genetic topics in the news, and HuGE NET
Galton Laboratory	http://www.gene.ucl.ac.uk	Teaching materials for undergraduate genetics courses
National Human Genome Research Initiative	http://www.nhgri.nih.gov/DIR/VIP/Glossary	A text and audio glossary of genetic terms and concepts
United States Department of Energy	http://www.er.doe.gov/production/ober/hug_top.html	Human Genome Project information, a primer on molecular genetics, and extensive links to genetic resources

with a combinatorial approach at primary, secondary, tertiary, and quaternary levels. This includes combining domains to create different proteins and combining proteins to make different complexes. Genome-wide efforts to determine 3-D protein structures, or at least one representative 3-D structure for all protein families, is in progress. Representative structures will allow modeling of related sequences and help infer the structures of all proteins. Structural insights can help in functional elucidation, which in turn can help to determine the biological role of the protein in normal physiological and pathological conditions.

V. DISEASE–TARGET GENE RELATIONSHIP

Over the last decade, a great deal of effort has been put into creating a physical map of the human genome—ordering genes within the genome by placing landmarks with which to navigate. In addition to providing an excellent framework for the complete sequencing of the human genome, the physical map has assisted directly in identifying about 100 disease-causing genes (positional cloning). One of the most difficult challenges ahead is to find genes involved in diseases that have a complex (nonsimple Mendelian) pattern of inheritance, such as those that contribute to diabetes, asthma, cancer, and mental illness. In all these cases, no one gene can determine whether a person will get the disease or not. It is likely that more than one mutation is required before the disease manifests. This is a well-known situation in cancers. A number of genes may each make a subtle contribution to a person's susceptibility to a disease. Genes may also affect how a person reacts to environmental factors. Unraveling these networks of events will undoubtedly be a challenge for some time to come. A list of selected databases pertinent to genetic epidemiology is given in Table 8. These databases are invaluable to epidemiological research and should be familiar to all epidemiologists because they contain (1) information regarding genes and molecular defects that contribute to human disease, (2) methods for detection of numerous disease-susceptibility mutations and polymorphisms, and (3) comprehensive descriptions of disease phenotypes [79].

An excellent source of information regarding genes and their relevance to diseases is the Online Mendelian Inheritance in Man (OMIM; http://www.ncbi. nlm.nih.gov/omim/) [80]. This is an electronic version of the catalog of human genes and genetic disorders edited by Victor McKusick. OMIM contains textual information from the literature published on most human genetic-related disease. It also has pictures illustrating the conditions or disorders. Since the online version of OMIM is housed at NCBI, links to MEDLINE and sequence retrieval sources like ENTREZ are provided from all references cited with each OMIM entry. The daunting task in this area is to integrate the diverse data—genomic sequence data, mutation information, disease conditions, mapping information, functional validation data, and experimental information. A good starting point to access the various types of data described above is given in Table 8.

VI. CONCLUSION

Over the last decade, bioinformatics has become an essential tool for biologists to mine the vast amount of genomic data. The current challenge is to identify genes from human genomic data. As genes are being identified, bioinformatics techniques can help us determine, or give insights into, the gene structure–function relationship. With advances in genomics, expression and SNP data associated with genes are being added to the human genomic database. Pathway relationships that connect various genes will be elucidated soon. Bioinformatics will play a key role in converting this data into knowledge. It is this knowledge which is of strategic value to commercial organizations. Already, the integration of bioinformatics, molecular biotechnology, and epidemiological methods of assessing disease risk is rapidly expanding our ability to identify genetic influences on complex human diseases. These technological advances are likely to have a profound impact on our knowledge of the etiology of complex diseases and reveal novel approaches to disease treatment and prevention. The era of personalized medicine, designing the right drug for the right patient, is not far away.

REFERENCES

1. US Human Genome Project: The first five years. FY 1991–1995. NIH Publication No. 90–1590, 1990.
2. D Benton. Bioinformatics—principles and potential of a new multidisciplinary tool. Trends Biotechnol 14:261–272, 1996.
3. J Drews. Drug discovery: a historical perspective. Science 287:1960–1964, 2000.
4. MC Capone, D Gorman, EP Ching, A Zoltnik. Identification through bioinformatics of cDNAs encoding human thymic shared AG-1/stem cell AG-2—a new member of the human Ly-6 family. J Immunol 157:969–973, 1996.
5. MR Fannon. Gene expression in normal and disease states—identification of therapeutic targets. 14:294–298, 1996.
6. W Bains. Company strategies for using bioinformatics. Trends Biotechnol 14:312–317, 1996.
7. PG Baker, A Brass. Recent developments in biological sequence databases. Curr Op Biotech 9:54–58, 1998.
8. RD Fleischmann, MD Adams, O White, RA Clayton, EF Kirkness, AR Kerlavage, CJ Bult, JF Tomb, BA Dougherty, JM Merrick et al. Science 269:496–512, 1995.
9. CJ Bult, O White, GJ Olsen, LX Zhou, RD Fleischmann, GG Sutton, JA Blake, LM Fitzgerald, RA Clayton, JD Gocayne. Complete sequence of the methanogenic archaeon. Science 273:1058–1073, 1996.

10. HW Mewes, K Albermann, M Bahr, D Frishman, A Gleissner, J Hani, K Heumann, K Kleine, A Maierl, SG Oliver et al. Overview of the yeast genome. Nature 387(suppl):7–65, 1997.

11. CJ Rawlings, DB Searls. Computational gene discovery and human disease. Curr Op Genetics Development 7:416–423, 1997.

12. JS Aaronson, B Eckman, RA Blevins, JA Barkowski, J Myerson, S Imran, KO Elliston. Toward the development of gene index to the human genome: an assessment of the nature of high-throughput EST sequence data. Genome Res 6:829–845, 1996.

13. P Hieter, DE Bassett Jr, D Valle. The yeast genome—a common currency. Nature Genetics 13:253–255, 1996.

14. E Lahad-Levy, W Wasco, P Poorkaj, DM Romano, J Oshima, WH Pettingel, CE Yu, PD Jondro, SD Schimdt, K Wang et al. Candidate gene for the chromosome 1 familial Alzheimer's disease locus. Science 269:973–977, 1995.

15. N Papadopoulos, NC Nicolaides, YF Wei, SM Ruben, KC Carter, CA Rosen, WA Haseltine, RD Fleischmann, CM Fraser, MD Adams et al. Mutation of a mutL homologue in hereditary colon cancer. Science 263:1625–1629, 1994.

16. CE Bronner, SM Baker, PT Morrison, G Warren, LG Smith, MK Lescoe, M Kane, C Earabino, J Lipford, A Lindblom et al. Mutation in the DNA mismatch repair gene homologue hMLH1 is associated with hereditary non polyposis colon cancer. Nature 368:258–261, 1994.

17. Advancing genomic research: The UniGene collection. NCBI News, 1996.

18. The South African National Bioinformatics Institute: Sequence Tag Alignment and Consensus Knowledgebase (STACK). http://www.sanbi.ac.za/ Dbases.html.

19. JM Claverie. Computational methods for the identification of genes in vertebrate genomic sequences. Hum Mol Genetics 6:1735–1744, 1997.

20. D Kulp, D Haussler, MG Reese, FH Eackman. A generalized hidden Markov model for the recognition of human genes in DNA. Proc Fourth Intl Conf Intell Sys Mol Biol 134–142, 1996.

21. X Huang, MD Adams, H Zhou, AR Kerlavage. A tool for analyzing and annotating genomic sequences. Genomics 46:37–45, 1997.

22. MQ Zhang. Identification of protein coding regions in the human genome by quadratic discriminant analysis. Proc Natl Acad Sci USA 94:565–568, 1997.

23. A Krogh, IS Mlan, D Haussler. A hidden Markov model that finds genes in *E. coli* DNA. Nucleic Acid Res 22:4768–4778, 1994.

24. S Salzberg, A Delcher, K Fasman, J Henderson. A decision tree system for finding genes in DNA. Technical Report, Department of Computer Science, Johns Hopkins University, Baltimore, MD, 1997.

25. J Henderson, S Salzberg, KH Fasman. Finding genes in DNA with hidden Markov model. J Comput Biol 4:127–142, 1997.

26. VV Solovyev, AA Salamov, CB Lawrence. Identification of human gene structure using linear discriminant functions and dynamic programming. Proceedings of the Third International Conference on Intelligent Systems for Molecular Biology, 1995, pp 367–375.

27. R Guigo, S Knudsen, N Drake, T Smith. Prediction of gene structure. J Mol Biol 226:141–157, 1992.

28. EE Snyder, GD Stormo. Identification of coding regions in genomic DNA se-

quences: an application of dynamic programming and neural networks. Nucleic Acids Res 21:607–613, 1993.

29. S Dong, DB Searls. Gene structure prediction by linguistic methods. Genomics 23: 540–551, 1994.

30. L Milanesi, N Kolchanov, L Rogozin, A Kel, I Tirov. Guide to human genome computing. Cambridge, UK: Academic Press, 1993, pp 249–312.

31. GB Hutchinson, MR Hayden. The prediction of exons through an analysis of spliceable open reading frames. Nucleic Acids Res 20:3453–3462, 1992.

32. AJ Viterbi, Error bounds for convolutional codes and an asymptotically optimal decoding algorithm. IEEE Trans Informt Theory 13:260–269, 1967.

33. MS Gelfand and MA Roytberg. Prediction of exon-intron structure by dynamic programming approach. BioSystems 30:173–182, 1993.

34. A Krogh. Two methods for improving performance of an HMM and their application for gene finding. In: T Gaasterland, ed. Fifth International Conference on Intelligent Systems for Molecular Biology. Menlo Park, CA: AAAI Press, 1997, pp 179–186.

35. M Borodovsky, J McIninch. GeneMark: parallel gene recognition for both DNA strands. Computers Chemistry 17:123–133, 1993.

36. AL Delcher, D Harmon, S Kasif, O White, SL Salzberg. Improved microbial gene identification with GLIMMER. Nucleic Acids Research 27:4636–4641, 1999.

37. A Thomas, MH Skolnick. A probabilistic model for detecting coding regions in DNA sequences. IMA J Math Appl Med Biol 11:149–160, 1994.

38. D Haussler. Computational genefinding. Trends Guide to Bioinformatics (Trends supplement) pp 12–15, 1998.

39. M Burset, R Guigo. Evaluation of gene structure prediction programs. Genomics 34:353–367, 1996.

40. N Brockdorff, A Ashworth, GF Kay, VM McCabe, DP Norris, PJ Cooper, S Swift, S Rastan. The product of the mouse Xist gene is a 15 kb inactive X-specific transcript containing no conserved ORF and located in the nucleus. Cell 71:515–526, 1992.

41. K Pfeifer, PA Leighton, SM Tilghman. The structural H19 gene is required for transgene imprinting. Proc Natl Acad Sci USA 93:13876–13883, 1996.

42. R Wevrick, U Francke. An imprinted mouse transcript homologous to the human imprinted in Prader-Willi syndrome (IPW) gene. Hum Mol Genet 6:325–332, 1997.

43. AY Liu, BS Torchia, BR Migeon, RF Siliciano. The human NTT gene: identification of a novel 17-kb noncoding nuclear RNA expressed in activated CD4+ T cells. Genomics 39:171–184, 1997.

44. GB Singh. Computational approaches to gene prediction. In: S Misener and SA Krawetz, eds. Methods in Molecular Biology, 132: Bioinformatics Methods and Protocols. Totowa, NJ: Humana Press, 2000, pp 351–364.

45. S Brenner. Practical database searching. Trends Guide to Bioinformatics (Trends supplement) pp 9–12, 1998.

46. SF Altschul, TL Madden, AA Schaffer, J Zhang, Z Zhang, W Miller, DJ Lipman. Gapped BLAST and PSI-BLAST: a new generation of protein database search programs. Nucleic Acids Res 25:3389–3402, 1997.

47. WR Pearson, DJ Lipman. Improved tools for biological sequence comparison. Proc Natl Acad Sci USA 85:2444–2448, 1988.

48. TF Smith, MS Waterman. Identification of common molecular subsequences. J Mol Biol 147:195–197, 1981.

49. SF Altschul. Fundamentals of database searching. Trends Guide to Bioinformatics (Trends supplement) pp 7–9, 1998.

50. DJ States, W Gish, S Altschul. Improved sensitivity of nucleic acid database searches using application-specific scoring matrices. Methods: A Companion to Methods in Enzymology 3:66–70, 1991.

51. JC Wootton, S Federhen. Statistics of local complexity in amino acid sequences and sequence. Comput Chem 17:149–163, 1993.

52. RL Tatusov and DJ Lipman. DUST program (manuscript in preparation).

53. S Henikoff, JG Henikoff. Amino acid substitution matrices from protein blocks. Proc Natl Acad Sci USA 89:10915–10919, 1992.

54. SE Brenner, C Chothia, TJP Hubbard. Assessing sequence comparison methods with reliable structurally identified distant evolutionary relationships. Proc Natl Acad Sci USA 95:6073–6078, 1998.

55. SR Eddy. Multiple-alignments and sequence searches. Trends Guide to Bioinformatics (Trends supplement) pp 15–18, 1998.

56. EL Sonnhammer, SR Eddy, E Birney, A Bateman, R Durbin. Pfam: multiple sequence alignments and HMM-profiles of protein domains. Nucleic Acids Res 26:320–322, 1998.

57. R Durbin, S Eddy, A Krogh, G Mitchison. In: Biological Sequence Analysis: Probabilistic Models of Proteins and Nucleic Acids. Cambridge: Cambridge University Press, 1998.

58. PFAM. Copyright © 1996–1999 The Pfam Consortium.

59. A Bateman, E Birney, R Durbin, SR Eddy, KL Howe, EL Sonnhammer. The Pfam protein families database. Nucleic Acids Res 28:263–266, 2000.

60. F Horn, G Vriend. G protein-coupled receptors in silico. J Molecular Medicine 76:464–468, 1998.

61. T Christiansen, N Torkington. Perl Cookbook. Sebastopol: O'Reilly and Associates, 1998.

62. K Hoffman. Protein classification and functional assignment. Trends Guide to Bioinformatics (Trends supplement) pp 18–21, 1998.

63. A Bairoch, P Bucher, K Hofmann. The PROSITE database, its status in 1997. Nucleic Acids Res 25:217–221, 1997.

64. A Bairoch. PROSITE: a dictionary of sites and patterns in proteins. Nucleic Acids Res 19(suppl):2241–2245, 1991.

65. J Schultz, F Milpetz, P Bork, CP Ponting. SMART, a simple modular architecture research tool: identification of signaling domains. Proc Natl Acad Sci USA 95:5857–5864, 1998.

66. EL Sonnhammer, D Kahn. Modular arrangement of proteins as inferred from analysis of homology. Protein Sci 3:482–492, 1994.

67. F Corpet, J Gouzy, D Kahn. Recent improvements of the ProDom database of protein domain families. Nucleic Acids Res 27:263–267, 1999.

68. EL Sonnhammer, SR Eddy, R Durbin. Pfam: a comprehensive database of protein domain families based on seed alignments. Proteins 28:405–420, 1997.

69. SR Eddy. Profile hidden Markov models. Bioinformatics 14:755–763, 1998.

70. S Pietrokovski, JG Henikoff, S Henikoff. The Blocks database—a system for protein classification. Nucleic Acids Res 24:197–200, 1996.
71. TK Attwood, DR Flower, AP Lewis, JE Mabey, SR Morgan, P Scordis, JN Selley, W Wright. PRINTS prepares for the new millennium. Nucleic Acids Res 27:220–225, 1999.
72. MA Hershkovitz, DD Leipe. Phylogenetic analysis. In: AD Baxevanis, BFF Ouellette, eds. Bioinformatics: A Practical Guide to the Analysis of Genes and Proteins. New York: Wiley-Interscience, 1998, pp 189–230.
73. DM Hillis, MW Allard, MM Miyamoto. Analysis of DNA sequence data: phylogenetic inference. Methods Enzymol 224:456–487, 1993.
74. JC Avise. Molecular Markers, Natural History and Evolution. New York: Chapman and Hall, 1994.
75. N Saitou. Reconstruction of gene trees from sequence data. Methods Enzymol 266:427–449, 1996.
76. WH Li. Molecular Evolution. Sunderland, MA: Sinauer Associates, 1997.
77. DL Swofford, GJ Olsen, PJ Waddell, DM Hillis. Phylogenetics inference. In: DM Hillis, C Moritz, BK Mable, eds. Molecular Systematics. Sunderland, MA: Sinauer Associates, 1996, pp 407–514.
78. JM Thornton. The future of bioinformatics. Trends Guide to Bioinformatics (Trends supplement) pp 30–31, 1998.
79. DL Ellsworth, TA Manolio. The emerging importance of genetics in epidemiologic research III: Bioinformatics and statistical genetic methods. Ann Epidemiol 9:207–224, 1999.
80. VA McKusick. Mendelian Inheritance in Man. Catalogs of Human Genes and Genetic Disorders. 12th ed. Baltimore: Johns Hopkins University Press, 1998.

18

The Evolution of Laboratory Automation

Jack Elands
IDBS Ltd., Guildford, United Kingdom

I. INTRODUCTION

Pharmaceutical and biotechnology companies are facing increasing global competition coupled with the need to maintain high market capitalization values. This has created a tremendous sense of urgency for them to look for ways to improve their efficiencies and reduce the cycle time of the drug discovery process.

Today, investments in workstations and robotic systems are justified not by the cost of the labor saved but by the financial impact of getting a new drug on the market a few months faster, or millions of dollars worth of opportunity cost. Laboratory automation is playing a major role in helping companies achieve their goals.

Laboratory automation is usually considered a relatively recent development. Although recent progress in automation has been spectacular, the introduction of automated technologies is not just a modern phenomenon. A vast range of equipment has been developed that could be considered laboratory automation. Hand-held pipettors with disposable tips are a great improvement over the glass pipettes and glass capillaries. Electronic balances have made weighing much easier, faster, and more reliable. Tube, flask, and bottle shakers, vortexers, etc. have eliminated the need to shake labware, etc. manually.

Although technology has allowed for laboratory automation to advance, in general economic drivers have caused laboratory automation to become reality. Examples can be found in the clinical diagnostics industry, which has been the source of many developments that are considered to be normal today. The recent wave of laboratory automation is clearly driven by the need for higher productivity and considerably faster cycle times both in drug discovery research and in drug development. Without those strong economic incentives the world would not have seen the current interest in automation.

From a broader industry perspective, drug discovery research and development has been very late in implementing automated processes. Since the pharmaceutical industry enjoyed high profit margins for a long time, there was no pressure to convert the academic research style to a more appropriate industrial R&D style. Also, there are huge differences between those industries that have automated their production and pharmaceutical research. Most industries that apply automation have established procedures and very well described practices. In addition, these procedures do not change often, and if they do, the nature of these processes does not change dramatically. In the process of gaining experience with automation, flexibility has been designed into those automated systems. A good example is today's car industry, where on a production line many varieties of the same car model are being assembled. Previously, the production of different varieties was done outside of the production line.

Drug discovery research has no well described best practices yet. Many companies, pharmaceutical and especially biotechnology, have proprietary approaches to discovery research in order to try to gain a competitive advantage over other companies. The need to shorten discovery time lines and the rapid developments in assay technology and related instrumentation cause these approaches to change almost continuously. There is no sign yet that those limits have been reached. The objective of this chapter is to address the evolution of laboratory automation—within the context of drug screening—in order to gain an understanding of where the industry is today, what the benefits and shortcomings are, and where the future promise may be.

II. EARLY HISTORY OF LAB AUTOMATION

In the early 1970s, analytical chemistry became a recognized and important discipline. Technologies had matured, and the instrumental analytical chemistry was considered an enabling technology for the pharmaceutical and chemical industry. While in the 1970s the industry operated with ever increasing R&D budgets, in the 1980s cost reduction and increased productivity became major drivers. In the mean time, sample preparation had become the bottleneck of the analytical chemistry. However, it was no longer acceptable to resolve this bottleneck by adding personnel. Shortly after its foundation, Zymark identified this unmet need in automated sample preparation. Its vision became to extend instrumental concepts into the sample preparation process. The analytical procedures at the time were very diverse and often depended upon internally developed methods. It was obvious that flexibility was required for sample preparation approaches. Zymark's founders came to the conclusion that sample preparation was a major bottleneck during a series of visits to laboratories when they were investigating what business strategy Zymark was to follow.

A close analysis of a wide range of sample preparation techniques revealed that they could be subdivided into a number of steps, such as weighing, liquid handling, separation, labware handling, etc. Zymark proposed the term Laboratory Unit Operation (LUO) to describe such a step in an application. A method thus consists of a series of appropriate LUOs together. Most LUOs are performed by different devices. These observations led to an architecture consisting of an array of instruments that each perform a LUO. Sample containers would be moved between instruments by a small-scale tabletop robot.

The Zymate® robot arm was specifically designed for moving laboratory consumables between storage racks and the different instruments. Payload and precision requirements were modest compared to what industrial robots could deliver. Moreover, the laboratory consumables themselves had widely variable dimensions and relatively high dimensional tolerances. A simple robot arm providing for vertical and horizontal movements would be largely sufficient and would simplify robot design and method programming. "Friendly" hardware, such as racks with tapered positions, would guide labware into their correct positions, regardless of their varying dimensions.

In order to control the robot arm and the different instruments a multitasking microprocessor-based controller was thought to be essential. No powerful and small microprocessor-based computer was available at the time. The Apple II was available, while the IBM PC was only announced. Both had primitive disk operating systems that only offered single tasking. A new and powerful microprocessor-based controller with a multitasking operating system was therefore developed. A user-friendly programming language (EasyLab®) provided intuitive programming for equipment interfacing and methods development. A self-configuring bus interface was developed so that new future modules could be easily connected to the controller. Together, this allowed the integration of up to 25 peripherals (instruments) and the Zymate robot arm into a circular array. This would become the basic architecture for laboratory automation for many years.

III. LAB AUTOMATION AND DRUG DISCOVERY RESEARCH

For many years, experiments using intact animal organs, tissues, or even whole animals were important tools for drug discovery research. Obviously these approaches were rather difficult to automate to the extent that an experiment could be undertaken without experimenter attendance. Today, however, many behavioral test models use sophisticated automated setups to study and analyze animal behavior. Enzymes, neurotransmitter (re-)uptake sites, and receptors for neurotransmitters became popular targets for drug discovery and development. These targets were studied amongst others using in vitro preparations. Thus the assay in which the direct enzymatic and neurotransmitter uptake activity or radioligand binding to receptors was measured took place in reagent vessels such as tubes. Enzymes were often analyzed using colorimetric reactions or radioactive substrates. Measurements were done with cuvette-based spectrophotometers and scintillation counters. Spectrophotometry was automated using autosamples in combination with flow through cuvettes. Radioligand binding was more difficult to automate. Many receptor-binding studies used filtration of the reaction mixture to separate the receptor preparation, to which the radioactive ligand was bound, from the rest of the test mixture.

Two main approaches have been in use; automating the manual procedure and implementing parallel sample handling. Several fully automated, single-tube-based approaches for assay automation existed. One example is the Filterprep FP102 system from Ismatec. It provided automated pipetting of reagents, incubation, filtration, filter punching into scintillation vials, and addition of scintillation cocktail in a conveyer belt fashion. They were rather complex single-application systems that automated all the steps of a manual procedure. Very successful was the parallel sample handling for the filtration of tube content, punching out of filters into scintillation vials, and the addition of scintillation cocktails to the vials. Examples of such parallel filtration devices are the Brandel multichannel cell harvesters. Parallel sample processing is often straightforward and intuitive, since it usually follows the same procedure as for processing single samples.

These examples are described only to illustrate some of the processes that were in use in R&D laboratories. Note that several of these technologies are still useful. Finally, it should be obvious that many more techniques have been used, and are still in use, that are not mentioned here.

IV. THE FIRST AUTOMATED SYSTEMS

The architecture developed by Zymark for automating sample preparation seemed to be ideal for the automation of processes in drug discovery research. Initially the developments derived from sample preparation automation were directly transferred to automate drug screening applications. In a way, the early integrated robotic systems were engineering challenges implemented by visionary technology adopters. Early automation projects were often awkward attempts that utilized the capabilities of the robotic arm to emulate the actions of human beings. For example, robots were used to pipette liquids as a human does with single-channel or multichannel pipette hands, to blot plates, or to tweeze filter paper from the bottom of a filter cup into a scintillation vial. The flexibility and throughput of these early systems was limited to the finesse and speed of the robot; they were often demanding and unreliable and offered no capability for multitasking.

In addition, most of the early screening methods were performed as single, discrete assays in test tubes or scintillation vials, and the automation was dedicated to one assay. Large volumes were required, typically 1 to 5 mL of reagents, and samples were consumed for each assay. Liquid handling devices offered some benefit as stand-alone workstations, but their productivity was limited in that they pipetted and dispensed liquids from test tube to test tube. Finally, most detection systems could only process one sample at a time; data acquisition rates were slow (30 seconds to several minutes per sample) and there was no data interface other than a hard copy print out. All things considered, automation added value by providing unattended operation.

Obviously, the role of automation through integrated robotic systems applied to the drug discovery process has evolved through the last 15 years. Today's laboratory automation is much more robust and reliable, the beneficiary of years of experience. Also, the standardization of assays into microplates has reduced the challenges for automation by providing a predictable environment to operate in.

V. THE MICROPLATE

The microtiter plate, now conveniently called "microplate" or "plate," deserves a special mention with respect to laboratory automation. In many respects it has been a major determinant of the direction of many products and solutions. And although chip-based technologies are rapidly gaining ground, the microplate will remain to play a prominent role in biopharmaceutical research. The microplate itself is lab automation. It introduced a concept of multiple reaction vessels that could be or needed to be treated in parallel. Rather than manipulating single vessels, a single container with a high density of vessels could be manipulated. This parallelization concept certainly contributed to the success of robotic systems and automated workstations.

The microplate has also been a standard for equipment development. Although the 96-well microplate has only recently been standardized (Society of Biomolecular Screening task force for "Standards in Automation and Instrumentation"), its physical dimensions and characteristics have always been roughly identical. The standardization was primarily required for enhanced reliability in today's automated environment. Recently, the 384-well plate also was standardized (Society of Biomolecular Screening task force "Standards in Automation and Instrumentation").

Apart from the 96- and 384-well microplates, there are no other microplate standards. The higher density plate formats (864-, 1536-, 3456-, and 9600-well plates) serve an important need to reduce assay costs by reducing the assay volume. The higher density also helps to increase throughput, although throughput is also determined by the speed of liquid handling devices and microplate readers. No standard is available for 1536-well plates. This may become a major issue, since the well-to-well distance is considerably smaller, and unacceptable tolerances may lead to difficulties in centering the wells (required for accurate pipetting and to obtain a high-quality signal during reading).

Although seldom thought of in the light of laboratory automation, the microplate is to be considered an important form of automation because it introduced a high degree of parallelization of reaction vessels. It is the most important platform for the development of standards that are available today in biopharmaceutical research.

VI. MICROPLATE INSTRUMENTATION

It is useful to differentiate between stand-alone products, workstations, and robotic systems. In general, a workstation is a device that performs multiple different functions (in our case mostly on microplates). As such, plate washers, dispensers, and readers are considered stand-alone devices, while liquid handlers that combine multiple tools, or that can perform multiple actions on a microplate, should be considered workstations. Robotic systems are a collection of stand-alone devices and workstations integrated into a functional environment, controlled by system management software and one or more robot arms to perform an application. In many cases the distinction between devices, workstations, and systems is difficult to make, e.g., a luminescence reader with a built-in injector and a microplate shaking function is not considered a workstation, while it clearly performs multiple actions on a microplate. Therefore these definitions may not be generally accepted, but they are sufficient for the purpose of this review.

Most of the devices that are available for microplate reading, scintillation counting, reagent dispensing, plate washing, etc. have been developed as benchtop, stand-alone devices and not as modules to be integrated into a robotic system. The result of the integration of such a module was therefore not always very reliable. This is especially true for the ability to control equipment by the system management software and the collection of data, generated by the equipment, in a way that is useful for the experimenter. More recently, the integration of devices in robotic systems has become significantly easier. Suppliers have made their equipment more robot friendly and easier to control by providing appropriate software interfaces. In spite of this, the industry is still far away from the standardization of device control and data output formats. In general, the combination of hand-held pipettors, etc. and stand-alone microplate equipment allow almost any microplate application to be performed manually.

VII. WORKSTATIONS

Earlier we defined a workstation as a device that performs multiple different functions. The first workstation was the Benchmate™ from Zymark. This versatile workstation has a dedicated robot arm with a gripper that can move containers to and from different locations on its workbench. Several devices are integrated, such as a precision balance, a single-channel canule for sample transfer, reagent addition, and dilutions, a vortexer, and a HPLC injector. It uses the cylindrical work envelope that is typical of the Zymate architecture that also enables the integration of additional external devices. The Benchmate has also been developed for areas in biopharmaceutical R&D where quality control and quality assurance are paramount. It has a built-in data audit trail and gravimetrically controlled liquid handling. The Benchmate is primarily designed for tube-based applications and is not well suited for microplate applications.

One of the first generally accepted workstations for microplate-related applications was the Biomek 1000 from Beckman Coulter. It combined single and multiple channel pipetting, reagent addition and washing, and single-channel reading (densitometry). Other workstations are single- or multichannel (4–8) liquid handlers, often gantry type systems with one or two arms carrying canules or different tools (e.g., tube or microplate grippers), and 96- and 384-channel disposable or fixed tip pipettors. There are two notable exceptions to this architecture, the Biomek and the Benchmate (see above). The Biomek is different since it uses exchangeable tools, such as different pipettors, dispensers, and a gripper. It offers thereby a great flexibility with respect to different applications. The Biomek 2000 workstation misses, however, the ability to adapt to nonmicroplate formats and individual tip liquid level sensing on the multichannel tools.

As workstations are designed to provide a certain capacity to perform an application without operator attendance, i.e., walk-away time, they must have a minimal capacity for consumable storage on the working deck. Thus the footprint is always larger than needed for the liquid handling technology itself. In a robotic system there is no benefit to the large surface that most workstations have. In fact, for integration into a robotic system the large decks are nonproductive, since the robotic system already has a consumable storage capacity itself. Smaller liquid workstations retaining all required functionality, and being optimal for a robotic system because of a small footprint, have become available only recently. One of the best examples is the Presto Liquid Handler™ from Zymark.

Workstations such as the ones described here have proven to be extremely useful for combinatorial chemistry, compound weighing, dissolution and distribution, screening, and many other applications. They add in a significant way to the functionality offered by the stand-alone devices, albeit in a very different way. Many of the stand-alone devices are critical to working with microplates, while the workstations enable repetitive work, provide more consistent results, and may also allow the experimenter to do other work, i.e., they create walk-away time. As such, workstations have taken a prominent role in laboratory automation. With increasing plate densities (384-well plates and beyond) that are almost impossible to pipet to and from manually, workstations are no longer a luxury but have become a necessity. It is therefore to be expected that application of the new and smaller footprint workstations will rapidly extend and become a set of personal tools for the scientist.

VIII. RECENT DEVELOPMENTS IN AUTOMATION

In recent years, different trends could be observed in the development of automated solutions for the drug discovery research community. First of all, there has been a further development of traditional robotic systems. Secondly, the workstations have expanded to include tools to transport consumables. These tools also enable interfacing with other devices such as readers and storage carousels. In addition, plate stackers or microplate autosamplers are now available for many stand-alone devices and workstations. Finally, solutions have become available that are more specifically developed for screening, an area where robotic systems and combinations of workstations and stand-alone devices (in combination with stackers) have mostly been used.

A. Modern Integrated Robotic Systems

These systems remain largely based upon the architecture that was originally developed by Zymark. Most progress has been made in the development of appropriate modules or devices and user-friendly graphical (mostly Windows-based) interfaces to system management and scheduling software. While in the original Zymate architecture a proprietary bus interface (the controller) was used exclusively for module interfacing, most systems today are based upon standard industry buses (ranging from RS-232 to standard industry network connections). When software interface specifications are available, modules can be easily integrated by in-house system integrators, which may decrease vendor dependence. However, not all suppliers of robotic systems have made interfacing specifications available.

Almost all robotic systems come with some form of scheduling software in addition to the required system management software. The very fact that a single robot arm serves many peripherals requires scheduling of the robot arm and of all the individual devices (and of required user intervention). The early software architectures were multilayered and reflected the development of the different components for the software architecture. Scitec Automation was the only company that developed a consistent software architecture for robotic systems in which scheduling was embedded and not just added onto an existing structure. For many years their software package, Clara, was the flagship of the industry. It was the first scheduler that incorporated true dynamic scheduling. Since then several companies have developed modern software architectures that include powerful schedulers. The Sagian division of Beckman Coulter deserves mention since it recently developed the first architecture in a Windows NT environment with dynamic scheduling in which modules can be exchanged very easily. In spite of all these developments, robotic systems remain very support intensive both from a user and supplier point of view. The degree of success of these systems is often directly related to the level of support. A significant part of the investment by suppliers in ''modern'' software architectures, based upon industry standard tools, has as goal a reduction of the complexity to build and support robotic systems.

Most integrators (vendors and ''in-house'' integrators) prefer to use standard equipment that is available on the market. When no products are available, custom modules are often developed. Such modules frequently become standard modules, or even are further developed to become standard products. The capability to develop those custom modules is a differentiating factor in today's automation industry.

The other factor that is often considered important is the robotic arm itself. The three main robot arms used today are the Zymate XP® arm, the Sagian Orca arm, and the family of CRS arms. Recently, Mitsubishi arms have also made their appearance. Although there are significant differences in positional accuracy and degrees of freedom, none of these elements plays a critical role in the success of integrated robotic systems. The fundamental task, to move containers from one location to another, can be carried out equally well by all arms. The approach of CRS Robotics, to move stacks of microplates rather than individual microplates, requires a certain minimum payload, for which the Zymate XP and the Sagian Orca arm are less appropriate. On the other side, the physical force of a robot arm can be disastrous when systems are programmed in an inappropriate way (during setup or after changes).

B. Integrated Systems as Standard Products

The time required for the integration of a robotic system is usually relatively long, especially for new configurations, when new modules need to be interfaced or when new modules need to be developed. Standard interface protocols have greatly facilitated the development of new interfaces, and most newer equipment and workstations are much more straightforward to interface. However, system integration includes not only setup of modules and teaching of robot positions but also optimization with respect to specific applications requiring precise fine-tuning of individual robot moves. As a result, system integration is and will remain delivery resource intense. An exception to this is the Allegro™ system from Zymark that is a unique combination of custom tailoring with standard modules. This approach eliminates configuration issues related to a specific application. Allegro will be discussed in more detail later.

Another attempt to reduce the complexity of these systems integration issues came from both Zymark [1] and Beckman Coulter [2], who have marketed systems that are specifically configured for an application or a family of applications. Obviously, focusing on standard configurations can significantly reduce the integration time for a system. First of all, components are used for which interfaces are already available, and secondly, because the application for which it will be used is well described, optimization can be done up front. Finally, it also represents an alternative way of product marketing.

C. Small or Large?

An even more important reason to propose systems that are specifically developed for an application or for a family of applications is to move away from the custom integration of large and complicated systems. In general, large systems are expensive, not just because of the hardware, but because of the additional amount of system integration time and project management resources that are required. Thus large systems take disproportionately more time to be implemented than small systems. If the application needs can be grouped in different families of corresponding applications, and each family can be implemented on a smaller system, the total cost, complexity, and delivery time can be significantly less than for a single large integrated system [1,2]. Furthermore, the argument has been made that large systems are economically less productive than several smaller systems [1]. There are many other advantages to having multiple smaller systems: maintenance scheduling has lower overall impact, system failures will not render the whole operation nonproductive, etc. In general, the time between ordering and delivery of a system should be as short as possible. Long delivery times often lead to revised objectives for a system, which in turn leads to expensive change order processes.

D. Expanded Workstations

Several workstations have seen their versatility increased by additional expansion capabilities. Modern liquid handling platforms (Tecan, Genesis; Packard Instruments, MultiProbe; Rosys Anthos, Plato) can also contain devices, such as readers, washers, and shakers, on their working surface that are integrated into the system. Using gripper tools, labware can be moved to and from different positions and devices. The Tecan Genesis also interfaces with devices that are external to the working surface, such as storage carousels, and devices such as readers and pipettors. Those external devices are in addition to many devices that can be integrated onto the working surface. When multiple devices are integrated to such a platform and with the additional devices integrated on the working surface, truly integrated robotic systems appear. The difference with traditional integrated systems is that the workstation is the core component, while it is only a peripheral in a traditional robotic system. Driving factors to add external modules are to enhance functionality or capacity. Integration of storage carousels is one way of adding capacity. Another way is to add plate stackers or plate autosamplers. It is interesting to note that the original design goals—to provide sufficient capacity for labware and reservoirs, etc. on the working surface—are no longer observed in these solutions.

E. Plate Stackers

Plate stackers originally had a single objective, to increase the capacity of a device or workstation. Many companies developed stackers for their products that were, and still are, specific to that product, or at best a line of products. Packard Instruments was one of the first to introduce a plate stacker for the TopCount scintillation counter. Many companies followed this example. Although it is relatively easy to remove a stack of plates of a stacker and to load another, there are several inconveniences. First, all stackers are not of the same size. Secondly, microplates may contain liquids, and these may be biohazardous. Titertek has developed one stacker model for several of their products. As a result, a stack can be carried from one device to another, further enhancing the concept of stackers and also reducing risks of errors.

Zymark has taken this concept a significant step further by introducing Twister™, a universal microplate stacker. Twister has been made universal through partnership and OEM agreements with many equipment suppliers, making it available for many readers, washers, dispensers, and pipettors. The success of Twister has led others to develop similar approaches. However, a significant investment for the development of partner relationships is required to be truly successful.

Several stackers exist now (the PlateStak from CCS-Packard, the PlateSilo from Hudson Control Group, and the Twister from Zymark) that can be used as devices that present a single microplate at a time to another device. As such they can easily interface to liquid handlers and pipettors. The Twister, PlateSilo, and PlateStak are ideal for feeding microplates to robotic systems, etc.

IX. NEW APPROACHES TO AUTOMATION

The original Zymate architecture, according to which most "integrated" robotic systems are built today, was developed initially for the automation of sample preparation for analytical chemistry. It is obvious that this architecture has proven to be exceptionally useful for many different approaches. The open character of the hardware side of the architecture facilitates flexibility and allows the tailoring of solutions, which has been simplified further by the newer generation of software architectures for these systems. However, it is only one approach to laboratory automation. Multiple approaches are possible. An alternative approach, as discussed already, is the liquid handler platform with a deck size that is large enough to store sufficient amounts of labware to run an application. However, the recent trend to expand those platforms with gripper tools, with devices on the liquid handling deck, and also with external devices has resulted in a solution that is in fact comparable to an integrated robotic system. The major difference is that the robot arm (the gripper tool) and the liquid handler are physically on the same platform, whereas on robotic systems they are separate devices.

Several new approaches for screening automation have been launched recently. Some of these appeared only very recently. These will therefore be mentioned without any comments with respect to their position, acceptance, utility, etc.

A. Allegro

Zymark developed the Allegro as a practical, pragmatic solution to achieve ultrahigh-throughput screening (UHTS) with the technology that was available in 1997/1998 and that was proven successful by many pharmaceutical companies [3]. In Allegro each step in an assay is executed by a separate module. Modules are standard, autonomous, independent, and self-contained. The modules are connected together in a linear fashion, and a dedicated plate handler, which is integrated in each module, passes plates from one module to another. Allegro's modular approach allows easy and very quick reconfiguration to adapt the system from one application to another. It is the only approach in which optimization for an application by the user is possible while using standard components. The Allegro architecture not only allows very high throughputs (one plate per minute) but is also very robust. The fact that each assay step is carried out by a single module contributes to the robustness but also renders this approach very hardware intensive. It is therefore relatively expensive and needs a large laboratory surface. Recently, Zymark launched the Allegro Combo™ to address those issues. Because Allegro Combo allows plates to also travel backwards, modules can be reused in an assay when needed. This approach reduces the number of modules required for an application. Allegro is the first system to offer a set of unique data handling and data organizing tools to interface to almost any data management system.

B. PlateTrak

CCS-Packard developed the PlateTrak originally as automation for the production of coated microplates. Reagent addition, microplate washing, and drying were the steps required. Obviously, in a production environment extremely high throughputs are required. This explains much of the design principles behind the PlateTrak, and its speed of processing microplates. However, the addition of reagents to and washing of microplates does not provide full assay automation. The original PlateTrak was clearly a very fast liquid handling workstation, but it had limited flexibility.

More recently, this technology has also been applied for UHTS. In order to provide more flexibility and versatility, CCS-Packard abandoned the PLC for a more sophisticated Windows-based PC control system and enhanced the modularity of the hardware components. The modular approach of the PlateTrak allows for flexibility during manufacturing but does not provide user reconfigurability. In addition, new modules have been developed, and the system can also be expanded to include on-line incubation and readers or other detectors. As such it is no longer only a liquid handling system but also an assay system.

C. TekCel

TekCel has launched a series of TekBench stations to automate microplate-based applications. A TekBench can be considered as an integrated system with a very small footprint. It comes with a novel gantry-type robotic arm to move consumables between the workbench and storage areas (capacity of 252 items), to which also pipetting heads can be attached. In this way the system can be equipped with up to four pipettors. A TekBench has several shelves. The top shelf is used to integrate devices such as readers, incubators, dispensers, etc. The lower benches provide the storage capacity. Several TekBenches have been developed to support different applications. A unique feature is the integrated exchange track, a belt-driven conveyer to transport microplates between two or more TekBenches. This track provides the possibility to combine multiple TekBenches, offering an integrated way of expanding capacity and functionality.

D. Others

Zeiss developed an integrated robotic system for high-throughput screening (HTS) and UHTS. This new approach, developed with the pharmaceutical industry [4], incorporates a new type of microplate reader, entirely based on micro-scope technology. In its final release, this detector has 96 parallel microscope optics with a 96-well microplate compatible spacing, enabling the detection of 96-, 384-, and 1536-well plates. Due to the close proximity of the optics to the source of the signal, the potential advantages are increased detection speed and high signal quality. When used with high optical quality microplates, this detection technology could offer significant opportunities.

Also integrated are the JobiWell from Jenoptik and the Cytomat 6000 from Kendro Lab Products (Heraeus). The system is a self-contained unit and consists of modules. It uses a novel plate transport mechanism and can be integrated with other units through a conveyer belt transport mechanism, thus providing true scalability.

Jenoptik Bioinstruments has built a system around the JobiWell pipettor, better known as the PlateMate. It uses the rail that existed already on the JobiWell to transport microplates to and from devices that are integrated into the system. A small rotating microplate platform transfers microplates from one stage to another. Several rails can be connected using this rotating platform. One or more axes (rails) can branch out of the main axis to include additional devices, such as different readers, incubators (Cytomat 6000), plate storage carousels, etc. to support a variety of applications. In a way this approach can be considered as an extreme form of an expanded workstation, since the heart of the system is still the JobiWell.

Panasonic introduced the Hornet in early 1999 in Japan. The Hornet is a compact, totally enclosed system with a professional appearance. It contains all devices required to run applications, except for a filtration station. The space below and above the ''workbench'' is also used in an efficient way, compared to traditional robotic systems. This concept is likely to be in an early phase, although at least one pharmaceutical company is using it. Conceptually, it resembles the TekBench approach from TekCel.

There are relatively few products for drug discovery research that are designed specifically for automating an application. This is true for screening as well as combinatorial chemistry. Most products are extensions or modifications of an existing platform. The new solutions listed above are certainly innovative approaches that break with this tradition. It is to be expected that some of these products will render the drug screening process more effective.

The speed with which the biopharmaceutical industry changes its approaches requires the supplier industry to focus to some extent on platforms that can be easily reconfigured, or customized, for keeping pace with these changes. The major platforms today are the robotic systems and the workstation-type liquid handlers.

Many new breakthrough developments are related to the assay detection technology. New and more sensitive readers, new detection technologies, and new assay platforms for which these detection systems are developed appear regularly. The second area of development has been liquid handling. The miniaturization of assays (384- and 1536-well microplates) requires lower and lower volumes to be dealt with. However, many of the microarraying applications have really pushed the limit and required existing technologies to be adapted to biopharmaceutical applications. Examples are combinations of classical syringe pump-based liquid handling and microsolenoid valve and piezo electric dispensing. The real breakthroughs are expected in the application of microchips in drug screening. In spite of these developments, the microplate-based platform will continue to play an important role for many years to come.

X. THE CHALLENGE OF LABORATORY AUTOMATION

The biopharmaceutical industry is not looking to automate; it is rather looking for tools to improve drug discovery processes. The goal of laboratory automation should be to facilitate this. Suppliers have traditionally been focusing on components with which the biopharmaceutical industry could build the tools they needed. More recently, the focus changed to better integration of components by the suppliers to deliver useful tools to the industry.

There are several areas where the laboratory automation industry could still provide major benefits to the biopharmaceutical industry. Two of them relate directly to existing products. The first is the integrated robotic system. Even today most systems are relatively difficult to use. Implementing applications on systems is a challenge and often the cause of deceptions and delays. There is much room for improvement to render systems more user friendly and to offer services to make systems more successful. Many companies have started with a vision of true multiuser, multitasking, "load my samples and walk away" assay systems. In spite of all the developments around integrated robotic systems and laboratory automation, such systems do not exist.

The second is the workstations. With the increased use of higher density microplates, workstations are going to be needed to assist scientists in their work. Most of today's workstations are too big and too expensive. Ideally workstations are small and are used by a single scientist, i.e., the Personal Workstation as part of the Personal Automation vision. The problem to solve is not the effective use of equipment but the effective use of the scientist's creativity. The parallel with the mainframe computer of the 1970s and 1980s and the PC of the 1990s seems to be appropriate.

AUTHOR'S NOTE

At the time of writing the author worked at Zymark Corporation. Several developments have taken place and new products have been launched since this chapter was written. The descriptions of the products may therefore no longer be accurate and complete. However, this does not affect the scope, intention, and conclusions of this review. Jenoptik Bioinstruments has changed its name to CyBio.

ACKNOWLEDGMENTS

Burleigh Hutchins, Frank Zenie, and Brian Lightbody are gratefully acknowledged for the information they provided for the historical overview and the insights shared with the author during many discussions.

Thanks also go to Nancy Leutert, who has provided the inevitable last-minute assistance to complete this chapter.

Andy Zaayenga (TekCel), Klaus Mlejnek (Carl Zeiss Jena), Dermot Boylan (Packard Instruments), Nader Donzel (Scitec Automation), and Carl Murray (Beckman Coulter) have been so kind as to provide essential information for this chapter about their companies' products.

Zymate and Easylab are registered trademarks, and Allegro, Allegro Combo, Twister, and Benchmate are trademarks from Zymark Corporation. All other trademarks are the property of the respective owners.

REFERENCES

1. J Elands. Automated screening: a review of existing automation approaches. Proceedings of the 2nd Annual Meeting of the Society of Biomolecular Screening, Basel, 1996, pp 291–303.
2. DW Brandt. Core system model: understanding the impact of reliability on high-throughput screening systems. Drug Discovery Today 3:61–68, 1998.
3. E Alderman, J Elands. Breaking a screening barrier. A novel approach to ultra-high-throughput screening. Genetic Engineering News 18:14–16, 1998.
4. K Mlejnek, M Gluch. The Zeiss Ultra High Throughput Screening (UHTS) system (Part 1): presentation of a novel system platform. Proceedings of the LabAutomation 99 Conference, San Diego, 1999, p 119.

19
Robotics and Automation

Robert F. Trinka
Robocon Incorporated, Plymouth Meeting, Pennsylvania

Franz E. Leichtfried
Robocon GmbH, Vienna, Austria

I. INTRODUCTION

As one of the first steps in the drug discovery process, high-throughput screening (HTS) has become a key tool in the discovery of pharmacological leads. Robotics and automation are now absolutely essential to perform the tremendous number of tests that a company conducts each year to discover new drugs in every step in the drug discovery process from lead optimization to clinical trials. Before robotics and automation were widely used in screening laboratories, it was customary for a researcher or laboratory technician to take one full year to test only a few hundred compounds [1]. Now, with robotics and other laboratory automation, it is possible to screen 100,000 or more compounds in a 24-hour period.

In addition to primary and secondary screens, robotic systems are now used in additional areas in the drug discovery process, such as early in vitro ADME (absorption, distribution, metabolism, and excretion) and toxicity studies. Because robotic systems can perform multiple assays simultaneously, early ADME and toxicity testing can now be performed at the same time as other HTS assays, allowing compounds to be profiled earlier in the drug discovery process. With the results of in-vitro ADME and toxicity tests, lead candidates can be selected that have the greatest chance of passing the later in vivo ADME and toxicity studies, ultimately leading to more successful drug candidates. Early drug profiling is useful because it allows companies to focus their efforts on lead compounds that have the greatest chance of becoming successful drugs.

Later in the drug discovery process, robotic systems again make a significant contribution by testing the data arising from the clinical trials. When a drug candidate goes to clinical trials, the high uptime and productivity of robotic systems provides fast sample analysis, quick turnaround time (TAT), accurate testing, and easy scale-up, without dramatic changes in the level of the human workforce. This provides more flexibility in the clinical trial and allows a higher trial population, because, within the system limits, sudden increases in the clinical sample load do not require a sudden increase in the number of laboratory technicians to analyze the samples. With more data reported and analyzed faster, the time for regulatory approval can sometimes be shortened, bringing the drug to market faster. One company recently estimated that by saving a year in drug development, sales increased by $580 million in the first year alone [2]. A typical robotic system is shown in Figure 1.

To put the term in perspective, automation is the technique of making a process automatic [3]. Fortunately, for those who work in today's laboratories, modern-day instruments are highly automated with sample changers, injectors, "sipping" cannulae, and so forth, which free the technician to perform other tasks. Indeed, the laboratory of today contains instruments that accept the sample,

Figure 1   A typical robot system.

perform the analytical test, reduce the data, and send the results onto the laboratory data network or laboratory information management systems (LIMS).

Although in recent years many laboratory analytical instruments have gradually become more automated through improvements in hardware and software, laboratory robots have been a relatively new introduction to pharmaceutical laboratories. Robots were generally first applied to laboratory tasks in the early 1980s, with Zymark Corp. (Hopkinton, MA) being credited as the first company to have developed a multifunctional robot specifically for widespread applications in analytical laboratories.

The classical guidelines for a task to be a candidate for robotic automation are the "four H's," i.e., hot, heavy, hazardous, and high-cost labor. Although laboratory tasks might be hazardous, depending on the chemical compounds being used, they usually are not hot or heavy, so on first glance it might have seemed unusual that robots would be used in the laboratory. However, although laboratory applications do not have much in common with industrial robotic applications, the laboratory has many elements that promote the use of robotics: high cost of labor, repetitive tasks, tasks that require accuracy, and the demand for higher productivity.

The high cost and limited supply of scientific and laboratory labor, coupled with the increasing workloads in drug discovery organizations, has required the use of many types of laboratory automation, including, of course, laboratory robotics. Technicians and scientists are too highly paid and too valuable to the organization simply to perform the simple, repetitive pipetting, incubation, and microplate transfer tasks in a typical screening assay.

Another reason for the widespread use of robotics is the need for precision. As microplate densities have increased and compound volumes decreased, it has become even more important to perform all steps in a laboratory method in the same manner. Liquid transfer and reagent addition, incubation times, washing or filter harvesting techniques are among a few of the steps that must be performed identically in order for the results to be comparable from run to run. Otherwise, erroneous conclusions may be drawn from the data, easily resulting in an incorrect conclusion about a compound, or a lost opportunity, both of which can be very costly in today's competitive drug discovery environment.

Unlike the automation found on typical analytical instruments, a robot is a reprogrammable mechanical device capable of gripping, for example, a test tube or microplate and moving it through the work envelop [4]. Of the other types of automation commonly used in drug screening, automated pipettors (or "liquid handlers" as they are sometimes called) are probably one of the most common automated instruments in the laboratory. These instruments usually do not have gripping capabilities but are often equipped with multiple fixed cannulae or, alternately, disposable pipetting tips. Automated pipettors began appearing in laboratories in the early 1980s and now have become standard equipment in most screening laboratories. There are many different types of automated pipettors from a range of suppliers, including Tecan, Inc. (Research Triangle Park, NC), Packard Instrument Company (Meriden, CT), Hamilton Company (Reno, NV), and Rosys, Inc. (Wilmington, DE) to name a few. A typical liquid handler is shown in Figure 2.

With a liquid handler, the operator will place the various test tubes, microplates, or other containers on the pipetting surface and the liquids will be transferred as programmed. After the liquid transfers are completed, the operator replaces the containers, possibly refills the reagent reservoirs and any other consumables, and the automated pipettor continues its tasks on the next set of containers. Thus, the liquid transfer steps are performed automatically, but the instrument relies on an operator to change the vessels on the deck and replenish reagents and other consumables.

On the other hand, laboratory robotic systems can run for hours or even days with little operator attention. The operator is only needed to replenish the

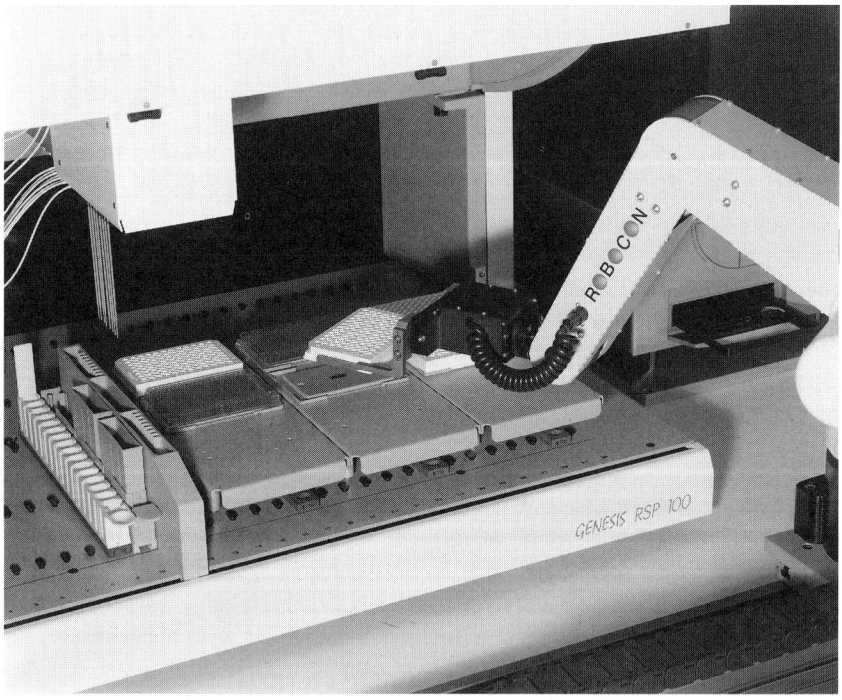

Figure 2   Robot arm placing a microplate onto a typical liquid handler.

consumables every few hours, and the robotic system can continue running, moving samples through the various steps of the method. Robotic systems for drug screening are usually built with a mixture of standard, commercially available components and some custom-engineered devices. In some robotic systems, some of the required mechanisms and devices may not be commercially available, and the manufacturer of the system, sometimes called a system integrator, must develop a device for a particular function. Examples of companies that provide complete robotic systems include Beckman Coulter Corp. (Fullerton, CA), Robocon, GesmbH (Vienna, Austria), and Zymark Corp. (Hopkinton, MA).

Within a range of capabilities, robotic systems for drug screening can usually perform several different but similar assays. The system will include all of the necessary equipment to perform the desired range of assays. Obviously, a system that can perform a wide range of different assays will have many more different instruments than a system that performs only one or two very similar assays. For this reason, the more versatile system is likely to cost more and require more space than the narrowly focused system. Not to be ignored, the rated throughput of the system will also be a major factor in the system size, with high-throughput systems requiring more and larger devices than low-throughput systems.

Robotic systems can be designed and assembled by system integrators and will contain all the devices, instruments, reagents, and storage capabilities for the particular assays that the system is being designed to perform. Such systems are typically sold as turnkey systems, with everything included, including programming, so the system need only be assembled at the user's site and will begin to perform the desired programs.

II. HIGH-THROUGHPUT SCREENING (HTS)

The term high-throughput screening (HTS) is used within the drug discovery field to denote the automated robotic systems that test thousands of compounds per week in a targeted assay [5]. Compound screening strategies can be somewhat random, that is, compounds are taken from a library without regard to the compound structure and tested against the selected target(s). Natural products and compounds from older libraries are frequently tested against a wide range of targets in the hopes of discovering the desired compounds. However, more frequently, compounds to be screened are selected in a directed manner. With the widespread use of combinatorial chemistry and synthetically produced compounds, a family of closely related molecules can be synthesized, to exploit the structure–activity relation.

Due to competitive pressures in the pharmaceutical industry, the emphasis on finding promising leads for new drugs has also increased. In the past few years, pharmaceutical researchers have responded to this pressure by utilizing automation and increased productivity to quicken the pace of drug discovery. Researchers are now expected to find new drug candidates at a faster rate, usually with little increase in personnel budgets to accomplish this task [6]. Robotic drug screening systems have provided the additional capacity to meet this need and fill the pipeline of drug candidates.

III. ELEMENTS OF A TYPICAL ROBOTIC SYSTEM

A typical robotic system consists of the robot (including arm, controller, and gripper), a servo track, a table, peripheral devices, and the system or robotic system controller. The specific peripherals depend on the assay(s) to be performed and the desired capacity and throughput for the robotic system.

A. Types of Robots

There are various types of robots in use in laboratories, some being better suited for a particular application than others. The primary types of robots used in today's laboratories are of the "articulated arm" and "cylindrical coordinate" style.

1. Articulated Arm Robots

Articulated arm robots are characterized by an arm with a bending joint in the middle, sometimes referred to as the "elbow" joint, with the obvious anthropomorphic reference to the human arm. This type of jointed arm allows the robot to work close to its base and also reach out to the furthest extent of the working range. Articulated arm robots typically have five axes, excluding the track that the robot may be mounted to. Typically there are three major axes, including a base swivel (the ORCA does not have a base swivel but can move the arm over the base to access both sides of the track), a shoulder, and elbow axes. There are typically two wrist axes. The roll axis moves the gripper or "hand" around the axis perpendicular to the gripper mounting plate (also sometimes called the "tool" mounting plate). The fifth axis is an angular axis and is called "pitch" when it is oriented to move the tool up and down, and "yaw" when it moves the tool from side to side. Some models of laboratory robots have both pitch and yaw axes and thus have six axes. (The linear track is usually termed an auxiliary axis and is rarely counted as one of the main robot axes). Typical suppliers of articulated arm robots are; CRS Robotics Corp. (Burlington, Ontario, Canada), Beckman Coulter, Inc. (Fullerton, CA), and Mitsubishi Robotics (Japan). A typical articulated arm robot is shown in Figure 3.

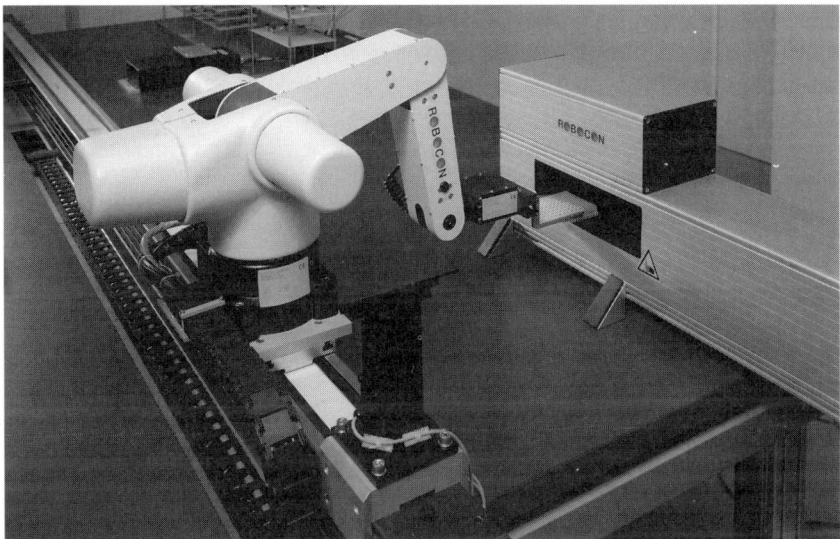

Figure 3   Articulated arm robot of the Robocon automation system.

2.   Cylindrical Coordinate Robots

Another style of robot commonly used in laboratories is the cylindrical coordinate style, such as the Zymate robot manufactured by Zymark Corp. (Hopkinton, MA). The arm on this style of robot is always horizontal with no joints and moves in simple horizontal and vertical movements in order to reach the programmed point within the work envelop.

The three main axes on the cylindrical coordinate robots include a base swivel, vertical movement of the arm, and horizontal arm movement. Unlike the articulated arm robots that move the elbow joint to fold up or extend the arm, cylindrical coordinate robots have a fixed length horizontal arm. Although these robots have fewer axes and are mechanically simpler than the articulated arm design, the fixed arm length is a disadvantage when considering the usable work envelope. Care must be taken when laying out the work area to assure that there is full clearance so the back of the arm does not hit an object on the other side of the table. The articulated arm style of robot avoids this problem by being able to fold up the arm so objects on the other side of the table are not hit when the base swivels.

This restriction serves to limit the robot's effective work envelope, but the simplicity of this type of robot has greatly contributed to its widespread use. Robot arms usually swivel at the base to allow the arm to move from side to side, allowing the robot arm to reach items on either side of the robotic system. Robots of this type are also oftentimes mounted on a linear track to expand the working range of the robot. Many of the devices used in robotic systems, such as plate storage devices, incubators, instruments, and so forth can be installed on either side of the track, which allows the robot working volume to be used more effectively.

3. Overhead Gantry Robots

Another style of robot is the overhead gantry. This design has a carriage that moves above the work surface, usually supported by a structure on one or more sides, depending on the weight and size of the carriage and overhead frame. From the carriage, the tools are raised or lowered as required. This vertical axis is sometimes referred to as the "Z" axis. The overhead robot uses linear rather than jointed or rotational axes, which makes the mechanical unit simpler than the articulated arm and cylindrical coordinate robot style.

B. Robot Controllers

The early robotic systems for drug screening applications were controlled by the robot controller, usually with the help of another controller for the "on-off" devices in the robotic system, such as a PLC (programmable logic controller). Quickly, however, robotic systems became more complex, and it became important to exchange data, such as the weight of a sample, the bar code identification, the volume of liquid to dispense, the reading from an analytical instrument such as a plate reader, etc. Laboratory robotic systems soon needed sophisticated data handling capabilities as well as data links via RS232 or network connections. Now, data handling and database interactions are a crucial part of any sophisticated drug screening robotic system, and a PC or PC-like platform is commonly used for the data handling tasks in the robotic system.

1. Servo Motors

As the name indicates, robot controllers send the signals to the motors and coordinate the movement of the motors in the robot arm. In order to move accurately and repeat the desired movements time after time, the motors and controller use bidirectional signals to position the axes motors accurately. Servo motors can be of either the AC or DC type, each having some differences in the motor construction and the type of robot controller (both still operate from the standard alternating current available in laboratories).

Servo motors are used in the articulated arm style of robot because they provide a broader range of torque output to accommodate the range of forces the arm experiences as the arm moves through the work envelop. With either AC or DC servo motors, an encoder sends signals or pulses as the motor moves. When the robot is being taught the movements and positions, the controller counts and stores the motor pulses. In the replay, or automatic mode, the controller tells the motors how many pulses to move, which translates to the speed and stopping position for each axis motor. By duplicating the taught program, the robot performs the tasks exactly as it was programmed, without operator attention.

2. Stepping Motors

Stepping motors are another type of motor that is used in robots. Stepping motors have a narrower range of torque output than servo motors and are more suitable for the cylindrical coordinate and gantry robots where the torque requirements are more uniform over the range of arm movement.

Stepping motors move in response to pulses sent by the controller, each pulsing causing a small amount of motor movement, called a "step." Because the controller counts and sends the appropriate number of steps to each motor, there is no need for an encoder to provide motor position information back to the controller. Controllers for stepping motor robots are simpler than for servo motors, resulting in reduced cost and controller complexity. For both price and performance reasons, stepping motors are used for simple robots, liquid handlers, and many other simple laboratory devices.

C. Gripper

The gripper, or robot "hand," is mounted on the tooling mounting plate at the end of the arm. The Zymate® and ORCA® robots commonly use a tool changer that allows them to change tools. This is a useful function, depending on the tasks that the robot will be performing in the robotic system. Grippers have an actuator or mechanism that opens and closes the "fingers" to grip the objects such as micro plates or tubes that will be carried within the robotic system. The gripper fingers are usually fabricated from metal or rigid plastic and should be designed to be easily replaced.

Replaceable fingers allow the gripper to accommodate changes to the robotic system, since the objects to be gripped may have a slightly different geometry. In addition, a gripper finger may become bent if the robot crashes, that is, accidentally hits something, and with replaceable fingers the gripper can be back in service very quickly.

The opening and closing mechanism is commonly pneumatically or electrically powered. Pneumatic mechanisms are fast and reliable, but they usually have only two positions, open and closed, and also offer little control over the speed by which the gripper opens and closes. Once closed, they grip the object quite tightly, because the compressed air exerts considerable force within the mechanism. While pneumatic mechanisms are simple, reliable, and powerful, they have some disadvantages. They cannot easily close to intermediate positions and cannot apply gentle gripping forces.

Electrically powered gripping mechanisms generally grip with less force than pneumatic mechanisms and are considerably more complex mechanically, but when powered by a servo motor they can easily close to any intermediate position and apply gentle pressure to grip even semirigid objects such as flexible PCR plates. Because a servo gripper closes to a pretaught position, not just until it stops, it can detect whether it has gripped the object, or whether the object was missing or had fallen.

Like the servo motors for the main robot axes, the gripper servo motor requires an encoder and servo motor controller, which increases the cost of a servo motor gripper over the cost of a pneumatic gripper. For many systems, however, the added utility of a servo gripper far outweighs the cost savings. Indeed, a robotic system that functions as a fully automated laboratory can be quite expensive, possibly costing a million dollars or more.

1. Digital Gripper

A digital gripper utilizes a servo motor to move the gripping fingers. With a digital servo motor, the gripper fingers can be opened or closed to virtually an infinite number of positions. In addition, digital grippers also have controlled speed and force in opening and closing the gripper fingers, so objects can be gripped as tightly or as loosely as desired. In a digital gripper, the fingers move in and out symmetrically, which means that the fingers move in and out by the same amount on each side. This is important so that the object will be gripped evenly and has less chance of being gripper off center. This symmetric movement is accomplished with a gear drive that includes the gear and linear rack. This mechanism moves the fingers smoothly and precisely, assuring reliable gripping. With the servo motor and encoder, positions can be taught and saved so the gripper can close to the same position without reteaching or resetting manual stops.

D. Linear Servo Track

The robot is usually mounted to a linear servo track that allows it to move to the various locations on the table. As mentioned previously, this greatly increases the effective work envelop of the robot and allows it to service more instruments and devices without human intervention. As with the robot joint movements, the

Figure 4   A robot mounted on a linear track.

servo motor moves the robot quickly and accurately, utilizing an encoder to provide the controller with positional information. The track is usually treated as an
auxiliary axis and is generally controlled by the robot controller. As with the
robot servo axes, the controller sends the track servo motor signals to accelerate,
move, decelerate, and stop. The speed is easily controlled, and, like the robot
axes, positions are saved as an integral part of the robot program. The robot track
usually extends for the length of the table for the robotic system to provide as
much robot work space as possible. Common track lengths are 2 m and 6 m,
although many times custom lengths are available from the system integrator.
The modern industrial-grade high-speed linear tracks can move the robot as fast
as 3 m/sec, which allows the robot to move quickly from one end of the table
to the other. A high-speed linear track is shown in Figure 4.

The fast industrial-grade linear track has several advantages over previous
track designs. The higher speed allows the long table to be traversed quickly.
This is important, as robotic systems have increased in size to accommodate more
instruments to provide higher throughput and a wider range of assays that can
be performed on the system. With a 6-m table, the high-speed track can be up
to ten times faster than tracks that had been previously supplied with robotic
systems.

E. Robotic System Table

The table on which the robotic system is mounted should be sized according to the space required for the assay(s) and the required instruments and devices to meet the desired throughput. It is common to use a table that is larger than the immediate needs, providing space for future expansion. Obviously, if the table is far too large for the present and anticipated needs, the robotic system will cost more than is necessary and take up much more laboratory space than is required. Table frames and supports are commonly fabricated from extruded aluminum members that have been cut to size. Once the aluminum pieces have been precisely cut and holes drilled for the mounting screws, the table frame can be assembled with common hand tools such as screwdrivers and Allen wrenches.

Tables of this design are very rigid and sturdy and offer more flexibility than traditional welded tables. The robot and track are usually mounted in the center of the table, and the various devices and peripheral instruments are mounted on either side of the track. In order to avoid mistakes in placing the devices and instruments on the system table, a scale CAD (computer-aided design) drawing should be made showing the reach of the robot arm. An engineering analysis can then be made to verify that the robot can reach all required locations on each of the instruments and devices. A drawing can then be printed to guide the system assemblers so that the devices are placed in the proper location and are reachable by the robot hand.

Minor mistakes in the table layout can usually be easily accommodated, but serious mistakes can be very costly to correct. Because the possible positions of the gripper are difficult to replicate accurately in the CAD drawing, the positions of the various instruments and devices may require minor adjustments when first setting up the robotic system. It is best to allow a little extra space between the various items for this adjustment.

Sufficient space is available under the table to mount the racks for the various controllers, wiring and tubing, reagent sources and waste, and so forth. Leveling screws are usually built into the table legs so the table can be easily leveled. This is important so the devices move and liquids flow easily. Tables of this design are easily disassembled for shipment and reassembled at the customer's site.

F. Typical Peripheral Devices

There are a number of devices and peripherals that are often used in different assays. Usually there is enough commonality between the various assays that the system can run any one of the assays without the operator making any hardware changes to the system. The operator need only call up the appropriate program, load the proper initial data, refill the consumables, and start the system. The robot will execute the assay as it had been programmed. There are a number of devices and instruments that can be part of a "core" robotic system.

G. Microplate Storage Unit

A key element of any system is the plate storage unit. These devices store microplates and provide incoming storage, in-process storage, room temperature incubation, and outgoing plate storage. For the most flexibility, storage devices should accommodate both standard well and deep-well plates as well as plates with covers. Storage units can be as simple as stationary vertical plate hotels or higher density units such as carousels that automatically present additional plate locations to the robot. Carousel storage units store more plates per foot of robot frontage than stationary plate hotels. One high-density unit from Robocon can store 216 standard well plates, with lids if desired, in a very compact cube that only takes up 400 square inches (Fig. 5).

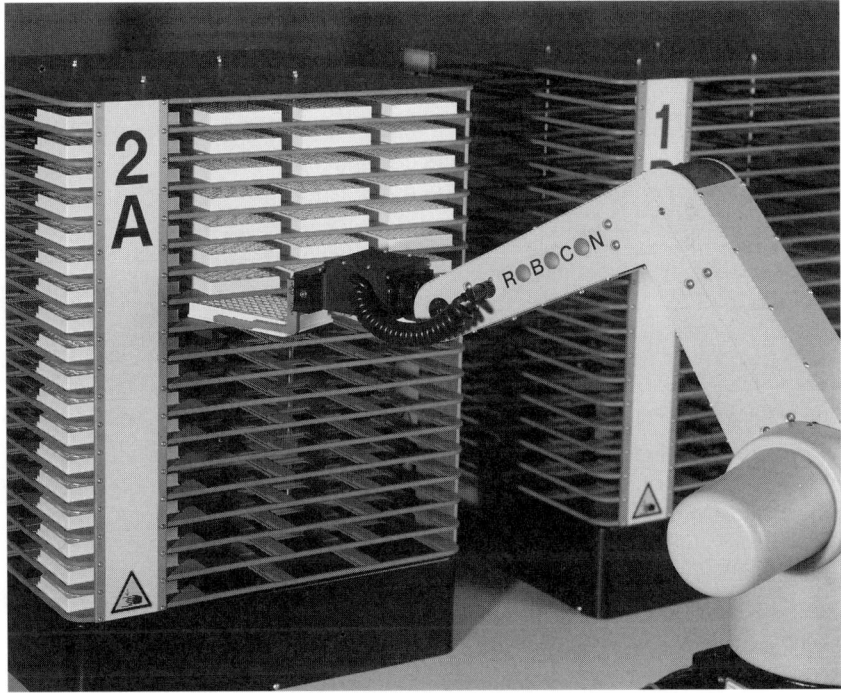

Figure 5   Robot loading plates in a high-density plate storage unit.

Plate storage units should provide random access to any plate. Sequential storage units such as plate stackers are adequate for plate feeding when the plates are in order, or if plate ordering is not necessary, but for a storage unit to serve also as an incubator, random access to any plate location is a requirement for properly timed incubation periods. Random access also gives the system more flexibility in storing plates in the middle of the process. The system controller remembers whether a plate is in each position and can place in-process plates in storage positions as they become available. Thus, random access to any plate storage location is very important in the efficient use of the system storage space.

H. Bar Code Scanner

Another important peripheral device for sample tracking is a bar code scanner. Since bar code scanners are small, they can be mounted at many different places in the systems. They can be mounted on the robot arm, above the gripper, and, as part of the robot, they can be taken to the plate for scanning at various locations in the robotic system. Alternatively, the scanner can be mounted in a fixed location on the table and plates can be taken to it for identification and then processed according to the desired method. The robotic system controller will record the plate identification (sometimes referred as the ''plate ID'') and will track the plate and the samples on the plate, through the various steps in the method. Because the robotic system computer controller also sends the move commands to the robot, the controller knows the location of any plate once the ID has been read. Without both a human-readable and a machine-readable plate identification, it would be very easy for the human operator to mix up plates during either loading or unloading of the plates from the system.

Any automated system requires consistency, so the location of the bar code label must be within the allowable tolerances of the reader for a proper reading. Bar codes must be printed on the label in a consistent manner, the labels must be placed on plates in the same location, and the operator must load the incoming plates into the storage unit with the label in the proper orientation. Some instruments have imbedded bar code readers for the automatic identification of plates. Some plate readers and liquid handlers have this feature. Fully integrated robotic systems, however, usually have an integrated stand-alone bar code reader so that bar codes can be read at virtually any step in the method.

I. Liquid Transfer

Liquid transfer is a key element of any drug screening method, and virtually every robotic system will have some provisions for the transfer of reagents and other liquids. There are three main types of liquid transfer modes: reagent dispensing from bulk reservoirs, single or multichannel transfer between micro plates and small reservoirs, and 96- or 384-multiwell transfer between plates. In each of these categories, syringe pumps are used to perform the liquid transfer because they are more precise than peristaltic pumps for drug discovery applications. Peristaltic pumps are used where the dispense accuracy is not as critical, such as dispensing media or in the washing of cannulae between reagents.

1. Bulk Reagent Dispensers

For the dispensing of bulk reagents, the reagent is drawn through the valve–syringe assembly and then flows through the liquid line to the dispensing cannula. Depending on the volume per dispense step and the capacity of the syringe, multiple aliquots usually can be performed without the syringe reloading. Because it is time-consuming to rinse the fluid path, including the valve–syringe assembly, and because the reagent in the liquid lines must be discarded, bulk reagents are not often switched from run to run.

Small amounts of reagent, or many different reagents, are more efficiently dispensed from reservoirs on the deck of the liquid handler. Even small amounts of reagent are easily aspirated from the reagent reservoir and dispensed into the appropriate microplate wells. Both bulk reagents and small reagent reservoirs are easily refrigerated to preserve the stability of the reagents during runs that can last 7 to 14 or more days.

2. Liquid Handlers

For single or multiple channel transfer between plates or small reservoirs, a liquid handler can be used. These instruments are most commonly used as stand-alone devices in the laboratory and have a specialized design and software for the transfer of liquids. As part of a robotic system, the liquid handler is useful for adding standards and controls to selected plate wells, or for the ''hit-picking'' operation to prepare plates for a secondary screen. When adding standards and controls, reagent reservoirs are located on the deck of the liquid handler and the pipetting probes transfer to the appropriate wells. The robotic system controller can send commands to the liquid handler to place any available reagents into any specific plate wells. The liquid handler can also perform serial dilutions and other liquid handling tasks that cannot be performed by the other devices.

Liquid handlers can usually be equipped with either disposable or fixed pipetting tips. Disposable tips minimize any cross-contamination between pipetting steps, but they have other limitations, including the cost of the disposable tips and the massive amount of space required for long, unattended runs. Fixed cannulae are usually coated with Teflon® or similar material (poly-tetrafluoroethylene or PTFE) on both inside and outside surfaces to minimize reagent surface wetting and aid in the rinsing process.

3. Multiple Channel Pipetters

Pipetting to 96 or 384 wells simultaneously is very useful in drug screening. As stated in other sections, the microplate with 96, 384, and most recently 1536 wells has become the standard testing format. These formats are all multiples of 96 ($384 = 4 \times 96$, $1536 = 16 \times 96$). Not only it is much faster to transfer liquid to 96 wells simultaneously compared with 1, 4, or 8 wells at a time as with a liquid handler, but for assays where a reaction is started or stopped with the addition of a reagent, it is possible to start and stop reactions at the same time. The 96- or 384-well pipettor will be capable of pipetting from one plate to another, or from a reservoir to a plate. Some models of pipettors use fixed tips and some use disposable plastic tips. Tips are washable by aspirating and dispensing wash fluid. Pipettors that use disposable tips are capable of changing tips. Most models can change tips automatically, but some require an operator to change the tips manually. For most applications, both fixed and disposable plastic tips can be washed with no detectable sample carryover.

J. Incubators

Many assays used in drug discovery methods utilize an incubation period at either room temperature or at an elevated temperature such as 37°C. Many assays use live cells that require humidity and a CO_2 atmosphere. Such incubators are a key requirement for robotic systems for drug screening. Some robotic systems use commercial incubators that have been modified for robotic use by adding an automatic door opener. But these designs are not optimal for robotic systems, because the single door faces the robot, making it difficult to service the incubator manually. More convenient designs have two incubator doors, opposite each other, one facing the robot and the other easily accessible by the operator. This allows the user to service the incubator safely and still provides convenient access for the robot.

   With easy access to the interior of the incubator, the operator can easily stock the incubator with incoming plates to augment the plate storage unit. Also, the operator can periodically clean or otherwise service the incubator. For safety reasons, when the incubator has separate doors for both the robot and the operator, there should be interlocks so that the robot cannot open a door if the operator door is open, and vice versa. In order to maintain reliably the interior atmosphere, the doors must have tight, flexible, and compliant elastomeric seals. Silicone rubber is an excellent material for the door seals, because it is flexible and compliant yet can easily withstand an elevated temperature. To supplement the mechanism to automatically open and close each of the doors, an additional automatic latch can be used to hold the door securely closed and maintain the integrity of the interior environment.

Figure 6 Robocon robot loading plates in tissue culture incubator.

With so many variables associated with the proper operation of the incubator, some manufacturers have found it useful to dedicate a PLC (programmable logic controller) to each incubator for distributed control of the many mechanisms, sensors, electrical contact switches, etc. These variables include the door open and closed sensors, the open and close actuators for each door, the door latches on each side, the heating, CO_2, and humidity controls, circulation fans, and so forth. Certainly these functions can be monitored at a central PLC as part of the controls for the robotic system, but the distributed control allows the incubator to be a self-contained module that can be easily integrated into a new system or retrofitted to an existing system. A typical robotic tissue culture incubator is shown in Figure 6.

K. Plate Turntable

Instruments such as plate readers and plate washers do not have the standardized orientation of the microplate; some require plates to be in the "landscape" orientation and others require the "portrait" orientation. In order to accommodate all types of instruments, the robotic system needs the capability of turning plates from the landscape to the portrait orientation and vice versa. The system controller will direct the robot to take the plate to the turning station as may be required for the next instrument or device. The plate orientation required for each instrument will be saved as an instrument parameter, and the system controller will orient the plate accordingly before the robot approaches the instrument plate nest. The user will do this automatically without any programming.

IV. PROGRAMMING LANGUAGES FOR ROBOTIC SYSTEMS

While there is not yet a common high-level programming language for robotic systems, and each supplier has developed its own language, suppliers have improved the ease-of-use of their programming languages, making it easy for operators to learn and use. In general, the languages have moved away from requiring the user to adhere to the programming, command, and syntax rules typical of programming languages. Instead, programming languages have become much more intuitive, allowing the user quickly to learn enough to write a usable program. In today's systems, programming might be reduced to a few simple commands, such as "get," "move," and "put." An array of microplate locations, such as for the incubator or plate storage units, will be pretaught by the integrator, and reference positions for a device can consist of three corner positions.

"Self-learning" third-generation software is now available from many integrators of robotic systems, which greatly reduces the time to write a new program. The software can anticipate the next move that the operator wants to program and presents it on the programming screen. This reduces the number of keystrokes required to write a new program for an assay. With this self-prompting capability, new programs can be written very quickly. Other features check syntax to reduce further the time to write and debug a program. Robotic systems from some vendors, such as Robocon, can run different assays simultaneously. This allows several different assays to be performed with the same or different compounds. With this capability, various different assays can be performed on a single robotic system that otherwise could not have been economically justified.

Other features of third-generation software include automatic cycle time calculating and "open-ended scheduling." The latter is a significant improvement over "finite scheduling," which had been used in robotic systems up until this time. Finite scheduling calculates the times of all of the steps up to the end of the run, which might be several days away. This is far too complex to be useful in today's robotic systems that might run nonstop for several days. For this reason, a "limited horizon" or "open-ended" scheduler is much more practical. With this type of scheduler, the system is only concerned with the next 30–

60 minute time horizon. Within that window, the scheduling software looks ahead to see what resources will be available and whether new plates can be introduced into the system.

All scheduling programs use approximations for the length of time to perform certain tasks. For shorter time horizons, such as 30–60 minutes, these approximations do not introduce significant error. However, for longer time periods, the error becomes so great that the forward scheduling is not useful in predicting exactly when resources will become available or if a scheduling conflict occurs. Open-ended scheduling eliminates the unnecessary calculations for the period beyond the next 60 minutes and easily accommodates changes in the middle of the run. Finite scheduling requires resetting and recalculating any time there is any operator intervention. For these reasons, which today's more complex robotic systems demand, limited horizon is the preferred scheduling method.

Automatic Maintenance Prompting is a recent software feature that prompts the user to perform various type of preventative maintenance. The system controller records the amount of time that the system has operated, even tracking usage for specific components, and alerts the operator when scheduled preventative maintenance should be performed. This assures that the operator is aware when the appropriate preventative maintenance must be performed in order to maintain the useful life of the robotic system.

Another feature of third-generation software is the ability to run multiple assays simultaneously. Several different assays with only a few thousand compounds each could be combined and run at the same time on a high-throughput robotic system. With earlier easy-to-use software, there was no provision to control and forward schedule multiple plates in multiple different assays with finite resources. Now, with third-generation software, this can be easily accomplished, allowing the high-throughput robotic system to be utilized processing assays with shorter runs.

One useful example of this capability is when secondary screens are performed to confirm "hits" from the primary screen. One might want to run several early in-vitro ADME and toxicity tests simultaneously on a selected group of compounds. It is more efficient to run assays of this type simultaneously because the compound plates need only be handled once to start the assay plates. If the assays were processed serially, one would be required to retrieve the compound plates from storage each time an assay was to be started. Since most compound plates are stored frozen, considerable time is lost in the steps of retrieving, thawing, unsealing, pipetting, resealing, and returning to storage.

With the hits selected from the compound plates and transferred to the various assay plates, the assays can be started. The system controller instructs the robot to take the appropriate plates through the various steps in each different assay, using the appropriate stations or system resources as required. Thus, the ability to process multiple assays simultaneously not only is a more efficient way to run assays with some common elements but also it increases the system utilization and increases the investment that the robotic system represents.

A. Software Operating System

Because there are multiple tasks to be performed simultaneously, most robotic systems for drug screening operate with software that operates in a multitasking environment. The controller issues device commands, receives data, and controls the system simultaneously. With today's increasingly complex systems, multitasking capability is a requirement. Due to their cost, availability, reliability, and user familiarity, standard personal computers are frequently used as system controllers. For a variety of reasons, at the time of this writing, most software for drug screening robotic systems operates in a Windows® NT environment. The Windows graphical user interface (GUI) has become widely accepted in drug screening laboratories, making it easy to train new users and assuring familiarity between different systems and different vendors.

V. AUTOMATED ASSAYS

In order to capitalize on the capabilities of the robotic system, it is highly desirable to run assays that require no human intervention. The manufacturer of the robotic system will work with the user to develop assays that can be performed by mechanical devices that, unfortunately, do not have the benefit of human dexterity or eye–hand coordination. Manufacturers of highly automated laboratory systems will recommend assay procedures that do not compromise the assay yet can be reliably performed by a robotic system without constant operator attention. Steps that might require human eye–hand coordination, such as placing a micro plate into a poorly designed nest on a device or instrument, will be difficult for the robotic system to perform accurately. At best, these difficult tasks can cause inaccuracies in the method, which reduces the repeatability and accuracy of test results between samples and between runs. More seriously, these poorly engineered devices may make the system more prone to faults that can interrupt the run, causing a delay in the test results or even the loss of the samples and expensive reagents.

For these reasons, it is important that assays be developed that can be performed as simply as possible without human intervention. If the robotic system is waiting for a step to be performed off the robotic table, such as an incubation or centrifugation step, then the system throughput will be greatly reduced. A "perfect screen" is one that can be run completely by a robotic system and requires no manual handling for at least 24 h [7]. A wide range of assays have been automated on robotic systems, including enzyme, ELISA, protein–protein interaction, protease, reporter gene, radioligand, and whole cell assays. Although these assays will certainly have different steps and reagents, they may use many of the same peripheral devices and will certainly use the basic robotic system.

A. Elements of Typical Assays

A basic robotic system for drug screening will usually have the following components: robot, servo track, digital gripper, plate storage unit, bar code reader, plate turning station, plate shaker, and liquid dispenser. Depending on the assays to be performed, additional devices will be required. Cell-based assays will require a 37°C temperature controlled CO_2 incubator with humidity, and, if cells are being plated by the robotic system, a gentle dispenser, usually with a stirred reservoir to keep the cells in suspension, will be required. If ELISA or similar assays will be performed, a microplate washer will be used to wash unbound reagents from the plate wells and prepare the plate for the next step in the assay. Plate readers are used to quantify the amount of reaction that has taken place, and assays use a variety of modes, including colorimetric, absorbance, fluorescence, luminescence, etc.

B. Typical Assays for Drug Discovery

Some of the drug discovery assays that run on automated robotic systems include receptor binding assays (heterogeneous phase receptor binding, such as filter binding), homogeneous receptor binding assays such as homogenous time-resolved fluorescence (HTRF) and scintillation proximity assay (SPA), enzyme assays, and enzyme-linked immunoabsorbent assays (ELISA).

1. ELISA

ELISA is frequently run on an automated system because it has many time-consuming steps including the addition of reagents, performing timed incubations, and subsequent plate washings before reading the plate in a plate reader. Most ELISA assays produce a color or fluorescence end product that, when the reaction is stopped, gives an end point reading. A robotic system is an ideal platform to perform ELISA assays because the forward scheduling software assures that the robot will perform the time-critical steps within the allowable time tolerance, so all steps are performed in exactly the same manner. The steps in a typical ELISA assay include (1) transfer of samples to be tested to an antigen-coated plate, (2) addition of reagent #1 (conjugate), (3) shaking, (4) incubation, (5) washing, (6) addition of reagent #2 (substrate), (7) shaking, (8) incubation, (9) washing, (10) addition of reagent #3 (stop solution), (11) shaking, (12) reading the plate in a plate reader.

A robotic system for an ELISA assay will need the basic system consisting of robot, servo track, gripper for micro plates, plate turning station, reagent dispenser (three different reagents are to be dispensed with minimum washing between reagents), liquid handler, incubator, plate washer, plate shaker, plate reader, and plate bar code reader. Separate locations in the plate storage unit will be dedicated to the sample plates and to the coated assay plates. The operator will load the sample and coated plates, refill the reagent reservoirs, replenish other consumables, and input the various run variables. Variables can include the number of sample plates to process, the volumes of reagents to be dispensed, the incubation temperature and times, the wavelength at which the plate should be read, etc. The robot will select the first sample plate from the plate storage unit, read the bar code if applicable, and place it onto the liquid handler. The samples will be transferred from the sample plate to the coated plate, with the pipetting tip being washed between samples and the original sample plate returned to the plate storage unit or placed in the waste container. The coated sample plate will be processed with the successive addition of reagents, incubations, and washing until it is ready to be read.

When the plate is read in the plate reader, the output from the reader will be a file of values, one for each plate well (96 values for a 96-well plate or 384 values if a 384-well plate is used). The file will be captured and stored by the robotic system controller. Although the samples to be tested are transferred from the bar coded sample plate to the antigen-coated assay plate, which may or may not be bar coded, third-generation software can track samples from plate to plate without bar code identification on each plate. The system controller merely remembers the position of the subsequent plates and transfers a virtual identifier to the subsequent plates, always knowing the location of any group of samples.

The readings from the plate reader will be linked with the identification (ID) of the original sample plate and can be outputted to a company-wide Laboratory Information Management System (LIMS). The data can also be reduced with reagent factors from the reagent manufacturer and presented in a spreadsheet format. Assay results are displayed with the plate bar code ID, and the resulting "plate map" shows the well locations for the samples, standards, controls, and blanks. With the robotic system's multitasking software and forward scheduling capabilities, multiple plates can be in process at one time. This ensures faster sample throughput, less robot waiting time, and higher system utilization.

2. Cell-Based Assays

Cell-based assays may use the same basic robotic system as described above, with the addition of a plate lid park station and a 37°C CO_2 incubator. Tissue culture plates are compatible, as the plate covers are easily removed to allow access to the plate wells, primarily for pipetting, and are easily replaced. Although the robot could simply remove the plate lids and place them on a surface while liquids are transferred to the well, the lid could be easily contaminated from the resting surface. A better approach is to suspend the plate cover so it does not touch any surface. One novel approach is to hold the cover from the top by vacuum, then releasing the cover back onto the plate. The first cover holding position can be above a hole in the table top with a waste container below it, so covers can drop directly into the waste when the cover is not needed for the balance of the assay. A typical robotic lid holding device is shown in Figure 7.

A practical design 37°C CO_2 incubator provides one access door for the robot on one side and another access door on the other side for the operator. If there is more than one set of incubation parameters (temperature, humidity, or CO_2) in an assay, then multiple incubators would be used. While it is possible for the system controller to change the temperature or CO_2 concentration remotely, it can take one or more hours for the incubator to reach the new conditions, which is not quick enough for a running assay.

3. Enzymatic Assays

Enzymatic assays are difficult to generalize and there will be customized differences for a specific enzymatic assay [8]. One enzymatic assay, for cytosolic phospholipase A2 (cPLA$_2$), is simple to perform on the basic robotic system described above. The steps include addition of ^{14}C labeled substrate in a microplate, addition of the compounds to be tested, addition of cPLA$_2$, and incubation for 30 min at 37°C. Reaction is stopped by the addition of a quench reagent, extracting the cleaved fatty acid, and separating the top organic layer in each well with an HPLC. The ^{14}C labeled fatty acid is detected by a flow scintillation counter.

Tyrosine Kinase Autophosphorylation Assay. This assay is useful in investigating the signal transduction process and is measured using a scintillation counter [9]. Coated plates are coated with antiphosphotyrosine monoclonal antibody, incubated at room temperature in the plate storage unit, and then washed with a standard plate washer. Appropriate reagents and sample can be added to the plate with either the liquid handler or the 96-channel pipettor. The plate is then incubated at room temperature, the liquid aspirated and discarded, and the plate washed again. The plates are then air dried, the scintillating reagent is added to each well with either the liquid handler or the 96-channel pipettor, and the plate is read in a scintillation counter. This assay can be performed with the basic robotic drug screening system using a scintillation counter as the plate reading instrument.

4. Receptor Binding Assays

Filter Binding Assay. Filter binding assays for receptor/ligand complexes are widely used in the drug discovery process; they utilize a plate harvester to transfer the cells from the assay plate to the filter plate. The filter plate is then

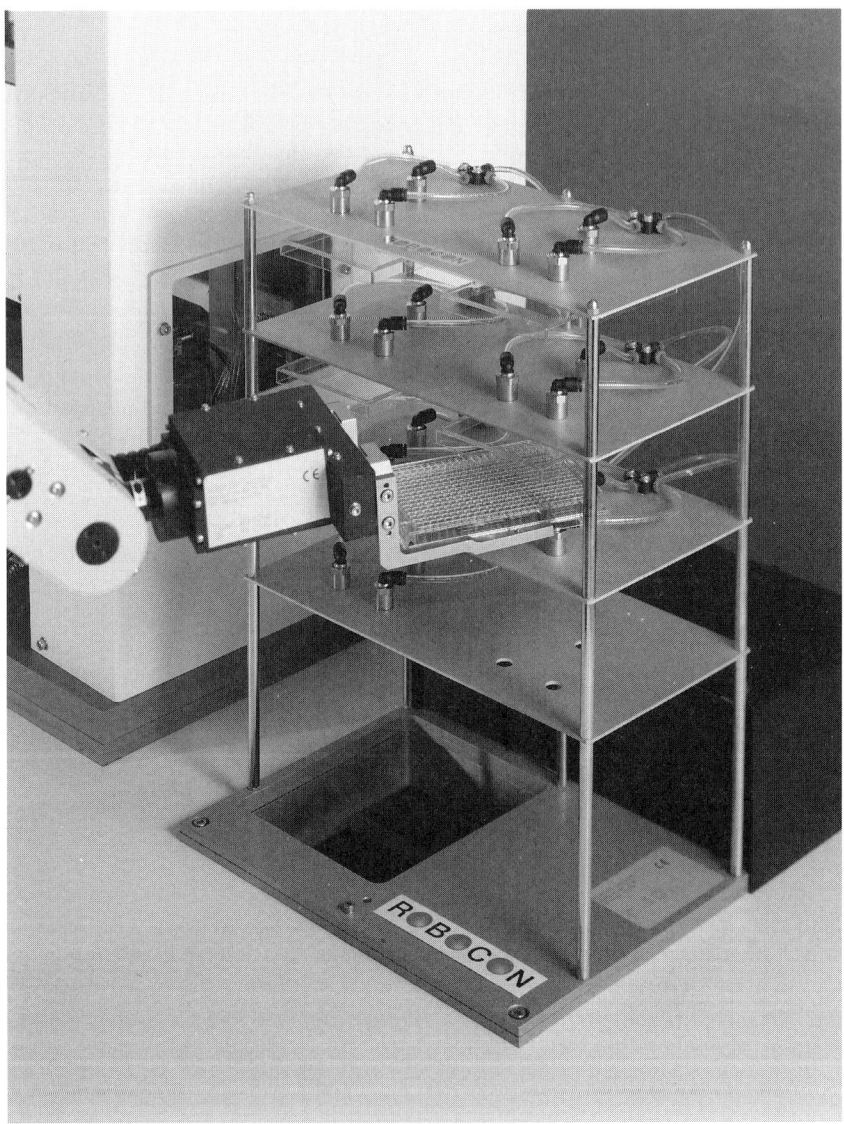

Figure 7 A typical plate lid storage device holding microplate lids from overhead by vacuum.

placed on a vacuum manifold and the excess liquid removed. The cells remain on the filter plate, which is then dried, and then the receptor/ligand binding is measured in a scintillation counter. Historically, plate harvesters can clog, which is unsuitable for unattended robotic operation. However, recently developed modifications allow the robotic system controller to monitor the performance of the plate harvester and alert the operator when the harvester performance deteriorates. With the system monitoring of the plate harvester, filter plate assays can now be automated for unattended laboratory operation.

5. HTRF

Homogenous time-resolved fluorescence (HTRF) is a homogeneous technique (discussed in Chap. 4) that can be applied to a variety of assays. It eliminates many of the time-consuming and difficult separation steps such as cell harvesting and plate washing. HTRF gives results comparable to ELISA, but because it is homogeneous and has fewer and shorter steps, the HTRF can be performed much quicker than a typical ELISA [10].

With the growing trend to miniaturize assays, there is always a concern whether the assay results are affected by the miniaturization. In one study, tyrosine kinase assays using HTRF have been performed on robotic systems in 96- and 384-well plates. Results were compared between the two plate formats, and results from 384-well plates with lower well volumes were found to be comparable to those from 96-well plates [11].

6. Cytochrome P450

The Cytochrome P450 assay is a relatively new assay to automate with an unattended robotic system. Because it is a valuable assay for early in-vitro metabolism studies and early drug profiling, there is considerable interest for automation. In this assay, liver hepatocytes, hepatic microsomes, and rat liver slices can be used. Because it is an enzymatic assay, it is processed robotically similar to other enzymatic assays.

A large number of exogenous substances are metabolized in liver (discussed in Chaps. 13 and 14). The cytochrome P450 assay is an indicator of the metabolism of the compound in question in the human liver. Assay readings are typically taken at various time points so the rate of metabolism can be estimated. The supernatant from each plate well is measured by LC/MS analysis. There are now available fully automated turnkey robotic systems for Cytochrome P450 analysis (Robocon) that can process plates at the rate of thirty 96-well plates per 24 hour period. With data from early in vitro metabolism testing, compounds can be evaluated and structures modified in the optimization studies, leading to more successful drug candidates.

7. Reporter Gene Assays

A typical reporter gene assay has adherent cells that contain the promoter of interest and a reporter protein such as luciferase. Adherent or suspension cells in microplate wells are incubated with the compound to be tested for typically 3–12 h at 37°C under tissue culture conditions. After incubation, the cells are washed. The most successful methods incorporate both the lysis buffer and reagents for the reporter gene product. With only one reagent to dispense, a step is saved and there is no chance of a variation in the time between the cell lysing and the addition of the reporter reagents. If a luminescence reagent such as luciferase is used, then the plates are read in a luminometer or Topcount. Some variations of this assay will use fluorescence for detection [12]. Some of the advantages of the reporter gene assay is that time-consuming steps that have historically been difficult to automate, such as filtration and centrifugation, are eliminated.

8. CaCO2 Assay

Another of the early in-vitro drug profiling assays is the CaCO2 cell assay. This assay correlates to the behavior of the mucosa cells in the human intestine and serves as a good predictor of the absorption of a compound into the human body [13]. In this assay, the CaCO2 cells, which are derived from human carcinoma cells, are grown on a filter membrane and inserted into 24- or 96-well plates. The CaCO2 cell monolayer develops on the top of the filter surface. The basal membrane is directed toward the filter, and the epitheleal surface points upward. The cell monolayer grows until it is continuous across the filter surface. The test compound is applied to the top of the cells and is absorbed by the cell and transported to the media below. The amount of compound that is transported (i.e., absorbed) by the cells is measured by LC/MS analysis of the media in the lower chamber of the plate. Readings can be repeated at various time intervals so the rate of absorption can be calculated.

9. SPA

Scintillation proximity assay (SPA) (detailed in Chap. 4) is an example of one of the newer HTS assays that are very robust and simple. Although the SPA assay uses a radioactive reagent that requires more careful handling, it has become a popular assay because there is no separation step. Frequently, SPAs are able to take the place of more complex and cumbersome assays such as ELISA, which has several steps of reagent addition, incubation, and washing, and the RIA (radioimmunoassay), which uses higher levels of radioactive reagents, which requires a more controlled system for waste collection and disposal. There are a number of variations in SPA assays, but the basic steps in the assay are very simple: pipetting of the sample and reagent into microplate wells, incubation for a short time at room temperature or at 5°C, and placing the plate into a scintillation plate counter.

A robotic system to perform this assay would consist of the basic robotic system, including robot, table, and track, as well as a storage device for incoming and outgoing microplates, a dispenser for the reagent, and a scintillation counter. Other devices might include a bar code reader to track the samples and correlate the results from the scintillation counter to the specific plate.

10. Polymerase Chain Reaction (PCR) Assay

Polymerase chain reaction (PCR) is a patented technique to amplify strands of DNA so that there will be sufficient quantity for analytical testing. PCR amplification is accomplished by adding the DNA to be replicated with oligonucleotide primers (amplimers or primers) and TAQ-polymerase into microcentrifuge tubes in a 96-rack or a well of a specialized PCR plate and then placing them into a thermal cycler where the samples are alternately heated and cooled, replicating the DNA. After a suitable number of thermal cycles, the DNA of interest has been amplified sufficiently to yield enough sample for testing. The DNA may be identified through a number of techniques, including gel electrophoresis, capillary electrophoresis, or mass spectroscopy.

A robotic system to perform PCR generally consists of the following items: basic robotic system table, including robot, robotic track of suitable length, storage for incoming microplates, PCR plates, thermal cycler, and pipettor. Because there are several different analytical techniques that may be performed on the DNA segments, the analysis of the sample is performed with manual intervention off the robotic table.

VI. INTERFACING TO DATABASES

The nature of drug discovery is to examine a large number of compounds for possible effectiveness as therapeutic agents. Because a large number of compounds are being analyzed, there are a tremendous number of data points created in the course of a screening run. In order to manage the data, it must be analyzed with the help of computerized bases. Thus, data handling becomes a key element of any robotic drug discovery system. The computer controller for the robotic system must be able to exchange data with a wide range of databases and laboratory information systems (LIMS).

VII. STANDARDIZATION OF MICROPLATES

Robotic systems combine many different types of instruments and devices, of which some are standard commercial products and some are custom designed. In order to integrate different types of instruments, some standardization is required.

Because the 96-well microplate has been widely used for a number of years, it has become the standard testing plate. The Society for Biomolecular Screening (Danbury, CT) has proposed standard dimensions for the microplate, including length and width dimensions (the height or thickness of the plate will vary depending on the capacity of the well and other functions of the plate).

The proposed standard dimensions of the microplate are approximately 127.8 mm (5.03 inches) long by 85.5 mm (3.36 inches) wide. This plate is fashioned after the Cooke microplate [14], which was subsequently trademarked as the Microtiter® plate, which is now marketed by Dynex (Chantilly, VA). The most common format contains 96 wells, but higher density plate formats have 384, 1536, and even more wells per plate. Microplates are available from a large number of sources and usually have only minor differences in dimensions.

VIII. SOURCING ROBOTIC SYSTEMS

Because robotic laboratory systems can be customized for the user, there are a number of considerations in selecting a vendor for a robotic system for drug screening and suggestions to new users to improve the implementation process. Users have made a number of recommendations to improve the success of robotic systems [15]. When evaluating vendors, it is recommended to visit the vendor's references, see the system, and ask specific questions regarding the vendor's performance. It is also recommended to see a demonstration of the proposed system before purchasing and look for specific proof of the quality of their systems, software, training, and support.

Both vendors and users agree that often there is not enough discussion of the system details at the beginning of the systems building phase. Some of the items that should be discussed include the users' expectations, the system's acceptance criteria, the desired amount of flexibility and modularity, the specific assays that can be performed on the system, the sample throughput rate, and the form of the operator interface, to name a few items.

Experienced users have identified a number of improvements that they have instituted within their companies in order to improve the success of robotic systems. These improvements include having a stable assay before attempting robotic automation, developing a complete list of system specifications before soliciting quotations from vendors, closer communication with vendors during the build phase at the vendor's site, a need to train multiple operators, and assigning a support technician to the robotic system. Although users reported minor problems in the initial days of their robotic systems, virtually all companies recently surveyed plan to increase their rate of purchase of robotic systems, or at least to maintain the same rate of purchase.

IX. COST JUSTIFICATION

The cost justification for automation systems is usually a calculation of the monetary payback over time. The total expenditure for the robotic system is balanced against the cost savings that can be quantified, which usually includes labor, materials, etc. To calculate the cost of the present method of performing the task, the actual time for all of the manual steps is measured and a cost is calculated based upon the cost of the person or persons who are involved with the task. In the case of drug screening systems, this might include many people, including supervisors and managers, and usually includes the total costs of each person, including benefits, which is sometimes referred to as a person's "loaded overhead cost."

After the full cost per sample is calculated, an estimate is made of the number of samples that will be processed in the future. The total out-of-pocket costs for the proposed robotic system is also added up and usually includes the cost to purchase the system, as well as any associated costs, such as installation, training, etc. Note that the acquisition costs are generally limited to the out-of-pocket costs, that is, those cash outlays that would be incurred only if the system were purchased. Other costs, such as personnel time that might be associated with the purchase, are not usually included in the acquisition costs unless they are significant.

In order to calculate the payback or return on the investment, the cost per sample is divided into the total cost of the system, to give the number of samples that need to be processed before the system is paid for. This is sometimes referred to as the break-even number. The break-even number is then compared with the number of samples that are forecasted to be processed on the system in future years. The length of time until break-even occurs would be the payback period. A short payback period, such as one or two years, indicates that there is considerable cost savings associated with the robotic system. A longer period, such as five to six years or more, would be considered a long time for payback.

Companies develop their own guidelines for the economic payback on capital expenditures, but generally two to three years is a common maximum length of time. Exceptions are often made when other factors are involved, such as when there is an overriding need to add screening capacity, when operator safety is involved, or when the "opportunity cost" of lost sales, for example, is considered.

The opportunity cost is the cost of the lost business opportunity because the expenditure was not made [16]. In the case of HTS, this would be the cost of not having leads that convert to drugs and in turn to products that can be sold. For a leading drug, the opportunity cost is quite high. Unfortunately, it is difficult to quantify the opportunity cost because there is considerable uncertainty when charting the path to a successful product. In instances where the path and cost are more predictable, there may be justification in including the opportunity cost with the cost justification to management for the purchase justification.

Some corporate accounting departments will adjust the payback analysis to express the future saving in dollars (or whatever local currency is being used), which has been adjusted for future inflation. The expenditures will be in current dollars, because the system is being paid for at the beginning, yet the savings will be in the future, after the system has been delivered and is operational. With inflation, current dollars do not have the same value as future dollars, so an adjustment is occasionally made to convert the future dollars into a value that is equivalent to the current dollars. This adjustment is made by discounting the future savings by estimating the amount of inflation over the payback period. (Future inflation reduces the value of future dollars when compared with current dollars, so the value in today's money—current dollars—must be calculated and reduce the future savings accordingly. When future cash flows are discounted to the present time to adjust for inflation, this is referred to as the net present value of the future cash flows from the robotic system [17].) The cash flows are of course the labor savings that are being realized with the system performing the tests rather than humans.

X. CONCLUSIONS

Sophisticated high-throughput screening techniques have been achieved through the development and refinement of robotic automation. At first, robotic screening systems were looking for the proverbial needle in the haystack, screening compounds from a variety of sources (compound libraries, natural product selections, and compound synthesis programs). However, with unique compound structures readily available through compound synthesis and combinatorial chemistry techniques, there is no shortage of compounds that can be screened against a target.

Now, however, the emphasis is to improve the chances of a lead candidate becoming a successful product. The objective is to select, as quickly as possible, the most promising structures for optimization to assure successful in-vivo testing and subsequent market introduction of a successful new drug. New assays are now being used for early drug profiling, including early ADME and toxicity studies to predict better the success of promising drug candidates.

In order to streamline HTS, users are seeking assays that are homogeneous and do not require separation steps such as centrifugation, filtering, washing, etc. Not only are these steps time-consuming but also they have been hard to automate and reduce the reliability of the system. Thus robotic systems for drug discovery have matured over recent years to meet users' needs of increased throughput, flexibility, and reliability and reduced costs.

REFERENCES

1. J Seega, J Delzer, P Eckard, F Emiling, R Janocha, C Markert, G. Paul, W. Wernet. Integrated use of robotic systems in high throughput random screening. Proceedings of the International Symposium on Laboratory Automation and Robotics, Boston, MA, 1994, pp 464–473.
2. Modern Drug Discovery, Feb 1999, p 24.
3. Webster's New Collegiate Dictionary. Springfield, MA: G & C Merriam Company, 1979.
4. R Trinka. Robotic workstations complement tracked robot systems. Lab Automation '97, San Diego, CA, 1997.
5. A Russo, L Heydt, K Ferriter, B Flam, B Lucoth, J Sredy, J Babiak. Measurement of para-nitrophenyl phosphatase (pNPPase) activity by an enzyme-colorimetric high throughput screening assay. Proceedings of the International Symposium on Laboratory Automation and Robotics, Boston, MA, 1994, pp 435–443.
6. D Hook. Team building and leadership in the successful implementation of automation for high throughput screening, Proceedings of the International Symposium on Laboratory Automation and Robotics, Boston, MA, 1995, pp 11–23.
7. P Fernandes. The new face of drug discovery with robotics and screening. Proceedings of the International Symposium on Laboratory Automation and Robotics, Boston, MA, 1995, pp 99–111.
8. M Palmer. High-throughput screening for accelerated drug discovery. Proceedings of the International Symposium on Laboratory Automation and Robotics, Boston, MA, 1995, pp 112–124.
9. G Nakayama, Z Parandoosh. A novel receptor tyrosine kinase autophosphorylation assay. Proceedings of Fourth Annual Conference and Exhibit, The Society for Biomolecular Screening, Baltimore, MD, 1998.
10. P Leitner, B Laskody, J Roth, A Shrago, G Tian, A Edison, P Domanico, Y He. Comparison of automated screening technologies: homogeneous time resolved fluorescence vs. ELISA. Proceedings of Laboratory Automation, San Diego, CA, 1998.
11. L Upham. HTRF® methods: optimum sensitivity for miniaturization in high throughput screening. IBC Third International Exposition and Symposium on Drug Discovery Technology, Boston, MA, 1998.
12. C Suto, D Ignar. Selection of an optimal reporter gene for cell-based high throughput screening assays. J Biomolecular Screening 2:7–9, 1997.
13. K Hwang. Using Caco-2 testing for absorption. Proceedings of Optimising Candidate Selection, Amsterdam, The Netherlands, 1998.
14. R Wallace. The microplate: a historical perspective. The Screening Forum 2:1, 1995.
15. R Trinka. Strategies for the successful sourcing of robotic systems. Proceedings of LabAutomation '99, San Diego, CA, 1999.
16. S Archer, C D'Ambrosio. Business Finance Theory and Management. New York: Macmillian, 1966, pp 202–203.
17. J Zeiman, P Morabito, R Tamilarasan. Economic justification of automation systems in the Dow Chemical Company. Proceedings of the International Symposium on Laboratory Automation and Robotics, Boston, MA, 1995, pp 75–85.

20

Assay Miniaturization
Developing Technologies and Assay Formats

Kevin R. Oldenburg, Ilona Kariv, Ji-hu Zhang, Thomas D. Y. Chung, and Siqi Lin
DuPont Pharmaceuticals Company, Wilmington, Delaware

I. INTRODUCTION

This chapter reviews the current state of the art in assay miniaturization, describes what types of experiments can currently be performed in miniaturized formats, and speculates on what will be possible in the near future. Before converting assays from the standard 96-well format into a miniaturized format such as the 384-, 864-, 1536-, 6144-, or 9600-well formats, several questions need to be answered in order to identify the format that will be most useful for the purposes at hand. First, why miniaturize an assay for drug discovery in the first place? Second, what are the advantages of each miniaturized format, and how low a volume is actually necessary to meet the need? Third, what are the inherent problems with each miniaturized format, and how will that limit the number and type of assays that can be performed in that format? These questions will be answered in the following sections.

The 96-well plate was first introduced over 20 years ago and has long been the standard in the pharmaceutical industry for assay development and high-throughput screening (HTS). Assays of nearly every conceivable type have been performed in 96-well plates [1–13], and the format has been proven to be a reliable means to discover lead compounds for further pharmaceutical development. The question then arises, why change a technology that has been so successful for so long? The miniaturization of assays for HTS is a change forced upon the industry by the invention of combinatorial chemistry [14–20] and other chemical methods that have significantly increased the number of compounds, and by the advances in genomics [21–24] that have significantly increased the number of targets that need to be tested.

Prior to the invention of combinatorial chemistry and the advances made in genomics, a typical pharmaceutical company may have from several tens of thousands to several hundred thousand compounds that would be tested against a small number of therapeutic targets per year. Biologists could easily clone, express, and purify the target of interest in the quantity needed and in a time frame that was acceptable for drug discovery. Similarly, HTS groups could easily screen compounds much faster than chemists could synthesize new ones. In short, the industry had reached an equilibrium where each group could meet the needs of the other groups involved.

With the invention of combinatorial chemistry and the advances made in genomics, this equilibrium was thrown out of balance. Chemists could now make large numbers of compounds very quickly and biologists could identify new therapeutic targets much more readily. In order to keep pace, HTS groups simply increased the throughput of the assays. That is, instead of testing one hundred 96-well plates per week, the groups increased the throughput to several hundred plates per day. This is only a short-term fix for the underlying problem and, as is usual, advances in technology can really only be met with concurrent technological advances.

The underlying problem with simply increasing the throughput of 96-well plates is that as the number of compounds increases, the amount of reagent needed for the biological testing increases as well. Carried to the extreme, the consequence of this is that the biologists spend the majority of the effort in target purification and very little time identifying new targets. The other inherent problem is that compounds made through combinatorial chemistry are usually synthesized in relatively small amounts. Typically, several hundred picomoles (for split and pool synthesis) to several micromoles (for parallel synthesis) are prepared. When these compounds are tested in 96-well plate format, a disproportionately large amount of the compound is consumed in each assay. Typically, these chemical reagents can be entirely consumed after testing several tens of targets.

The goal then is to develop a technology that would allow all the targets that are obtained from the genomics efforts to be screened against all the compounds synthesized by combinatorial and medicinal chemistry. The obvious solution is to miniaturize the assay so that each assay consumes only a small amount of both the biological and the chemical reagents. In this way, with the same amount of effort put into isolating and purifying the biological reagents, 10 to 100 times more assays could be performed. Miniaturization would also extend the ''life'' of the chemical reagents, and hence they could be used for many more assays as well.

How far can an assay be miniaturized and how far is it practical to miniaturize? In theory, an assay could be miniaturized down to single molecules of en-

Table 1 Cost–Benefit Analysis of Miniaturization (Dollars)

	96-well	384-well	1536-well	6144-well
Plates	4,000	2,000	1,600	1,000
Peptide reagent	25,000	12,500	1,000	150
Enzyme production	50,000	100,000	50,000	50,000
Time (screening)	30,000	25,000	3,500	2,500
Total	209,000	139,500	56,100	53,650
Savings/assay		69,500	152,900	155,350
~30 assays/yr		2.09M	4.59M	4.66M

zyme and substrate or receptor and ligand. Unfortunately, today this is neither practical nor desirable since the technology is not available nor would it be cost-effective to develop it. Currently, a 96-well assay is usually performed in 100–200 µL volumes. A number of groups have reported miniaturization of assays to 384-well [25–28] (50 µL) plates, and a few groups have reported miniaturization to 1536-well [29,30] (3–10 µL) or 9600-well [31] (0.2 µL) plates.

Table 1 shows a cost–benefit analysis for assay miniaturization. These numbers are based on screening a 100,000 member library in a protease assay that uses a peptide substrate containing a donor/quencher pair. This is one of the simplest of assays, so the savings observed will likely be conservative estimates. Also, these numbers do not take into account the savings/benefits of chemical library conservation. However, it is apparent from Table 2 that significant savings can be seen with miniaturization from the standard 96-well (100 µL) assay to the 384-well (50 µL) assay. Additional savings are observed upon miniaturization to 1536-well (3–10 µL) formats, but little additional savings are seen upon miniaturization to the 6144-well (0.5 µL) format.

This point of diminishing returns in savings accrued through miniaturization is due to a number of factors. The plate cost is relatively constant for an assay independent of the format. The most significant savings are seen in the

Table 2 Well Volume Vs. Savings

Plate format	% 96-well cost	% 384-well cost	% 1536-well cost
96-well (100 µL)	100		
384-well (50 µL)	67	100	
1536-well (5 µL)	27	40	100
6144-well (0.5 µL)	26	38	96

biological reagents. In the instance described in Table 1, a 20-fold reduction in volume from the 96-well (100 µL) assay to the 1536-well (5 µL) assay results in a 20-fold reagent savings. However, this degree of savings is not seen with the enzyme production, since the majority of that cost is in the labor to produce the reagent. That is, a given amount of labor is required to clone and express the enzyme. This fixed cost is independent of the number of batches that need to be prepared in order to obtain the material needed to complete the screen. Consequently, one batch of this reagent is sufficient to complete an entire screen in 1536-well format and so, in this case, there are no additional savings by moving to a smaller 6144-well format.

An additional factor in determining the degree that an assay can or should be miniaturized is the time (labor) that it will take to complete the screen. As can be seen in Table 1, there is very little time savings between 96-well and 384-well or between 1536-well and 6144-well formats, while there is a large difference between the 96-well and the 1536-well. These differences occur primarily due to differences in plate processing and signal detection times. Since very different technologies are employed, as discussed below, to process 96-well or 384-well plates vs. 1536-well or 6144-well, significant reductions in the length of time it takes to run an assay can be achieved.

Significant savings that are achieved through miniaturization but were not included in the calculations in Table 1 are the reductions in the amount of the test compounds that are used. When synthesized by traditional medicinal chemical methods, the syntheses typically produce on the order of several tens of milligrams of each compound. A typical assay in a 96-well format may consume on the order of 0.5 µg of the test compound per test (at 10 µM test concentration in 100 µL). The consumption of this amount of material is not significant when compounds are made through traditional methods in large quantities. However, in combinatorial chemistry the amounts of material synthesized are significantly smaller. For example, in a split and pool synthesis the amount of material synthesized is only about 0.1 µg, and in a parallel synthesis the amount of material may only be 50–500 µg. The small amounts of material synthesized by these methods are consumed rapidly when testing is performed in the 96-well format, whereas the number of tests can be extended 20- or 30-fold by miniaturization to a 1536-well format.

Three critical factors involved in assay miniaturization include plate design, liquid handling, and signal detection. Table 3 shows a comparison of well density to the well volume and subsequent reagent savings. Miniaturization from the 96-well format to the 1536-well format is achieved by a volume reduction of from 20- to 100-fold. With the reduced volumes that are used in the 1536-well format, new methods of liquid handling had to be developed.

The ability to move liquid rapidly, accurately, and precisely is a key component of any HTS effort. Fully automated devices capable of either single-channel

Table 3 Well Volume and Savings
With Miniaturization

Well density	Well volume (μL)	Fold reagent savings
96-well	200	0
384-well	50	4
864-well	20	10
1536-well	1–10	20–200
6144-well	0.2–0.7	250–1000

or multichannel (8, 12, 96 or 384) pipetting have been developed over the years. For a recent review covering state-of-the-art high-speed conventional pipetting devices see Stevens et al. [32]. Conventional pipetting devices usually work on a positive displacement principle. That is, movement of a piston creates a vacuum that liquid is then pulled into. These mechanical pipettors are generally quite accurate and precise when pipetting volumes greater than 1 μL. The loss of accuracy and precision at lower volumes is due to the viscosity of the liquid, which causes adherence to both the pipet tip and the well. As the pipetting volume decreases, the amount of liquid that remains adhered to the tip becomes a large percentage of the total pipetted volume.

An assay run in a 1536-well plate typically has a volume in the range of 3 to 5 μL. Mechanical pipettors are quite good at pipetting volumes down to 1 μL but tend to lose accuracy and precision when pushed below this range. Consequently, these mechanical pipettors can be used for much of the liquid handling in 1536-well plates. However, most of the test compounds are solubilized in neat DMSO, and the majority of biological reactions cannot tolerate greater than 1% DMSO. This means that the test compounds must be pipetted into each well in a volume of no more than 50 nL for a 5 μL assay, and these volumes cannot be achieved by mechanical pipettors.

For assays miniaturized to 1536-well formats, pipetting volumes from 10 nL up to a few μL is necessary. In order to achieve these volumes, a switch to noncontact dispensing is preferable. Devices based on either ink-jet or piezoelectric technology have been developed that are capable of accurately and precisely pipetting volumes between 100 pL and 5 μL. These devices dispense liquid as a series of small droplets between 10 picoliters and several nanoliters (Fig. 1) and thus never come in direct contact with the well itself. There are several advantages to this method of dispensing liquid. First, the effect of viscosity is minimized, since the liquid is dispensed as microdroplets in a noncontact fashion. Second, there is no need to exchange or wash tips between dispense cycles, since

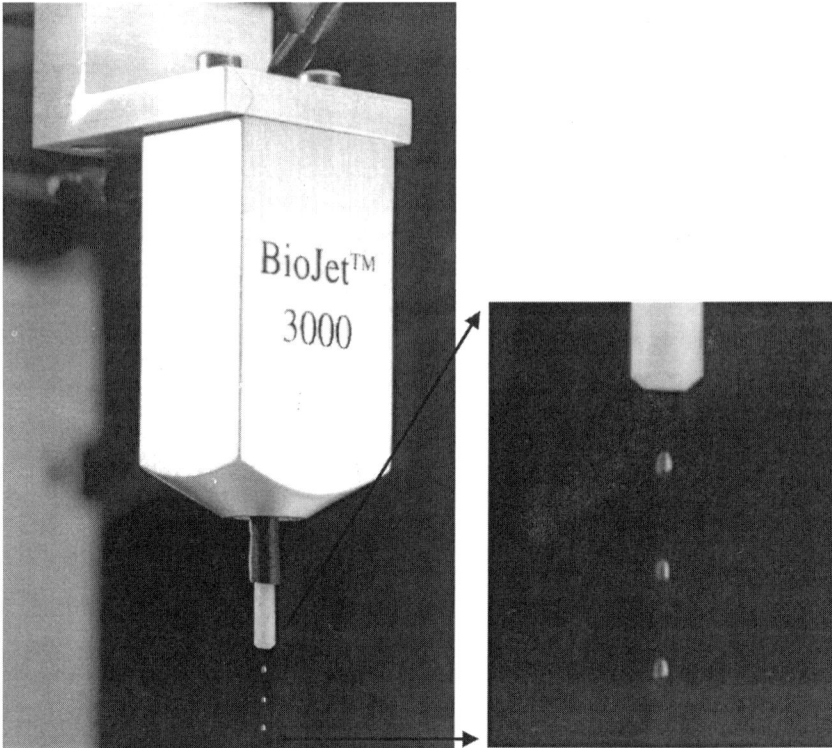

Figure 1   Microliquid handling using a device based on ink-jet printer technology. The figure displays such a device dispensing 10 nL drops in rapid succession. Devices based on this principle are capable of dispensing volumes from several hundred picoliters up to the microliter range. (Courtesy of Cartesian Technologies, Irvine, CA.)

the tips never come in contact with reagents in any of the wells. Carryover from one well to the next is essentially eliminated.

A technological hurdle had to be overcome in liquid handling in order for miniaturized assays to become practical. Similarly, a conceptual hurdle needed to be overcome in terms of plate design and construction. 96-well plates come in several formats, and one of these has become the standard set by the Society of Biomolecular Screening [33]. These plates are typically constructed of polystyrene or polypropylene, have a round or square shape, and have a volume of between 50 and 300 μL (for screening plates; volumes up to 3 mL for compound storage plates).

Miniaturization creates a number of challenges in plate design and construction. As can be observed from Table 4, a volume decrease from 200 μL in a standard 96-well plate to 4 μL in a 1536-well plate results in a 6.5-fold increase in the surface-area-to-volume ratio and greater than a 7-fold increase in the plastic-area-to-volume ratio. The increased surface-area-to-volume ratio translates into a much greater rate of evaporation from the 1536-well plate (Fig. 2). Consequently, reagent addition to these plates has to be rapid in order to avoid concentration changes via evaporation. Also, during incubations, the plates may need to be kept in a humidified chamber so that the volume in the wells does not change during the reaction.

The plastic-area-to-volume ratio is also a concern, since it is known that many proteins adhere to and denature when in direct contact with certain materials. As a well is miniaturized, the plastic-area-to-volume ratio increases from 0.55 mm²/μL for a 96-well plate to almost 4mm²/μL for a 1536-well plate. This means that if one of the reagents binds to the plastic, the proportion of that reagent bound to the plastic will be much greater in the 1536-well plate than in the 96-well format. This has serious consequences for assay development. For example, if one were looking for a kinase inhibitor and wanted to avoid compounds that would compete with ATP, then one would usually run the reaction at several fold the Km for ATP. However, if the ATP bound nonspecifically to the plastic, then the actual concentration of ATP in solution would be much lower than expected. Consequently, all the inhibitors that may be found in this assay format could potentially be competitive with ATP, exactly the opposite of what was searched for. Since the plastic-area-to-volume ratio is much higher in a 1536-well format than in the 96-well format, it will be imperative that Kms and other enzymatic parameters be determined directly in the new format.

Table 4 Specifications for Various Density Microtiter Plates

Well density	Surface area (mm²)	SA/volume ratio (mm²/μL)	Plastic area (mm²)	PA/volume ratio (mm²/μL)	Volume (μL)
96-well	38.5	0.19	110	0.55	200
384-well	9.6	0.19	98	1.95	50
864-well	2.5	0.17	33.3	2.2	15
1536-well[a]	5	1.25	15.5	3.91	4
1536-well[b]	1.3	0.13	36.8	3.7	10
6144-well	1	2	3	61	0.5

[a] Inverted pyramidal shaped "Vision" plate.
[b] Square well "Greiner" plate.

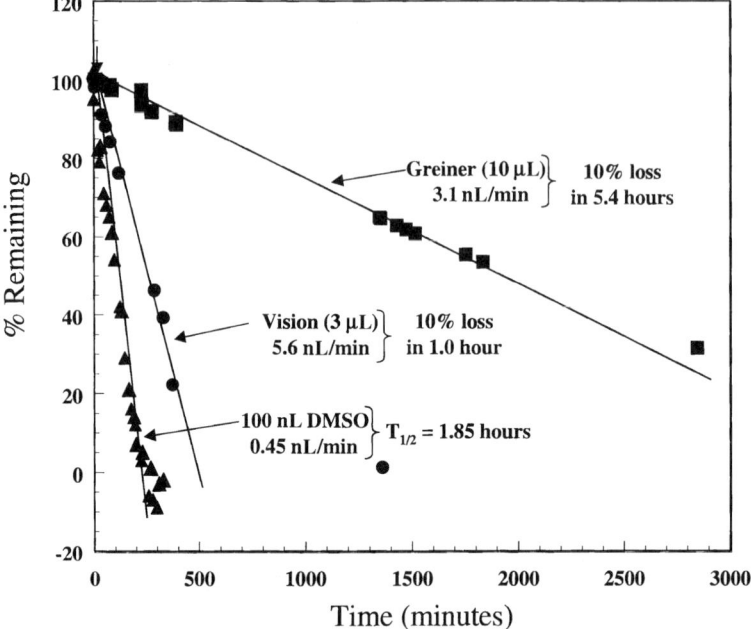

Figure 2 Evaporation from microwell plates presents a significant challenge for developing assays in miniaturized formats. The figure shows the evaporation rate of dH_2O from a Greiner 1536-well plate (10 μL volume) and a Vision 1536-well plate (3 μL volume). The evaporation rate in the Vision plate is approximately 5.6 nL per minute and almost twice as rapid as the evaporation rate, 3.1 nL per minute, in the Greiner 1536-well plate. The more rapid loss of liquid from the Vision plate is a direct function of the greater surface area. Since most HTS assays are complete within a matter of a couple of hours, evaporation may not pose a significant problem. However, should longer incubations be required, incubation can be performed in a humidity-controlled environment. Relative humidity was 42%, and the lab temperature was 24°C.

The shape of the well, as well as the material from which it is constructed, can serve to ameliorate some of these potential problems. For example, the data in Table 4 for 1536-well plates assumes that the well shape is an inverted pyramid. If a standard cylindrical 96-well is miniaturized to the format of a 1536-well then the plastic-area-to-volume ratio is nearly four times larger. This inverted pyramidal shape can also aid in liquid handling.

The inverted pyramidal shape confers two additional advantages in assay miniaturization. First, since the walls separating the wells come to a knife edge, liquid can be "sprayed" into the wells much as a spray painter can paint a wall.

Any liquid that hits the apex or intersection between two adjacent wells must, by well design, run into one of the two wells. This lack of interstitial spaces between wells allows reagent to be added very quickly to all the wells on the plate. Oldenburg et al. [31] report that using an ink-jet device they were capable of adding 100 nL of liquid to each of the wells of a 9600-well plate in under 70 seconds with CVs of less than 4% using this method. This ''angled'' well will also be important for mechanical pipetting as well. Since the surface area to be hit by the pipettor is quite small, the angled well will help guide the pipet into the center of the well. Thus, the tolerance required of the disposable pipet tips can be more generous. If one were to miniaturize the shape of the current 96-well plates into a 1536-well format, this would not be the case, and problems will result when the pipet tips hit the flat interwell spaces. Second, as will be discussed further below, the angled well will help increase the image quality when data is collected from these plates by CCD technology.

II. SIGNAL DETECTION

As should be obvious from the above sections, as an assay is miniaturized from a 96-well (100 μL) format to a 1536-well (3–5 μL format), the signal from each individual well will decrease proportionately. In most cases, 20- or 30-fold less signal would doom a 96-well assay to failure, since a suitable signal window could not be achieved [34]. Using this analysis, it would be easy to conclude that it is impossible to perform an assay in a 1536-well format. However, new technologies as well as modifications and improvements to old technologies have overcome this potential problem.

It is impossible to avoid the fact that when the volume of an assay is reduced from 100 μL to 3 μL the resultant signal will be 33-fold less. However, the signal per unit surface area is not necessarily much different between the 96-well plate and the 1536-well plate. For example, assume that in a 96-well plate the total signal is 100. Then, for a standard well, the signal per unit area is 2. The same assay run in a 1536-well plate will generate a total signal of about 3, but the signal per unit area is about 1. Thus, upon miniaturization, there is only about a 50% loss in signal *if* the signal can be measured as signal per unit area rather than the total signal.

Charge coupled devices, CCD cameras, are devices that are capable of taking images of assay plates based on signal per unit area rather than on total signal. This is possible because the silicon chip within the camera is sectored into an array of about 1000 × 1000 pixels. Light (photons) striking each of the pixels is measured independently of the other pixels. If an entire 1536-well plate is imaged at once, then the signal generated by each well of the plate will be analyzed by approximately 150 pixels. As long as the volume imaged per pixel remains constant, identical signal should be obtained independently of the format. In this case, the theoretical limit for assay miniaturization would be the size of an individual pixel.

One of the hurdles in developing a CCD-based system for HTS was the development of a telecentric lens for the camera. Typical lenses such as those found on 35 mm cameras exhibit spherical aberration. That is, the light entering the lens is bent more with increasing distance from the center of the lens (Fig. 3). The consequence of this for imaging an assay plate would be that the signal from each well would decrease from the center of the plate toward the edges, leading to an unacceptably high plate %CV. This hurdle was overcome with the development of a large telecentric lens. In a telecentric lens, the angle of bending of the light is nearly independent of the distance from the center of the lens. Consequently, a nearly uniform image can be taken of the plate, leading to acceptably small %CVs.

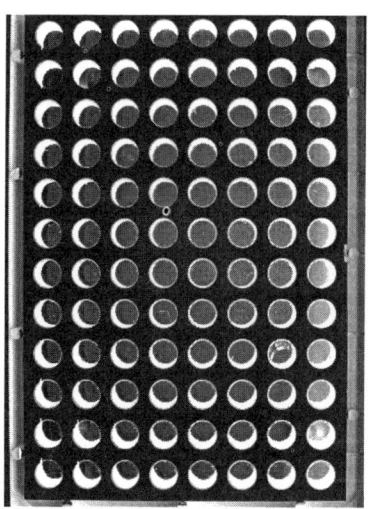
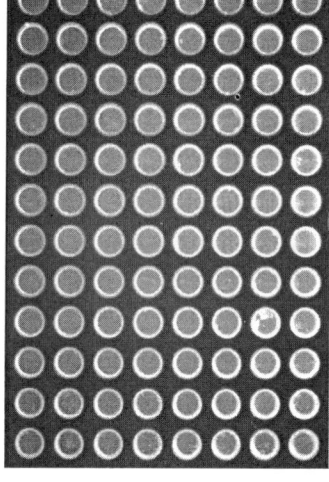

Standard Lens   Telecentric lens

Figure 3 The image on the left was taken with a standard 35 mm lens, while the one on the right was taken with a telecentric lens on the Tundra imaging system. As can be observed, the amount of light captured by the 35 mm lens from each well decreases as the distance from the center of the plate increases. However, with the telecentric lens, the amount of light capture is nearly equal across the entire surface. (Courtesy of Amersham Pharmacia Biotech, Cardiff, Wales.)

Another advantage of CCD technology is that it images an entire plate at once. Detectors based on photomultiplier technology are usually ganged together to increase throughput. For 96-well plates, photomultiplier-based detectors usually come in some multiple of 8 or 12. Increasing the number beyond 12 usually is not practical due to size and cost constraints. This means that between 8 and 12 separate "detections" need to be performed for each individual 96-well plate or from 128 to 192 detections for a 1536-well plate. Since evaporation is a significant issue for these miniaturized formats, whichever system can image the plate faster has a significant advantage. Head to head, a CCD-based system can image a plate approximately 10 to 100 times faster than a photomultiplier-based system. However, a photomultiplier-based system is approximately 10 to 100 times more sensitive than a CCD-based system. What you gain in speed you lose in sensitivity and vice versa.

The earlier section reviewed the advantages, disadvantages, and pitfalls of miniaturizing assays along with the technologies, i.e., hardware that is currently available to address some of those problems and challenges. The following section will deal with the assay formats that have either already been successfully used in miniaturized format or can be used, at least theoretically, in the miniaturized format.

III. FLUORESCENCE TECHNOLOGIES

Since fluorescent molecules can be excited many times, the total number of photons obtained from a sample is simply a function of the quantum yield, the number of fluors present, and the number of excitation photons presented. This effect amplifies the signal and improves the measurement statistics. Sufficiently sensitive detection of fluorescence emission permits even single-molecule detection. Over the years, several fluorescence methods have been developed to address a wide range of biological assays. These range from those of biological origin, such as green fluorescent protein (GFP), to those of a chemical nature, such as fluorescein, coumarin, and the cyanine-based dyes. These fluors have been utilized in a number of different fluorescent technologies such as standard fluorescence, fluorescence resonance energy transfer (FRET), fluorescence polarization (FP), and time-resolved fluorescence (HTRF and TRF). These and related technologies as they relate to miniaturization are reviewed and discussed below.

Green fluorescence protein (GFP) has recently been used in live-cell imaging by monitoring fluorescence intensity redistribution. GFP can signal physiological activation when it is fused to a host protein that translocates from one cellular compartment to another upon stimulation. For example, GFP, glucocorticoid receptor fusion protein, translocates from the cytoplasm to the nucleus upon hormone stimulation [5,35]. With the aid of a high-performance digital imaging system, GFP has also been used as a reporter in an antimicrobial assay in a miniaturized mode [31]. In this assay, *E. coli* expressing GFP was plated into 9600-well plates. The bacterial growth is monitored by GFP fluorescence intensity. Bacterial growth studies in both the 9600-well and the 96-well plate yield similar MIC values for known antimicrobial agents.

Fluorescence quenching is another technique widely used in fluorometric enzymatic assays for proteases, peptidases, nucleases, reverse transcriptases, etc. In the protease system, peptide substrates are double labeled with fluor and quencher chromophores [36]. The fluorescence of the fluor on this type of substrate is initially suppressed by a nearby quencher through intramolecular resonance energy transfer. When an enzyme cleaves the substrate, quencher is removed from the proximity of the fluor, and the fluorescence signal is observed. Fluorescence signal increase is directly proportional to the amount of substrate hydrolyzed. This is a highly sensitive homogeneous assay that is suitable for miniaturization. Peptide or DNA substrates are of particular utility, since these substrates are relatively simple to make synthetically and are usually well tolerated by the enzymes.

Good donor/quencher pairs (EDANS/DABCYL and Mca/Dpa are two commonly used pairs) should have spectral overlap between the fluorescent emission of the former and the absorption of the latter. Efficient fluorescence quenching translates into a very low fluorescence background, a large dynamic range, and excellent assay sensitivity. Both of these pairs have been adapted to ultra-high-throughput screening. In particular, the Mca/Dpa donor/quencher pair has been used with a peptidyl substrate to measure metalloprotease (MMP) activity in both 1536- and 9600-well formats [31] (Fig. 4).

Even though fluorescence is an inherently sensitive technique, it does not always provide adequate sensitivity for HTS or uHTS. The major limitation is that the readout is a change in fluorescence intensity. The fluorescence intensity change is subject to many different measurement artifacts. These include susceptibility to background fluorescence, both from the biological materials in the sample and from the plate itself or test compound quenching or fluorescence, and from photophysical effects such as light scattering and inner filter effects that are dependent upon geometry and sample composition.

A. Fluorescence Resonance Energy Transfer (FRET)

One way to increase sensitivity, or perhaps more appropriately, selectivity, is to utilize fluorescence resonance energy transfer (FRET) technology. This technique has seen increasing use in cell imaging, flow cytometry, and confocal microscopy. FRET is the physical process by which energy is transferred from an excited chromophore (donor) to another chromophore (acceptor) by means of intermolec-

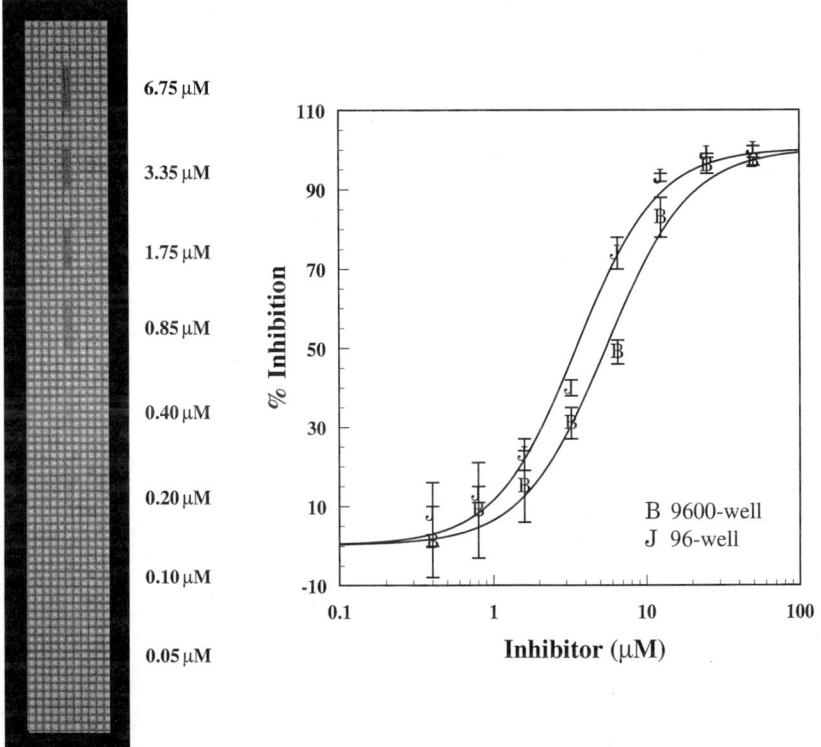

Figure 4 Titration curves of a compound found to inhibit MMP3 were prepared in a 96-well plate (100 µL) and in a 1536-well plate (3 µL). The figure on the left shows the image of the titration in the 1536-well plate, and the data is displayed graphically on the right. Nearly identical IC50 data is generated in either format.

ular long-range dipole–dipole coupling. The efficiency of energy transfer depends on the inverse sixth power of the distance between the dyes. The extent of energy transfer can be measured by measuring the fluorescence intensity decrease of the donor and the increase of the acceptor. FRET is highly sensitive to intra- and intermolecular distance changes (10–100 Å) between two fluorophores and can therefore sense relatively small changes in protein conformation, protein–ligand, or protein–protein interactions. Consequently, FRET-based biosensors and bioassays provide high sensitivity, specificity, and relative simplicity.

A FRET-based sensor, BFP-GFP dual chimeras, has been designed as a fluorescent reporter protein to ''sense'' intracellular activation [37,38]. The two protein-based fluorophores are linked by a flexible calcium binding site derived from calmodulin. Binding of Ca^{2+} to the spacer changes its conformation into an extended rod. This increases the distance between the two fluors and therefore alters the energy transfer between the two fluorescent proteins. The altered FRET spectrum, measurable by fluorescence-ratio change, is an elegant indicator for free intracellular Ca^{2+} concentration.

In addition to protein-based fluorophores like GFP, the two points of interest can also be labeled with small organic fluorophores. To increase the sensitivity, the Em_{max} of the donor should match the Ex_{max} of the acceptor (Fig. 5). There should be little or no overlap of the emission spectra from the two fluors, and both the donor and the acceptor should have good quantum yields in order to maximize the emission signal. There are several chromophore pairs that have been used as FRET pairs; these include fluorescein/rhodamine, fluorescein/coumarin, and terbium/rhodamine.

β-lactamase has been used as a reporter protein in a FRET based system. This system, in which fluoroscein and coumarin are connected by the β-lactamase substrate, cephalosporin, has been reported to show enhanced sensitivity over other biological reporter systems [39]. In the intact substrate, excitation of the coumarin donor at short wavelength leads to energy transfer to the longer wavelength fluorescein acceptor and reemisson of green light. When β-lactamase-catalyzed hydrolysis of the substrate separates the donor and acceptor, the donor then emits blue fluorescence, whereas the acceptor is not excited. The ratio of the

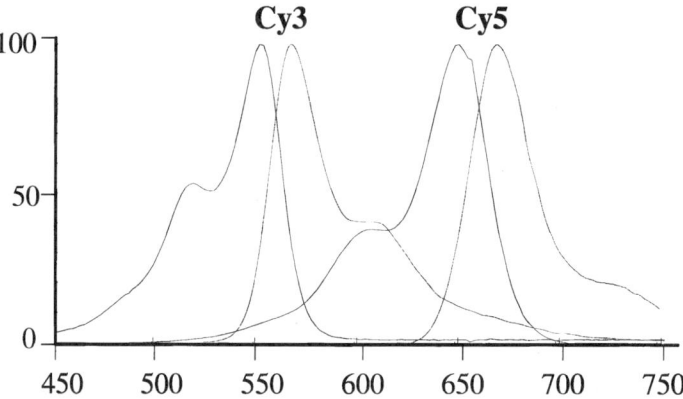

Figure 5 Excitation and emission wavelengths are shown for a FRET pair composed of Cy3 and Cy5. In order for a pair of fluors to function as a FRET pair, the emission wavelength of one member of the pair (Cy3) has to overlap the excitation spectrum of the second member of the pair (Cy5). It is also important that the excitation band of the second of the pair (Cy5) not overlap with the excitation band of the first of the pair (Cy3). (Courtesy of Amersham Pharmacia Biotech, Cardiff, Wales.)

emission intensity of blue to green light is a sensitive indicator for transcriptional readout.

B. TRF and HTRF

Separation or wash steps may not be acceptable for miniaturized assays. Ideal assays for miniaturized formats will be in a simple mix-and-read, homogeneous format. Recently developed FRET-based approaches combine the benefits of a highly sensitive fluorescent label and a homogeneous assay protocol under a variety of HTS experimental circumstances. One of these systems, called homogeneous time-resolved fluorescence (HTRF), involves fluorescence resonance energy transfer between a slow-decay fluor and a short-lived fluor [9,40–42].

Processes involving light emission do not occur on the same time scale. When excitation is performed in a repetitive timed fashion, fluorescence from any long-lived (or slow decay) species can then be detected selectively by delaying measurement until after such time that all short-lived species have decayed. Various lanthanide ion cryptates (particularly Eu^{3+}, Tb^{3+}, Sm^{3+}, and Ru^{2+}) are a source of long-lived fluorescence. Organic chromophores (the acceptor), when in proximity to a lanthanide donor, emit long-lived fluorescence resulting from the energy transfer from the lanthanide. Free acceptor, distant from the donor, emits only short-lived fluorescence, that is not measured in a time-resolved fashion (Fig. 6).

The HTRF principle has been demonstrated with the donor–acceptor pair europium cryptate (EuK) and allphycocyanin (APC) [9,40]. APC has an absorption band overlapping the EuK emission. APC has a typical broadband organic emission monitored at 665 nm. Because of the energy transfer process, the APC is continuously reexcited, and the apparent emission decay is delayed to match the lifetime of EuK; consequently, the APC emission can be measured after a 50 µs time delay. The emission of an unbound APC-labeled biomolecule at 665 nm can be disregarded because of its fast decay. Free EuK-labeled biomolecule, that is also present during the measurement, emits a slow-decay fluorescence at 620 nm and can be captured in a time-resolved fashion as well, and used as an internal reference. This reference intensity, as well as the bound APC emission, may vary due to interfering absorption by the serum component or turbidity. However, the ratio (intensity at 665 nm/620 nm) is unaffected and depends only on the concentration of analytes.

Typical background emission in biological samples is due to organic fluorophores and is short-lived. The long-lived fluorescence from the energy transfer to APC allows removal of essentially all background fluorescence by time resolution. Labeling of biological reagents with either APC or cryptate can be achieved either directly or by using labeled antibodies directed toward a nonobtrusive site in the target. The development of APC- and cryptate-labeled secondary antibod-

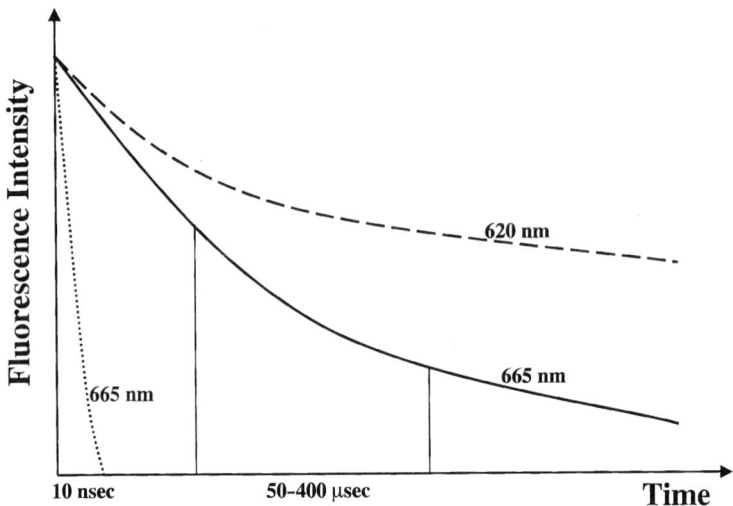

Figure 6 Time-resolved fluorescence. This figure shows the rapid excitation of the "background" fluorescence and its rapid decay. It also shows the rapid excitation of the donor fluor and its long lifetime (slow decay). Readings are taken after the rapid decay of the background but before the complete decay of the sample.

ies for HTS will not only allow a more generic approach but also further reduce the cost of this technology regardless of specific application.

So far, FRET- and HTRF-based assays have been demonstrated in 96-, 384-, and 1536-well microplates. Since conventional plate readers for fluorescence typically read one well at a time, for a miniaturized format (1536 or higher), it would take at least sixteen times longer to read, which is unacceptably slow for high/ultrahigh-throughput requirements. A more promising approach is to use fluorescence imaging technology to collect data from all the wells at once. This detection system, similar to FLIPR (fluorescence imaging plate reader) [12], should permit parallel laser pulses to each well and capture dual-wavelength fluorescence intensity imaging simultaneously.

C. Fluorescence Polarization

Although the theoretical principles of fluorescence polarization (FP) were described at the beginning of the century [43], the application of FP to drug discovery, and especially to HTS, became popular only recently [10,44–46]. FP measurements permit analysis of molecular orientation and motion of intrinsically fluorescent or fluorophore-labeled molecules in solution. Therefore FP methodology can be applied to different assays measuring direct binding, such as antibody–antigen interaction, compound–protein interaction, DNA–protein interaction, protease activity, and others [3,47–49].

Several phenomena need to be taken into consideration to characterize the basic principle of fluorescent polarization, FP. First, a time interval exists between the excitation of the fluorescent molecule and the emission of light. This fluorescent lifetime (τ) is usually in the nanosecond range. Second, molecules absorb and emit light based on the transition moment of the molecular electronic framework—where transition moment is defined as the direction in which the molecule most easily absorbs the light; therefore the linear direction of the emitted light is dependent on the direction of the excitation vector. The excitation polarized light and the emitted light will be parallel only if the molecule remains stationary during the fluorescent lifetime. However, if the molecule is freely rotating in solution during its excited state, then the emitted light will be depolarized. The ability to distinguish between polarized and randomly scattered light is the basis of FP. The fluorescence polarization values (P) are determined by the ratio between the difference of perpendicular and parallel emitted light measurements and the sum of those measurements, where parallel direction is defined as a direction of a polarized excitation light vector. The molecules with their excitation dipoles oriented perpendicular to the excitation light vector cannot absorb light and therefore will not emit light when excited. The parallel alignment will result in 100% light absorbtion. Molecules at other orientations will absorb light at different efficiencies depending on their alignment to the excitation light plane. Therefore the above maximum theoretical polarization value is equal to 0.5P (or 500 mP) [3,50,51]. A polarization value greater than 500 mP suggests an experimental artifact, due either to an additional source of scattered light or to incorrect instrument calibration.

Fluorescent probes can be selected based on their lifetime, but molecules with more extensive τ values have an advantage because they can rotate more during the excited state, thus causing greater depolarization. Currently, fluorescein and its derivatives are the fluorophores of choice. However, new fluorophores, that possess long lifetimes, have been recently developed, e.g., BODIPY (Molecular Probes, Eugene, OR) or the Cy dyes (Amersham Pharmacia Biotech, Little Chalfont, UK). Because the emission of those fluores is in the red portion of the spectrum, the interference from fluorescent organic compounds is minimal, thus making this fluorophore more suitable for HTS.

The advantage of FP is that the assay is homogenous, eliminating the necessity to separate bound versus unbound fluorescent probes. The reaction usually reaches equilibrium in a few minutes, thus making it even more attractive for assay miniaturization. Moreover, FP is independent of fluorescence intensity and therefore less sensitive to quenching than other fluorescent detection methods. It also is not affected by solution turbidity, because the FP values are obtained as single-wavelength ratiometric measurements. Because FP measures the changes in molecular orientation and motion, there are two major considerations in designing a successful FP assay. First, high binding affinity/avidity between the "low" molecular weight fluorescent-labeled ligand and a relatively large receptor, and, second, the interaction between receptor and ligand has to be rigid in order to eliminate the "propeller effect" [10].

FP has multiple advantages for use in the miniaturized format. The assay itself is homogenous, without the necessity to separate bound versus unbound fluorescent probes. The measurement is independent of the fluorescent signal intensity, and the limit of FP sensitivity is dependent only on the affinity of the binding between assay components. Accordingly, in FP, as compared to other assay formats, the decrease in assay volume does not translate to a decrease in P values, although the noise may increase. Figure 7 provides an example of a FP assay formatted in a 1536-well plate [52]. In this experiment, the binding between a fluorescein-labeled 8-mer peptide and an 81.1 kD CDK2/E protein complex was measured. The Z' factor of 0.73 indicates that it is an excellent assay to use in uHTS. The %CV across the 1536-well plate of the positive control (maximum signal with protein complex) and negative control (minimum signal without protein complex) are 2.9% and 3.6%, respectively. The specificity of interaction is determined by the competition experiment with nonlabeled tracer

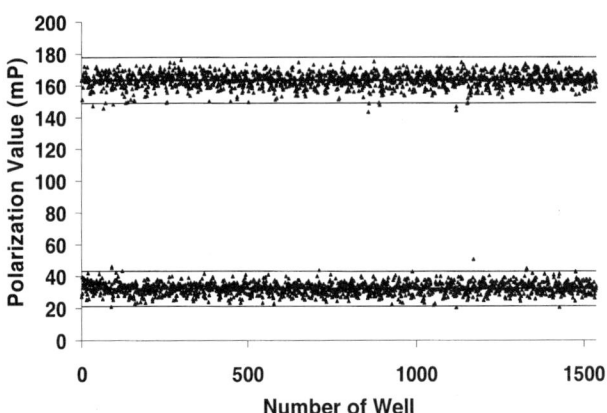

Figure 7 Flourescent polarization measurements of CDK2/E complex and fluorescein-labeled 8-mer peptide (tracer) interaction. Well-to-well variation for positive and negative controls (with and without CDK2/E, respectively) was performed on two separate plates ($N = 1536$). Solid lines represent the mean $\pm$ 3 standard deviations. The coefficient of variation (CV) is 2.9% for the positive control and 11.1% for the negative control. The z' factor is 0.75.

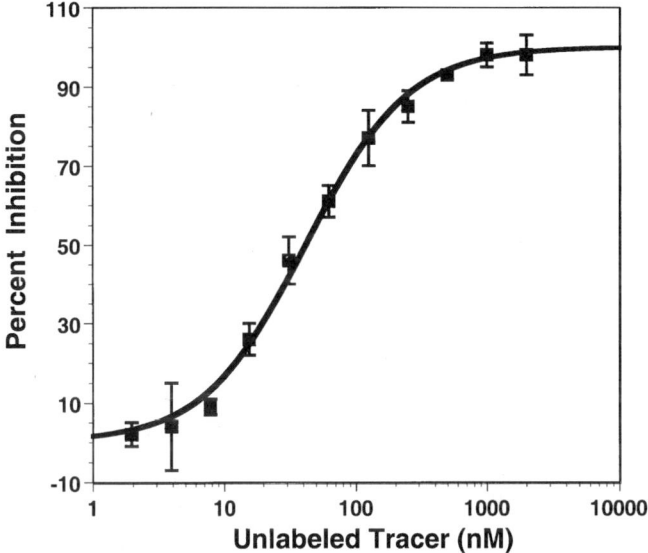

Figure 8 Demonstration of fluorescent polarization assay specificity, showing competition binding of unlabeled and fluorescently labeled tracer to CDK2/E. The IC_{50} is 42 nM, and the hill slope is 1.1.

(Fig. 8). These data indicate the suitability of FP for use in the miniaturized format.

IV. MINIATURIZATION OF CELL-BASED ASSAYS (LUMINESCENCE AND FLUORESCENCE)

In vitro mammalian cell-based assays constitute an important part of any biomolecular screening effort. The main advantage in using whole-cell-based assays is that a cell model may be predictive of the desired physiological response. Data obtained from a cell-based assay not only relates to the target of interest but can on occasion predict such parameters as compound membrane permeability, general cell toxicity, and cell type specificity and selectivity. The disadvantage of using a cell-based assay is that it is hard to determine the exact mechanism of compound activity without a battery of parallel assays. For example, if the primary screen detects the expression of the gene of interest during the course of the assay, a parallel cell-based assay measuring nonspecific inhibition of transcription/translation machinery is required to eliminate nonspecific compound effects. The major benefit of miniaturizing cell-based assays is the decrease in the number of cells required to complete the screen. However, multiple criteria have to be met in order to accomplish cell-based assay miniaturization.

Miniaturization of the mammalian cell-based assay poses a number of unique challenges as compared to the other assay types utilized in HTS. Frequently, engineered immortalized cell lines are used to monitor the activity of the gene of interest in the cell-based screens. The most common cell-based formats currently used in HTS are the receptor binding assays and the reporter gene assays. Homogenous assays, where all the components are mixed and the signal is read directly without the necessity of washing cells, are preferable for miniaturization, because technology is only now becoming available that is capable of washing 1536 plates. Additional requirements have to be considered to miniaturize cell-based assays successfully, including signal strength and detection, efficient cell dispensing, evaporation control, and the concentration of the organic solvents in the assay.

Signal detection is limited by the available optical reader systems and therefore has to be either fluorescent or luminescent based. When the detector is CCD-camera-based, the signals obtained from 96-well and 1536-well formats are comparable, because signal per unit pixel is measured. However, when a photomultiplier-based detection system is used, the transition from 96-well to higher density plates displays a concurrent decrease in signal. This is partly because only a small portion of the photomultiplier tube is used for the measurement, and also because the system measures total signal and not a signal per unit area. However, photomultipliers tend to be one or two orders of magnitude more sensitive than CCD cameras and consequently can usually detect much lower signals. Therefore cell lines that express high levels of the reporter gene, or where the signal can be amplified, are usually more suitable for miniaturization.

Different reporter gene models have been characterized for use in HTS [53–58]. The firefly luciferase reporter gene is an enzyme of choice due to the high detection sensitivity (λem = 510 nm) and the amplification provided by the luciferase enzyme. The luminescence readout signal is generated by the light production resulting from the ATP and Mg^{2+}-dependent cleavage of the luciferin substrate. In addition, different systems have been recently described that lead to significant signal amplification and production of steady light "glow" rather than the light "flash" previously used [59]. Other luciferase reporter genes can also be utilized, including the enzyme cloned from *Renilla reniformis*, which is ATP and Mg^{2+}-independent for dual glow signal in the same cell (Packard Instruments Co., Meriden, CT). The use of the later luciferase genes is limited by the short half-life. The disadvantage of both methods is that both require cell lysis and therefore are not amendable for kinetics studies.

The β-lactamase reporter gene is also potentially suitable for miniaturization. Several commercial β-lactamase substrates, that are cleaved to produce either fluorescent or chemiluminescent products by the enzyme, generate a reasonable signal and should be adaptable to a miniaturized format (Clontech Laboratories, Inc., Palo Alto, CA; Molecular Probes, Eugene, OR; Tropix-PE Biosystems, Bedford, MA). However, these methods also require cell lysis and are thus limited in utility. Recently a β-lactamase fluorogenic substrate has been described for use in live cell kinetic studies, applying FRET technology [60]. This substrate has been shown to be membrane permeable and displays low cellular toxicity. This substrate should overcome many of the disadvantages of the previously described substrates.

Green fluorescent protein (GFP) and GFP derivatives are alternatives for the noncatalytic fluorescent reporter [61,62]. This system allows noninvasive tagging and monitoring of cellular proteins. The use of a GFP reporter in miniaturization requires a very high level of GFP expression because of the low quantum yield of GFP. This requirement limits the choice of the gene promoter to strong viral promoters such as SV40 or CMV and is usually less suitable for "native" mammalian promoters.

The secreted form of human placental alkaline phosphatase (SEAP) has previously been shown to be an efficacious reporter gene [63,64] and should be amenable to assay miniaturization. Since SEAP is secreted by the cell into the culture supernatant, there is no need to prepare cell lysates in order to monitor gene expression. The same cell preparations can be sampled multiple times and used for kinetic studies. The concern in using SEAP in miniaturized format is that it may be difficult to eliminate background due to endogenous alkaline phosphatase (AP), which is usually abrogated in standard methods by heating the test samples to 65°C (SEAP unlike other APs is heat-stable). However, chemical inhibitors of AP, which are utilized to decrease background, have been recently introduced (Tropix-PE Biosystems, Bedford, MA). Both chemiluminescent and fluorescent substrates are available to monitor SEAP enzymatic activity (Clontech Laboratories, Inc., Palo Alto, CA and Tropix-PE Biosystems, Bedford, MA).

The time that it takes to complete a cell-based assay is also a major consideration when contemplating miniaturization. A well volume must be used that will allow for sufficient cell density to produce the total signal needed for detection, but not so small that cell death is induced due to overcrowding. With miniaturization, the surface-area-to-volume ratio increases approximately 6.5-fold (Table 4). This provides both a benefit and a limitation to the cell-based assays. The advantage of the increased surface-area-to-volume ratio is that it allows better gas exchange between the cell culture media and the surrounding environment. However, this increased surface-area-to-volume ratio also increases the rate of evaporation, and therefore special precautions must be taken to ensure that the volume change during the assay does not affect the results.

The organic solvents that are often used to maintain compound libraries, both discrete medicinal chemistry and combinatorial in origin, cause unique problems in a cell-based assay. Most mammalian cells show high sensitivity to the solvents, such as DMSO, commonly used for compound storage and can only be maintained in less than 1% solvent concentration. Thus test compounds must be delivered in a way that minimizes solvent concentration in the final assay, requiring either very precise delivery of nanoquantities of tested compounds or drying of compounds prior to testing.

One of the primary concerns in miniaturization of cell-based assays is the ability to dispense the cells at a high density in a relatively small volume (1–2 µL) while maintaining cell integrity. Piezoelectric or ink-jet liquid dispensers may be suitable for certain cell-based assays, where the pathway under study is not associated with stress response(s). When the activity of the gene of interest is implicated in a stress response, the use of more conventional mechanical liquid handlers, based on either positive or negative displacement, is recommended in order to avoid stressing the cells.

Taking all the above concerns and considerations into account, a cell-based assay has been adapted to the 1536-well plate format using a total volume of 3 µL per well [29]. This system utilizes a human T cell line transfected with the luciferase reporter gene under the control of the promoter of the gene of interest.

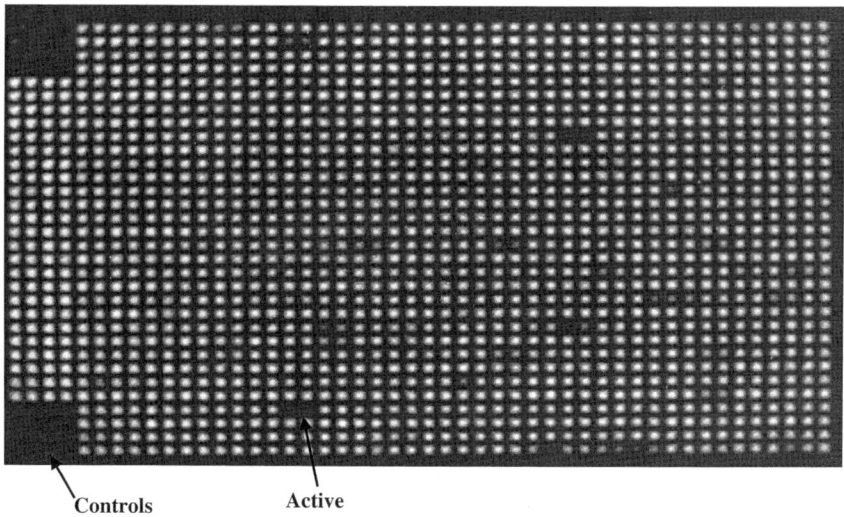

Controls   Active

Figure 9 Miniaturization of cell-based assay. Representation of screening data of random compounds from eight 96-well compound plates in duplicate consolidated into one 1536-well plate. Inhibitors of the expression of the gene of interest were identified using a T cell line transfected with the luciferase reporter gene under the control of the promoter of the gene of interest.

Figure 9 represents the screening plate, where eight 96-well compound plates in duplicates were consolidated into one 1536-well plate. The data indicates that identification of inhibitors of the gene of interest can be successfully performed in a miniaturized format.

V. SPA MINIATURIZATION

The scintillation proximity assay (SPA) format has been applied to a variety of HTS assays including antibody–antigen binding, receptor–ligand binding, protein–protein interactions, protein–DNA interactions, and a number of enzyme-based assays [65]. SPA is a nonseparation radioisotopic technique. In many cases, the SPA bead or plate is coated with a "generic" reagent that is capable of sequestering the "target" of interest near the surface of the bead. When the target binds to its corresponding radiolabeled ligand, the SPA bead is brought within the mean free path distance of the β-particle, and the energy from the radioactive decay is transferred from the ligand to the scintillant in the SPA bead. Upon absorption of this energy, the scintillant emits a photon of light (for a review of SPA technology see Refs. 6 and 65–67).

The main driving forces behind SPA miniaturization are reagent (both biological and chemical) conservation, cost, and reduction in radioactive material usage and disposal. Since this is a "homogeneous" assay format it should be amenable for performing SPA assays in miniaturized HTS formats (384-well, 1536-well, or beyond). The scintillants used in the standard SPA beads have been optimized for use in photomultiplier-based detectors. Consequently, the scintillant was designed to emit light with an emission maximum at approximately 450 nm—the region of maximum sensitivity for this type of photomultiplier. CCD cameras, however, are maximally sensitive in the 600–700 nm region and thus required the development of an appropriate scintillating particle. The recent development of the Leadseeker™ SPA beads (Amersham) enables the emitted light to be detected by field imaging CCD cameras.

SPA assays have currently been performed in 96- and 384-well formats [68] with well volumes ranging from ~200 µL in the 96-well format to ~50 µL in the 384-well format. Miniaturization beyond the ~5 µL/1536-well format is currently limited by detection sensitivity as well as the compatibility with systems for liquid handling.

To evaluate the possibility of miniaturization of SPA in the 1536-well format, a SPA-RT assay has been optimized for substrate, primer/template, enzyme, reaction time, and SPA Leadseeker™ beads used in the assay. Figure 10 shows an enzyme titration curve of the miniaturized SPA-RT assay performed in a 1536-well plate. The total reaction volume was 5 µL, with an additional 3 µL of stop solution containing SPA Leadseeker™ beads added after the required incubation.

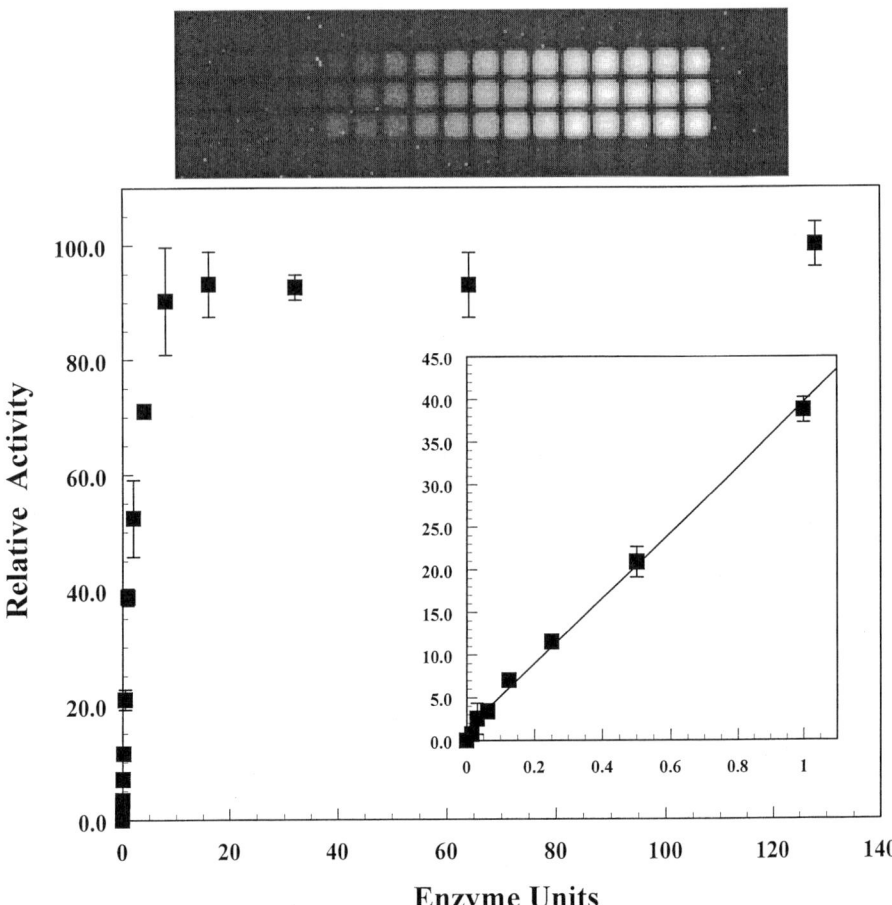

Figure 10 Miniaturization of a SPA assay into a 1536-well plate. Advances in SPA technology have demonstrated the feasibility of miniaturizing a SPA format. In this example, an assay for HIV reverse transcriptase was adapted for SPA. The figure demonstrates a typical enzyme titration curve for HIV RT. As the amount of enzyme in the reaction is increased, the amount of product formed increases linearly up to 2.5 units of enzyme. The figure on top is the image from the assay as detected by the Tundra. The data is presented graphically on the bottom.

The signal was acquired with a Tundra™ imaging system with a cooled CCD camera. The results indicate that up to $1 \times 10(-3)$ unit enzyme concentration, the enzyme titration curve is linear and can be used for inhibitor screening. At the selected conditions, a titration with a known HIV-1 reverse transcriptase inhibitor yields an IC50 of 72 nM, compared to the reported value of 46 nM determined under similar conditions in a 96-well plate. Although the signal obtained from the miniaturized format has a relatively low signal-to-background ratio compared to the 96-well plate format, the assay quality estimated by the z-factor (see the statistical analysis section of this chapter) shows that it is still suitable for use in HTS. Recently, a miniaturized SPA assay for MMP (matrix metalloproteinase) has also been performed in the 1536-well plate format with similar assay quality to the aforementioned SPA-RT assay.

In comparison with the regular SPA assays, the miniaturized SPA has several advantages: reagent saving, increased throughput, and reduced radioactive waste. The miniaturized assay also has the potential for improving hit confidence by performing the assay in duplicate or triplicate without consuming significantly more reagent. The miniaturized assay also displayed some disadvantages. For example, in the miniaturized format, the detection system is a low-light CCD camera instead of a photomultiplier-based system. The sensitivity of miniaturized SPA with the currently available CCD cameras is somewhat less than with a similar fluorescent method. The SPA assay will still remain a simple and efficient assay when the ligand or substrate is too small to be fluorescently labeled without loss of activity.

VI. A STATISTICAL PARAMETER FOR USE IN ASSAY QUALITY EVALUATION

Among other possible advantages of assay miniaturization, one less well addressed issue is the gain of assay quality in drug screening from a statistical point of view. Assay miniaturization makes it possible for each data point to be measured multiple times without consumption of significantly more key reagents.

The active compounds (called hits) from a high-throughput random screening process are those statistical ''outliers'' of the screened compounds. The quality of a HTS assay will directly affect the quality (e.g., confirmation rate) of the active compounds that come out of the screen. In the past several years, the quality or suitability of a HTS assay has been loosely defined. Terminology such as signal-to-background ratio (S/B) or signal-to-noise ratio (S/N) borrowed from electronic engineering have been widely used as the major criteria for evaluating the quality of a HTS assay. However, in many cases these ratios are inadequate or even improper for use in assay quality evaluation.

In validating a typical HTS assay, unknown samples are assayed along with reference controls. The sample signal refers to the measured signal for a given test compound. The negative control (usually referred to as the background) refers to the set of individual assays from control wells that give the minimum signal. The positive control refers to the set of individual assays from control wells that give the maximum signal. In assay validation, it is critical to run several assay plates containing both positive and negative controls in order to assess the reproducibility and signal variation at the extremes of the activity range. The positive and negative control data can then be used to calculate their respective means and standard deviations (SDs). The difference between the mean of the positive controls and the mean of the negative controls defines the dynamic range of the assay signals. The variation in signal measurement for samples, positive controls, and negative controls (i.e., SDs or CVs) may be different. All these parameters together define the quality of a specific biological assay. The inherent problem with using either the S/N ratio or the S/B ratio is that neither of them takes into account both the signal dynamic range and the variability in the sample and reference control measurements. To define a HTS assay, μ_s, μ_{c-} and μ_{c+} are denoted for the mean of the library sample signal, negative control signal and positive control signal, respectively. The SD's of the signals are denoted as σ_s, σ_{c-} and σ_{c+}, respectively. Figure 11 illustrates a typical assay structure.

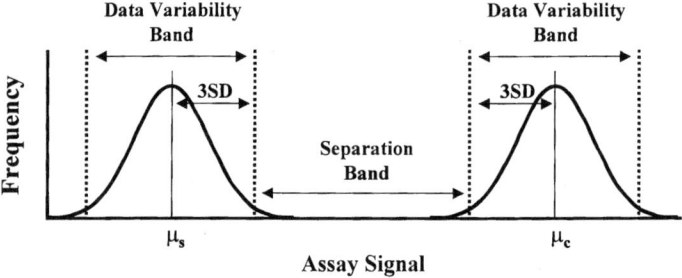

Figure 11   Signal-to-noise ratio and signal-to-background ratio are two terms that have been loosely used to describe HTS assay quality. However, a more mathematically correct term has been described by the z-factor equation. This equation is based on the simple premise that the quality of an assay is a function of the standard deviations of the unknowns, the standard deviations of the controls, and the separation band between the unknowns and the controls. An assay suitable for HTS has either a large separation band between the samples and controls, or small data variability bands for the samples and controls, or preferably, both. For example, an assay with an average signal of 1 and an average background of 0.9 may be an excellent assay if the data variability bands are on the order of 0.01. Conversely, an assay with an average signal of 100 and an average background of 1 may be a poor assay if the data separation band is very small or nonexistent.

Recently, Zhang et al. [34] have introduced a simple statistical parameter for use in any biological assay quality evaluation, especially for high-throughput drug screening assays. This *dimensionless* screening window factor (denoted the z-factor) is defined as the ratio of the separation band to the assay signal dynamic range (Fig. 11):

$$z = \frac{|\mu_s - \mu_c| - (3\sigma_s + 3\sigma_c)}{|\mu_s - \mu_c|} \tag{1}$$

where $|\mu_s - \mu_c|$ defines the *usable* dynamic range for the screen. This expression for the separation band is suitable for either the activation type assay (where $\mu_c > \mu_s$) or the inhibition type assay (where $\mu_c < \mu_s$). Obviously, μ_c and σ_c need to be replaced by μ_{c+} and σ_{c+} for agonist/activation type assays and by μ_{c-} and σ_{c-} for antagonist/inhibition type assays, respectively. The z-factor takes into account the assay signal dynamic range, the data variation associated with the sample measurement, and the data variation associated with the reference control measurement. The z-factor thus defines a characteristic parameter of data quality for hit identification for each given assay at the defined screening conditions. It is therefore suitable to use the z-factor for evaluating the quality (or performance) of HTS assays [34].

Rearranging Eq. (1) yields

$$z = 1 - \frac{3\sigma_s + 3\sigma_c}{|\mu_s - \mu_c|} \tag{2}$$

or substituting Eq. 2 with

$$Z_{act} = 1 - \frac{3\sigma_s + 3\sigma_{c+}}{|\mu_s - \mu_{c+}|} \quad \text{for activation type assays} \tag{2a}$$

$$Z_{inh} = 1 - \frac{3\sigma_s + 3\sigma_{c-}}{|\mu_s - \mu_{c-}|} \quad \text{for inhibition type assays} \tag{2b}$$

In Eq. (2), the ratio $(3\sigma_s + 3\sigma_c)/|\mu_s - \mu_c|$ can be regarded as the data *variation factor* of the screening assay. It is clear that this variation factor by itself also takes into consideration all of the information necessary for HTS assay characterization.

The z-factor can be used to evaluate the quality or performance of any given HTS assay. It is sensitive to the data variability as well as the signal dynamic range. For example, as $3\sigma_s + 3\sigma_c$ approaches zero, i.e., very small standard deviations, or as $|\mu_s - \mu_c|$ approaches infinity, the z-factor approaches unity (maximum value of z), and the HTS assay approaches an ideal assay. This is in agreement with one's intuition that a "good" assay has a large dynamic range and small data variability. The z-factor can be any value less than or equal to one ($-\infty < z \leq 1$). Since the z-factor is a dimensionless quantity, it is also suitable and useful for comparison of assays. The larger the value of the z-factor of an assay, the higher the data quality (or the suitability) of the assay for HTS. Table 5 lists a simple comparison of screening assays by the value of their respective z-factor. The z-factor is sensitive not only to the assay procedure and the instrumentation used but also to the composition of the compound library and the compound concentration at which the assay is performed. Different libraries and different compound concentrations will give different distribution profiles of the sample signals. Consequently, the μ_s and σ_s values will change with these conditions, which causes a subsequent change of the z-factor.

Similar to the screen window coefficient defined in Eq. (2), a *z'-factor* can be calculated using only control data:

$$z' = 1 - \frac{3\sigma_{c+} + 3\sigma_{c-}}{|\mu_{c+} - \mu_{c-}|} \tag{3}$$

The z'-factor is a characteristic parameter for the quality of the assay itself, without intervention of test compounds. Therefore, the z'-factor is a statistical characteristic of any given assay, not limited to HTS assays. Since in every case $z \leq z'$ for all large data sets, the z'-factor can be utilized for quality assessment in assay development and optimization. If the z' value is negative or close to zero, it usually indicates that the assay conditions have not been optimized or the assay is not feasible as it is configured. If the z' value is large and the z value is relatively small (where both z' and z are > 0), it may indicate that the compound library

Table 5 A Simple Categorization of Screening Assay Quality By the Value of the z-Factor

$$z = 1 - \frac{3\text{SD of sample} + 3\text{SD of control}^{a}}{|\text{mean of sample} - \text{mean of control}|}$$

z-factor value	Structure of assay	Related to screening
1	SD = 0 (no variation), or the dynamic range	An ideal assay
$1 > z \geq 0.5$	Separation band is large	An excellent assay
$0.5 > z > 0$	Separation band is small	A doable assay
0	No separation band, the sample signal variation and control	A "yes/no" type assay
<0	No separation band, the sample signal variation and	Screening essentially impossible

[a] For agonist/activation type assays the control data in the equation is substituted with the positive control (maximum activated signal) data; for antagonist/inhibition type assays the control data in the equation is substituted with the negative control (minimum signal) data.

and/or the compound concentration for screening need to be further examined or optimized. It is important to note that the z-factor as defined in Eq. (2) is a statistical parameter that requires relatively large data sets. Furthermore, the z-factor is a plug-in formula useful for evaluating HTS assay quality. It has no preset requirement for setting a "hit limit," and therefore the hit limit is still a floating parameter of the screen. However, the z value of an HTS assay will provide a useful guideline for where to set the most reasonable hit limit from the quality of the assay.

In most HTS programs, each compound is tested only in singlet. A high degree of accuracy and sensitivity in the assay is therefore critical for identifying active compounds (called hits). Due to the nature of each assay methodology and the perturbation introduced by instrumental and human associated random error, all of the measurements from an assay contain a degree of variability, yet hits need to be identified in the presence of and despite such signal measurement variation. For large unbiased chemical libraries, the vast majority of the compounds have little or no biological activity. Therefore error analysis by statistical models will apply, and an activity histogram of a typical HTS assay approximates a normal distribution model. The hit threshold or hit limit is usually expressed as standard deviations away from the mean of the library sample signals. Due to the variation associated with each measurement, compounds with activity close to the hit limit always have some probability of crossing over the hit limit upon remeasurement. Conceivably, compounds with an activity near the hit limit only have a certain probability of being declared a hit [34,69]. From a statistical point of view, the number of false positive and false negative hits can only be minimized by reducing the data variability. This can be achieved either by improving the assay quality itself or by assaying each data point multiple times. By averaging multiple (n) measurements for each data point, the variability (standard deviation) of its mean value decreases by $\sqrt{n}$.

Assay miniaturization has several potential advantages over the standard 96-well plate screening scheme. First, since in miniaturized format the well density on each plate is significantly higher, many more unknown samples and reference controls can be assayed in one plate. For example, 16 times more unknowns and controls can be loaded in a 1536-well plate than in a 96-well plate. The data generated from each plate should display less fluctuation for statistical analysis. The intra- or interplate data variation can be assessed with a higher degree of certainty. Secondly, in a miniaturized, high-density assay format, each unknown sample can be tested multiple times without consuming a large amount of critical reagents. Therefore, assay miniaturization should enable data averaging and hence decrease the variability of the mean value and subsequently increase the confidence of the screening results. A 1536-well plate will allow each compound to be assayed in duplicate or triplicate and still maintain a compound density greater than that in a 96-well plate. In addition, by being able to assay many more compounds within each plate, assay miniaturization can also shorten the overall screen as downstream processes such as "cherry picking" and reconfirmation of initial hits need not be performed. This will alleviate some assay variation caused by day-to-day random error. These statistical considerations combined with other features (such as reagent and cost savings) of assay miniaturization make it one of the most preferred formats for use in high-throughput drug screening.

VII. DATA HANDLING AND DATA VISUALIZATION

Handling the massive amounts of data generated by a high-throughput screening lab is a continuing challenge that will be further exacerbated by the move towards assay miniaturization, with its potential to generate data at a 10- to 100-fold greater rate. Whether the physical format of test devices are 1536-well, 9600-well, or chip-based, one of the potential rate-limiting activities is the transfer of data (raw or processed) acquired from measurement devices, association of this data with some unique indexing or cataloging bits, matching of the data back to compounds tested, and loading of this data into a database that can be queried. Once the data is in the database, the additional bottleneck is data presentation as an aid to supporting decisions about the quality and significance of the data accumulated. Even with current production runs of 100–400 96-well plates per day, reviewing reams of tabular information or plate-by-plate graphical reports is untenable.

The HTS community would do well to learn from the progress towards automation of the DNA sequencing and transcriptional profiling being done by functional genomics companies. All data loading is done automatically from sequencing readers directly to ORACLE relational tables as raw fluorescent traces, which are then provisionally assigned A, C, T, or G by automatically launched processes. Sequencing run sets, associated primer sequences, filenames, and actual raw traces are automatically linked and available for use for other downstream queries, such as searches against public databases. The end user can monitor the progress of any sequencing effort in real time, drill down to primary data to verify data quality, and edit these data files all from a client–host environment. Removal of all human intervention except at the point of data review streamlines data handling.

Handling and processing of these increased volumes of data and automatically loading both raw and processed data into a robust relational database management system (RDBMS) is central to a HTS process. All processed data is derived from automatically launched algorithms that calculate them from the underlying raw primary data. This paradigm uncouples the data loading process from the data review and analysis process. Some HTS groups load only validated data, but this creates a bottleneck for data capture and exposes the system to misplaced or lost data. The loss of a link back to the primary data obviates verification or modification during subsequent analyses. For instance, while a test plate may look out of specifications, upon subsequent review it is found that a spurious control well has thrown off all the calculated percentage inhibitions; masking this value (not deletion from the database) and recalculation yields an acceptable data set from this test plate. Loading only a processed data set of percentage inhibitions would have led to a loss of data. More importantly, at any later date another scientist could see the underlying data in order to verify the rejection of the "spurious" control. These quality control decisions are made after the data is acquired and already in the database. HTS is a highly iterative and dynamic process. Certain criteria such as hit rejection thresholds are provisionally set at the start of an HTS campaign; it needs to be updated throughout the entire campaign, and the final screen selections need to be reviewed as a whole after the screening collection is complete. This cannot be done appropriately in a manner based on sound statistical analysis if data is rejected too early in the process.

Regardless of the particular RDBMS, any HTS group should give considerable thought at the outset in defining their work flow to highlight the kinds of data types being generated and used, and more importantly, in defining what kinds of queries they might want to ask of their database and their processes.

It is particularly important in image-based technologies that the primary raw data, e.g., pixel intensity, be directly stored to tables, rather than the processed image that shows up as a grey scale or color-coded image on a CCD camera's terminal. These visual images can be readily distorted by changing gain and contrast, whereas the primary absolute intensity values do not change. Link back from a processed image and its setting files to the table of raw intensity values would be indispensable.

Another key to data analysis is that human beings see patterns better than computers in the absence of considerable investment in algorithm and code generation. It is better to have good visual representations of the data rather than simple tables of data. Rapid graphical representation of data is critical during two main HTS activities, the initial data review to uncover experimental artifacts and the final review of library performance against an assay and across assays as an assessment of library quality.

Data from runs represented as grids of plate data are useful to obtain a feel for how a particular plate did on a given day. However, as run sets routinely have become larger, the value in reviewing each data set as individual plates becomes less, except as an alert that there may be some assay artifact for that plate. Therefore an aggregate histogram or scattergram becomes more useful, with the ability to drill back to an individual plate that may be associated with a section of that scattergram. Visualization is very useful in the area of recognizing plate artifacts, such as edge effects or pipetting errors. If during a day's run, all plates run on that day collapsed onto a "virtual" aggregate plate with some measure of hit frequency, it would be obvious if a statistically anomalous number of hits were located in one area on the plates. Potential artifacts of this type could be noted readily.

A dynamically updated histogram of all aggregated data for a screen that can be broken into constituent parts and presented as subset histograms according to a day's run, last batch run, etc., would facilitate decision support for data rejection or inclusion. It would also give an indication of whether the library was truly random in its distribution of activity or whether it was more active for a subset of the library (i.e., bimodal).

During final review, data visualization as a multidimensional surface of biological activity of the library against several assays would uncover hot spots of activity for certain target types. These could then be parsed out for analysis against chemical space parameters to uncover additional pharmacophores. There is a need for displaying these n dimensions simultaneously on a 3-D projection. There are several software companies that have enabled these features. The data visualizer from Spotfire allows dynamic display of up to six dimensions or characteristics for up to 150K samples.

VIII. NEW DETECTION TECHNOLOGIES

This section reviews a few nascent technologies that can be easily adapted to the CCD imaging detector employed in assay miniaturization efforts and that use the existing optics of a field imager. Therefore, methods such as Evotec's (Hamburg, Germany) fluorescence correlation spectroscopy (FCS) and Kairos' (Santa Clara, CA) fluorescence imaging microspectrophotometry (FIMS) requiring a confocal microscopic system are not discussed.

A. Fluorescent Nanocrystals

There has been a recent interest in the use of biocompatible nanocrystals as in situ fluorescent labels. These microcrystalline particles have interesting fluorescent properties that vary with their composition (CdSe, InP, InAs) and size [70,71]. Their emission spectra are narrow (25–35 nm at half-height) and symmetrical with peak maxima shifting from blue to infrared (460–1800 nm) as the sizes of the particles increase from 2 to 6 nm in diameter. Their emission is long lived (>100 ns), which allows for time gating to reduce autofluorescent backgrounds. While these nanocrystals can be efficiently excited at any wavelength below their emission peak, all nanocrystals share common absorption bands in the 350–450 nm range. Therefore, an ensemble of different nanocrystals can be simultaneously excited with light of only one wavelength, to yield emission peaks that can be tuned in 1–2 nm increments across the entire 460–1800 nm range. This allows considerable choice in simultaneous multicolor labeling [72] to maximize non-overlap. There has been considerable progress on functionalizing the surfaces of nanocrystals for water solubility, biological compatibility, and enablement of bioconjugation chemistries. These derivatized nanocrystals have already been used for receptor internalization and microstructure studies in cells [72]. The ability to tune the emission of nanocrystals to red wavelengths and their long emission lifetimes will allow them to be applied to difficult FRET configurations where inner filter, optical quenching, and autofluorescence are problematic in HTS.

B. Liquid Crystal

Recently Gupta et al. reported on a device that uses distortions in a liquid crystal (LC) to amplify and transduce a receptor-mediated binding event into optical outputs [73]. They sandwiched a 1–20 μm LC layer of 4-cyano-4′-pentylbiphenyl between a pair of patterned anisotropic gold films that align the LC, so that the plane of polarized light transmits through this layer uniformly with no resultant features. These gold films were coupled to self-assembled monolayers (SAMs) containing biotin. Specific binding of avidin to the biotin SAMs disorders the surface, and this disorder is propagated within seconds throughout the LC fluid phase. The result is a 5-order-of-magnitude amplification that can be observed by the naked eye as a darkening of the uniform background. The SAMs can be micropatterned on the millimeter scale, so such devices could be used in assaying the effects of spatially arrayed chemical libraries on protein–ligand interactions. These devices require no electroanalytical apparatus and can be imaged with simple illumination with plane polarized light.

C. Thin Film Technology

There are several techniques that report on the "thickness" of a layer above a reflective surface. Surface plasmon resonance (SPR), which measures the changes in refractive index of a monolayer above a gold surface, has been commercialized as the BIACore system (Pharmacia, Sweden), but the device has only one channel and is of limited use for HTS [74]. The detection of binding events by reflectometric interference spectroscopy (RIFS) is based on the interference that occurs at thin transparent films [75]. As a light beam passes through a weakly reflecting thin film deposited on a reflective surface, it will be reflected in part at each of the interfaces. Therefore each of the reflected light beams will traverse different optical paths, and a phase shift results that will modulate the resultant reflected light intensity. Any change in this film thickness will cause a change in the reflectance spectrum. Such reflectance shifts have been demonstrated for immobilized antigens bound by antibodies. Additionally, various techniques are employed to enhance the local deposition of mass into this reporter layer of binding (e.g., TNB deposition from a localized alkaline phosphatase reaction). Recently, BIOSTAR (Boulder, CO) introduced a variant of RIFS, ellipsometry, which utilizes plane polarized light and the attendant depolarization of reflected light from the thin film. The extent of depolarization can be measured by an imaging system through a cross-polarizer. Their simple visually readable devices comprise a thin-film antireflective material deposited on solid substrate to a depth of 500 Å to create a gold-colored background. Additional deposition of a layer 10–25 Å results in a change in color from the gold background to an intense purple in direct proportion to the layer thickness. Biostar corporation has patents on these materials and their fabrication according to semiconductor technologies. Much of the current activities of such devices are in the diagnostic market, where immunoassays to detect the presence of antigens are being researched, but there is a logical extension to any kind of ligand–receptor assay. The technologies have already been demonstrated for enzyme assays wherein a protease is used to digest a thick molecular layer.

IX. CONCLUSIONS

We hope that this review of technologies that are currently available and those that should be available in the near future has provided a basis for implementation of these new technologies. Our group and others have shown that essentially any assay that can be performed in a 96-well plate can be performed in a miniaturized format such as a 1536-well plate. It is readily apparent that with the increases in library size due to combinatorial chemistry and the increased number of targets from genomics efforts, HTS groups will need to increase capacity to meet this growing need.

REFERENCES

1. E Altermann, JR Klein. Synthesis and automated detection of fluorescently labeled primer extension products. BioTechniques 26:96–101, 1999.
2. TS Angeles, C Steffler, BA Barlett, RL Hudkins, RM Stephens, DR Kaplan, CA Dionne. Enzyme-linked immunosorbent assay for trkA tyrosine kinase activity. Anal Biochem 236:49–55, 1996.
3. T Burke, R Bolger, W Checovich, R Lowery. Measurement of peptide binding affinities using fluorescence polarization. In: Phage Display of Peptides and Proteins. A laboratory manual. BK Kay, J Winter, J McCafferty, eds. New York: Academic Press, 1996, pp 305–326.

4. D Burns et al. Development of fluorescence-based high throughput screening assays: choice of appropriate instrumentation. SPIE 3259:188–196, 1998.
5. KL Carey, SA Richards, KM Lounsbury, IG Macara. Evidence using a green fluorescent protein-glucocorticoid receptor chimera that the Ran/TC4 GTPase mediates an essential function independent of nuclear protein import. J Cell Biol 133:985–996, 1996.
6. ND Cook. Scintillation proximity assay: a versitile high-throughput screening technology. Drug Discovery Today 1:287–294, 1996.
7. A Dhundale, C Goddard. Reporter assays in the high throughput screening laboratory: a rapid and robust first look? J Biomol Screening 1:115–118, 1996.
8. S Froidevaux et al. A microplate binding assay for the somatostatin type-2 receptor (SSTR2). J Recept Signal Transduction Res 19:167–180, 1999.
9. JM Kolb, G Yamanaka, SP Manly. Use of a novel homogeneous fluorescent technology in high throughput screening. J Biomol Screening 1:203–210, 1996.
10. ME Jolley. Fluorescence polarization assays for the detection of proteases and their inhibitors. J Biomol Screening 1:33–38, 1996.
11. E Sarubbi, SD Yanofsky, RW Barrett, MA Denaro. A cell-free, nonisotopic, high-throughput assay for inhibitors of type-1 interleukin-1 receptor. Anal Biochem 237:70–75, 1996.
12. KS Schroeder, BD Neagle. FLIPR: A new instrument for accurate, high throughput optical screening. J Biomol Screening 1:75–80, 1996.
13. T Winkler, U Kettling, A Koltermann, M Eigen. Confocal fluorescence coincidence analysis: an approach to ultra high-throughput screening. Proc Natl Acad Sci USA 96:1375–1378, 1999.
14. JJ Burbaum, JHJ Ohlmeyer, JC Reader, I Henderson, LW Dillard, G Li, TL Randle, NH Signal, D Chelsky, JJ Baldwin. A paradigm for drug discovery employing encoded combinatorial libraries. Proc Natl Acad Sci USA 92:6027–6031, 1995.
15. JC Chabala. Solid-phase combinatorial chemistry and novel tagging methods for identifying leads. Cur Opin Biotech 6:632–639, 1995.
16. R Fathi, J Rudolph, RG Gentles, R Patel, EW MacMillan, MS Reitman, D Pelham, AF Cook. Synthesis and properties of combinatorial libraries of phosphoramidates. J Org Chem 61:5600–5609, 1996.
17. EM Gordon, RW Barrett, WJ Dower, SPA Fodor, MA Gallop. Application of combinatorial technologies to drug discovery. 2. Combinatorial organic synthesis, library screening strategies and future directions. J Med Chem 37:1385–1401, 1994.
18. EM Gordon, MA Gallop, D Campbell, C Holmes, J Bermak, G Look, M Murphy, M Needels, J Jacobs, J Sugarman, J Chinn, B Ruhland-Fritsch, D Jones. Combinatorial organic synthesis: application to drug discovery. Eur J Med Chem 30:S237–S248, 1995.
19. DAM Konings, JR Wyatt, DJ Ecker, SM Freier. Deconvolution of combinatorial libraries for drug discovery: theoretical comparison of pooling strategies. J Med Chem 39:2710–2719, 1996.
20. L Wilson-Lingardo, PW Davis, DJ Ecker, N Hebert, O Acevedo, K Sprankle, T Brennan, L Schwarcz, SM Freier, JR Wyatt. Deconvolution of combinatorial libraries for drug discovery: experimental comparison of pooling strategies. J Med Chem 39:2720–2726, 1996.

21. ST Cole, IS Girons. Bacterial genomics. FEMS Microbiol Rev 14:139–160, 1994.

22. JR Korenberg, XN Chen, MD Adams, JC Venter. Toward a cDNA map of the human genome. Genomics 29:364–370, 1995.

23. MG Palfreyman. Functional genomics conference: from identifying proteins to faster drug discovery. Expert Opin Invest Drugs 7:1201–1207, 1998.

24. A Paulus, S Williams, R Tan, P Kao, A Sassi, I Cruzado, V Burolla, W Levine, G Alongo-Amigo, HH Hooper. Microfabricated devices for high throughput separation in genetic analysis and drug screening. SO Book of Abstracts, 217th ACS National Meeting. Washington, D.C.: American Chemical Society, 1999.

25. JCW Comley, A Binnie, C Bonk, JG Houston. A 384-HTS for human factor VIIa: comparison with 96- and 864-well formats. J Biomol Screening 2:171–178, 1997.

26. B Janzen, P Domanico. The 384-well plate: pros and cons. J Biomol Screening 1: 63–64, 1996.

27. AJ Kolb, K Neumann. Beyond the 96-well microplate: instruments and assay methods for the 384-well format. J Biomol Screening 2:103–109, 1997.

28. P Pasini, M Musiani, C Russo, P Valenti, G Aicardi, JE Crabtree, M Baraldini, A Roda. Chemiluminescence imaging in bioanalysis. J Pharm Biomed Anal 18:555–564, 1998.

29. AM Maffia, I Kariv, KR Oldenburg. Miniaturization of a mammalian cell-based assay: luciferase reporter gene readout in 3 µL 1536-well format. J Biomol Screening 1999, in press.

30. G Knebel. High-throughput screening with 1536 wells. Application Note 22–24, 1998.

31. KR Oldenburg, J Zhang, T Chen, A Maffia III, KF Blom, AP Combs, TDY Chung. Assay miniaturization for ultra-high throughput screening of combinatorial and discrete compound libraries: a 9600-well (0.2 µl) assay system. J Biomol Screening 3: 55–62, 1998.

32. ME Stevens, PJ Bouchard, I Kariv, TDY Chung, KR Oldenburg. Comparison of automation equipment in high throughput screening. J Biomol Screening 3:305–311, 1998.

33. TW Astle. Microplate standardization report. J Biomol Screening 3:3–8, 1998.

34. J Zhang, TDY Chung, KR Oldenburg, A simple statistical parameter for use in evaluation and validation of high throughput screening assays. J Biomol Screening 4:67–73, 1999.

35. H Htun, J Barsony, I Renyi, DL Gould, GL Hager. Visualization of glucocorticoid receptor translocation and intranuclear organization in living cells with a green fluorescent protein chimera. Proc Natl Acad Sci USA 93:4845–4850, 1996.

36. M Taliani, E Bianchi, F Narjes, M Fossatelli, A Urbani, Steinkühler, R De Francesco, A Pessi. A continuous assay of hepatitis C virus protease based on resonance energy transfer depsipeptide substrates. Anal Biochem 240:60–67, 1996.

37. VA Romoser, PM Hinkle, A Persechini. Detection in living cells of Ca^{2+}-dependent changes in the fluorescence emission of an indicator composed of two green fluorescent protein variants linked by a calmodulin-binding sequence. A new class of fluorescent indicators. J Biol Chem 272:13270–13274, 1997.

38. A Miyawaki, J Llopis, R Heim, JM McCaffery, JA Adams, M Ikura, RY Tsien. Fluorescent indicators for Ca^{2+} based on green fluorescent proteins and calmodulin. Nature 388:882–887, 1997.

39. G Zlokarnik, PA Negulescu, TE Knapp, L Mere, N Burres, L Feng, M Whitney, L Roemer, RY Tsien. Quantitation of transcription and clonal selection of single living cells with β-lactamse as reporter. Science 279:84–88, 1998.

40. AJ Kolb, PV Kaplita, DJ Hayes, Y Park, C Perrnell, JS Major, G Mathis et al. Tyrosine kinase assays adapted to homogeneous time-resolved fluorescence. Drug Discovery Today 3:333–342, 1998.

41. K Stenroos, P Hurskainen, S Eriksson, I Hemmila, K Blomberg, C Lindqvist. Homogeneous time-resolved IL-2-IL-2R alpha assay using fluorescence resonance energy transfer. Cytokine 10:495–499, 1998.

42. G Mellor, MN Burden, M Preaudat, Y Joseph, SB Cooksley, JH Ellis, MN Banks. Development of a CD28/CD86 (B7-2) binding assay for high throughput screening by homogeneous time-resolved fluorescence. J Biomol Screening 3:91–99, 1998.

43. P Perrin. Polarisation de la lumière de fluorescence. Vie moyenne de molécules dans l'état excité. J Phys Radium 7:390–401, 1926.

44. GR Nakayama. Microplate assays for high throughput screening. Curr Opin Drug Discovery Dev 1:85–91, 1998.

45. MV Rogers. Light on high-throughput screening: fluorescence-based assay technologies. Drug Discovery Today 2:156–160, 1997.

46. JR Sportsman, SK Lee, H Dilley, R Bukar. Fluorescence polarization in high throughput screening: the discovery of bioactive substances. In: JP Devlin, ed. High Throughput Screening. New York: Marcel Dekker, 1997, pp 389–399.

47. WB Dandliker, SA Levison. Investigation of antigen–antibody kinetics by fluorescence polarization. Immunochem 5:171–183, 1968.

48. R Devlin, RM Studholme, WB Dandliker, E Fahy, K Blumeyer, SS Ghosh. Homogenous detection of nucleic acids by transient-state polarized fluorescence. Clin Chem 39:1939–1943, 1993.

49. R Bolger, W Checovich. A new protease activity assay using fluorescence polarization. Product Application Focus 17:585–589, 1994.

50. KL Bentley, LK Thompson, RJ Klebe, PM Horowitz. Fluorescence polarization: a general method for measuring ligand binding and membrane microviscosity. Bio-Techniques 3:356–366, 1985.

51. E Colett. Polarized Light—Fundamentals and Applications. New York: Marcel Dekker, 1993.

52. SS Pin, I Kariv, NR Graciani, KR Oldenburg. Analysis of protein–peptide interaction by a miniaturized fluorescence polarization assay using cyclin dependent kinase 2/cyclin E as a model system. Anal Biochem 275:156–161, 1999.

53. W Scheirer, C Roelant, DA Burns, BV Groningen. Introduction to LucLite: a bioluminescent reagent system for reporter gene assays. Vienna, Austria: Packard Instruments, 1995.

54. W Scheirer. Reporter gene assay applications in high throughput screening: the discovery of bioactive substances. In: JP Devlin, ed. High Throughput Screening. New York: Marcel Dekker, 1997, pp 401–412.

55. JL Moreira, M Wirth, M Fitzek, HJ Hauser. Evaluation of reporter genes in mammalian cell lines. Meth Mol Cell Biol 3:23, 1992.
56. J Alam, JL Cook. Reporter genes: application to the study of mammalian gene transcription. Anal Biochem 188:245–254, 1990.
57. I Bronstein, J Fortin, PE Stanley, GSAB Stewart, LJ Kricka. Chemiluminescent and bioluminescent reporter gene assays. Anal Biochem 219:169–181, 1994.
58. PC Gailey, EJ Miller, GD Griffin. Low-cost system for real-time monitoring of luciferase gene expression. BioTechniques 22:528–534, 1997.
59. CH Roelant, DA Burns, W Scheirer. Accelerating the pace of luciferase reporter gene assays. BioTechniques 20:914–917, 1996.
60. G Zlokarnik, PA Negulescu, TE Knapp, L Mere, N Burres, L Feng, M Whitney, K Roemer, R Tsien. Quantitation of transcription and clonal selection of single living cells with β-lactamase as reporter. Science 279:84–88, 1998.
61. H Tanahashi, T Ito, S Inouye FI Tsuji, Y Sakaki. Photoprotein aequorin: use as a reporter enzyme in studying gene expression in mammalian cells. Gene 96:249–255, 1990.
62. M Chalfie, K Steven. Green Fluorescent Protein. Properties, Applications, and Protocols. New York: Wiley-Liss, 1998.
63. J Berger, J Hauber, R Hauber, R Geiger, BR Cullen. Secreted placental alkaline phosphatase: a powerful new quantitative indicator of gene expression in eukaryotic cells. Gene 66:1–10, 1988.
64. BR Cullen, MH Malim. Secreted placental alkaline phosphatase as a eukaryotic reporter gene. Meth Enzymol 216:362–368, 1992.
65. M Picardo, KT Hughes. Scintillation proximity assays. In: JP Devlin, ed. High Throughput Screening. New York: Marcel Dekker, 1997, pp 307–316.
66. ND Cook et al. SPA: a revolutionary new technique for drug screening. Pharm Manuf Int 1:49–53, 1992.
67. N Bosworth, P Towers. Scintillation proximity assay. Nature 341:167–168, 1989.
68. Scintillation Proximity Assay. Amersham International, C.L.,.
69. GS Sittampalam, PW Iversen, JA Boadt, SD Kahl, S Bright, JM Zock, WP Janzen, MD Lister. Design of signal windows in high throughput screening assays for drug discovery. J Biomol Screening 2:159–169, 1997.
70. MA Hines, P Guyot-Sionnest. Synthesis and characterization of strongly luminescing ZnS-capped CdSe nanocrystals. J Phys Chem 100:468–471, 1996.
71. M Bruchez Jr, M Moronne, P Gin, S Weiss, AP Alivisatos. Semiconductor nanocrystals as fluorescent biological labels. Science 281:2013–2015, 1998.
72. WCW Chan, S Nie. Quantum dot bioconjugates for ultrasensitive, nonisotopic detection. Science 281:2016–2020, 1998.
73. VK Gupta, J Jskaife, TB Dubrovsky, NL Abbott. Optical amplification of ligand–receptor binding using liquid crystals. Science 279:2077–2080, 1998.
74. A Brecht, R Burckardt, J Rickert, I Stemmler, A Schuetz, S Fischer, T Friedrich, G Gauglitz, W Goepel. Transducer-based approaches for parallel binding assays in HTS. J Biomol Screening 1:191–201, 1996.
75. G Gauglitz et al. Chemical and biochemical sensors based on interferometry at thin (multi-) layers. Sensor Actuator B 11:21–29, 1993.

21

Screening in the NanoWorld

Single-Molecule Spectroscopy and Miniaturized High-Throughput Screening

Rodney Turner, Dirk Ullmann, and Sylvia Sterrer
EVOTEC BioSystems AG, Hamburg, Germany

I. DEMANDS ON ULTRAHIGH-THROUGHPUT SCREENING (uHTS) SYSTEMS

A. A Rationale for New Methods in Drug Discovery

The dramatic increase simultaneously, in the number of targets and the potential hit compounds is the reason for what has been called a change in paradigm in drug discovery [1]. To capitalize fully on the promise of both genomics and combinatorial chemistry, the drug discovery process must change in a fundamental way.

Rather than screening a limited number of chemical compounds against a particular target, combinatorial approaches often require screening thousands of compounds. Because of the preponderance of targets, screening must take place in days rather than months. Out of necessity, high-throughput screening (HTS) will be a key factor in realizing the potential of new drug discovery strategies by building a bridge between the increased number of targets and the vast number of compounds to be screened.

Not only does screening throughput need to be increased but also novel assay techniques and detection technologies have to be employed in order to provide more information about the suitability of specific compounds earlier in the screening process. This combination of quality and quantity must be achieved in order to take full advantage of the increased potential for target identification offered by genomics and proteomics and the potential for direct rapid access to novel chemical compounds offered by combinatorial chemistry and combinato-

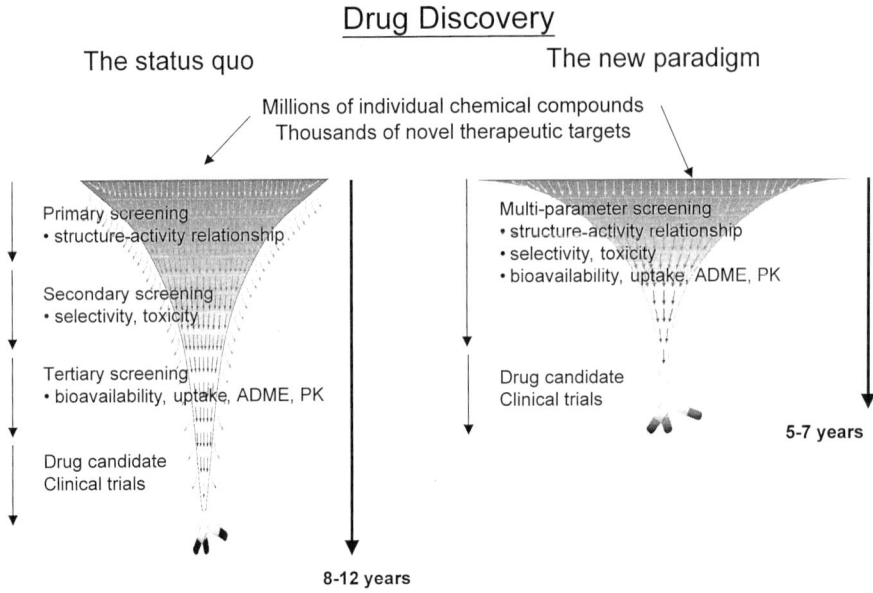

Figure 1   A comparison of today's and tomorrow's drug discovery paradigm.

rial biology [2]. Figure 1 illustrates the time and cost savings to be achieved by using novel technologies and miniaturization.

B. Chemistry, Biology, and Technology

The technologies such as (combinatorial) chemistry, robotics, cell-based assays, miniaturization, and genomic sciences is starting to change the drug discovery process significantly. Thus, the effective use of HTS depends largely on the effective synergy between these disciplines. Screening, therefore, is the link between biology (targets) and chemistry (druglike molecules). In the age of genomics and combinatorial chemistry, that link must achieve high throughput. Figure 2 illustrates the unification of targets, compounds, and assays in screening applications for drug discovery.

    At the screening rates enabled by HTS and ultra-HTS (uHTS) applications, reagent consumption is a critical factor. Cost reduction will not come from economics of scale of the microminiaturized systems but from reductions in the costs of reagents and manpower [3]. Approximately three-quarters of the costs of screening derive from the costs of both chemical and biological reagents. Miniaturization of HTS results in three significant benefits that address the pharma-

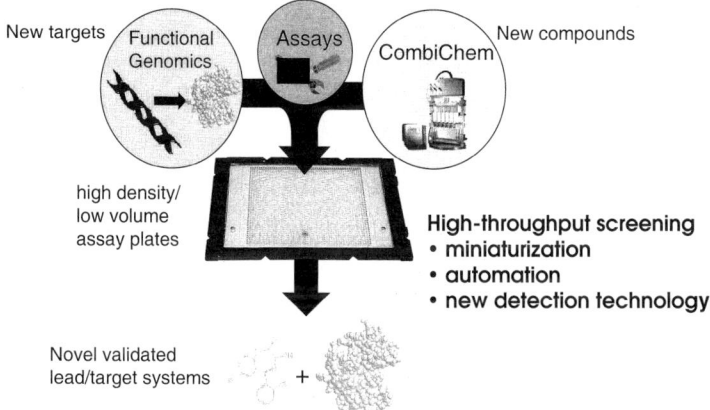

Figure 2 The integration of the three primary components of drug discovery screening: compounds, targets, and assays.

ceutical industry's needs: lower costs, faster turnaround, and reduced space requirements. Miniaturization reduces the costs of chemical and biological reagents in proportion to the reduction in volume [4,5]. Additionally, given the number of compounds to be screened, often of similar chemistry, novel assay techniques and detection technologies must be employed in order to provide better characterization of specific compounds, and this earlier in the screening process. A collection of synthetic compounds is likely to be used for several different assays—another reason to keep reagent consumption to a minimum.

Multiplexing strategies, whereby multiple parameters can be measured by employing multiple assays and/or readouts in a single sample, will also improve screening efficiency. This combination of quality and quantity must be achieved in order to take advantage of the increased potential for target identification offered by genomics and proteomics and the potential for direct, rapid access to novel chemical compounds offered by combinatorial chemistry and combinatorial biology [5].

Although miniaturization is an inherent characteristic of most modern HTS systems, the detection technology is often largely unchanged from many of the readout technologies used in conventional assays. To provide the most effective synergy between screening technology and assay development, screening systems must provide detection technology that is matched to the demands of miniaturization, on the one hand, and offers significant flexibility of readout methods, on the other. Fluorescence-based detection technologies are by far the choice of most modern HTS systems.

The assays themselves present a significant demand on HTS systems. Established biochemical assays must be readily adapted to miniaturized format. Additionally, assay throughput can be increased by multiplexing strategies: the rapid sequential or simultaneous detection of multiple parameters. This multiplies the amount of information which can be obtained from a single assay—and thereby the information that can contribute to the characterization of a single compound. This also means that the lines between primary and secondary screening will become increasingly blurred. A variety of readout modes must be available in any HTS system, not only to enable assay multiplexing and thereby the rapid and efficient characterization of compounds but also to allow for flexibility in assay design. This flexibility also contributes to the ease with which specific assays can be translated from standard manual formats to miniaturized automated formats. This translation of assay protocols is a challenge not to be underestimated: a fact to which many of those who are already involved in screening will most likely testify grudgingly. For this reason, those same scientists would also be glad to have a variety of readout modes with which to test different assay strategies.

Thus, a screening system must be capable of achieving throughput of at least 50,000 compounds/day, in volumes of less than a microliter, possess multiple readout modes, master multiplexed assays in which the results are tracked by a database capable of correlating synthesis information with assay results, and be cost effective.

C. Using Primary Cells

Another challenge to the screening machines of the future is the ability to work with living cells. Because of the difficulty of working with living cells, as well as the difficulty in automating functional assays (single-step, cell-based, homogeneous assays are preferred because of their speed and ease of use but are technically challenging to develop), cellular assays have been relegated to smaller screens much later in the discovery process. The need to increase the screening speed and the type of information derived means that assays using living cells are now even being used in primary screens. Cellular assays provide a functional approximation of the in vivo biological context and therefore offer biologically relevant information.

Most assay strategies for use in living cells rely on genetically manipulated, otherwise known as recombinant, cells. However, such overexpressing recombinant cell lines frequently have altered physiological properties compared to primary or nonrecombinant cells, particularly with regard to regulatory mechanisms of signal transduction. Thus they cannot be considered to be ideal for this purpose.

For cellular assays, miniaturized systems such as EVOscreen are capable of providing a superior solution: the combination of sensitivity and miniaturization enables primary cells to be employed. The restrictions resulting from the availability of primary cells are overcome by the very small number of cells needed for each assay (100–1000 cells/nanowell).

Apart from primary cells (e.g., stem cells) as therapeutic agents, the power of developmental biology for drug discovery has recently begun to be recognized. In addition to the use of model developmental organisms for functional studies, certain pluripotent cell populations and gene products that are present during embryogenesis are candidate therapeutic tools for use in the treatment of human diseases [6].

D. Blurring the Lines Between Primary, Secondary, and Tertiary Screening

Drug screening has traditionally been divided into three phases: primary, secondary, and tertiary. As a result of the dramatic increase in targets and compounds to be screened, traditional screening is currently undergoing a re-evaluation. In addition to the move to miniaturization and increases in throughput, increased efficiency is also called for. One way to increase efficiency is to obtain more information at an earlier stage of drug screening.

Primary screening is largely the establishment of a structure–activity relationship (SAR). At this stage of screening, several strategies are typically employed to manage the biological testing of large chemical compound sets against multiple biological targets. A single compound per bioassay per well is the most direct, and this is precisely why throughput must be increased. The advantages are clear: no deconvolution is required, and minimal potential for ''masking'' of bioactivity exists. With one compound per well, the primary bioassay provides extensive structure–activity relationship data (SAR), while negative bioassay data also provide additional information for subsequent lead optimization activities.

For example, using fluorescence correlation spectroscopy (FCS), the structure–activity relationship between the compound and the target as well as the concentration can be determined in a single measurement. This enables dose–response to be evaluated immediately during screening. Miniaturization and automation are required to reduce the cost of this approach compared to the alternative strategy of compound pooling. While the pooling of 3–10 compounds per bioassay has been utilized quickly and efficiently to assay large compound sets, the primary disadvantage is the need for subsequent deconvolution of positive readouts and the potential for the masking of one compound's activity by others. By maintaining the one-compound-per-well configuration, the maximum amount of information can be obtained per compound.

In secondary screening, multifunctional testing for parameters such as selectivity, toxicity, and dose–response is conducted. By combining miniaturization, automation and multiple detection modes, this process can be moved forward in the drug screening process. New readout technologies will enable several parameters to be measured simultaneously, meaning that assays, rather than compounds, will be pooled in a single sample. Just as conventional strategies lead to the pooling of compounds to increase throughput, new detection technologies enable far more information to be obtained in a single measurement, leading to an increase in information and a shortening of screening time. This will make lead evaluation feasible in early stages of the drug discovery process.

Novel types of assay can also be designed for tertiary screening: the profiling of compounds to obtain valuable biological information, such as bioavailability and cellular uptake. To predict whether a molecule has the potential for in vivo pharmacological activity, it is necessary to determine whether the molecule persists in the body long enough at concentrations sufficient to exert the intended pharmacological effect.

Such evaluations hinge on assessments of absorption, distribution, metabolism, excretion, and pharmacokinetics (ADME/PK). If these evaluations are made in the early stages of drug discovery, resources can be appropriately channeled to those molecules that are likely to succeed in terms of in vivo efficacy. Moreover, where a series of lead compounds in identified through in vitro pharmacological testing, AMDE/PK provides a means of rank-ordering the compounds. Such tertiary screens are typically carried out in test animals or in living cells. The new detection technologies described in this report are being used in functional cellular assays to explore signal transduction, transcriptional control, and even cellular viability. The availability of human liver tissues, cell culture systems (e.g., Caco-2) as a model of the intestinal mucosa, recombinant drug metabolizing enzymes, and cytochrome P450 isoform-selective substrates and inhibitors have provided the tools for effective conduct of such in vitro studies.

II. MULTIDIMENSIONAL READOUT TECHNOLOGY

Miniaturization requires changes in detection strategies, away from radioactive methods and towards methods based on fluorescence. Presently, radioactivity is used in about half the assays employed in HTS. Many if not most of these assay types have an analogous fluorescence approach, although it is inevitable that particular assays will prove to be difficult or impossible to adapt. Based on currently available data, it is expected that fluorescence methods will be feasible in > 90% of HTS assays [4].

A. Confocal Technology

The use of fluorescent dyes has effectively replaced radioactivity as the tagging method of choice for biological assays. The ability to make full use of the fluorescent molecule and all its properties should be the benchmark of any fluorescence-based assay technology. While most assay strategies make use of only a single fluorescent property, the broad applicability of a screening system requires that flexibility of assay design be extended to the readout technology by including a variety of detection modes. While most scientists may associate the use of fluorescence in biological assay systems solely with the measurement of fluorescence intensity, the measurement of additional fluorescent properties such as lifetime, polarization, and quenching can yield a wealth of information from a single measurement. This ability to collect multiple data points per measurement not only provides an internal control but also contributes to screening efficiency by enabling rapid multiparameter evaluation of compound–target interactions.

Confocal microscopic optical systems form the foundation of the readout technologies employed in the EVOscreen system. Using confocal optics, submicroliter miniaturization can be accomplished without any loss of signal quality (Fig. 3). The highly focused confocal optics reduce the detection volume to one femtoliter. This means that the detection volume, roughly the size of a bacterial cell, is much smaller than a eukaryotic cell or the standard miniaturized assay volumes of one microliter. Whereas other methodologies often suffer from the increased contribution of surface effects and from drastically reduced signal-to-noise ratios when assay volumes are reduced, the confocal optics employed for all EVOscreen readout technologies eliminates such concerns.

Furthermore, confocal detection technology enables the use of homogeneous assay methods, i.e., those that eliminate washing or so-called ''sandwich'' strategies. This is a great advantage in screening in that it allows biological systems to be evaluated in close to their natural in vivo state. Building on this confocal fluorescence detection platform, the EVOscreen® system described fulfills all the demands made of a HTS system in the combinatorial age.

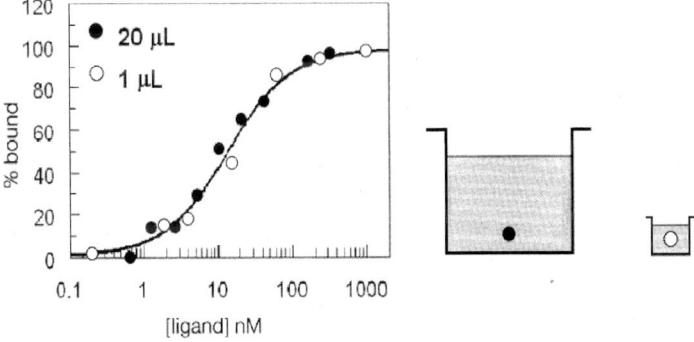

Figure 3 A comparison of binding curves derived from a DNA–peptide binding assay in 1 μL and 20 μL volumes.

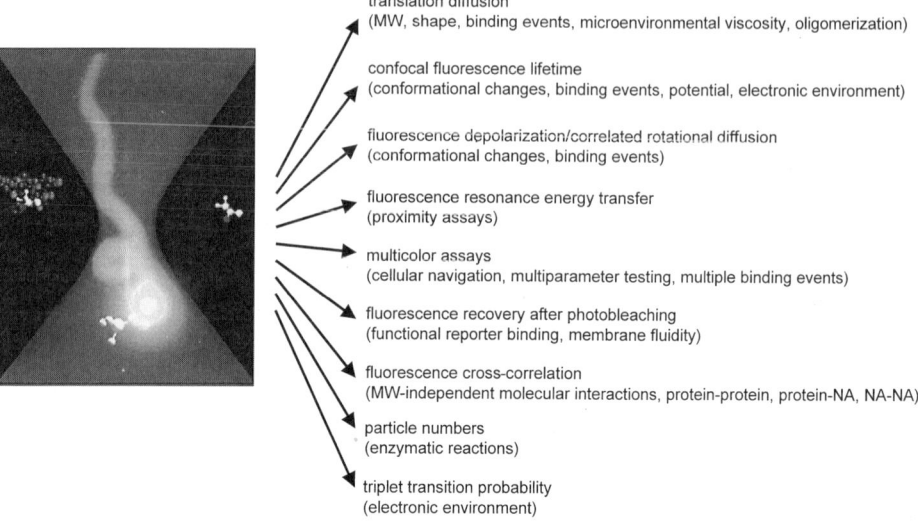

translation diffusion
(MW, shape, binding events, microenvironmental viscosity, oligomerization)

confocal fluorescence lifetime
(conformational changes, binding events, potential, electronic environment)

fluorescence depolarization/correlated rotational diffusion
(conformational changes, binding events)

fluorescence resonance energy transfer
(proximity assays)

multicolor assays
(cellular navigation, multiparameter testing, multiple binding events)

fluorescence recovery after photobleaching
(functional reporter binding, membrane fluidity)

fluorescence cross-correlation
(MW-independent molecular interactions, protein-protein, protein-NA, NA-NA)

particle numbers
(enzymatic reactions)

triplet transition probability
(electronic environment)

Figure 4   A survey of readout modes for confocal fluorescence detection technologies and their applications to biological systems.

Figure 4 lists a variety of different fluorescent parameters used as readout modes for biological assays and their areas of application. Though this list is by no means exhaustive, it gives an indication of the wide variety of information accessible through fluorescent measurement methods.

B.   FCS⁺ Detection Technologies

Fluorescence correlation spectroscopy (FCS) takes advantage of differences in the translational diffusion of large versus small molecules [7,8]. Each molecule that diffuses through the illuminated confocal focus gives rise to bursts of fluorescent light quanta during the entire course of its journey, and each individual burst is registered. The length of each photon burst corresponds to the time the molecule spends in the confocal focus. The photons emitted in each burst are recorded in a time-resolved manner by a highly sensitive single-photon detection device. This detection method achieves single-molecule sensitivity, but because diffusion is a random process, the diffusion events for a minimum ensemble of molecules must be averaged to achieve statistically reliable information. The detection of diffusion events enables a diffusion coefficient to be determined. This diffusion coefficient serves as a parameter to distinguish between different fluorescent species in solution, for example between free and bound ligand. In screening, the diffusion coefficient can be used to determine such factors as concentration or degree of binding. Confocal optics eliminate any interference from background signals and allow homogeneous assays to be performed. FCS measurements are conducted in a matter of seconds, which makes the technology ideally suited for HTS application.

While FCS is based on the diffusion properties of molecules, thereby detecting significant changes in molecular weight upon molecular interaction, new analytical methods have been developed at EVOTEC that enable interactions to be evaluated on the basis of a wide range of fluorimetric properties, independent of changes in molecular weight. These new methods, collectively dubbed FCSplus (FCS$^+$), evaluate fluorescence signals from single molecules in biological interaction on the basis of changes in fluorescence lifetime, brightness, depolarization, or spectral shift or on the basis of transferred fluorescence energy or quenching characteristics. All these parameters utilize the same optical and electronic configuration employed for FCS analysis; they merely employ a different algorithm for analysis. This means that all parameters noted above can be read using the same detection unit.

During the assay development and adaptation phase, prior to the scree process itself, the entire complement of FCS and FCS$^+$ methods stand ready to be employed in analysis of the biological system of interest. The variety of readout options available using a single detection device means that the nature of the biological interaction itself will determine the best readout method. This aspect is central to EVOTEC's strategy to simplify and improve the drug discovery process. By using a single detection device capable of evaluating multiple fluorimetric parameters, assay development is no longer a process of fitting an assay to a specific detection method but rather one of selecting the method best suited for the characterization of the biological interaction. The screening run can then be rapidly implemented utilizing EVOscreen's multiple readout options (Fig. 4). This rapid initial assay prototyping leads to a substantial reduction of assay development times.

C. Case Study of Multidimensional FCS$^+$ Readout

A novel assay based on a multiparameter fluorescent readout is described below. A theophylline/antitheophylline antibody interaction was chosen as a pharmacologically relevant assay system in order to demonstrate the full potential of EVOTEC's multidimensional fluorescence analysis based on its proprietary FCS$^+$ detection platform.

Theophylline therapy has been a cornerstone of asthma therapy through the years. Theophylline inhibits the breakdown enzyme phosphodiesterase with a resultant increase in cAMP concentrations, which results in smooth muscle cell relaxation of the bronchial tree. In view of the very small margin between therapeutic effects and toxicity, individualization of dose is mandatory with the-ophylline therapy. Therefore a strong demand for highly sensitive assays and detection technology exists in order to fine tune the theophylline level in serum [9–11].

The antibody used for this study was a polyclonal antitheophylline serum. The most interesting antigens were chosen to be those bearing TAMRA labels. Reference K_ds for antibody and ligand titration series were determined by FCS and compared with those obtained by the multidimensional fluorescence analysis (Fig. 5).

The mode of interaction is depicted in Figure 6. For a precise description of this antigen–antibody interaction, a model was chosen assuming two identical and independent binding sites. L represents the ligand and R the antibody with two binding sites. Three reaction diagrams show the reaction participants, which can be discriminated from one another by two-dimensional analysis, FCS, and one-dimensional brightness analysis.

Using FCS, one is able to discriminate between the bound and the non-bound fraction of the ligand based on the translational diffusion properties of the molecules but not between the state where either one or two ligands are bound by

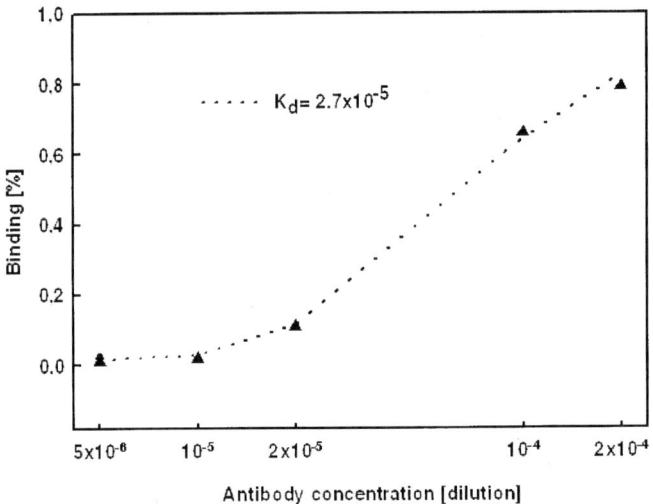

Figure 5 FCS measurement of theophylline and antitheophylline interaction. The pre-liminary FCS studies indicated well-defined binding characteristics of the ligand and reinforced the suitability of the system for a homogeneous assay. The data were fitted according to a single binding site model.

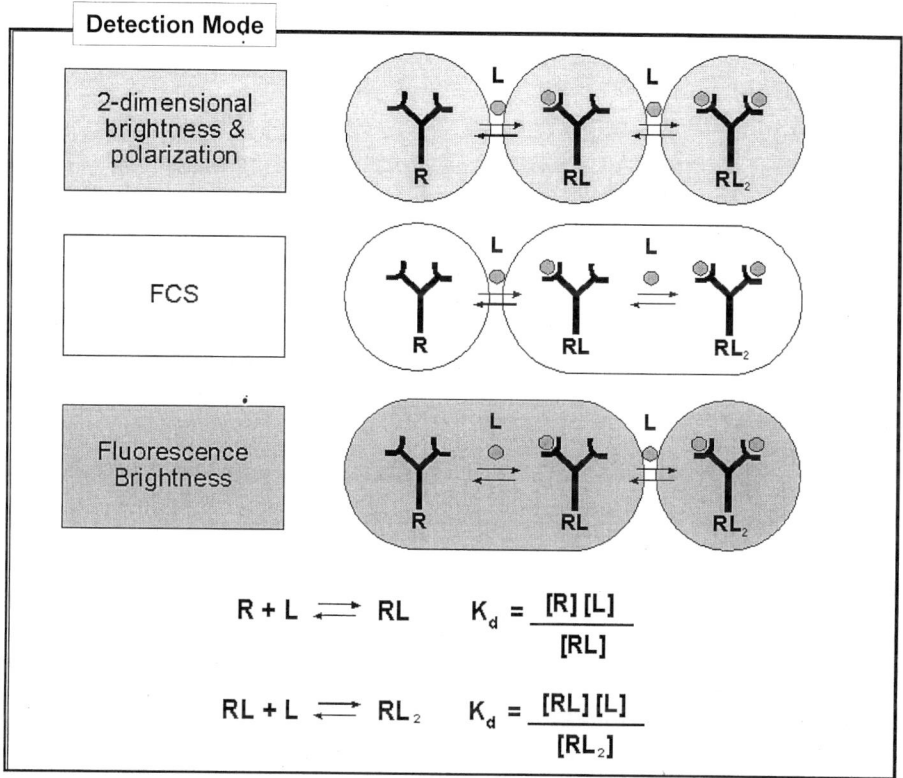

Figure 6 Differentiation of information provided by different fluorimetric readout parameters.

one antibody. A one-dimensional brightness analysis allows for a differentiation between antibodies carrying either one or two ligands but not between free ligand and a single ligand bound by an antibody.

Using EVOTEC's multidimensional fluorescence analysis, whereby the polarization ratio and the brightnesses of the particles are determined in parallel, each individual step of such a multiple binding reaction can be analyzed. The results of these studies are shown in Figure 7, where the individual components of a multicomponent system can be resolved in a single measurement. Here it can be seen that the concentration of free ligand decreases with increasing antibody concentration. The population of antibodies bearing two ligand molecules reaches a maximum at an antibody dilution of 2×10^{-5} and begins to decrease as antibody becomes limiting. At this point the number of antibodies bearing a single ligand

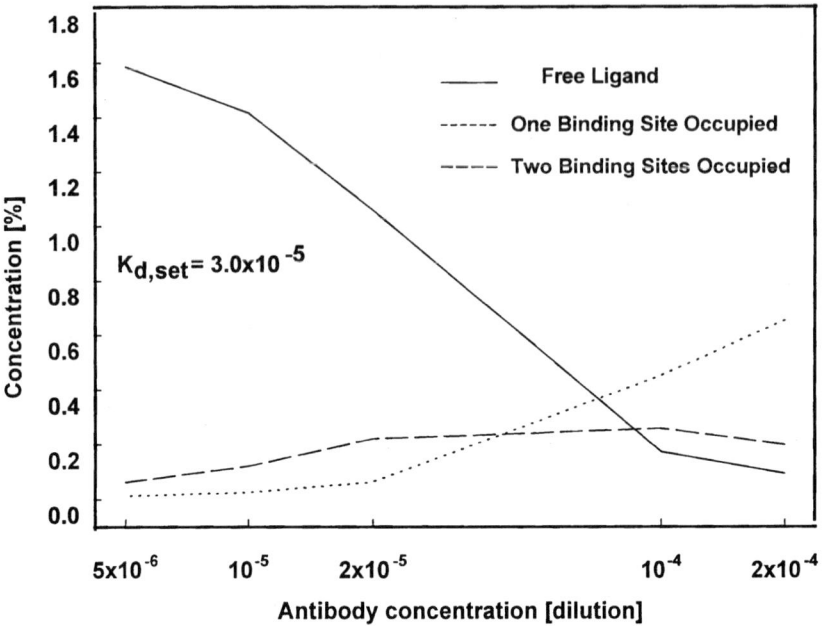

Figure 7 2D fluorescence analysis showing all three species in solution at differing concentrations of antibody. The model for this simulation assumes two identical independent binding sites. FCS analysis enables differentiation between free and bound ligand. One-dimensional fluorescence brightness analysis allows discrimination between single free ligand molecules and those antibody molecules with a single site occupied by a fluorescent ligand, and those antibody molecules with two sites each occupied by a fluorescent ligand. Two-dimensional analysis of fluorescence brightness and polarization allows discrimination between free ligand, those antibody molecules with a single site occupied by a fluorescent ligand, and those antibody molecules with two sites each occupied by a fluorescent ligand.

molecule begins to increase, and these dominate as the concentration of antibody increases. As is to be expected, when the antibody is limiting, both sites on all molecules are occupied, and as the ligand becomes limiting, antibodies with a single ligand bound become more prevalent. The unique feature of multidimensional analysis is that all three species, ligand and antibodies bearing one or two ligand molecules, can be resolved and their relative concentrations determined in a single measurement.

III. SCREENING TECHNOLOGY

A. EVOscreen™

The EVOscreen™ is a unique platform integrating highly sensitive detection technology and high-precision liquid handling systems in a modular system capable of 100,000 assays per day. A schematic layout of the EVOscreen™ system is shown in Figure 8a, and a view of the system in operation is shown in Figure 8b. The core elements of the modular system are (1) a high-performance confocal fluorescence detection unit usable in either single-channel or multichannel mode. Proprietary signal processing protocols are based on FCS and FCS$^+$ single-molecule-based confocal fluorescence methodologies. (2) A miniaturized, automated liquid handling system for nano- to low microliter volumes (including pipetting, dispensing, and compound retrieval). Four reagent dispensing modules are shown

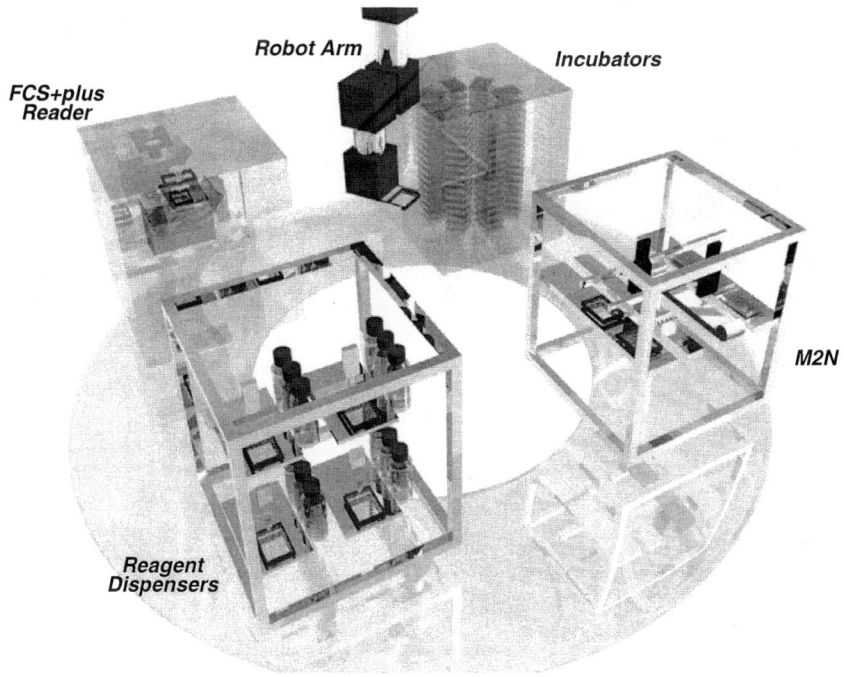

(a)

Figure 8 (a) Schematic diagram of the modules comprising EVOscreen® Mark 1. (b) Picture of EVOscreen® Mark 1 in operation.

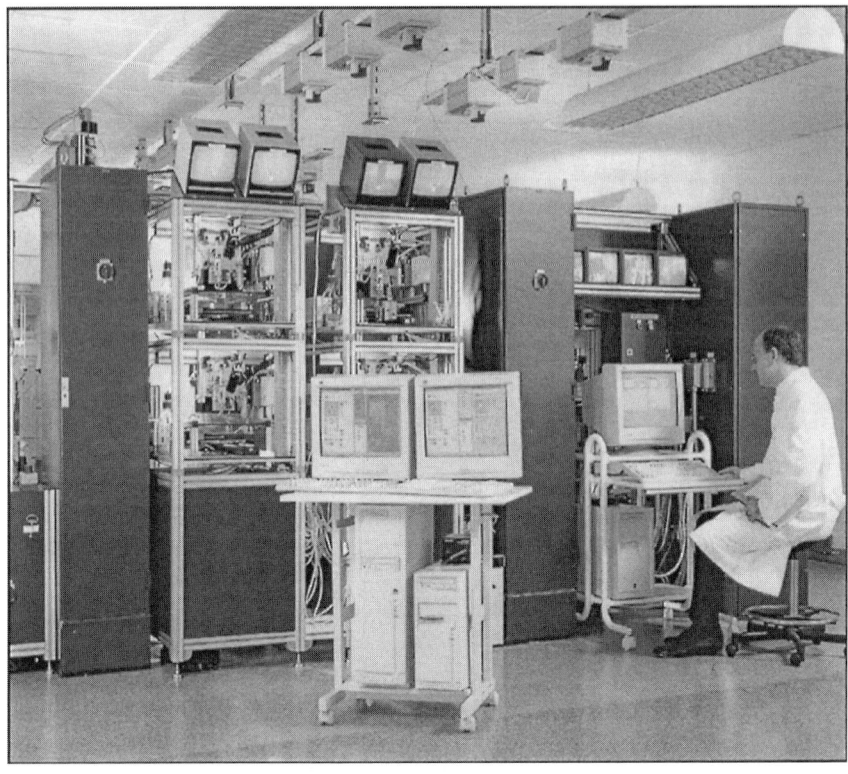

(b)

Figure 8 Continued

in Fig. 8b. (3) NanoStore, a rapid, miniaturized multireplica compound repository providing the link between traditional single-compound library formats and EVOscreen™. (4) A microseparation device (HPLC) coupled to the detection system. (5) A scanner and picker device for the analysis of combinatorial libraries and for functional genome analysis (described below).

1. Liquid Handling

The ability to miniaturize assays to the submicroliter level is largely dependent on the precision of liquid handling systems. Modern pipetting and dispensing systems enable components to be reproducibly added in volumes in the nanoliter range of thousands of samples per plate (Fig. 9). The EVOscreen™ system in-

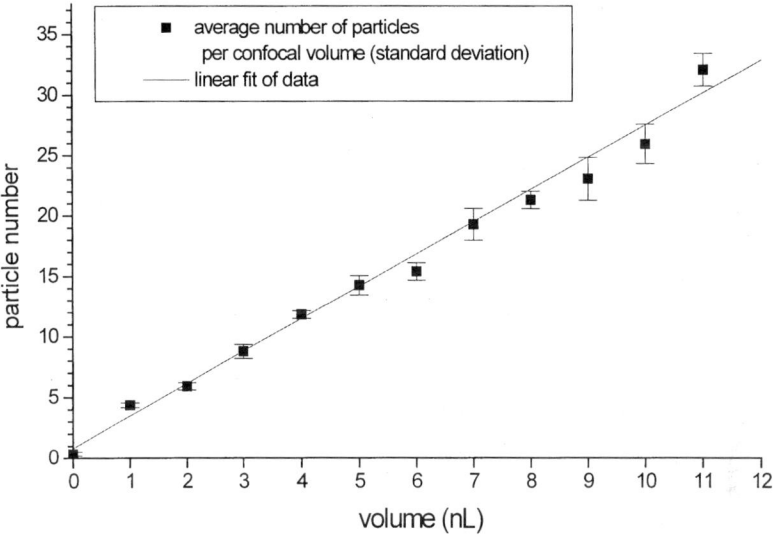

Figure 9 A plot demonstrating the linearity of performance of the liquid dispensing system in the low nanoliter range.

cludes a specially designed M2N unit which is able to reformat samples stored in any conventional format, such as 96-well plates, to the high-density 2000-well NanoCarriers (see Fig. 10) used for miniaturized assays. Also included is an automated storage and retrieval system providing a compound repository, as well as transport mechanisms for the many millions of compounds to be used in the screening runs. One specific feature of the EVOscreen™ liquid handling system is single-drop monitoring and on-line quality assurance.

2. Scanner/Picker

EVOTEC has developed a unique assay device specifically designed to address the need for on-bead screening of combinatorial libraries. The system is designed to enable tens of thousands of beads to be screened per day. Using the confocal optics and detection technologies described above, the Bead Scanner/Picker enables homogeneous assays to be conducted without the need to cleave the chemical compound from the bead surface. Additionally, the system allows individual beads to be recovered for further analysis. Figure 11 is the image of a scan of 12,000 TentaGel™ resin beads.

b

a

Figure 10   (a) A comparison of the reduction in space achieved by using miniaturized sample carriers. 2000 samples may be carried by either the 23 microtiter plates seen in this picture or a single 2000-well nanocarrier seen in the lower right. (b) The head of a nanodispenser over one of the wells of a 2000-well nanocarrier.

B. Case Study: Multiparameter Analysis of Screening Results

An important component of a successful HTS strategy in drug discovery is the ability to assess HTS structure–activity data and to distinguish between promising drug leads and the many useless false positives (where inactive compounds score a hit in the assay) that can plague screening efforts [12]. Pursuing false positives is a drain on the time and resources of drug development, requiring secondary assays for their subsequent elimination. EVOTEC's multiparameter analysis of screening results allows for an early and unequivocal identification of false positives: from a single measurement, three different parameters (e.g., % complex, count rate, particle number) can simultaneously be determined and be used for the evaluation of the screening data.

In order to demonstrate the multiparameter FCS$^+$ readouts of the EVO-screen platform, a simple model system based on streptavidin-biotin has been

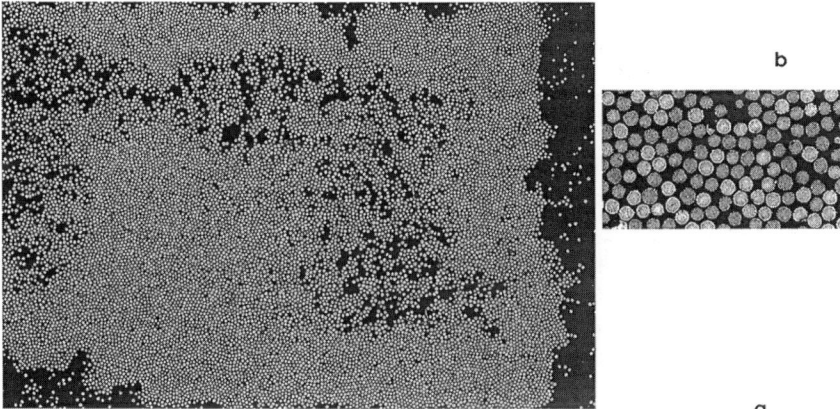

Figure 11 (a) A scan of more than 12,000 TentaGel™ beads using EVOTEC's Bead Scanner/Picker module in the presence of 1 μM fluorescent protein conjugate. (b) A close-up of a region of the scan, showing the bright fluorescent ring on the positive beads.

used. The assay was run in a miniaturized (1536-well) format in an HTS mode (2 sec read time per well). Biotin is an essential compound acting as a cofactor for several enzymes. One of its most striking features is its tight binding to avidin and streptavidin with dissociation constants of 10^{-15} M and 10^{-14} M, respectively. The high affinity of streptavidin for biotin has been used in many biotechnical applications exploiting the fast (approx. 10^7 M^{-1} s^{-1}) and almost irreversible (dissociation rate approx. 10^{-6} sec^{-1}) binding.

Figure 12 shows the readout of fluorescence count rate (a measure of the average fluorescence intensity in the confocal volume, denoted concentration 1) for all wells of a 1536-well NanoCarrier. In this control plate, the alternation of high and low controls is visible in the alternation of low columns (registering a value of 2 kHz) and high columns (registering a value of 5 kHz). One well is notable in that it registers a value of approximately 7 kHz.

A closer examination of the row in the NanoCarrier containing this very high sample is shown in Figure 13. Here the additional information is provided by the additional parameters of particle count (a measure of the concentration of fluorescent molecules present in the confocal volume) and percent complex formed (complex %, a measure of the number of large complexes formed upon the binding of streptavidin and biotin). As can be seen, the average fluorescence intensity (count rate) and the concentration of fluorescent molecules (particle number) increase in the well in question. However, the number of large com-

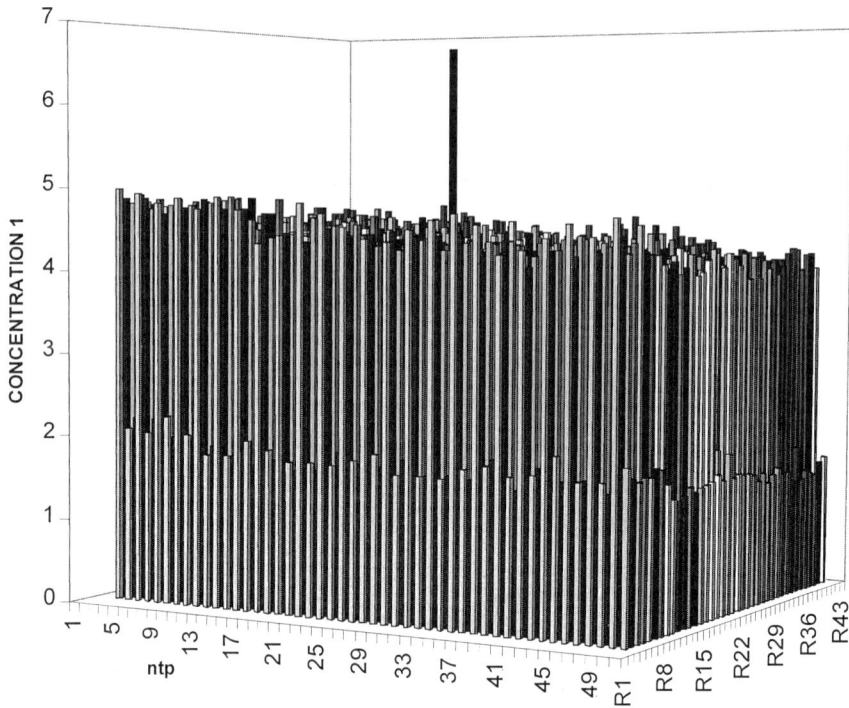

Figure 12   The FCS readout of fluorescence count rate for 1536 wells of a NanoCarrier. Alternating high- and low-control wells are filled with reagents for a streptavidin–biotin binding assay.

plexes formed upon binding of biotin and streptavidin decreases (complex %). This value provides the key additional information that allows the high value in this well to be discarded as a false positive. Namely, the high value can be attributed to the failure of a dispenser to deliver the appropriate amount of streptavidin to this well. The resulting lower volume in this well results in the increase in count rate and particle number (the final concentration of biotin is high due to the lower volume), while the lower amount of streptavidin results in a decrease in the amount of complex formed. This example shows how the simultaneous analysis of multiple parameters helps improve the efficiency and precision of screening runs by helping to eliminate false positive results rapidly and effectively.

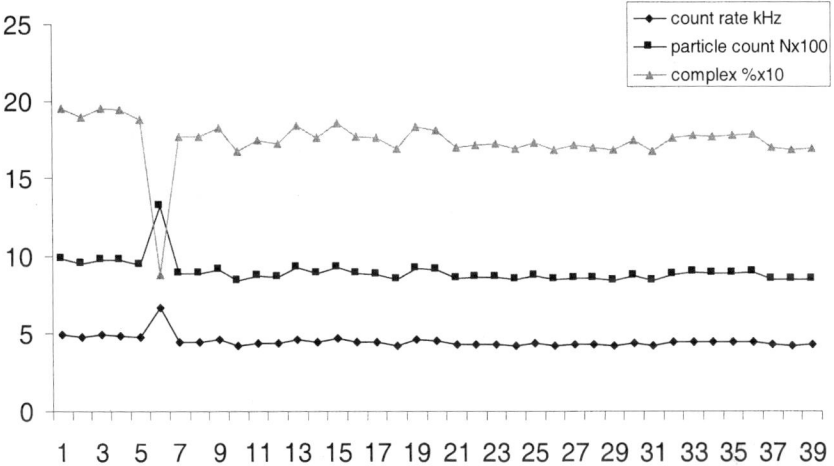

Figure 13 An extract of the data from a single column of the NanoCarrier described in Fig. 11. The data shows the results obtained from analysis of three different fluorimetric parameters demonstrating the cause for the abnormally high count rate found in well 6.

IV. CONCLUSIONS

The nature of screening for new drugs is undergoing a fundamental change. The exponential increase in the number of compounds available through the application of combinatorial strategies coupled with the increases in therapeutic targets made possible through genomics means that screening must reach a new level of efficiency. The need to shorten the drug discovery process and simultaneously allow millions of new compounds to be tested against hundreds of new targets has led to a complete re-evaluation of screening technology. Miniaturization and automation cannot be avoided if screening is to remain economical.

A maximum of information must be gleaned from each assay. This means that multiple readout modes must be used to assess a wide variety of parameters in a single sample. As in vitro assays lend themselves to be miniaturized and integrated in a HTS format, the new compounds can be directly tested for biological significance and data obtained to help design the next iteration of lead optimization.

EVOscreen meets the requirements of HTS: to achieve a degree of sensitivity and specificity through high spatial resolution in miniaturized formats, a multiplicity of readout modes all operating in high-throughput mode, with an integrated database system capable of correlating the data required to obtain useful information.

ACKNOWLEDGMENTS

The authors would like to thank the microfluidics, biophotonics, screening, and downscaling teams at EVOTEC for contributing to the data presented herein. Especially, they are indebted to Dr. Michael Busch and Dr. Sabine Schaertl for providing data from screening test runs. We would also like to thank Dr. Keith Moore and Dr. Andrew Pope of SmithKline Beecham for providing comparative data on FCS analysis in different volumes.

REFERENCES

1. AW Lloyd. Modifying the drug discovery/drug development paradigm. Drug Discov Today 2:397–398, 1997.
2. GL Verdine. The combinatorial chemistry of nature. Nature 384 (suppl):11–13, 1996.
3. D Hook. Ultra high-throughput screening—a journey into Nanoland with Gulliver and Alice. Drug Discov Today 1:267–268, 1996.
4. JJ Burbaum. Miniaturization technologies in HTS: how fast, how small, how soon? Drug Discov Today 3:313–322, 1998.
5. S Borman. Combinatorial chemistry. C&EN 76:47–67, 1998.
6. SJ Rhodes, RC Smith. Using the power of developmental biology for drug discovery. Drug Discov Today 3:361–369, 1998.
7. M Eigen, R Rigler. Sorting single molecules: application to diagnostics and evolutionary biotechnology. Proc Natl Acad Sci USA 91:5740–5747, 1994.
8. R Rigler. Fluorescence correlations, single molecule detection and large number screening. J Biotechnol 41:177–186, 1995.
9. SM Poncelet, JN Limet, JP Noel, MC Kayaert, L Galanti, D Collet-Cassart. Immunoassay of theophylline by latex particle counting. J Immunoassay 11:77–88, 1990.
10. JA Hinds, CF Pincombe, RK Kanowski, SA Day, JC Sanderson, P Duffy. Ligand displacement immunoassay: a novel enzyme immunoassay demonstrated for measuring theophylline in serum. Clin Chem 30:1174–1178, 1984.
11. W Mallin, E Eber, HJ Semmelrock, M Zach. Determination of theophylline serum level: comparison between a rapid test (enzyme-immunochromatography) and conventional fluorescence polarization immunoassay. Pneumologie 44:967–969, 1990.
12. GM Rishton. Reactive compounds and in vitro false positives in HTS. Drug Discov Today 2:382–384, 1997.

Index

Acqueyst, 79, 408
Actinomadura, 362
Actinomycetes, 154, 357, 359, 361, 362, 371, 373–377
Action potential, 317
Activating domain, 54, 142, 421
Activity base, 21
ADME (absorption, distribution, metabolism, and excretion), 61, 403, 493, 568
Affinity chromatography, 212
Aggregation, 250, 292, 294
Agonists, 100, 189, 208–210, 214, 239, 255, 256, 272, 417
Agricultural chemistry, 153
Alginate, 373, 374, 376
  Ca-alginate, 370, 376
  Na-alginate, 370, 376
α- factor, 169
Alternate splice variant, 416
Amantadine, 145
Amphetamine, 391
Amplification, 108
Analyst, 79, 408
Anisotropy, 77
Annotation, 434, 442, 443, 453

Antagonist, 106, 136, 189, 192, 208–210, 212, 214, 239, 249, 272, 417
Antibiotic response, 172
Antibiotics, 52, 171
Antibody-antigen complex, 114, 572
Antimicrobial agents, 11
Apoaequorin, 56, 241
Array scan, 98, 250
Aryl-hydrocarbon receptors, 157
Autophosphorylation, 242, 516
Assay details, 255, 518
  acceptor, 86–88, 91, 93, 94, 106, 108
  close proximity, 88, 91, 111, 113, 229
  development, 571
  donor, 86–88, 91, 93, 94, 106, 109
  efficacy, 426
  energy transfer, 93, 96
  heterogeneous, 39–41
  homogeneous, 39, 42, 51, 52, 61, 62, 70, 237
  optimization, 13, 31
  secondary, 233, 236
  signal-amplification, 266
  signal dynamic range, 550, 551
  signal-generation, 266

[Assay details]
  signal to background, 38
  signal to noise, 38, 54, 62, 91, 104,
    407
  validation, 13, 300
Assays (*see also* Fluorescence; Recep-
    tors)
  absorbance, 9, 39, 42, 44
  adsorption, 25, 110
  agar based, 174, 175
  agar diffusion assays, 174, 176
  amperometric systems, 148
  bead-based, 39, 48
  biochemical, 8, 70
  cell-based, 36, 39, 51–54, 313, 514,
    543- 546
  cell-based functional, 226
  cell-free, 35, 39
  cell-proliferation, 51, 52
  cellular, 565, 566
  centrifugation, 218, 219
  charcoal adsorption, 214, 216
  chemiluminescence, 9, 13, 40, 59,
    102, 105
  chromogenic, 39, 42, 43, 251
  colorimetric, 6, 9, 13, 19, 148, 303
  competition immunoassay, 82
  enzyme linked sorbent assay
    (ELISA), 11, 39, 42, 51, 70, 82,
    103, 216, 230, 231, 244, 245,
    277, 303, 512–514, 517, 518
  filtration, 39, 41, 51, 52, 55, 246
  gel-filtration, 110, 226
  homogeneous, 214, 236, 256, 303
  homogeneous cell-based, 112
  immunoassays, 40, 45, 244
  mass-dependent, 94
  mass-independent, 94
  microbial-based, 51, 177
  precipitation, 39, 41
  protein fragment complementation,
    247
  quench, 45
  quench relaxation, 45
  spectrophotometric, 144
Association rate constant, 205, 206

ATP-binding cassette, 132, 133
ATP-transporters, 133
Auto correlation techniques, 48
Auto fluorescence, 76
Automatically launched algorithms, 554
Automatically launched processes, 554
Automatically linked, 554
Automatically loaded, 554
Automation, 2, 13, 18, 31, 33, 34, 40,
    52, 57, 64, 253, 493–495, 498,
    512, 517, 520, 554
Auxilary proteins, 157
Auxotrophic markers, 160
Avidin, 83

Bacterial artificial chromosome (BAC),
    366
β-particle, 113
Biased combinatorial libraries, 363
Binding domain, 54
Binding equilibrium, 11, 205, 206
Binding sequence, 160
Binding sites, 226
Bioavailability, 423
Biochemical pathway, 417, 451
Bioinformatics, 441,424, 447, 456, 471
Biological database, 449
Bioluminescence, 105, 238, 239
Biosensors, 148
Biosynthetic pathway, 361, 362, 366,
    367, 368
Biotechnology, 449
Biotin, 41, 83, 84, 91, 104, 109, 216,
    249, 575, 79, 580
  biotin peptide, 41, 237, 244, 246
  biotinylation, 220
B_{max}, 230

Calcium (Ca^{2+}), 235, 238, 271, 272,
    314, 323
  bound, 273
  calcium sensitive dyes, 323
  channel, 143, 145, 198
  cytoplasmic, 272, 273
  free, 273
  intracellular, 226, 272

Caloric restriction, 418
cAMP-responsive elements (CRE), 57,
    233
Camptothecin, 139
Capillary electrophoresis, 320, 321
Captopril, 1
Carbonic anhydrase like-domain, 197
Catabolite gene activation protein, 147
Catalytic domain, 195
Cells
    adherent, 211, 216
    cycle, 165
    permeability, 2, 132
    recombinant, 204, 213
    suspension, 211
    microbial cells
        Aequorea victoria, 60
        Asperigillus fumigatus, 173
        Bacillus subtilis, 378
        Bordetella pertussis, 147
        Candida albicans, 173, 175
        Drosophila melanogaster, 156,
            158, 419, 433, 442
        Escherichia coli, 53, 59, 97,133,
            135,140, 141,146, 148, 175,
            214, 247, 249, 366, 373, 375,
            376, 378, 420, 422
        Helicobacter pylori, 439,444
        Mycobacterium tuberculosis, 421
        Saccharomyces cervisiae, 133–135,
            138, 142, 155, 161, 174, 175,
            419, 421, 433
        Sarcina aurantica, 378
        Schizosaccharomyces pombe, 142
        Staphylococcus aureus, 378
        Streptomyces, 361, 362, 366
        Streptomyces avellaneus, 372
        Streptomyces lividans, 36, 368,
            369, 372
        Vibro cholerae, 145
    mammalian cells, 52, 58, 70, 132,
        210, 256
    CaCO-2, 60, 518
    CHO-K1, 57, 109, 232, 236, 241
    CV1, 269
    D98/raji, 269
[Cells]
    fibroblasts, 385
    HEK293, 241, 250, 269
    hepatocytes, 383, 386, 394
    hepatocytes primary, 390, 391, 394
    HEPG2, 394
    human hepatocytes, 61, 385, 389,
        395–399
    SKNMC, 218
    SMS-KAN, 218
    T-cells, 425
    Caeranobdibtis elegans, 419, 433,
        443,
    Helicoverpa zea, 377
    Plutella xylostella, 377
    Xenopus laevis, 233, 234
    Yeast, 52, 155, 157, 168
Cell wall biosynthesis, 173
Cephalosporin, 1, 133, 139, 140, 142,
    148
Channel blockers, 137
Chaperones, 202
CHAPS, 211
Charge coupled device (CCD) camera,
    99, 117, 119, 120, 274, 533–535,
    544, 547, 549, 555, 556
Chemical genomics, 423
Chemical library, 107, 121, 122, 217,
    303, 423
Chemiluminescence imaging plate
    reader (CLIPR), 56, 120
Chemokines, 271, 272
Chemotypes, 70
Chimeric gene, 57, 241
CHIP technology, 35
Chlorzoxazone, 397
Cirrhosis, 394
cis-acting elements, 276
cis-trans assay, 159
Clathrin coated pits, 250
Clinical microbiology, 148
Clones, 367–370, 372, 379, 380
Cloning, 268, 452
Coactivator, 159, 203
Coactivator binding, 254
Coexpression, 161, 213, 267

Coelenterazide, 56, 241
Coelenterazine, 56, 241
Collagen-gel sandwich, 386
Combinatorial biology, 360–363, 367–371, 379
Combinatorial chemistry, 2, 26, 31, 69, 148, 483, 490
Combinatorial library, 163, 168, 279, 423, 577
Competition binding, 207
Complementation, 135
Complement function, 134
Compound deck, 31
Computational methods, 453
Computational tool, 453
Configuration, 285, 286, 296
Confocal fluorescence microscopy, 71
Confocal focus, 570
Confocal technology, 568
Confocal volume, 94
Corepressors, 159, 201
Coumarin, 397
Counter-selectable marker, 161
Cryopreservation, 384, 387–399
Crystallization, 424
Cyclic AMP (cAMP), 51, 55–57, 60, 62, 108, 109, 111, 147, 167, 168, 229–234, 274
Cyclic nucleotides, 314
Cyclosporin, 135, 169, 421
Cytochrome P450 (CYP), 61, 385, 387, 388, 403–413, 423, 434
 CYP 1A2, 61, 385, 387, 388
 CYP 2C19, 61, 385, 387, 388, 399
 CYP 2C9, 61, 385, 387, 388
 CYP 2D6, 61, 385, 387, 388, 399
 CYP 2E1, 61, 385, 387, 388
 CYP 3A4, 61, 385, 387, 388
 CYP 2A6, 61, 385, 387, 388
CYP induction, 389, 393
CYP inhibition, 389, 393
Cytofluor 4000, 72, 407
Cytokines, 50, 197, 251, 279
Cyto loop, 194

Cytomegalovirus (CMV) promoter, 9, 268
Cytoplasmic domains, 213
Cytoplasmic receptors, 157, 158
Cytosensor, 241
Cytosol, 285
Cytotoxic, 60, 129, 384, 398

Data analysis, 464
 profile-based, 464
 sequence-based, 464
 structure-based, 464
Data base, 449
 primary, 450
 protein information resource, 451
 Swissport, 451
 knowledge, 451
 EcoCyc, 451
 PDB, 451
 SCOP, 451
 others
 BLAST, 456, 457, 459
 BLOSUM62, 459
 DUST, 459
 EGAD, 452
 Enterz, 455
 FASTA, 456, 457, 459
 Hidden Markov Model (HMM), 455, 461, 462, 464
 Online Mendelian inheritance in man (OMIM), 470
 SEG, 459
 SMART, 465
 UniGene, 452
 XREFD, 452
 profile-based, 460
 Pfam, 461, 466
 secondary, 451
 BLOCKS, 451, 465
 OWL, 451
 PROSITE, 465, 466
 PRINTS, 451, 465, 466
Depolarizing-gated channel, 323
Depolarization, 314
Desensitization, 166, 194, 250, 314, 315

Detection technology, 565, 568, 569, 577
Detection unit, 575
Dextromethorphan, 397
2D-gel electrophoresis, 422
Diacyl glycerol (DAG), 235–237
Differential gene expression, 418
Diffusion coefficient, 95, 570
Digitalis, 153
Digitonin, 211
Dimerization, 145, 201, 247, 248, 251, 255
Dimethylnitrosamine, 394
Dissociation buffer, 211
Dissociation enhanced lanthanide fluorescence immuno assay (DEL-FIA), 17, 82, 83, 216
Dissociation rate constant (Kd), 205, 206–210, 215, 220, 247, 572
DNA binding, 254
DNA-binding domain (DBD), 142, 158, 201, 421, 426
DNA-binding protein, 193
DNA chips, 61, 71
DNA-cleavage fragments, 140
DNA-damaging agents, 140
DNA loop, 147
DNA-response elements, 202
Dominant phenotype, 141
Dopamine, 135
Drug discovery, 3, 60, 105, 213, 256, 257, 270, 447, 449, 478–480, 487, 513, 519, 563, 564
Drug-DNA fragments, 139
Drug-drug interaction, 383, 391, 392, 397, 398, 403, 404, 406, 408, 412
Druggable targets, 6
Drug intervention, 267
Drug metabolism, 390, 392, 404
Drug screening, 31, 122, 212, 478, 484, 516
Drug targets, 415–417
Dyes
  alamar blue, 61
  1-anilino naphthalene 8-sulfonic acid (ANS), 221–223

[Dyes]
  calcium green, 238, 241
  calcium-green-2, 273, 274
  Cy5, 17, 48, 91, 100, 101, 222, 223, 541
  Fluo-3, 238, 239, 273, 274
  Fluo-4, 238, 240
  Fura-2, 238
  oligreen, 44, 74
  picogreen, 75
  ribogreen, 44, 74

EC_{50}, 208, 230, 233
Efflux system, 132
Electrical potential, 314
Electrochemiluminescence (ECL), 50, 83, 102, 104, 123, 224, 231, 245
Electro-osmosis, 122
Electrophysiology, 144, 313, 318, 320, 322, 326
Encapsulate, 370, 371, 374, 376, 378
  coencapsulation, 373, 376–378
  encapsulation, 373, 375
Endogeneous ligands, 193
Endothelin, 193
Enterohepatic recirculation, 384
Enzyme details
  action, 83, 85
  activation, 4, 11, 38
  allosteric modulators, 315, 316, 324
  allosteric sites, 226, 321
  Beer-Lambret, 293, 295, 297
  cofactor, 159, 162, 286, 389
  competitive, 288
  coupled enzyme assay, 302,
  coupled system, 302,
  coupling reaction, 302,
  competitive inhibition, 411
  curvature, 292
  stability, 295
  substrate interaction, 104
Enzymes
  adenyl cyclase (AC), 55, 57, 60, 147, 163, 169, 228, 229, 239, 234

[Enzymes]
angiotensin converting enzyme, 140, 416
ATP-citrate lyase, 44
biotin ligase, 220
cAMP-dependent protein kinase (PKA), 230, 235
chitin synthase, 173
dihyrofolate reductase, 267
DNA gyrase, 420
DNA helicase, 9, 417
glucan synthase, 173,
glucose-6-phosphate dehydrogenase, 408
β-galactosidase, 420
β-glucuronidase, 44, 73–75
GTPase, 166, 227
HCV protease, 303
helicase, 80, 107
HIV protease, 303
kinases, 9, 110, 422–424
lanosterol 14 α methylase, 173
MAP kinase, 163, 284
methylase, 426
nuclease, 72, 74, 80
oxygenase, 404,
phosphatidyl inositol kinase 3 (PI3 kinase), 242
phosphodiesterase, 228
phospholipase, 228
phospholipase A2, 226, 516
phospholipase C, 169, 226
phospholipase C (plc), 163,235–237
phospholipase D, 226
polymerase, 72, 74, 286
polypeptide synthase, 421
proteases, 9, 47, 80, 85, 87, 107, 140, 225, 246, 417, 422, 424, 513
protein kinase C (PKC), 83, 226, 234–238
protein kinase JAK, 197
protein kinase STAT, 251
protein phosphatase 2A, 135
protein phosphatases, 80, 82, 142
protein tyrosine kinases, 2, 40, 80, 81, 107, 142, 249, 275, 333

[Enzymes]
restriction enzymes, 426
RNA polymerase, 72, 146, 422
sphatases, 9, 10, 163
TAQ-polymerase, 519
threonine-serine kinases, 101, 107
thymidilate synthase, 423
tyrosine kinase, 62, 101, 517, 517
Epidermal growth factor, 193
Episomes, 267, 269, 270
Equilibrium potential, 314, 316, 322
Ergosterol pathway, 134
EST database, 451
Estrogen, 86, 221
EvoScreen, 94, 575,577, 581
Exo loop, 194, 225
Exon, 463, 464
Expressed sequence tags (EST), 283, 423, 451, 456, 460, 464
Expression cassette, 268
Expression libraries, 168
Expression profile, 449
Expression vectors, 160
Extracellular loops, 229

False positives, 578, 580, 585
Fermentation, 154
Fgenes, 454
Fgeneshm, 454
FGF, 195
Fibrinectin like-domain, 197
Fibrosis, 3, 94
5-fluoro orotidic acid (FOA), 142
Flash plate, 11, 43, 55, 59, 71, 110–113, 322
Flow cell, 104
Flow cytometer, 50
Fluomicrospheres, 113
Fluorescein labeled ligand, 204
Fluorescence, 6, 9, 16, 19, 25, 26, 36, 39, 40, 42–44, 45, 47, 50, 54, 59, 61, 63, 70, 71, 75, 87, 89, 96, 97, 122, 217, 319, 323–325, 518, 568, 569, 571–573
dyes, 72, 273, 274
emission, 273

[Fluorescence]
excitation, 273
fluorogenic, 71, 73
imaging, 71, 98,123
intensity, 44
life times, 44, 48, 121
quench, 71, 72, 536
quench relaxation, 71
Fluorescence correlation spectroscopy
(FCS), 17, 18, 44, 48, 94–96,
567, 571, 572, 574, 575
Fluorescence linked immunosorbent
assays (FLISA), 49, 101, 231
Fluorescence microvolume assay technology FMAT), 49, 100, 101,
222
Fluorescence polarization (FP), 17, 44,
45, 47, 56, 71, 76–86, 95, 117,
122, 217, 204, 221, 226, 231,
238, 244, 245, 249, 251, 256,
303, 535, 540–543
Fluorescence resonance energy transfer
(FRET), 13, 45, 47, 56, 57, 71,
86–88, 96, 97, 108, 117, 122,
217, 223, 303, 319, 324, 535,
536–540, 545
Fluorescent DNA probe, 425
Fluorescent probe, 253
Fluorogenic assay, 71, 221, 303
Fluorogenic substrates, 423
Fluorometric assay, 405, 409
Fluorophores, 71, 76–78, 100, 102, 113,
121
allophycocyanine (APC), 47, 88, 91
BODIPY, 72, 78, 85, 541
europium, 88, 91–93, 117, 223
fluorescein, 72, 77, 78, 85, 88
lanthanide, 47, 88, 93
rhodamine, 72, 75, 88
rhodamine green, 78, 95
rhodamine red, 78
terbium, 88, 89, 91–93
tetramethylrhodamine, 95, 572
Texas red, 72
Fluoroscent substrate, 410, 411, 413
Forskolin, 233

Freez-thaw, 296
Functional assignment, 463
Functional homology, 134
Functional proteomics, 147
Fungicidal, 153
Furaylfylline, 394, 397
Fusion tags, 212
glutathione s-transferase (GST), 212,
214, 217, 226
histidine (His), 212, 214, 217, 227
maltose binding protein (MBP), 212,
214

Galanin, 216
Gal promoter, 140
Gal4 transcriptional activator, 140
Genecentric drug discovery, 449
Gene discovery, 455
Gene disruption, 442
Gene expression, 105, 253, 449
Gene finding, 453
Gene function, 416
Gene identification, 453
Gene prediction, 454
Gene shuffling, 360
Gene structure, 454
Genetic epidemiology, 468
Genewise program, 463
Genome, 69, 257, 450
Genome sequencing, 453
Genomic data, 464, 477
Genomics, 35, 62, 69, 419, 433, 447,
564
Genomic sequences, 415
Genotyping, 426
Ghrelin, 200
Glutathione, 16, 113, 117
Glycoprotein, 225
Glycosylation, 132, 194
G-protein coupled receptor kinase
(GRK), 194
G-proteins, 53, 55, 56, 165, 166, 193,
194, 221, 250, 314, 325
Granulocyte colony stimulating factor,
170
Graphical user interface (GIU), 512

Green fluorescent protein (GFP), 57, 58, 60, 1, 96, 98, 170, 250, 276, 535, 537, 538, 545
Gripper, 487, 488, 498, 501
Growth factors, 279
Growth promoter, 419
GTP-binding, 227, 229
Gutamate, 316

Halothane, 394
Heat shock proteins, 202
Hepatotoxicity, 383
Heterodimer, 145, 146, 158, 202
Heterodimerization, 91, 139, 142. 248
Heterodimerizing protein, 422
Heterogeneous binding, 216
Heterogeneously expressed, 132
Heterogeneity, 212, 423
Heterologous expression, 164, 166, 213
Heteromeric channels, 315, 318
Hetero oligomer, 198
Hetero trimeric G-proteins, 169
Hexose transporters, 133
High affinity, 266
High affinity ligand, 204
High content screening (HCS), 98
High density plates, 33, 34, 39, 74
High flux synchtron radiation, 424
High impact database, 451
High-resolution receptor structure, 214
High-throughput genome sequence, 450
High-throughput microbial assays, 178
High-throughput screening (HTS), 2, 3, 5, 6, 9, 11, 13–19, 21–25, 27, 32–38, 40, 44, 47, 48, 54, 60–62, 69, 70, 86, 94, 97, 110, 115, 117, 123, 147, 148, 162, 215, 221, 226, 229, 233, 239, 244, 250, 253, 256, 265, 266, 269, 273, 275, 313, 315, 317, 320–326, 359, 360, 363, 395–399, 412,489. 493, 497, 518, 521, 522, 525, 526, 532, 563, 564, 581
Hill coefficient, 207, 297
Histogram, 555, 556
Hits, 129

Homodimer, 139, 142, 146, 158, 422
Homodimeric protein, 422
Homogeneous receptor binding, 217
Homogeneous technology, 78
Homogeneous time-resolved fluorescence (HTRF), 11, 44, 47, 88, 89, 91–93, 223, 244, 245, 251, 303, 513, 517, 539, 540
Homo oligomer, 198
Homologous, 135, 441
Homology, 147, 422, 460
Homology screening, 200
Homomeric channels, 318
Hormone responsive element (HRE), 159
Hormones, 313
HPLC, 404, 420, 483, 516
HTS 7000, 72
HTS-cell-based, 266
Human genome, 164, 453
Human genome project, 453
Human liver microsomes, 405, 406, 410
Hybridization, 426, 436
Hydrophilic, 403, 404
Hydrophobic pocket, 194, 221
Hydrophobic segments, 200
Hydrophobic transmembrane, 198
Hygromycin B, 139
Hyperpolarized, 155
Hypertension, 140, 416
Hypothalamus, 201
Hyperpolarization, 325

IC_{50}, 20, 123, 207, 208, 218, 242, 300, 404, 408
IMAGE consortium, 451
Imaging systems, 63, 273
Imaging technology, 119, 123
Immobilized ligand, 216
Immunoassay, 107, 122
Immunocomplex, 80, 244
Immunofluorescence, 99, 442
Immunoglobulin like-domain, 197
Immunosuppressants, 135
Inactivation, 300

Inhibitors, 25, 82, 172, 173, 246, 278, 304, 407
Initial burst, 289
Initial rate, 288, 300
Initial velocity, 288–291
Inositol triphosphate (IP3), 226, 235
Insect growth regulator, 158
Instrumentation, 78
Insulin receptor substrate-1 (IRS-1), 242
In situ hybridization, 99
Integrated data base system, 451
Interhepatocyte concentration, 389
Internalization, 250
International human genome, 451
Intracellular loops, 227, 229
Intracellular pH, 273
Intron/exon, 454
In vitro methodology, 407
Inward ion flux, 316
Inwardly rectifying potassium channel, 137
Ion channels, 189, 228, 267, 279,313–323, 325, 326
Ion flux, 279, 314, 316, 317, 324
Ionic strength, 287, 295
Ionotropic, 315
Isoproterenol, 250, 275
Isotopic detection, 16

Jacques Monod, 13
JNK pathway, 226

Ketoconazole, 394, 397, 398
K_i, 404,
Kinetically predominant form, 287
K_m, 287, 288, 296–299, 301, 393
Knockout mutation, 416

Laboratory automation, 178, 477, 491
Laboratory information management system (LIMS), 514, 519
Laboratory unit operation, 479
LacZ tet proteins, 422
Lag period, 289
LC/MS, 396, 518
Lead discovery, 6

Lead identification, 280
Lead optimization, 130
Leadseeker technology, 71, 117, 118, 122
Leukocytes, 271
Lidocaine, 393
Ligand binding, 54, 91, 94, 110, 168, 190, 194, 202, 213
Ligand binding domain (LBD), 194, 195, 214, 215
Ligand binding site, 168
Ligand receptor binding, 190, 203, 205
Ligand-gated-ion channel, 97,193, 198
Ligands, 38, 157, 189
Linear interpolation, 408
Lipocytes, 394
Lipophilic, 403
Lipopolysaccharide, 134
Liposome, 236
Liquid crystal, 121, 557
Liquid dispensing, 36
Liquid handlers, 487, 488, 495, 496, 407
Liver microsomes, 391
Low throughput, 51
Luciferase, 53, 57, 58, 99, 102, 105, 106, 255, 275, 276, 544, 546
Luminescence, 6, 19, 36, 42, 54, 79, 239
Luminescent semiconductor quantum dots, 121
Luminometer, 34, 56, 241

Macrodroplets, 363, 370–372, 375–379
Macrolides, 1
Magnetic bead, 50, 102, 104
Mammalian hormones, 171
Mammalian proteins, 134
Mammalian steroid receptor, 159, 162
MAPK pathway, 226
Mass spectrometry, 28, 420, 422
Mating factor, 135
Mating phermone, 164
Matrigel, 386
Mechanism-based, 166, 177, 178
Mechanism based inhibitors, 411, 412

Medicinal chemistry, 265
Medium-throughput, 33,51, 52, 215, 236
Melanocorticoid receptor, 158
α-melanocyte stimulating (a-MSH), 233
Melanophores, 57, 233, 234
Melanosomes, 233
Melatonin, 233
Membrane filtration, 11
Membrane potential, 155, 253, 273,
        314, 316, 318, 319, 323–325, 424
Membranes, 94, 210, 284–286, 301
Metabolic stability, 383, 396, 398
Metabolism, 403, 441
Metabolite, 408
Mevacor, 1
Michaelis-Menton kinetics, 296, 297
Micro-beta, 41, 57, 110, 112, 114, 215
Microbial genomes, 447
Microbial systems, 135, 148
Microchannel, 23
Microchips, 27, 62, 122, 148, 424
Microfluidics, 21, 122
Micromonospora, 362
Microphysiometer, 241, 321
Microplate reader, 482
Microscopic channels, 122
Microsolenoid valve, 491
Microsome, 61
Microtiter plate, 2, 26, 27,34, 35, 59,
        99, 118, 229, 405, 412, 481–484,
        490, 508, 519, 520, 1536
    1536 well plate, 26, 33, 35, 74, 86,
        94, 100, 120, 222, 482, 489,
        508, 579
    3456 well plate, 33, 35
    384 well plate, 2, 6, 20, 26, 33, 34,
        59, 61, 69, 72, 74, 75 79, 81,
        94, 99–101, 115–118, 120,
        230, 481–484, 490, 508,
        514, 517
    96 well plate, 2, 6, 27, 33, 34, 41,
        58, 61, 69, 72, 75, 79, 81,
        86, 94, 99–101, 115–117,
        120, 230, 236, 246, 481–
        483, 490, 508, 517
    9600 well plate, 2, 26, 33, 35

Micro volume fluorometry (MVF), 49
Miniaturization, 2, 16, 31, 33–35, 40,
        54, 61, 62, 69, 74, 94, 177, 253,
        257, 279, 313, 326, 491, 517,
        525–532, 543, 544, 546–549,
        553–556, 564, 565, 568, 575,
        579, 581
Mitochondrial respiration, 120
Molecular surface analysis, 424
Molecular targets, 189
Monoclonal antibody, 223
Monomers, 159
Motif, 2, 422, 437, 438, 451
Multichannel pipettes, 407, 480
Multidomain protein, 464
Multidrug resistance, 132
Multiplexing, 49, 101, 565, 566
Mutagenesis, 418
Myoinositol, 236

NADP/NADPH, 408
Na/K pump, 321
Naladixic acid, 376
Nanocarriers, 577, 579
Nanocrystals, 121, 127, 556, 557
Nanodroplets, 148
Nanaodrop technology, 148
Nanoparticles, 121
Nanostore, 576
Nanoworld, 563
Natural products, 1, 171, 177, 357,
        361
Negative cooperativity, 207
Neuropeptide Y, 193
Neurotransmitter, 480
Neurotransmitter transport, 322
New drug entity, 403, 408
NFAT-transcription factor, 56, 97, 241,
        275
NMR, 422, 424, 426
Non-competitive, 296
Nonidet P-40, 211
Nonpolyadenylated, 421
Nonradioactive, 39, 43, 51
Nonspecific binding, 117
Novel gene discovery, 452

Novel targets, 447
Nuclear hormone receptors, 60, 86, 157, 201, 257

Oligomeric, 193
Oligonucleo tides, 62, 91, 92, 271
Open reading frame (ORF), 439
Orphan gene, 447
Orthologs, 419
Over expression, 11, 161, 210, 211, 419, 420, 566
Oxidation-reduction reaction, 102

p21 protein, 419
Patch clamp, 320
Pathobiology, 11
Pattern database, 451
PCR, 519
Pencillin, 1, 357, 374
Peptide ligands, 136
Periplasmic space, 134
pH, 205, 287, 290, 295, 297–299
Pharmacokinetics, 412
Pharmacological agents, 154
Phase I metabolism, 384
Phase II metabolism, 384
Phermone, 165, 166
Phermone response, 164, 165
Phosphatidyl inositol (PI) turnover, 236
Phosphoproteins, 272
Phosphorylation, 41, 162, 237, 247, 315
Phosphorylation/dephosphorylation, 194
Phosphoserine, 83, 425
Phosphoserine antibody, 83, 236
Phosphothreonine, 83
Phosphotyrosine, 138, 249, 425
Phosphotyrosine antibody (PY antibody), 80, 81, 88, 104, 115
Phosphotyrosine binding domain (PTB), 242
Photocleavable resin, 374, 378
Photolithogaphy, 26, 27
Photon counting, 25
Photosensitizer, 106
Phylogenetics, 466, 467
Piezo electric dispensing, 491

Piezo electric pulse ink jet, 62
Pigment aggregation, 57, 234, 235
Pigment dispersion, 57, 233–235
Pipettor, 100, 477, 483, 489, 490, 495, 516, 519
Plasmid, 160
Plasmid library, 168
Plate harvester, 517
Plate readers, 34, 72
Plate stackers, 506
Platinum electrode, 102
PMT, 59, 79, 1000
Point mutations, 213
Polarization, 76–78, 88, 91
Polarized light, 76
PolarStar, 79
Poly A, 267
Polybrominated biphenyl, 393
Polycistronic message, 420
Polyethylene glycol, 214, 215
Polyethylimine, 220
Polymorphism, 449, 470
Polysomes, 418
Polytron homogenizer, 210, 211
Polyvinyl tolune (PVT) bead, 113, 114
Porins, 134
Post-translational mechanism, 315
Post-translational modification, 52, 132, 162, 194
Potassium channel(s), 137, 155–157, 198, 321
Potassium channel opener, 137
Potassium channel-two pore, 156
Potassium transporter, 137
Potassium uptake, 155, 156
Preincubation, 289, 290, 295, 298, 299
Primary database, 449
Progess curve, 288, 289, 294, 295
Promoter, 51, 58, 142, 145, 165, 255, 271
  AraBAD, 147
  AraC, 147
  CTX, 145
  elements, 172
  inducible, 275
  Lac, 139

[Promoter]
LexA, 147
metalthionein, 159
Protease inhibitor, 140
Protein A, 11, 55, 117
Protein expression, 418, 436
Protein expression maps, 437
Protein finger printing, 422
Protein fold, 424
Protein function, 416, 433–435, 442
Protein interaction maps, 415, 421, 438
Protein interactions, 49, 129, 142, 147
Protein kinase, 48, 275, 279
ERK, 279
JAK, 251
JNK, 279
MAPK, 279
p38 kinase, 279
Protein–protein interaction, 11, 91, 101, 104 105, 107, 246, 248, 416, 439–441, 512
Protein–protein interaction data, 440
Protein–protein interaction map, 440
3D-protein structure, 423
Proteomics, 31, 147, 422, 433, 434
Proximity assay, 106
Purity, 286, 287

Quinidine, 394, 397

Radioactive, 39, 42, 54, 110,123, 132, 236
Radioactive waste, 123, 220
Radioimmunoassay(s) (RIA), 42, 43, 51, 52, 111, 114, 230, 231, 518, 539
Radioisotopes, 112
Radiolabeled ligand, 190
Radioligand, 113, 114
Radioligand binding, 205, 214, 513
Rapamycin, 167, 169
Ratiometric methods, 78, 123
Reagent dispensing, 482
Real-time kinetics, 239
Receptor binding, 11, 14, 54, 107, 205, 214, 226

Receptors, 267
adenosine A2, 167
β-adrenoceptor (β -AR), 164, 192, 221, 233, 275, 435
cell surface, 221
competitive antagonist, 208, 209
cytoplasmic, 225
EGF, 195, 250
erythropoietin, 234
estrogen, 158
GHRH, 168
G-protein coupled receptor (GPCR), 2, 8, 11, 53, 55–57, 60, 97, 108, 109, 135, 136, 148, 163, 193–195, 201, 211, 213, 221, 224, 225, 227, 229, 232–238 250, 257, 279, 417, 422, 424, 435, 462, 463
growth hormone, 143
insulin, 195, 249
internalization, 249
membrane, 115, 249
neuropeptide Y, 218, 219, 222
nicotinic receptor, 315
NMDA, 314, 321
nuclear, 254–257
orexin receptor, 417, 418
orphan GPCR, 136, 164, 200, 201, 257
orphan nuclear, 203
orphan receptor, 158, 189, 200, 224, 435
PDGF, 195
peroxisome proliferator activatior (PPAR), 203
phermone, 164
protein tyrosine phospatase, 195, 250, 251
recombinant, 212
retenoid, 139
RXR, 139
serotinin, 135
somatostatin, 167
TNF, 197, 277–279
tyrosine kinase, 57, 143, 193, 195, 417

Recessive phenotype, 141
Recombinant protein, 266, 267
Regulated gene expression, 416
Relational database management system
    (RDBMS), 554, 555
Renal hypertrophy, 418
Repetitive elements, 470
Reporter assays, 170
Reporter gene, 8, 51, 57, 58, 163, 271,
    274, 544
Reporters, 51
    β-galactosidase, 53, 60, 106, 248,
        256, 276
    β-glucuronidase, 60, 72–74, 276
    β-lactamase, 60, 71, 42, 256, 538,
        544, 545, 597
    chloramphenicol acetyltransferase, 53,
        58, 59, 110, 276
    secreted alkaline phosphatase
        (SEAP), 545
Repressor protein, 422
Response elements, 275
Rifampin, 393
RNA expression, 441
RNA-protein interaction, 144
RNase, 75, 83, 84
Robotic arm, 480, 485, 500, 506
Robots, 19, 31, 53, 482, 486, 493, 495,
    498–506, 508–517, 519–522
Robot system, 482, 493, 495, 498–506,
    508–517, 519–522
Rotational relaxation, 76, 77
Ruthenium chelate (Ru^{2+}), 102, 224

Sample registration, 22
Satiety, 198, 417
Scatchard pPlot, 206, 207
Scintillant, 215, 216, 229
Scintillation counter, 34, 41, 42, 113,
    116, 190, 517
Scintillation counting, 482, 487, 517
Scintillation proximity assay (SPA), 11,
    16, 17, 42, 48, 55, 71, 113–117,
    123, 217–219, 226, 230, 231,
    237, 238, 244, 246, 251, 513,
    518, 547–549

Screening, 33
Screening program, 154
Screen platforms, 33
Screens, 189
Screens amplified luminescent proximity
    homogeneous assay (ALPHA),
    49, 106–110, 217, 224, 230, 231
    cell based, 276, 278, 280
    functional assay, 267
    mechanism-based, 276, 278
SDS-PAGE, 295
Selective drug, 417
Sequence alignment, 460
    multiple, 460
    pairwise, 460
Sequence data, 450
Sequence homology, 201
Signal transduction, 6, 11, 55, 97, 105,
    189, 192, 212, 213, 221, 249, 257
Signaling pathway, 226, 284, 285, 301,
    302
Silicon layer, 425
Single analysis gene expression
    (SAGE), 436
Single channel pipette, 480
Single nucleotide polymorphism (SNP),
    433, 471
Singlet cluster, 464
Small molecule agonist, 204
Small molecule ligand, 225
SNP, 426
Sodium channel, 198
Solubility, 297
Solubilized, 211, 212,
Soluble, 256
Solvent, 287, 295,
Specific binding, 190, 204, 205
Spectramax Gemini, 72, 408
Splice site prediction, 455
Src homology 2 (SH2), 242
Stability, 295, 296, 298, 299, 300
Stacker, 20
Standard deviation (SD), 15
Standard industry busses, 484
Staurosporin, 82
Stokes equation, 76

Streptavidin, 11, 41, 43, 84, 89, 91, 92, 95, 101, 104, 109, 110–112, 117, 119, 220, 244, 246, 575, 579, 580
Stress response genes, 418
Structure activity relationship (SAR), 283, 284, 567
Sulfaphenazole, 397
Super family
   cytokine receptor, 197
   GPCR receptor, 193
   nuclear receptor, 201
   RPTK receptor, 195, 233, 234
Surface plasmon resonance (SPR), 118, 119, 123, 557

T-antigen, 270
Terfenadine, 394
Testosterone, 397
Tetracycline, 141, 372, 374–376
   efflux protein, 140
   resistance, 140
Theophylline, 571
   antitheophylline, 572
Thermalcycler, 519
Thermogenic, 120
Thermography, 120
Thin film, 557
Thrombin, 225
Thrombin receptor activating peptide, 225
Time resolved fluorescence (TRF), 47, 77, 88–90, 117, 216, 535
Titertek, 72
Tolbutamide, 397
Topcount, 41, 57, 110, 112, 114, 215, 220, 230, 236, 237, 246, 249
Toxicity, 155, 423, 426
Toxicology, 62
Toxins, 318, 323
TPA, 275
Transactivation domain, 139
Transcription, 6, 57, 96, 172, 202, 203, 227, 253, 254, 255, 274, 441
Transcription activity, 146, 201

Transcription based reporters, 266
Transcription complex, 201
Transcription factors, 221, 271
Translocation, 255
Transmembrane, 193–195, 197, 198, 225, 226, 229
Transmembrane helix, 227
Transphosphorylation, 251
Transport(s), 267, 322, 325
Transposon, 421, 442
Transverse relaxation optimized spectroscopy, 426
Tranylcypromine, 397
Tripropylamine, 102, 104, 224
Triton X-100, 211
Two-hybrid system, 162, 170, 171
   bacteria, 145, 421
   mammalian, 421
   reverse, 143, 171
   yeast, 53, 54, 142, 143, 168, 169, 424

Ultrahigh-throughput screening (UHTS), 3, 26, 27, 33, 35, 48, 69, 94, 96, 105, 116, 118, 123, 177, 488, 489
Uncultivable, 359

Validation, 81
Vector, 268, 363, 366, 369
Vesicle-like particle technology, 224
Vesicles, 94
Victor 1420, 72
VIPR, 325
Visualization, 555, 556
V_{max}, 297–299, 393,
Voltage-gated calcium channels, 323
Voltage-gated ion channel(s), 193, 198, 314, 317, 321
Voltage sensors, 318, 324, 326
Wheat germ agglutinin (WGA), 217, 218, 220, 229
Windows NT, 512
Workstation, 484, 486, 491

Xenobiotic, 61, 384, 395, 403
Xenobiotic biotransformation, 383

X-ray crystallography, 214
X-ray diffraction, 424

Yeast genome, 435,
Yeast reporter, 160,
Yittrium silicate (Ysi) bead, 114

Z-factor, 38, 550–553
Zinc fingers, 201
Z prime factor (Z′), 38, 542, 552
Zymate robot, 479, 484, 485, 499,
    501